AF669515

Sustainable Development and Environmental Science

Sustainable Development and Environmental Science

Editors
Professor (Dr.) Arvind Kumar
FLS (London), FASc (Swiss), FISEC, FSESc, FAZ, FZA (Gold Medalist)
Post Graduate Department of Zoology,
S.K.M. University, Dumka, Jharkhand

Dr. R. Sivakami
Lecturer, Department of Zoology,
A.A. Govt. Arts College, Musiri, Tiruchirapalli (Tamil Nadu)

2013
Daya Publishing House®
A Division of
Astral International Pvt. Ltd.
New Delhi – 110 002

© 2013 ARVIND KUMAR (b. 1953 –)
R. SIVAKAMI (b. 1969–)
ISBN 9789351241966

Despite every effort, there may still be chances for some errors and omissions to have crept in inadvertently. No part of this publication may be reproduced in any form or by any means, electronically, mechanically, by photocopying, recording or otherwise, without the prior permission of the copyright owners.

The views expressed in various articles are those of the authors and not of editor or publisher of the book.

Published by : **Daya Publishing House®**
A Division of
Astral International Pvt. Ltd. –
ISO 9001:2008 Certified Company –
4760-61/23, Ansari Road, Darya Ganj
New Delhi-110 002
Ph. 011-43549197, 23278134
E-mail: info@astralint.com
Website: www.astralint.com

Laser Typesetting : **Classic Computer Services**
Delhi - 110 035

Printed at : **Salasar Imaging Systems**
Delhi - 110 035

PRINTED IN INDIA

Preface

India has successfully crossed 50 year mark of its independence marching ahead with covetable progress scored through unflagging commitments and tireless efforts of the people and the government in past and could not produce even a needle at the time of independence, and had to start from the scratch, is running neck to neck with most advanced nations of the world in the field of frontier technologies and we are proud to say that in information technology, on which the future of human race rests, it is surging ahead, specially in software technology. During the last five decades, India's achievement in science and technology seem to be very impressive which would reveal expertise built up in space research, nuclear engineering, production of steel, fertilizers, petroleum, chemical, machine tools, thermal power plants, construction of big dams etc. Miraculous achievement has been made in the agricultural production through Green Revolution during last three decades which converted India of the fifties, as an importer of food grains to an exporter. The technological advancement in agriculture is brought about through the increased production of new high yielding varieties of crops by means of the application of chemical fertilizers and pesticides bringing more and more land under crop cultivation for the ever increasing population has no doubt saved mankind from hunger and pestilence. On the other hand various developmental activities such as construction of huge dams, unscientific mining of coal for power plants and industrial units have changed the man-nature relationship may they have changed not only the economic and socio-cultural life of the people but also their values, systems, ideas, believes and indeed their entire life style. Destruction of more forests for extension of land for agriculture purpose, for coal mining activities as well as for making multi-storied building, roads and other constructions has led to the extinction of plant and animal species and is also responsible for ecological imbalance. Apart from these, the indiscriminate disturbance of the freshwater ecosystems leads to the disturbance in corresponding recycling system. The storm of modernization and industrialization has not only uprooted man but in fact has destroyed his habitat and environment too. The increase in the discharge of toxic gases from the industrial unit and carbon dioxide liberated from animals and human beings and from burning of fossil fuels is as sharp as decrease in release of oxygen by the trees and plants, as a result of which the biosphere equilibrium maintained since time immemorial has been affected.

The relationship between human health and his environment is a two-way process. If one disturbs, other will be automatically disturbed. Therefore, the present work has been undertaken to illustrate the impact of ill environmental health on the health of human beings.

This book is a unique compilation various 64 research articles of eminent environmental scientists, which will be useful for academicians, researchers, scientists, administrators and ecoplanners and it would also be used as an invaluable reference book for the generations to come.

My special thanks and appreciation go to the scientists whose contributions have enriched this volume. I wish to express my sincere gratitude to Dr. B. A. Khan, Hon'ble Vice Chancellor, S. K. M. University, Dumka, Professor M. C. Das, Hon'ble former Vice Chancellor of Sambalpur University, Professor N. C. Datta of Calcutta University, Professor K. C. Pandey of Lucknow University, Professor S. K. Konar of Kalyani University, Professor D. K. Belsare of Bhopal University, Professor Professor U. G. Goswami of Gauhati University, Professor P. Natarajan of Kerala University, Professor P. S. Murthy of Bangalore Universtiy, Professor Ajit Varma of JNU, Professor A. K. Mittal of BHU, Professor S. P. Hosmani of Mysore University, Professor K. Kapoor of Udaipur University, Professor K. B. Reddy of Nagarjuna University, Professor M. Vikram Reddy of Pondicherry University, Professor B. K. Tiwari of NEHU, Professor G. Tripathi of Jodhpur University, Professor R. Ramalingam of Annamalai University, Professor Sharif U. Ahmad of Nagaland University, Professor G. K. Kulkarni of Aurangabad University, Professor S. K. Battish of PAU, Professor B. M. Sharma of Manipur University, Professor B. D. Joshi of Hardwar University, Professor G. C. Pandey of Faizabad University, Professor K. C. Sharma of Ajmer University, Professor M. Raziuddin of Hazaribag Universtiy, Professor U. S. Bagde of Mumbai University, Professor Gurdeep Singh of I. S. M., Dhanbad, Dr. P. K. Goel of Karad, Professor S. P. Roy and Shri Tribhuwan Poddar of Bhagalpour University, Bhagalpur for encouragements.

I also express my deep sense of gratitude to my parents whose blessings have always prompted me to pursue academic activities deeply. I am also thankful to my wife, Kumari Bimla and my two lovely sons, *Kumar Pallav Shivshankaran and Kumar Prasun Ramakrishnan* whose natural smiles extended to me relief all through this tiresome endeavour.

Last but not the least, I am also thankful to Mr. Anil Mittal, Profrietor, Daya Publishing House, New Delhi for taking keen interest in bringing out of this book. Finally, I will always remain a debtor to all my well-wishers for their blessings, without which this book would not have come into existence.

Arvind

Professor Arvind Kumar

Contents

Preface *v*

1. Environment and Sustainable Development Indicators 1
I. Sundar and Arvind Kumar

2. Effect of Urban and Agricultural Waste Composts on Yield and Nutrient Uptake by Different Crops 9
K. Jeevan Rao and Ch. S. Rama Lakshmi

3. Microbial Fuel Cells: A Bioelectrocatalytic Technology for Bioelectricity, Biohydrogen, Biosensors, Wastewater Treatment and Biomedical Applications 20
M. Vijay and Rm. Meyyappan

4. Limnological Studies of Ooty Lake in Tamil Nadu 35
M. Manimegalai, S. Umavathi and S. Binukumari

5. Nutrient Dynamics and Phytoplankton Abundance in the Marine Environment of Kollam (Quilon), South West Coast of India 47
Wilma Cyril, S.T. Shibu Joseph and B.S. Sharma

6. Pre-impact Studies of the 2000MW Lower Subansiri Dam on Certain Hydrobiological Aspects of Down Stream of River Subansiri with Special Reference to Plankton and Fishes 66
Debojit Baruah, Lakhi Prasad Hazarika and Sabita Borah

7. Impact of Paper Mill Effluent on Agroecosystem: A Review 84
P. Dhevagi, C. Udayasoorian and G. Oblisami

8. **PCP Degradation by *Pseudomonas aeruginosa* and *Pseudomonas putida*: A Comparative Study** 98
T. Byragi Reddy, K. Swapna Vahini and M. Sumanth Kumar

9. **Study of Spoilage Causing Pathogenic Microorganisms in Indian Cheese (Paneer) Sold in Allahabad City** 104
Om Prakash Verma, Ankit Kumar, Nirmala Singh, Ajay Kumar Singh, Purushottum Kumar and Birendra Kumar Singh

10. **Comparative Analysis of Variability in Rice (*Oryza sativa* L.) Under Different Land Situations** 113
Monalisa Manna, Hossain Ali Mondal and Biman Gopal Sasmal

11. **Alterations in Major Nutrient Molecules of Albino Mice Under Monocrotophos Stress** 120
P. Madhaveelatha, K. Sankaraiyah, R. Jayakumar and K. Jayantha Rao

12. **Non-keratinizing Carcinoma in the Sputum of Occupational Workers (Stone Crusher Mill Workers) as an Indicator of Health Hazard** 124
Puspanjali Parida

13. **Correlation and Path Analysis of Some Economic Traits in Potato (*Solanum tuberosum* L.)** 129
K.G. Santosh, H.R. Bhoomika, N.S. Mavarkar and S.L. Biradar

14. **Genetic Variability Studies in Potato (*Solanum tuberosum* L.)** 133
K.G. Santosh, K.M. Indiresh, H.R. Bhoomika, S.L. Biradar and N.S. Mavarkar

15. **Effect of Different Packing Materials on the Shelf Life and Quality of Acid Lime (*Citrus aurantifolia* Swingla.) at Room Temperature** 139
V. Bali Reddy, G. Bindu Madhavi, V. Chenga Reddy, D. Venkata Reddy and B. Srinu

16. **Seasonal Periodicity and Pollution Level of a LNMU Campus Pond, Darbhanga** 147
Ashwini Kumar, Onkar Mishra and S.B. Shashi

17. **Variations in the Primary Productivity of Singanallur Pond in Coimbatore, Tamil Nadu** 150
S. Umavathi and K. Logankumar

18. **Effect of Water Activity on the Germination of Certain Fungi Isolated from Working Environments in Imphal, Manipur** 155
A. Premila Devi, S. Mukta Singh and N. Irabanta Singh

19. **Acute Toxicity of Some Indigenous Plants against Pulse Beetle, *Callosobruchus chinensis* Linn. (Coleoptera : Bruchidae)** 161
S.S. Patole and R.T. Mahajan

20. **Electron Spin Resonance Spectroscopic Technique for Dating of Fossil Bones** 166
G. Velraj and K. Janaki

21. **Electricity Generation and Effect of Catholyte in a Dual Chambered Microbial Fuel Cell Using *Klebsiella pneumoniae* as Biocatalyst** 174
M. Vijay and Rm. Meyyappan

22. **Selection of Potential Site for Solid Waste Disposal through GIS Technique: A Case Study of Tambaram Municipality, Kancepuram District, Tamil Nadu, India** 182
R. Annaduri

23. **Studies on Growth and Biomass Production of *Alnus nepalensis* D. Don, as Affected by Crop and Mulches** 189
P.C. Vanlalhluna and U.K. Sahoo

24. **Cryopreservation of Spermatozoa of an Endangered Fish *Osphronemus goramy* and a Common Table Fish *Puntius sarana*** 200
L. Bindu

25. **Bacteriological Assessment of the Marine Environment of Kollam (Quilon), Southern Kerala Coast of India** 205
Wilma Cyril and B.S. Sharma

26. **Effect of Seed Stratification and Nursery Raising of *Quercus robur* (English Oak)** 212
Gh. Mohi-u-din Bhat, A.H. Mughal, P.A. Sofi, J.A. Mugloo, A.A. Mir and T.A. Rather

27. **Environmental Impact of Mine Spoils** 216
M. Baskar, A. Solaimalai, P.T. Ramesh and T.S. Muthukumar

28. **Growth Studies in Groundnut (*Arachis hypogaea*) Under the Influence of Fluctuating Groundwater Table and Levels of Irrigation** 248
J.M.L. Gulati, M.M. Mishra and D. Lenka

29. **Field Culture of Fairy Shrimp *Streptocephalus dichotomus* Baird** 256
B. Jesi Selvarani

30. **RAPD Analysis of F_1 Diallel Cross Population of Chickpea for Drought Tolerant Trait** 261
P.K. Lokhande, U.S. Dalvi, S.V. Damame, R.M. Naik and L.B. Mhase

31. **Identification of Molecular Marker Associated with Chickpea for Drought Tolerance Using RAPD Analysis** 265
P.K. Lokhande, U.S. Dalvi, S.V. Damame, R.M. Naik and L.B. Mhase

32. **Modelling on Risk of Vulnerability to STDs/HIV Infection Among Pre-marital Female Migrants in Urban Area** 270
Himanshu Pandey, J.N. Mishra and Umashankar Shukla

33. **Diversity and Impact of Macrophytes in Bhandardaha Beel, Murshidabad, West Bengal** 275
Anita Chakraborty, B.C. Jha and R.K. Bhakat

34. Ecology and Biodiversity of Belloon Beel, Murshidabad, West Bengal with Special Emphasis on Macrophytes 280
Anita Chakraborty, B.C Jha and R.K. Bhakat

35. Effect of Farm Yard Manure and Lime on Soil Adsorption and Plant Uptake of Phosphorus 288
H.K. Senapati, A.K. Dash, B.B. Dash, D.K. Behera and P.K. Das

36. Mineral Characterization Studies of Sediment Samples of Tuna, Gujarat by Spectroscopic Studies 296
R. Ravisankar A. Rajalakshmi, D. Sahayathna Selvi, K. Vijay Anand and Prince Prakash Jeba Kumar

37. Seasonal Analysis of Sanganer Handmade Paper and Cardboard Industries Effluent 310
Shweta Kulshreshtha, Nupur Mathur and Pradeep Bhatnagar

38. Water Quality Assessment of Ayroor River (Kerala) with Special Reference to Zooplankton Diversity 321
C. Shaji, S. Suvidha and L. Bindu

39. Influence of Silicate Solubilizing Bacterial (SSB) Inoculation to Rice in Different Soils 328
P. Kalaiselvi and S. Anthoni Raj

40. Yield and Economics of Intercropping of Pulses in Pigeonpea 339
D.D. Dudhade, G.P. Deshmukh, S.S. Nawake and P.N. Harer

41. Observation on the Avifaunal Diversity in and Around Shringarbandh Lake, Bondgaon (Surban), District Gondia, Maharashtra 343
S.V. Bhandarkar and R.N. Chavan

42. Correlation Coefficient of Physico-Chemical Properties with Zooplankton in Sager Lake Bhandara, Maharashtra 347
S.V. Bhandarkar and G.N. Gopale

43. Comment on Rotifer Diversity in Two Water Bodies of Brahmapuri, Maharashtra 352
S.V. Bhandarkar, A.B. Jagne and W.R. Bhandarkar

44. Studies on Population Fluctuations and Abundance of Predators Fauna in Rice Ecosystem 359
K. Verma and M.A. Khan

45. Biosuppression of Scirpophaga incertulas (Walker) 364
K. Verma and M.A. Khan

46. Effect of Soil Solarization on Damping Off Disease Management in Vegetable Nursery 372
V. Bali Reddy, G. Bindu Madhavi, D. Venkata Reddy, V. Chenga Reddy, K. Gurava Reddy and M. Chandra Sekhar Reddy

47. Screening of Eight Plants for their Repellent Action Against Adult Khapra Beetle, *Trogoderma granarium* (Everls) Infesting Wheat Grain 377

S.C. Dwivedi and Santosh Yadav

48. Some Aspects on Vision in Relation to Light in Birds 382

P. Jasmine, M. Surendran and D.J. Prakash

49. On Some Ciliates (Protozoa) of Bor Wildlife Sanctuary, Maharashtra, India 389

L. Bindu

50. Effect of Silicate Solubilizing Bacterial (SSB) Application Along with Rice Residues on Growth and Yield Attributes of Rice 394

P. Kalaiselvi and S. Anthoni Raj

51. Effect of Wheat Straw and Sugarcane Trash on Yield and Nutrient Status of *Agaricus bitorquis* 404

D. Venkata Reddy, V. Bali Reddy, G. Bindu Madhavi, V. Chenga Reddy, K. Gurava Reddy and M. Chandra Sekhar Reddy

52. Sustainable Development of Environmental Conservation on Natural Land Resources and Population through Carrying Capacity and Footprints 410

S.R. Singh, Th. Manimala Devi and N. Sangbanbi Devi

53. Biomedical Waste Management: A Prophalactic Measure for Future Health Hazards–A Case Study of Warangal District of Andhra Pradesh 421

G. Benarjee, K. Srikanth, G. Laxman Prasad, G. Ramu and K. Narasimha Ramulu

54. Assessment of Irrigation Water Quality Parameters of Certain Tank Located Around Tumkur 430

C. Vijaya Bhaskar, Mahadevaiah and G. Nagendrappa

55. Assessment of Impact of Human Activity on Mangrove Ecosystem South East Coast of India 437

C. Sivasubramanian, S. Raveendran and P. Kennedy

56. Impact of Weeds on Some of the Morphological Attributes of a Little Millet Crop 442

K.L. Barik

57. Nutrient Dynamics in an Annual Fire Influencing Humid Subtropical Grassland of Indra Gandhi Wildlife Sanctuary, the Western Ghats, India 448

S. Paulsamy, S. Manian and D. Suresh

58. Prospect of Biomonitoring of Pollution in Aquatic System by Algal (Chlorophyceae) Assessment 458

Sushma Das Guru

59. Antimicrobial Activity of *Merremia tridentata* (L.) Hall. f. 464

S. Gopalakrishnan, M. Kamalutheen and T. Syed Ismail

60. *In vitro* Studies on Antibacterial Effect of Aqueous Crude Extract of Bryophytes on *Xanthomonas citri* 470

G.S. Deora, Durgesh Bhati and Narendra S. Jain

61. Larvicidal and Ovicidal Effect of a Neem Product on *Aedine* and *Culex* Mosquitoes 477

Kiran Joseph, R.S. Mohanraj, B. Dhanakkodi and K. Sathya

62. Comparative Effect and Efficacy of Fenbendazole and Tetramisole on *Ascaridia galli* 486

K. Surender Reddy, V. Bikshapathi and G Benarjee

63. Postharvest Fungal Rot of Chilli (*Capsicum annuum* L.) Fruits 492

V. Bali Reddy, G. Bindu Madhavi and D. Venkata Reddy

64. Evaluation of Toxicity of Yellow Oleander (*Thevetia peruviana*) and Tomato (*Lycopersicon esculentum*) as Fish Toxicants on some Important Blood Parameters of an Air Breathing Fish (*Heteropneustes fossilis*) 496

Tapan Chandra Das and Samir Banerjee

Index 509

Chapter 1

Environment and Sustainable Development Indicators

I. Sundar[1] and Arvind Kumar[2]

[1]Department of Economics, Directorate of Distance Education, Annamalai University, Annamalainagar – 608 002, Tamil Nadu

[2]Vice Chancellor, Vinoba Bhave University, Hazaribag, Jharkhand

ABSTRACT

Environmental and sustainable development indicators are a relatively new phenomenon. The Rio Conference on Environment and Development in 1992, and other similar environmental milestone activities and happening, recognized the need for better and more knowledge and information about environmental conditions, trends, and impacts. To achieve this, it is not only necessary to collect new and better data; new thinking and research with regard to indicator frameworks, methodologies, and actual indicators are also needed. This chapter deals with indicators of pollution, indicators of natural resource management, benefits and problems of sustainable development indicator and biodiversity indicators. It outlines problems and prospects of developing and measuring indicators of environment and sustainable development. This chapter concludes with policy measures to develop effective indicators of environment and sustainable development.

Keywords: *Indicators, Natural resource management, Pollution indicators, Benefits and problems of sustainable development indicator and biodiversity indicators, Baselines, Thresholds, Targets and comparators.*

Introduction

In working with indicators of natural resource management and pollution, the geographic and the time scale of the indicator playa significant role. The geographic scale is partly an issue of globally significant indicators versus locally significant indicators. For example, the extinction of a species in a region within a country may be of serious concern for that region, even if the species is not near extinction globally. At a global level, there may even be completely different species that matter.

The role of indicators in decision-making processes that concern natural resource management and pollution aspects is also potentially linked to the geographic scale. Indicators of pollution and natural resource management do not always coincide with politically defined areas, to which decision making is commonly delimited. The time scale of an indicator also affects the usefulness and interpretation of indicators. The issue of time becomes important in the design of mitigating measures and to enable prevention rather than reaction to a problem.

Pollution Indicators

Apart from these common considerations in the practical work with indicators, there are a few that are specific for pollution and natural resource management monitoring respectively. For example, while several pollution issues may be more difficult to interpret than some natural resource issues due to their commonly global characteristic, more and more practitioners agree on the indicators to use in their monitoring. The World Health Organization has developed global standards of water quality, air quality indicators have been used in many cities around the world, and their monitoring is fairly easily motivated due to their link to human health. Impacts from many pollutants also commonly take longer before they are detectable and last for a long time indicators of pollution therefore have to consider the time scale.

Natural Resource Management Indicators

Many indicators of natural resource management, on the other hand, show an impact directly even though the importance of that impact may be of the long-term nature. Deforestation is apparent the same instance it happens, so is land use change. There are of course other indicators that may not show a clear trend immediately–indicators of water use may belong to that category since it may be difficult to establish a threshold for sustainable use of fresh water resources. Effective management, and thus quality, is therefore a central aspect in the monitoring of natural resources. This is easily missed when only monitoring number of hectares of deforestation or people trained in environmental management. Some natural resources are furthermore difficult to monitor due to their characteristics. For example, it is difficult to estimate marine fish stocks and the richness of a country's biodiversity due to the vast areas utilized by the resource. Unlimited monitoring areas are not the only reason why issues such as biodiversity are complicated to monitor. Political, conceptual, practical and institutional factors all contribute to the complexity of some issues.

Sustainable Development Indicators

Developing sustainable development indicators at an international level cannot be categorized as an easy task. In 1996, the Development Assistance Committee of the Organization for Economic Cooperation and Development initiated such an effort, the indicators for the then called International Development Goals (IDGs) initiative, inviting the United Nations, the World Bank and the International Monetary Fund to become partners. Over the four years that followed, five working groups discussed indicators for issues such as poverty, education, gender, infant and child mortality, maternal health, HIV/AIDS, malaria and other diseases, environment, and global partnership.

At a later stage, the name of the targets changed from the IDGs to the MDGs (the Millennium Development Goals). Each goal has a number of targets identified, which are more specific in their character. The targets for environmental sustainability and regeneration are: (*i*) Integrate the principles of sustainable development into country policies and programmes and reverse the losses of environmental resources; (*ii*) halve, by 2015, the proportion of people without sustainable access to safe drinking water; and (*iii*) have achieved, by 2020, a significant improvement in the lives of at least 100 million slum dwellers. The working group on environmental indicators agreed on nine indicators to be included in the set of indicators to monitor the MDGs:

1. Proportion of land area covered by forest
2. Ratio of area protected to maintain biological diversity to surface area
3. Energy use (kg oil equivalent) per $1 GDP (PPP)
4. Carbon dioxide emissions (per capita)
5. Consumption of ozone layer depleting CFCs (ODP tons)
6. Proportion of population using solid fuels
7. Proportion of population with sustainable access to an improved water source, urban and rural
8. Proportion of urban population with access to improved sanitation
9. Proportion of population with access to secure tenure (owned or rented)

The nine indicators are selected according to a number of selection criteria, of which the most important were that they captured important environmental problems, were relevant for the majority of the countries in the world including developed countries, and that data were available collection of new data was beyond the scope of the project. Furthermore, there was a limit to how many environmental indicators could be included in the total set of indicators for the MDGs.

The core set of indicators will be used–at a global level-to monitor performance and adjust development strategies as required. The World Development Indicators published annually by the World Bank, will report on the nine core indicators, making the information accessible world wide, and putting it in the front of the World Bank's priorities in its collaboration with other international partners.

Biodiversity Indicators

Biodiversity is here used as an example of the difficulties in defining and developing indicators for some issues such as biodiversity and several crosscutting issues including sustainable development. The monitoring of biodiversity has traditionally been difficult due to several factors. These factors can be divided into political, conceptual, practical and institutional factors:

Political Factors

Factors that affect the development of biodiversity indicators are often political. Political willingness to prioritize an issue such as biodiversity, and to spend money on the monitoring of it, is commonly low since the benefits to the general population sometimes can be difficult to identify. This in turn is often a result of ignorance the economic benefits of conserving biodiversity, or the economic losses of destroying biodiversity, may not be analyzed enough. The ignorance can also be a result of a traditionally greater focus on biodiversity's non-utilitarian values as opposed to its utilitarian values, which may result in a lower priority given to biodiversity conservation.

Conceptual Factors

The monitoring of biodiversity is extremely complex as a result of various aspects. First, the diversity within the concept of biodiversity demands a flexible approach to the selection of indicators. Different aspects such as genetic diversity, species diversity, diversity in untouched areas, and diversity in agricultural areas all require different indicators. Second, the ambiguity associated with the term "diversity" itself creates confusion in how to monitor and measure it. Third, the multidimensional roles mentioned above contribute to the complexity. This intricacy clearly creates problems when one tries to express it in a few selected parameters.

Practical Factors

The general problem of a lack of data plays a significant role for the few identified biodiversity indicators that are currently commonly used. Difficulties in being exhaustive in the measuring of biodiversity indicators are noticeable. In addition, practical measuring problems result in a lower reliability in the data that do exist. For example, "number of species" is a commonly proposed and used indicator of biodiversity. However, with the current knowledge gaps in, for example, how many species actually exist, such an indicator does not necessarily reflect reality.

Institutional Factors

Capacity to develop and, perhaps even more important, to interpret and analyze the developed indicators is commonly lacking in many countries. The capacity needed is both of the human and of the technical kind. Hammond *et al.* (1995) state, for example, that "national measures of biodiversity of use to policymakers may be impossible to compile unless they are based on spatially referenced data–essentially digital maps." While the technology to achieve this is becoming more and more developed, it is still far from available to all involved parties.

Many concerned with sustainable development would like to see a single indicator to compete with the enormous political power of the Gross Domestic Product (GDP). In addition, it is becoming more and more common for decision makers to ask for a single, powerful number that is easy to understand and use in decision making. But many are skeptical that a single number could assess something as complex as sustainable development, especially if one considers all the criticism that has been raised against GDP. The skeptics are instead promoting sets of indicators that include numbers on the most important aspects of society's well-being.

Benefits of Sustainable Development Indicators

1. The aggregation of indicators makes it possible to obtain clear messages, gain an overview of sustainable development, and show where performance is especially weak or strong.
2. An aggregated indicator of sustainable development can compete with the GDP indicator in the communication to the general public about material well-being.
3. Aggregated indicators explores the relationship among the variables, which lies at the heart of the linkages intrinsic to sustainable development.
4. Quality assurance of the individual indicators is made easier.
5. A set with several indicators provide countries with more flexibility with regard to which indicators to include according to variations in conditions, activities and priorities.
6. A set with several indicators can be based on indicators that are conceptually accepted and familiar to developers and the public.

Problems of Sustainable Development Indicators

1. An aggregated number is more difficult to quality assess since the aggregation hides the individual parts of the indicator.
2. The actual aggregation may become a bit ambiguous when items, whose units differ, are added together for example, life expectancy, educational attainment and adjusted income as is done in the Human Development Index.
3. The aggregation is, in general, difficult to achieve in a clear and unambiguous manner since it demands weighting of items that are difficult to value.
4. An aggregated number may be good for comparisons between countries, or regions, but does not necessarily have a functional value as a policy tool.
5. The linkages between the economic, social, and environmental trends in society as well as between and among systems are not always illustrated properly for the sets to be measuring sustainable development.
6. It is not uncommon that sets of sustainable development indicators include multiple indicators for essentially the same issue while hardly (or not) including indicators of other important issues.
7. Their assessment is made difficult by the diversity of issues that different indicators measure, and the different directions in which the indicators move.

Baselines

For any indicator to be meaningful, either baselines, thresholds and/or targets for the indicators need to be established. A baseline is a value that is determined before an indicator initiative starts in order to show a positive or negative environmental change as a result of an initiative. Without a baseline, the indicator values detected as the monitoring continues can not be compared to anything meaningful; it is not possible to say whether things have improved or become worse. It is then difficult to establish what kind of impact an activity has had, or what kind of changes are occurring in a region, country, or area.

If the project's objective is not about improving the environmental conditions, but rather about another development initiative which needs to be monitored so that it does not degrade the environment, the baseline is best established through the monitoring of the project area at the initiation of the project to establish the current state of environment. The same method to establish the baseline is used in those initiatives that wish to illustrate how that state of environment is improving. For those indicator initiatives that have as their objective to illustrate how the state of environment has changed in total, the baselines are established as zero for example, extent of protected areas or total CO_2 emissions.

It can be worthwhile to point out that, for the interpretation of the change within the area to be correct, the baseline may not be enough to compare with. To be able to determine whether, for example, the project was the best alternative, the resulting values should be compared to the alternative development scenarios. That is, the alternative development path for the monitored area without the project may not have been a path of non-existing change. Whatever would have happened in the area without the project should therefore be kept in mind during the interpretation of the monitoring result.

Thresholds

For some monitoring systems the establishment of thresholds may be of even greater importance. than that of baselines as a tool for analysis. Thresholds are useful in initiatives that do not necessarily have environmental improvements, or sustainable development, as their main objective. Instead, there may be possible negative impacts on the environment or the development could in some other way be unsustainable, and thresholds should therefore be included for those aspects. This is, for example, true for a monitoring system that is based on alarm and diagnostic indicators. Without such levels, there is no way of knowing when the alarm sounds or when to react to what the indicator reveals. As a result of years of research, such thresholds already exist for several indicators. However, there are still many indicators that are not currently being monitored in the world, and thresholds therefore need to be established before those indicators become meaningful. Those thresholds should be established in an objective manner so that the current management practices and resource uses can be related to them.

Targets

The use of targets is very similar to the use of thresholds. However, targets are used to improve the state of environment through actions such as improved natural resource management, reduced pollution levels, or increased institutional efficiency. In short, targets are useful in initiatives that have environmental or sustainable development improvement as one of their objective. Performance indicators are used to monitor the progress towards these targets. The Millennium Development Goals, as developed by the world community, are an example of such targets. Each of the eight areas monitored (poverty, education, gender equality, infant and child mortality, maternal health, HIV/AIDS, malaria and other diseases, environmental sustainability, and global partnership for development) has a related development goal established. In addition, each of the indicators used for the monitoring of the areas has individual goals. For many of the most common indicators at the national, regional and international levels, targets already exist in the form of international commitments or consensus. Examples include the World Health Organization's guidelines on availability of safe drinking water and water for sanitary means of excreta disposal (WHO estimates this amount to 20 to 40 litres of water per person per day located within a reasonable distance from the household (WHO, 1996) or the World Conservation Union's target for protected areas (10 per cent of a country's surface). Other targets are best determined through the monitoring of an "unspoiled" part of the monitoring area, or even of another area (if it seems likely that the area that is to be monitored already has been changed from its original state).

Comparators

While baselines, thresholds and targets can be seen as a sort of comparator–the result of monitoring the indicator is compared to a predetermined baseline, threshold or target for that indicator–the concept of comparators is much wider than that. Comparators belong to the most basic but important tools for facilitating indicator analysis. In order to decide which comparator to use and what it should look like, the indicator which one wishes to analyze needs to be thought about. The first choice for this number is whether to present it as a relative or absolute number. The difference between the two is easily illustrated by the indicator "protected areas." If the number of hectares of protected land in a country is presented–for example 47,000 hectares in Argentina-it probably does not say much to most people. If, on the other hand, the hectares of protected land are compared to the total land area, the interpretation of the indicator is simplified as it can now be expressed as 1.7 per cent protected land of total land area. For some people, this may still not say much–is this a small or a large share of protected land? To answer that question, a comparator is necessary. The comparator should, quite clearly, be presented in the same way as the original indicator, that is as an absolute or relative number. The assessment of what comparator to use can further highlight the relative use of these two types. For example, if the absolute number of protected hectares discussed above is compared to the regional average of protected hectares, the extent of biodiversity protection in Argentina would not be well illustrated. The main reason for this is that some countries are so much larger than others-the number of protected hectares in Argentina can therefore be much higher than the protected hectares in another smaller country, even if the latter has protected a larger share of its land. If, on the other hand, the relative number of protected hectares is compared to the regional average (7.3 per cent) or the average of the income group that Argentina belongs to (5.7 per cent), the picture becomes a lot clearer–Argentina is protecting a relatively small share of its land. These types of considerations are taken into account in indicator publications such as the World Development Indicator, World Resources Report or The Little Green Data Book which present some of the indicators as absolute numbers and some as relative.

A second choice concerns the actual comparator. This is very much an issue of the message that the indicators are supposed to convey. In some cases, like in the case of some Argentinean indicators, a comparison with most of the other countries in the region may work more as an excuse not to work on the issues monitored since the Argentine values look above average (for example, Argentina's annual freshwater use is 2.8 per cent of its total resources compared to 3.6 per cent in Chile). In that case an average for a couple of developed countries may be more useful as comparators since Argentina is relatively highly developed for its region (using the same example as above, Canada's annual freshwater use is 1.6 per cent of its total resources). However, one should keep in mind to select comparators that are relevant, that is comparators that face more or less the same circumstances.

Conclusion

Sustainable development is a process of change with the intention of achieving harmony between social, economic and ecological objectives and system requirements in short and long run. It is an economic problem of scarcity and choice. However, analysis must go beyond the conception of conventional economic problem. It calls for a trans-disciplinary approach. For integrating basic ecological and economic concepts of sustainability concepts, of needs and wants, as well as absolute and related limits of the overall system. It is a normative problem and' it involves social choice about issues of intra and inter-generational equity. A proper study of sustainable development depends on the following criteria.

1. Development of a set of indicators on economic sustainability along with operational definition.
2. Development of a set of environmental and ecological indicators of sustainability and proper operationalization of the key concepts.
3. Development of a set of socio-cultural indicators of sustainability with suitable operational definition.
4. Proper weighing of the chosen indicators of sustainable development.
5. Construction of various types of index of sustainable development along with construction of composite index of sustainable development.
6. Attempts should be made to study sustainable growth on the basis of empirical findings.

References

Hammond, A., Adriaanse, A., Rodenburg, E., Bryant, D., and Woodward, R., 1995. *Environmental Indicators: A Systematic Approach to Measuring and Reporting on Environmental Policy Performance in the Context of Sustainable Development*. World Resources Institute: Washington, DC, USA.

OECD (Organisation for Economic Co-operation and Development), 1994. *Environmental Indicators*. OECD Core Set. OECD, Paris, France.

Pretorius, R., Grobler, L., and Moore A. (Eds.), 1998. *Results from Testing of CSD Indicators of Sustainable Development in South Africa*. Report to the United Nations Commission on Sustainable Development. Department of Environmental Affairs and Tourism, Pretoria, South Africa.

Reid, W.V., McNeely, I.A., Tunstall, D.B., Bryant, D.A. and Winograd, M., 1993. *Biodiversity Indicators for Policy-Makers*. World Resources Institute, Washington, DC, USA.

Segnestam, L., 1999. *Environmental Performance Indicators: A Second Edition Note*. Environment Department Papers No. 71. The World Bank, Washington, DC, USA. Segnestam, L., in collaboration with Winograd, M., and Farrow, A. 2000. Developing Indicators: Lessons Learned from Central America. The World Bank, Washington, DC, USA.

UNDSD, 2000. *Report of the Consultative Group to Identify Themes and Core Indicators of Sustainable Development*, March 6–9. UN, New York, USA.

UNEP, 1999. *GEO–2000: UNEP's Millennium Report on the Environment*. Earthscan Publications Ltd., London, UK.

WHO (World Health Organization), 1996. *Water and Sanitation*. Fact Sheet No. 112.

Winograd, M., Fernandez, N.F. and Farrow, A., 1998. *Tools for Making Decisions in Latin America and the Caribbean*. CIAT: Cali, Colombia; UNEP, Mexico City, Mexico.

World Bank, 2000. *The Little Green Data Book from the World Development Indicators*. The World Bank, Washington, DC, USA.

World Bank, 2002. *Millennium Development Goals 2002 from the World Development Indicators*. The World Bank, Washington, DC, USA.

World Economic Forum, 2000. *Pilot Environmental Sustainability Index. An Initiative of the Global Leaders for Tomorrow Environment Task Force*. World Development Forum, in collaboration with Yale Center for Environmental Law and Policy, and Center for International Earth Science Information Network, Annual Meeting, Davos, Switzerland.

WRI (World Resources Institute), 2000. *Developing Environmental Indicators: Materials Ecology*. http://www.wri.org/matecology/index.html.

WWF, NEF, and WCMC, 1999. *Living Planet Report*. WWF International, Gland, Switzerland.

Chapter 2

Effect of Urban and Agricultural Waste Composts on Yield and Nutrient Uptake by Different Crops

K. Jeevan Rao* and Ch. S. Rama Lakshmi

Department of Soil Science and Agricultural Chemistry,
College of Agriculture, ANGRAU, Rajendranagar, Hyderabad - 30

ABSTRACT

Increase in the yields of vegetables, cereals and pulses crop was observed where treatments received 40 tonns urban waste compost per hectare when compared to lower doses of compost application. Dry matter production was also followed the same trend. Tomato recorded highest yield of 18.30 t/ha, maize 4.5 t/ha, bhendi 5.56 t/ha and soybean 1.99 t/ha to the levels of 40 t/ha USW compost application. Second highest yields were observed with treatments received 40 t/ha of agricultural waste composts. The mean N uptake in vegetable (tomato)–vegetable (bhendi) crop sequence varied from 14.62 kg/ha (control) to 60.39 (T_{17}: 40 tonns of 100 per cent urban waste compost enriched with fungus and chemical additives) by tomato and 10.68 (control) to 69.00 kg/ha (T_{17}). In cereal-pulse crop sequence (maize and soybean) highest N uptake of 114.4 and 38.6 kg/ha was observed, respectively. Though the maize crop is exhaustive crop highest NPK uptake was recorded in this crop. The treatment with 40 t ha^{-1} UW compost showed maximum N, P, K and S uptake by edible parts of different crops were 70.22, 24.37, 37.20 and 6.03 kg ha^{-1} in tomato fruit, 106.6, 36.45, 80.55 and 10.85, kg ha^{-1} in maize cobs, 32.33, 11.22, 21.23 and 3.89 kg ha^{-1} in bhendi fruit and 105.47, 10.94, 36.21 and 6.16 kg ha^{-1} in soybean grains, respectively.

---

* E-mail: kjeevanrao@yahoo.co.in

Phosphorus accumulation in edible parts was more as compared to haulm portion. Unlike incase of other nutrients, potassium uptake was higher in haulms compared to fruit/grain indicating that the adsorbed K had greater role to play in vegetative parts of the plants.

Keywords: *Urban and agricultural composts, Yield, Nutrient uptake, Crops.*

Introduction

The recycling of organic wastes for increasing soil fertility has gained importance in recent years due to high cost of fertilizers and reduced availability of organic manures. A wide variety of wastes are generated which are known to contain appreciable amounts of organic matter and plant nutrient elements and can be effectively used as a source of organic manure. Urban solid waste compost is a rich source of organic matter and plant nutrients. Therefore, it is a potential resource for soil amelioration and crop production. Both beneficial and adverse effects of USW were reviewed by Gallardo-Lara and Nogales (1987) and He *et al.* (1992). Due to energy crisis, prohibitive cost of fertilizers and poor purchasing power of marginal and small farmers, it is imperative to develop strategy to use wastes to its maximum potential with proper technology to meet the shortage of fertilizers and for improving soil fertility.

Materials and Methods

Different levels and combinations of Agricultural and Urban Waste composts were evaluated under field conditions to evaluate workable standards for maximum loading rates and wastes (composts) carrying capacity of soils and effect on nutrient uptake by plant in Vegetable-Vegetable and Cereal-Pulse cropping sequence during *Kharif*, 2002 (Tomato and Maize) and *Kharif*, 2003 (Bhendi and Soybean). Experiments were conducted at Agricultural Experimental Station, College of Agriculture, Rajendranagar, Hyderabad by taking Tomato (Pusa Ruby) and Maize (Pusa Early) as first season crops (*Kharif*, 2002). During *Kharif*, 2003–Bhendi (PK) and Soybean (PK 472) were taken up as second season crops to study the direct and cumulative effects of composts applied during *Kharif*, 2002 and *Kharif*, 2003. The experiment was laid out in a randomized block design (Factorial) with 18 treatments consisting of different levels and combinations (10, 20 and 40 tons) of AW and USW composts (aerobic and anaerobic), RDF and control.

In the field experiment the treatments imposed are: T_1: 10 tons of 100 per cent agricultural waste compost enriched with microbe and chemical additives (compost treatment-1) + 50 per cent RDF, T_2: 10 tons of 100 per cent urban solid waste compost enriched with microbe and chemical additives (compost treatment-2) + 50 per cent RDF, T_3: 10 tons of 75 per cent AW and 25 per cent USW compost enriched with microbe and chemical additive (compost treatment-3) + 50 per cent RDF, T_4: 10 tons of 50 per cent AW and 50 per cent USW compost enriched with microbe and chemical additive (compost treatment-4) + 50 per cent RDF, T_5: 10 tons of 25 per cent AW and 75 per cent USW compost enriched with microbe and chemical additive (compost treatment-5) + 50 per cent RDF, T_6: 10 tons of 100 per cent agricultural waste compost without enrichment (compost treatment-6) + 50 per cent RDF, T_7: 10 tons of 100 per cent urban solid waste compost without enrichment (compost treatment-7) + 50 per cent RDF, T_5: 20 tons of T_1 compost, T_9: 20 tons of T_2 compost, T_{10}: 20 tons of T_3 compost, T_{11}: 20 tons of T_4 compost, T_{12}: 20 tons of T_5 compost, T_{13}: 20 tons of T_6 compost, T_{14}: 20 tons of T_7 compost, T_{15}: 40 tons of T_1 compost, T_{16}: 40 tons of T_2 compost, T_{17}: 100 per cent RDF and T_{15}: Control.

Results and Discussion

The experimental soils were slightly alkaline (7.78 and 8.15 in maize and tomato fields, respectively), medium in OC (0.66 and 0.62 per cent), available N (240 and 220 kg ha^{-1}), P (15 and 12 kg ha^{-1}) and K (269 and 254 kg ha^{-1} in maize and tomato fields, respectively). The urban and agricultural waste composts used in this experiment are slightly alkaline in reaction. The highest N, P and K content was recorded in urban solid waste compost than agricultural waste compost.

Dry Matter Production

The difference in the dry matter production of different crops *viz.*, tomato, maize, bhendi and soybean due to different levels and types of composts were significant (Table 2.1). Among all the composts tested urban waste compost (T_{16}, 40 t ha^{-1}) recorded highest dry matter production, which was significantly differed from other treatments. The highest dry matter production of tomato, maize,

Table 2.1: Effect of Different Levels of Urban and Agricultural Waste Composts on Plant Dry Matter (q ha^{-1}) in Vegetable–Vegetable and Cereal–Pulse Cropping Sequence

Treatments	*Kharif, 2002*						*Kharif, 2003*					
	Tomato			*Maize*			*Bhendi*			*Soybean*		
	A	*An*	*M*	*A*	*An*	*M*	*A*	*An*	*M*	*A*	*An*	*M*
T_1	24	25	25	61	62	62	31	30	31	14	14	14
T_2	26	28	27	68	70	69	38	39	39	16	16	16
T_3	25	26	26	63	63	63	34	35	35	14	14	14
T_4	26	26	26	65	66	66	36	35	36	15	15	15
T_5	23	27	25	66	67	67	36	37	37	15	16	16
T_6	25	24	25	60	61	61	31	31	31	14	14	14
T_7	23	28	26	66	67	67	37	38	38	15	15	15
T_8	25	24	25	60	62	61	31	32	32	14	13	14
T_9	24	29	27	68	69	69	39	40	40	17	19	17
T_{10}	24	25	25	62	61	62	35	38	37	15	14	15
T_{11}	25	26	26	64	65	65	37	38	38	16	16	16
T_{12}	22	28	25	67	68	68	38	39	39	16	16	16
T_{13}	24	23	24	58	59	59	31	32	32	14	14	14
T_{14}	24	27	26	66	67	67	37	39	38	15	16	16
T_{15}	29	31	30	71	73	72	39	38	39	17	18	18
T_{16}	32	34	33	79	81	80	45	46	46	19	21	20
T_{17}	22	21	22	60	61	61	28	29	29	13	14	14
T_{18}	16	17	17	49	48	49	12	11	12	9	8	9
M	24	26	25	64	65	65	34	35	35	15	16	16

	SEm±	CD	SEm±	CD	SEm±	CD	SEm±	CD
Treatments	0.2185	0.4310	0.6125	1.1825	0.3769	0.7692	0.0825	0.1658
Methods	0.0958	0.1852	0.2895	0.5120	0.1239	0.2555	0.0393	0.0652
Treatments × methods	0.3125	0.5892	0.7982	1.4015	0.4834	0.9871	0.1246	0.2321

bhendi and soybean were 33 q ha^{-1}, 80 q ha^{-1}, 46 q ha^{-1} and 20 q ha^{-1} respectively. Highest dry matter production at T_{16} might be ascribed to increased availability of nutrients in the urban solid waste compost which enhanced the uptake of nutrients resulting in improved dry matter production. With in the treatments plots received enriched composts recorded significantly highest dry matter production when compared to untreated composts. Treatments with 50 per cent compost + 50 per cent RDF gave higher dry matter production than plots with 100 per cent compost application, might be ascribed due to the immediate availability of nutrients from chemical fertilizers than composts.

Yield

The fruit/corn yields were significantly increased with increasing levels of composts and 40 tons of USW composts recorded significantly more yields (44 and 46 q ha^{-1} in aerobic and anaerobic composts of maize and 18 and 19 t ha^{-1} in aerobic and anaerobic composts of tomato, respectively) over other treatments (Table 2.8). For second season crops the yields were 55.12 and 56.00 q ha^{-1} in aerobic and anaerobic composts of bhendi and 19.80 and 20.00 q ha^{-1} in aerobic and anaerobic composts of soybean was observed.

The yields of all four crops *i.e.* tomato, maize, bhendi and soybean were higher at higher levels of compost application, may be due to presence of several plant growth hormones which have positive beneficial effect on all yield parameters. However, the treatments applied with composts had significantly enhanced the yields of crops in both the years of investigation over RDF and control. This might be due to prevailing of better soil conditions and increased availability of nutrients which enhanced the uptake of nutrients resulting in improved yields.

Plant Uptake

Uptake of nutrients by different crops showed significant difference under different treatments. The total uptake of nutrients in the plant was computed from the dry matter accumulation and nutrient content (Tables 2.2, 2.3 and 2.4). The maximum N, P and K uptake by different crops were 60.39, 26.73 and 44.22 kg ha^{-1} in tomato, 114.4, 67.20 and 141.60 kg ha^{-1} in maize, 69, 32.66 and 64 kg ha^{-1} in bhendi and 38.60, 9.0 and 36.72 kg ha^{-1} in soybean, respectively.

Table 2.2: Effect of Different Levels of Urban and Agricultural Waste Composts on N Uptake (kg ha^{-1}) by Plant in Vegetable–Vegetable and Cereal–Pulse Cropping Sequence

Treatments	*Kharif, 2002*						*Kharif, 2003*					
	Tomato			*Maize*			*Bhendi*			*Soybean*		
	A	*An*	*M*	*A*	*An*	*M*	*A*	*An*	*M*	*A*	*An*	*M*
T_1	30.48	31.25	30.75	59.78	62.62	62	41.54	40.2	41.00	24.08	24.36	24.22
T_2	35.36	39.76	38.07	87.72	90.3	89.01	52.82	54.6	54.00	27.75	30.08	29.92
T_3	32.75	34.84	34.58	70.56	71.82	71.19	45.56	47.6	47.25	24.92	25.20	25.06
T_4	34.58	35.10	34.84	78.00	80.52	79.86	48.96	47.95	49.32	27.15	27.15	27.15
T_5	31.05	36.45	33.75	79.86	80.4	81.07	48.96	50.69	50.69	27.60	29.28	29.44
T_6	28.75	27.84	29.00	56.40	58.56	57.95	40.92	41.23	41.23	23.80	24.08	23.94
T_7	29.67	36.96	33.80	80.52	81.07	81.74	53.28	52.06	53.58	27.75	28.05	27.90
T_8	32.00	32.00	32.00	61.20	65.10	61.61	42.47	43.52	43.52	24.36	22.88	24.50
T_9	36.48	44.37	41.31	89.76	93.84	92.46	55.38	57.2	57.2	31.62	35.72	31.79

Contd...

Table 2.2–Contd...

Treatments	*Kharif, 2002*						*Kharif, 2003*					
	Tomato			*Maize*			*Bhendi*			*Soybean*		
	A	*An*	*M*	*A*	*An*	*M*	*A*	*An*	*M*	*A*	*An*	*M*
T_{10}	32.88	34.75	34.50	75.64	73.2	75.02	47.60	53.96	51.43	26.85	25.20	27.00
T_{11}	35.25	36.66	36.66	79.36	83.2	81.9	51.06	53.58	53.2	29.28	29.28	29.28
T_{12}	32.56	41.44	37.00	87.10	87.72	88.4	53.58	56.16	55.77	29.60	29.60	29.60
T_{13}	29.28	28.52	29.52	58.00	60.18	59.59	43.40	45.44	45.12	24.22	24.08	24.22
T_{14}	35.52	40.23	38.74	84.48	86.43	86.43	52.54	55.77	54.34	27.75	29.44	29.60
T_{15}	44.37	48.05	46.20	96.56	100.74	98.64	56.94	55.86	57.33	31.79	33.84	33.84
T_{16}	57.92	62.56	60.39	112.18	115.83	114.4	67.05	69.46	69.00	36.67	40.53	38.60
T_{17}	28.16	25.20	26.62	74.40	73.20	74.42	37.80	38.28	38.86	21.45	22.96	23.10
T_{18}	13.60	14.62	14.62	41.16	40.32	41.16	9.60	9.79	10.68	10.89	9.84	10.98
M	32.16	35.36	33.75	76.26	78.05	77.60	47.20	48.50	48.10	26.55	28.48	27.54

	SEm±	CD	SEm±	CD	SEm±	CD	SEm±	CD
Treatments	0.4578	0.9510	0.9982	1.9758	0.4586	0.9552	0.2654	0.5102
Methods	0.1520	0.3110	0.4120	0.7210	0.1562	0.3202	0.1035	0.2015
Treatments × methods	0.6570	1.3420	1.3825	2.7820	0.6589	1.3510	0.3450	0.7210

Table 2.3: Effect of Different Levels of Urban and Agricultural Waste Composts on P Uptake (kg ha^{-1}) by Plant in Vegetable–Vegetable and Cereal–Pulse Cropping Sequence

Treatments	*Kharif, 2002*						*Kharif, 2003*					
	Tomato			*Maize*			*Bhendi*			*Soybean*		
	A	*An*	*M*	*A*	*An*	*M*	*A*	*An*	*M*	*A*	*An*	*M*
T_1	11.52	12.50	12.01	31.11	33.48	32.60	13.95	13.2	13.95	4.76	4.76	4.76
T_2	17.94	19.32	18.63	48.28	51.80	50.37	23.18	23.79	23.79	5.70	6.24	6.24
T_3	14.50	15.60	15.05	41.58	40.95	41.58	17.68	17.50	17.85	4.76	4.90	4.80
T_4	15.25	16.12	15.81	44.85	46.20	45.85	20.52	19.25	20.16	5.25	5.25	5.25
T_5	17.42	18.09	17.76	46.86	47.57	47.36	21.60	22.57	22.57	5.25	5.76	5.76
T_6	10.58	11.28	11.05	28.80	29.89	29.65	13.64	13.95	13.95	4.62	4.76	4.76
T_7	16.50	18.76	18.09	44.88	47.50	45.89	22.20	23.18	23.18	5.55	5.70	5.60
T_8	13.34	14.16	13.87	36.00	37.20	36.60	13.64	14.08	14.08	4.90	4.68	5.04
T_9	18.75	22.04	20.52	53.72	55.20	54.11	24.57	24.00	24.8	6.63	7.79	6.80
T_{10}	16.56	17.25	16.91	43.40	42.70	43.05	17.50	20.14	19.24	5.25	5.04	5.40
T_{11}	16.80	18.46	17.75	44.80	46.80	45.80	21.09	20.52	21.28	5.92	5.92	5.92
T_{12}	18.50	20.72	19.61	48.91,	51.68	49.95	22.42	23.01	23.01	5.92	6.24	6.08
T_{13}	12.10	13.11	12.60	31.90	33.04	32.76	13.02	14.08	13.76	5.04	5.04	5.04

Contd...

Table 2.3–Contd...

Treatments	Kharif, 2002						Kharif, 2003					
	Tomato			Maize			Bhendi			Soybean		
	A	An	M	A	An	M	A	An	M	A	An	M
T_{14}	16.80	19.17	18.02	48.18	50.25	49.21	22.20	23.40	22.80	5.70	6.40	6.40
T_{15}	20.59	22.01	21.30	51.12	54.75	53.28	22.62	21.28	22.23	6.63	7.20	7.20
T_{16}	25.60	27.54	26.73	65.57	68.04	67.20	32.40	32.20	32.66	8.55	9.45	9.00
T_{17}	10.78	10.29	10.54	30.60	30.05	30.86	11.20	10.73	11.31	4.16	4.34	4.48
T_{18}	5.12	5.78	5.45	14.70	15.36	15.04	3.36	3.30	3.48	2.52	2.32	2.61
M	14.88	16.38	15.75	40.99	42.25	41.94	17.00	18.20	17.85	5.40	5.92	5.72

	SEm±	CD	SEm±	CD	SEm±	CD	SEm±	CD
Treatments	0.0961	0.1910	0.3854	0.8015	0.0992	0.1972	0.0386	0.0654
Methods	0.0315	0.0645	0.1385	0.2715	0.0368	0.0710	0.0139	0.0282
Treatments × methods	0.1320	0.2760	0.5325	1.1855	0.1382	0.2820	0.0543	0.1195

Table 2.4: Effect of Different Levels of Urban and Agricultural Waste Composts on K Uptake (kg ha^{-1}) by Plant in Vegetable–Vegetable and Cereal–Pulse Cropping Sequence

Treatments	Kharif, 2002						Kharif, 2003					
	Tomato			Maize			Bhendi			Soybean		
	A	An	M	A	An	M	A	An	M	A	An	M
T_1	21.12	22.50	21.81	76.25	77.50	76.88	29.45	29.7	30.07	22.68	22.96	22.82
T_2	27.30	29.96	28.62	102.68	105.00	104.19	41.8	41.34	42.12	30.3	32.8	32.64
T_3	24.75	26.78	25.76	85.05	85.05	85.05	33.32	34.65	34.65	24.5	25.2	24.92
T_4	25.00	26.52	25.76	91.00	93.06	92.36	36.0	35.7	36.36	27.15	27.15	27.15
T_5	26.26	28.08	27.30	96.36	96.48	96.72	37.80	37.74	38.48	30.0	32.64	32.32
T_6	19.55	21.60	20.68	76.80	76.25	76.23	29.14	30.07	29.76	22.4	22.68	22.54
T_7	25.00	29.12	27.54	99.00	98.49	99.09	39.96	40.28	40.66	30.0	30.0	30.0
T_8	22.77	24.00	23.50	81.00	81.84	81.74	30.38	31.36	31.36	24.36	22.88	24.5
T_9	30.00	35.09	32.67	108.80	112.47	110.29	43.68	42.8	44.00	35.36	39.14	35.19
T_{10}	27.36	29.00	28.18	87.42	86.01	86.72	35.35	39.14	37.74	27.0	25.48	27.15
T_{11}	27.84	30.16	29.00	92.80	92.95	92.88	39.96	39.9	40.66	30.72	31.04	30.88
T_{12}	29.75	33.60	31.80	10 1.84	102.00	101.93	41.04	41.73	42.12	32.48	32.64	32.64
T_{13}	21.78	22.54	22.28	75.98	77.29	76.64	30.07	30.08	30.72	23.8	23.52	23.66
T_{14}	28.32	31.59	30.09	105.60	107.87	107.07	40.70	39.78	40.28	30.0	32.32	32.16
T_{15}	33.64	35.96	34.80	117.86	121.18	119.52	44.85	42.94	44.46	34.68	36.54	36.72
T_{16}	42.56	45.56	44.22	141.41	142.56	141.60	63.9	63.94	64.4	42.94	47.88	45.40
T_{17}	18.70	18.69	18.71	81.60	81.13	81.07	25.2	27.84	26.97	20.54	22.4	22.26

Contd...

Table 2.4–Contd...

Treatments	Kharif, 2002						Kharif, 2003					
	Tomato			Maize			Bhendi			Soybean		
	A	An	M	A	An	M	A	An	M	A	An	M
T_{18}	12.00	12.58	12.38	54.39	52.80	53.35	7.80	6.38	7.44	8.55	7.36	8.46
M	24.90	27.56	26.25	91.59	92.95	92.28	35.02	35.7	36.05	27.45	29.44	29.44

	SEm±	CD	SEm±	CD	SEm±	CD	SEm±	CD
Treatments	0.3290	0.7150	0.5942	1.1540	0.3841	0.8120	0.3358	0.7282
Methods	0.1215	0.2410	0.2415	0.4110	0.1541	0.2980	0.1310	0.2542
Treatments × methods	0.5210	1.1050	0.7860	1.3920	0.6120	1.1420	0.5410	1.1150

The mean N uptake by tomato plant increased significantly from 14.62 (control) to 60.39 kg ha^{-1} (40 t ha^{-1} UW compost) with increase of compost levels. Mean N uptake by bhendi plant was significantly increased from 10.68 in control (T_{18}) to 69 kg ha^{-1} in the treatment with 40 ton of USW compost ha^{-1} (T_{16}). There was significant difference between enriched composts and untreated composts. The mean N uptake by soybean ranged from 10.98 (control) to 38.6 kg ha^{-1} (T_{16}). To the same level of compost application highest N uptake was observed with UW compost treated plots than AW compost treated plots. Further more, Treatments with 50 per cent compost + 50 per cent RDF exhibited highest nutrient uptake than plots with 100 per cent compost application, might be ascribed due to the immediate availability of nutrients from chemical fertilizers than composts.

Table 2.5: Effect of Different Levels of Aerobic and Anaerobic Composts on Uptake of N (kg ha^{-1}) by Grain/Fruit in Vegetable–Vegetable and Cereal–Pulse Cropping Sequence

Treatments	Kharif, 2002						Kharif, 2003					
	Tomato			Maize			Bhendi			Soybean		
	A	An	M	A	An	M	A	An	M	A	An	M
T_1	41.80	43.01	42.61	59.4	60.39	60.06	22.53	24.42	23.53	60.73	61.36	61.04
T_2	50.39	50.62	50.62	77.14	82.00	79.56	26.89	26.16	26.53	69.61	69.23	69.50
T_3	44.10	45.15	44.73	63.24	65.80	65.45	23.52	24.62	24.13	61.96	63.59	62.77
T_4	45.58	45.79	45.79	67.68	70.30	69.93	25.11	25.61	25.42	65.92	65.09	65.50
T_5	47.30	48.17	47.95	74.10	74.86	74.48	26.33	25.18	25.83	67.85	67.84	67.84
T_6	36.20	40.60	38.38	51.00	53.01	53.01	22.55	23.5	23.16	58.50	60.80	59.64
T_7	43.68	44.52	44.31	70.30	71.41	71.04	24.68	25.20	24.95	68.68	67.20	67.94
T_8	41.78	42.61	42.40	59.84	62.37	62.04	25.49	24.62	24.14	63.76	65.60	64.71
T_9	49.22	50.81	50.13	82.08	83.22	82.84	26.05	27.10	26.57	79.00	79.58	79.29
T_{10}	44.42	46.42	45.53	64.94	65.62	65.28	24.49	24.43	24.53	68.80	68.68	68.74
T_{11}	46.59	47.29	47.05	69.30	71.28	71.28	25.28	24.94	25.12	73.28	73.04	73.16
T_{12}	48.56	48.79	48.79	74.00	75.48	74.74	25.76	26.61	26.19	77.49	77.61	77.55

Contd...

Table 2.5–Contd...

Treatments	*Kharif, 2002*						*Kharif, 2003*					
	Tomato			*Maize*			*Bhendi*			*Soybean*		
	A	*An*	*M*	*A*	*An*	*M*	*A*	*An*	*M*	*A*	*An*	*M*
T_{13}	38.69	39.08	39.08	54.60	58.88	56.73	23.28	24.32	23.80	61.60	63.96	62.77
T_{14}	45.90	47.69	47.01	69.30	70.56	70.92	26.32	26.65	26.49	78.20	78.99	78.77
T_{15}	56.64	57.60	57.12	85.20	88.97	88.15	28.48	28.92	28.70	85.86	86.08	85.97
T_{16}	68.89	71.28	70.22	103.8	107.1	106.6	32.09	32.57	32.33	102.96	108.0	105.47
T_{17}	40.00	39.37	39.79	63.72	63.36	63.72	20.84	20.93	20.90	52.85	50.95	51.90
T_{18}	25.56	26.25	26.06	30.80	3 1.24	31.02	7.12	7.53	7.32	27.36	27.42	27.39
M	45.20	48.83	47.09	66.85	70.20	69.48	25.29	25.91	25.66	67.14	68.04	67.59

	SEm±	CD	SEm±	CD	SEm±	CD	SEm±	CD
Treatments	0.5120	0.9890	0.6358	1.2015	0.2985	0.5612	0.6321	1.1910
Methods	0.1730	0.3410	0.2668	0.5012	0.1065	0.2210	0.2627	0.4995
Treatments × methods	0.6850	1.3815	0.8951	1.7254	0.3584	0.6845	0.8895	1.7212

The variation in N, P and K uptake among A W composts and USW composts might be attributed to their inherent capability to supply nutrients in the crop growth period, which in turn effected the dry matter production and hence N uptake by the plants. During second season also the nutrient uptake was significantly maximum with 100 per cent compost level (*i.e.* 40 t/ha) than RDF and control. This prompts that the applied nutrients through organic sources will not be exhausted in one season, and will have residual effect on succeeding crop also. Although heavy metals have been tied with Urban Solid Waste compost as a potential source of pollution, present study showed that urban waste compost addition as high as 40 t/ha apparently did not have any deleterious effect on plant growth (Bhattacharya *et al.*, 2001).

Uptake of Nutrients by Edible Parts of Different Crops

Results reveal that nutrient uptake by tomato fruit, maize cobs, bhendi fruit and soybean grain was significantly improved with the increasing levels of composts used (Tables 2.5, 2.6 and 2.7). The uptake was significantly more in 40 tons of USW compost per hectare over control, followed by 40 t. of A W compost ha^{-1}. The treatment with 40 t ha^{-1} UW compost showed maximum N, P, K and S uptake by edible parts of different crops were 70.22, 24.37, 37.20 and 6.03 kg ha^{-1} in tomato, 106.6, 36.45, 80.55 and 10.85, kg ha^{-1} in maize, 32.33, 11.22, 21.23 and 3.89 kg ha^{-1} in bhendi and 105.47, 10.94, 36.21 and 6.16 kg ha^{-1} in soybean, respectively. Phosphorus accumulation in edible parts was more as compared to haulm portion. This may be attributed to the fact that the phosphorus accumulates as phytic acid, which is most important portion of any economic part. Hence, most of the absorbed phosphorus gets translocated to edible portion at the time of maturity. Unlike incase of other nutrients, potassium uptake was higher in haulms compared to fruit/grain indicating that the adsorbed K had greater role to play in vegetative parts of the plants because of its involvement in translocation of photosynthates from source to sink.

Table 2.6: Effect of Different Levels of Urban and Agricultural Waste Composts on Uptake of P (kg ha^{-1}) by Grain/Fruit in Vegetable–Vegetable and Cereal–Pulse Cropping Sequence

Treatments	*Kharif, 2002*						*Kharif, 2003*					
	Tomato			*Maize*			*Bhendi*			*Soybean*		
	A	*An*	*M*	*A*	*An*	*M*	*A*	*An*	*M*	*A*	*An*	*M*
T_1	14.21	17.77	14.63	19.14	19.14	19.14	6.13	6.35	6.30	6.25	6.44	6.42
T_2	17.56	17.85	17.71	27.36	29.14	28.47	8.48	8.24	8.36	8.02	7.88	7.95
T_3	14.70	15.12	15.05	23.12	22.75	23.45	6.84	7.37	7.10	6.81	6.82	6.82
T_4	15.62	15.98	15.91	24.84	25.90	25.90	7.69	7.73	7.71	7.52	7.58	7.63
T_5	16.50	16.73	16.85	26.98	27.36	27.36	8.25	7.98	8.18	7.55	7.84	7.69
T_6	11.58	13.40	12.54	15.00	15.19	15.50	5.94	5.92	6.00	6.00	6.23	6.19
T_7	15.54	16.19	15.90	25.53	25.90	25.90	7.52	7.89	7.77	7.77	7.68	7.72
T_8	14.54	14.67	14.60	22.08	22.11	22.44	6.96	6.87	6.65	6.85	6.72	6.87
T_9	17.33	17.96	17.79	29.64	30.02	30.02	8.47	8.86	8.73	8.72	8.99	8.94
T_{10}	15.68	16.67	16.27	24.48	24.82	24.82	7.17	7.24	7.27	7.36	7.61	7.56
T_{11}	16.43	16.83	16.67	25.90	27.36	27.00	7.67	7.57	7.62	7.62	7.63	7.63
T_{12}	17.04	17.33	17.33	27.38	28.49	28.12	8.08	8.65	8.43	8.29	8.51	8.48
T_{13}	13.02	13.464	13.28	18.30	19.20	18.91	6.30	6.57	6.50	6.46	6.70	6.66
T_{14}	16.12	16.88	16.64	24.85	25.92	25.92	8.56	8.58	8.57	8.5	8.52	8.60
T_{15}	19.04	19.68	19.36	30.00	31.16	31.16	8.25	8.30	8.28	8.51	8.75	8.72
T_{16}	23.83	24.55	24.37	35.20	36.45	36.45	11.04	11.55	11.22	10.89	11.00	10.94
T_{17}	13.20	13.124	13.20	19.80	19.80	19.80	5.48	5.73	5.61	5.73	5.36	5.55
T_{18}	7.24	7.25	7.34	9.46	9.68	9.68	1.61	1.48	1.55	2.73	2.63	2.73
M	15.41	16.80	16.20	23.45	24.48	24.48	7.05	7.24	7.20	7.17	7.20	7.19

	SEm±	CD	SEm±	CD	SEm±	CD	SEm±	CD
Treatments	0.1012	0.1995	0.3886	0.8102	0.0589	0.1019	0.0588	0.1017
Methods	0.0356	0.0685	0.1012	0.1985	0.0202	0.0415	0.0200	0.0411
Treatments × Methods	0.1412	0.2880	0.3612	0.9210	0.0785	0.1412	0.0782	0.1408

Table 2.7: Effect of Different Levels of Urban and Agricultural Waste Composts on Uptake of K (kg ha^{-1}) by Grain/Fruit in Vegetable–Vegetable and Cereal–Pulse Cropping Sequence

Treatments	*Kharif, 2002*						*Kharif, 2003*					
	Tomato			*Maize*			*Bhendi*			*Soybean*		
	A	*An*	*M*	*A*	*An*	*M*	*A*	*An*	*M*	*A*	*An*	*M*
T_1	21.11	21.40	21.41	43.23	42.90	43.23	14.75	15.45	15.16	24.72	25.15	24.93
T_2	28.50	28.70	28.60	56.24	59.60	58.11	17.35	17.14	17.25	28.50	27.85	28.25
T_3	23.73	24.28	24.17	45.90	47.25	47.25	15.36	16.00	15.68	26.02	26.21	26.19
T_4	24.82	25.06	24.94	49.32	50.32	50.69	16.40	16.35	16.44	27.20	26.86	27.03

Contd...

Table 2.7–Contd...

Treatments	Kharif, 2002						Kharif, 2003					
	Tomato			Maize			Bhendi			Soybean		
	A	An	M	A	An	M	A	An	M	A	An	M
T_5	26.18	26.98	26.64	52.82	53.20	53.20	17.16	16.46	16.88	27.65	27.84	27.74
T_6	18.10	20.00	19.00	39.00	40.61	40.61	14.74	15.04	14.88	24.30	25.08	24.76
T_7	26.04	26.84	26.50	53.28	54.75	54.02	16.04	16.93	16.55	27.86	27.68	27.85
T_8	21.58	22.31	22.00	44.16	45.87	45.87	16.68	15.50	15.52	26.77	27.20	27.00
T_9	28.25	28.47	28.43	56.62	57.00	57.00	16.95	17.85	17.46	30.09	30.62	30.44
T_{10}	24.32	25.32	24.93	47.94	48.62	48.28	15.95	15.96	16.02	27.20	27.86	27.53
T_{11}	26.21	27.26	26.80	50.70	52.56	52.60	16.60	16.38	16.50	28.51	28.55	28.53
T_{12}	27.26	28.03	27.82	54.76	53.65	54.39	16.92	17.44	17.25	29.61	29.95	29.86
T_{13}	20.09	20.38	20.38	41.40	43.52	42.47	15.16	15.14	15.21	25.41	25.89	25.73
T_{14}	25.50	27.01	26.42	51.10	51.84	52.20	16.99	17.68	17.40	29.58	30.45	30.10
T_{15}	28.56	29.28	28.92	66.00	68.06	68.06	19.08	19.33	19.27	31.04	31.43	31.32
T_{16}	36.52	37.75	37.20	78.32	80.55	80.55	22.09	21.61	21.85	35.83	36.40	36.21
T_{17}	20.00	19.49	19.90	49.68	50.04	50.04	11.62	12.12	11.88	25.21	25.00	25.20
T_{18}	12.35	13.20	12.82	26.40	26.62	26.62	4.02	4.029	4.05	9.30	9.44	9.37
M	24.16	26.43	25.27	50.05	51.48	51.48	15.44	15.71	15.64	26.63	27.05	26.84

	SEm±	CD	SEm±	CD	SEm±	CD	SEm±	CD
Treatments	0.2570	0.5490	0.4425	0.9412	0.0974	0.1998	0.2610	0.5510
Methods	0.1090	0.2120	0.1510	0.3112	0.0364	0.0682	0.1108	0.2158
Treatments × Methods	0.4110	0.8410	0.6258	1.2985	0.1382	0.2910	0.4210	0.8468

Table 2.8: Effect of Different Levels of Urban and Agricultural Wastes Composts on Plant Yield (q ha^{-1}) in Vegetable–Vegetable and Cereal–Pulse Cropping Sequence

Treatments	Kharif, 2002						Kharif, 2003					
	Tomato			Maize			Bhendi			Soybean		
	A	An	M	A	An	M	A	An	M	A	An	M
T_1	146.6	149.8	148.2	33	33	33	45.40	46.10	45.75	15.26	15.34	15.30
T_2	159.9	158.5	159.2	38	40	39	47.10	47.80	47.45	16.38	16.10	16.24
T_3	146.7	149.0	147.9	34	35	35	46.00	46.28	46.14	15.49	15.51	15.50
T_4	149.8	151.5	150.7	36	37	37	46.85	46.91	46.88	16.00	15.80	15.90
T_5	154.3	156.0	155.2	38	38	38	47.00	47.40	47.20	16.08	16.00	16.04
T_6	136.7	140.0	138.4	30	31	31	45.21	46.00	45.61	15.00	15.20	15.10
T_7	146.7	149.0	147.9	37	37	37	46.90	47.10	47.00	16.20	16.00	16.10
T_8	138.3	141.0	139.7	32	33	33	46.12	46.24	46.18	15.94	16.00	15.98

Contd...

Table 2.8–Contd...

Treatments	*Kharif, 2002*						*Kharif, 2003*					
	Tomato			*Maize*			*Bhendi*			*Soybean*		
	A	*An*	*M*	*A*	*An*	*M*	*A*	*An*	*M*	*A*	*An*	*M*
T_9	150.0	153.0	151.5	38	38	38	48.08	48.22	48.15	17.10	17.30	17.20
T_{10}	141.0	147.5	144.3	34	34	34	47.00	47.14	47.04	16.00	16.20	16.10
T_{11}	145.3	149.0	147.2	35	36	36	47.75	47.85	47.80	16.58	16.60	16.59
T_{12}	149.0	150.0	149.5	37	37	37	48.00	48.10	48.05	16.92	17.02	16.97
T_{13}	129.9	131.0	130.5	30	32	31	46.20	46.60	46.40	15.40	15.60	15.50
T_{14}	142.7	148.0	145.4	35	36	36	48.00	48.10	48.05	17.00	17.40	17.20
T_{15}	166.7	168.0	167.4	40	41	41	52.65	52.85	52.70	17.74	17.86	17.80
T_{16}	181.0	185.0	183.0	44	45	45	55.12	56.00	55.56	19.80	20.00	19.90
T_{17}	140.0	135.0	137.5	36	36	36	42.80	43.00	42.90	15.10	14.90	15.00
T_{18}	99.3	101.5	100.4	22	22	22	16.12	15.82	15.97	9.12	9.08	9.10
M	145.8	156.6	151.2	35	36	36	45.68	46.05	45.87	15.95	16.01	15.98

	SEm±	CD	SEm±	CD	SEm±	CD	SEm±	CD
Treatments	0.815	1.644	0.3772	0.7702	0.4472	0.9219	0.0808	0.1649
Methods	0.273	0.555	0.1257	0.2567	0.1386	0.3088	0.0271	0.0550
Treatments × methods	1.146	2.331	0.4834	0.9875	0.6581	1.3478	0.1142	0.2331

References

Bhattacharya, P., Chakrabarti, K., Chakraborty, A.A. and Bhattacharya, B., 2001. Microbial biomass and activities of soils amended with municipal solid waste compost. *Journal of the Indian Society of Soil Science*, 49(1): 98–104.

Jackson, M.L., 1973. *Soil Chemical Analysis*, Prentice Hall of India Pvt. Ltd., New Delhi.

Gallardo-Lara, F. and Nogales, P., 1987. *Biol. Wastes*, 19: 35.

He, X.T., Traina, S.J. and Logan, T.J., 1992. *J. Environmental Quality*, 21: 318.

Tester, C.F., Sikora, L.J., Taylor, J.M. and Parr, J.F., 1977. *Journal of Environmental Quality*, 6: 459.

Chapter 3

Microbial Fuel Cells: A Bioelectrocatalytic Technology for Bioelectricity, Biohydrogen, Biosensors, Wastewater Treatment and Biomedical Applications

M. Vijay* and Rm. Meyyappan
Department of Chemical Engineering, Annamalai University, Annamalai Nagar – 608 002, Tamil Nadu

ABSTRACT

Microbial fuel cells (MFCs) are similar to traditional fuel cells, except the metallic electro catalyst is replaced with a biological electro catalyst. MFCs are not yet commercialized but they show great promise as a method of wastewater treatment and as power sources. The power produced by these systems is currently limited, due to high internal resistance. However, improvements in the system architecture will soon result in power generation that is dependent on the capabilities of the microorganisms. The detection of new electronic properties of microorganisms/enzymes makes them even more attractive for the growing fields. The results of various experimental studies shows that increased power densities in BFC will be obtained, only when the internal resistance of the system is reduced. Methods are needed to control the dissolved oxygen flux into the anode chamber in order to increase overall columbic efficiency.

* Corresponding Author: E-mail: vjtech98@yahoo.co.in.

Here we discussed the multidisciplinary role of MFC in the fields of wastewater treatment, biosensors, bio-hydrogen production and biomedical applications, thus solving all the current global issues.

Keywords: *Fuel cell, Electro catalyst, Microbial fuel cell, Enzymes, Wastewater*

Introduction

Technological advancement made mankind, to an increased energy consumption which leads to a global energy deficit. Researches are attempting to produce energy from renewable sources to overcome our dependency on oil. Techniques have been developed to produce ethanol from corn and biomass but after ethanol is produced, methods are needed to produce power from ethanol. Moreover, the demand on the power supplies of portable electronic equipment such as laptops and cellular phones has substantially increased due to additional technical features and miniaturization efforts. Well-developed lithium ion battery technology currently powers most of these applications. It is unlikely, however, that improvements in battery technology will keep pace with the accelerating power demand. In order to enable further advances, and other alternatives such as fuel cells were considered which offer an environmental friendly alternative to fossil fuels. Scientists attempted to develop ethanol fuel cells but traditional platinum based catalysts are unable to catalyze the oxidation of ethanol efficiently and are easily passivated. In addition the high cost associated with the development of new small scale fuel cell power supplies are surpassed by the growing portable power needs of the mobile work force and the major industries that now supports them. Direct methanol fuel cells exhibit relatively high power and energy densities. There are several problems associated .with fuel cells, for instance, methanol cross over and slow anode kinetics reduce efficiency and open-circuit voltage. When considering energy supply for applications, fuel fells have some disadvantages because the reaction conditions are usually harsh: high temperature and pressure may be strongly acidic or alkaline solutions. In many applications, platinum is used as catalyst, which is expensive and resources are limited. Further it is easily poisoned by CO, which will be formed if the used fuel is not pure hydrogen (Larmine *et al.*, 2003 and Katz *et al.*, 2003).

Basic Concepts

Fuel Cell-Functions

Fuel cells generate electricity from a simple electrochemical reaction in which an oxidizer, typically oxygen from air, and a fuel, typically hydrogen, combine to form a product, which is water for the typical fuel cell. Oxygen (air) continuously passes over the cathode and hydrogen passes over the anode to generate electricity with by-products as heat and water. The fuel cell itself has no moving parts-making it a quiet and reliable source of power.

Type of Fuel Cells

Fuel cells are generally categorized by their electrolyte-the material sandwiched between the two electrodes. This material's characteristics determine the optimal operating temperature and the fuel used to generate electricity. Each comes with its particular set of benefits and shortcomings. Table 3.1 shows types of fuel cells differing with their electrolytes (Katz *et al.*, 2003).

Table 3.1: Types of Fuel Cells (Katz *et al.*, 2003)

Fuel Cell Type	*Electrolyte*	*Anode Gas*	*Cathode Gas*	*Temperature*	*Efficiency*
Proton Exchange Membrane (PEM)	Solid polymer membrane	Hydrogen	Pure or atmospheric oxygen	75°C (180°F)	35–60 per cent
Alkaline (AFC)	Potassium hydroxide	Hydrogen	Pure oxygen	Below 80°C	50–70 per cent
Direct Methanol (DMFC)	Solid polymer membrane	Methanol solution in water	Atmospheric oxygen	75°C (180°F)	35–40 per cent
Phosphoric Acid (PAFC)	Phosphorus	Hydrogen	Atmospheric oxygen	210° C (400°F)	35–50 per cent
Molten Carbonate (MCFC)	Alkali Carbonates	Hydrogen, methane	Atmospheric oxygen	650°C	40–55 per cent
Solid Oxide (SOFC)	Ceramic Oxide	Hydrogen, methane	Atmospheric oxygen	800–1000°C (1500–1800°F)	45–60 per cent

Recently fuel cell technology has been challenged by emerging microbial fuel cell technology that utilizes biocatalysts such as enzymes and microorganisms, as the electro-catalysts, instead of transition metal catalysts that are utilized in the conventional fuel cell systems. Several organic compounds can be utilized in biofuel cells, especially biomass based compounds can be used as the fuels such as sugars and several alcohols. Higher energy conversion can be achieved when compared with the transition metal based fuel cells theoretically.

Biofuel Cell–Mechanism

A biofuel cell is a device that directly converts microbial power into electricity using electrochemical technology. Chemical energy is converted to electric energy by coupling the biodegradative oxidation or organic or inorganic substrates to the chemical reduction of an oxidant at the interface between the anode and the cathode (Willner *et al.*, 1998).

A Biofuel cell consists of anode and cathode separated by a proton exchange membrane(optional). In the anode compartment, microorganisms/enzymes oxidize fuel (substrate) generating electrons and protons. These protons are transferred through the membrane and the electrons through the external circuit. They are consumed by reducing oxidant at the cathode, usually oxygen supplied by aeration.

Equation (1): $2H_2 + 4OH^- \rightarrow 4H_2O + 4e^-$ (At anode)

Equation (2): $O_2 + 2H_2O + 4e^- \rightarrow 4OH$ (At Cathode)

Types of Biological Fuel Cells

There are two types of biological fuel cells, namely microbial fuel cells and enzymatic fuel cells.

Microbial Fuel Cells

The use of microorganisms in fuel cells eliminates the isolation of individual enzymes, thereby providing cheaper substrates for biological fuel cells. Microorganisms can be used in four ways for producing electrical energy (Katz *et al.*, 2003 and Shukla *et al.*, 2004).

1. Microorganisms can produce electrochemically active substances through fermentation or metabolism. For the purpose of energy generation, fuels are produced in separate reactors and transported to the anode of a conventional fuel cell. In this configuration, the microbial bioreactor is kept separated from the fuel cell (Figure 3.1A).

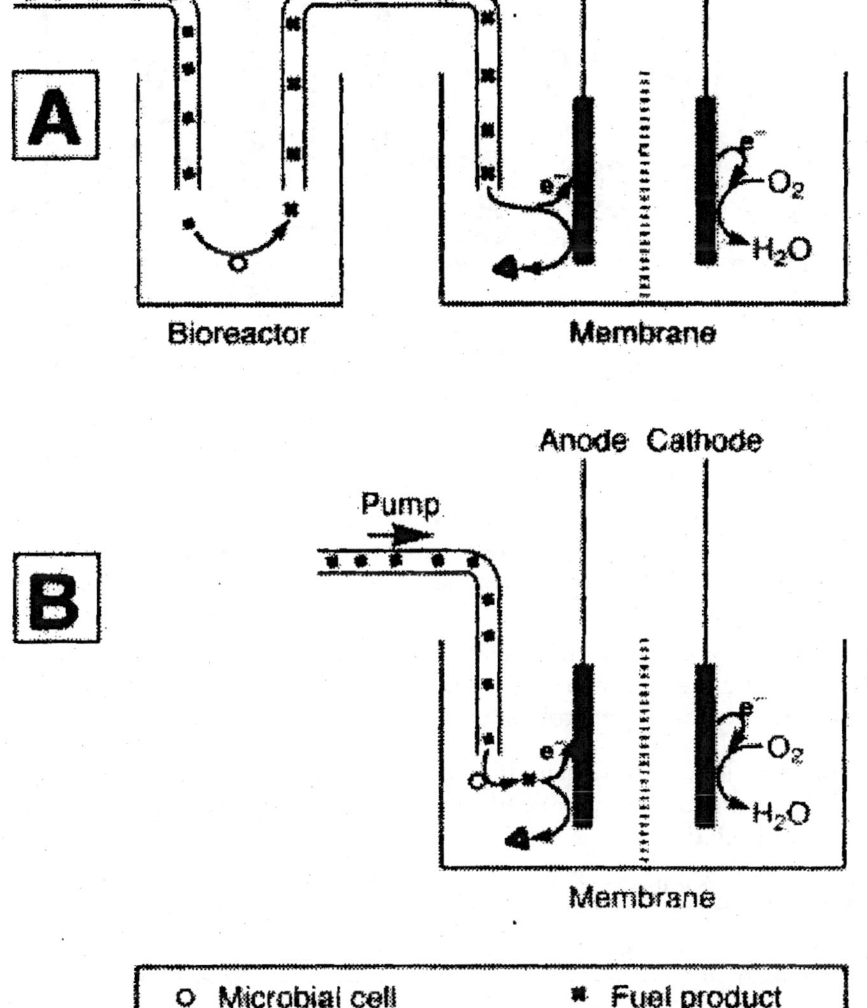

Figure 3.1: Schematic Configuration of a Microbial Fuel Cell:
(A) Bioreactor Providing Fuel Separated from the Anode Compartment;
(B) Bioreactor Providing Fuel Directly in the Anode Compartment (Shukla *et al.*, 2004)

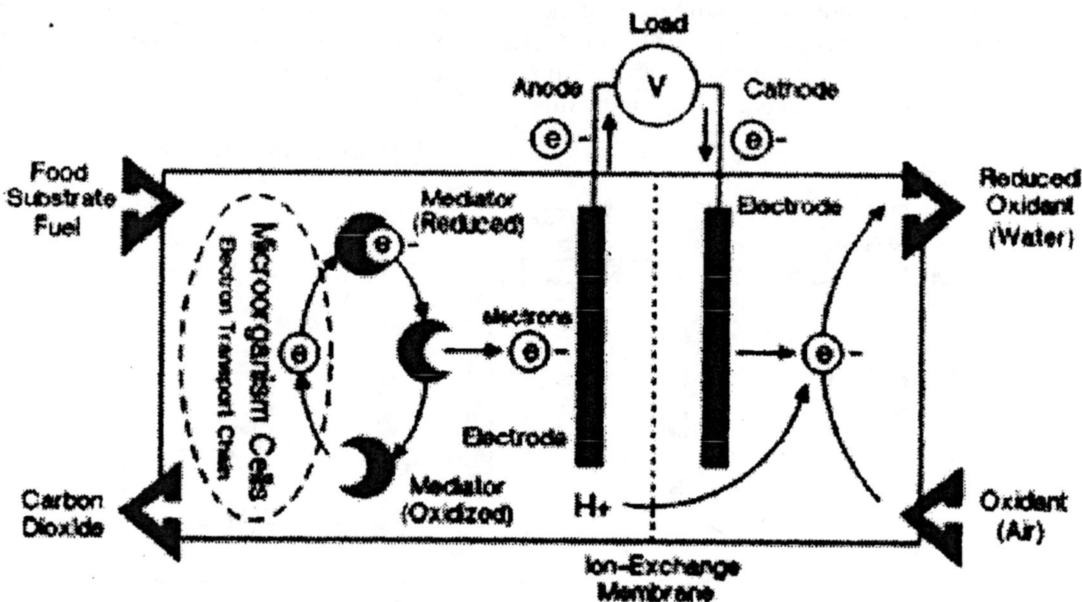

Figure 3.2: Schematic Configuration of a Mediator Microbial Fuel Cell (www.ncbe.reading.ac.uk)

2. In the second configuration, the microbiological fermentation process proceeds directly in the anodic compartment of the fuel cell (Figure 3.1B).
3. In the third configuration, electron-transfer mediators shuttle electrons between the microbial biocatalytic system and the electrode. The mediator molecules accept electrons from the biological electron transport chain of the micro-organisms and transport them to the anode of the biological fuel cell (Figure 3.2)
4. In the fourth configuration, the metal-reducing bacterium having cytochromes in its outer membrane and the ability to communicate electrically with the electrode surface directly result in a mediator-less microbial fuel cell.

The conversion of carbohydrate to hydrogen is achieved by a multienzyme system. In practice, improvement in H_2 production is possible by genetic engineering techniques and screening of new hydrogen-producing bacteria.

$$\text{Glucose} + 2\,\text{NAD}^+ \xrightarrow{\text{Multenzyme Embden-Meyerhof Pathway}} 2\,\text{Pyruvate} + 2\,\text{NADH}$$

$$\text{Pyruvate} + \text{Ferredoxin}_{ox} \xrightarrow{\text{Pyruvate-ferredoxin Oxidoreductase}} \text{Acetyl–CoA} + CO_2$$

$$\text{NADH} + \text{Ferredoxin}_{ox} \xrightarrow{\text{NADH-ferredoxin Oxidoreductase}} \text{NAD}^+ + \text{Ferredoxin}_{red}$$

$$\text{Ferredoxin}_{red} + 2H^+ \xrightarrow{\text{Hydrogenase}} \text{Ferredoxin}_{ox} + H_2$$

$$H_2 \rightarrow 2H^+ + 2e^- \text{ (at anode)}$$

Microbial Fuel Cell (MFC) can be classified as two types depending on how electrons are transferred from bacteria to the electrode (anode). Mediator MFC where electron shuttles or mediators are added into the system and Mediator less MFCs where no mediators need to be added (Park *et al.*, 2000). Mediators are generally toxic compounds (*e.g.* Thionine, Neutral Red and Azure A); therefore using them is impractical for economic electrical production (Bond *et al.*, 2002). In some cases, bacteria of the type present in mediator less MFCs have electrochemically active redox enzymes on their outer membrane that can transfer electrons to external materials and therefore, do not require exogenous chemicals for electron transfer to an electrode (Leang *et al.*, 2003).

Mediator–Coupled Microbial Fuel Cells

A microbial membrane isolates reductive species generated by metabolic processes inside microbial cells. Thus, the contact of the microbial cells with an electrode usually results in only a diminutive electron-transfer across the membranes of the microbes, except in some special cases. The electro-active groups responsible for the redox activity of enzymes present in the microbial cells are deeply buried inside their prosthetic groups, which leads to poor electrode surface. The cells can, however, be wired to the electrode surface with the help of mediators. Low molecular weight redox species may assistance the shuttling of electrons between the intracellular bacterial space and an electrode, and are referred to as mediators (Park *et al.*, 2000).

Requirements for a Mediator

1. The oxidized mediator should easily penetrate through the bacterial membrane to reach the reductive species inside the bacteria.
2. The redox potential of the mediator should match the potential of the reductive metabolite.
3. None of the oxidation states of the mediator should interfere with other metabolic processes.
4. The reduced mediator should easily escape from the cell through the bacterial membrane.
5. Both the oxidized and reduced states of the mediator should be chemically stable in the electrolyte solution, should be easily soluble, and should not adsorb on the bacterial cells or electrode surface.
6. The electrochemical kinetics of the oxidation process for the mediator reduced state at the electrode should be fast.

Varieties of organic compounds have been studied in combination with bacteria to test the electron-transfer efficiency of the mediator between the microorganism and anode surface. Thionine and organic dyes have been frequently used as mediators in biological fuel cells (Shukla *et al.*, 2004).

Mediator Less Microbial Fuel Cells

Fe(III) reducing microorganisms are found to be electrochemically active as they have cytochromes in their outer membranes. It was first demonstrated with the Fe(III) reducers, *Shewanella putrefaciens* that these can be used as a catalyst in a mediator less microbial fuel cell. Recent studies have demonstrated that Fe(III) reducing micro-organisms of the family *Geobacteraceae* can directly transfer electrons on to electrodes (Bond and Lovley, 2003). Metal-reducing bacteria are the most used species in this type of fuel cells. Studies have been conducted to understand the mechanism of Fe(III) reduction by bacteria such as *Shewanella putrefaciens* and *Geobacter metallireducens.* Development of mediator-less microbial fuel cells using *Shewanella putrefaciens* IR-1 has been reported by Kim *et al.* In these fuel cells, the electro-chemical activity of the micro-organism has been confirmed by cyclic voltammetry when grown under anaerobic conditions, although no activity was found when they were grown under aerobic conditions. Table 3.2 shows various metal reducing bacteria so far utilized in MFC. Recently, there is a report on the bacteria, *Rhodoferax ferrireducens* that can be used in microbial fuel cells effectively without a mediator (Chaudhuri and Lovley, 2003). *R. ferrireducens* was isolated from anoxic sub-surface sediments of Oyster Bay, Virginia, USA as a dissimilatory Fe(III) reducing microorganism. *R. ferrireducens* grows on glucose in the presence of Fe(III) reduction can be explained as below:

$$C_6H_{12}O_6 + 6H_2O + 24Fe(III) \rightarrow 6CO_2 + 24Fe(II) + 24^+ + 24e^-$$

Table 3.2: Outline of Metal Reducing Bacteria Applied in MFC

Microorganism	*Authors*
Shewanella putrefaciens	Bond and Lovley 2003; Kim *et al.*, 1999a; Kim *et al.*, 1999b; Kim *et al.*, 2002; Schroder *et al.*, 2003
Geobacter sulfurreducens	Bond and Lovley 2003
Geobacter metallireducens	Bond *et al.*, 2002
Desulfuromonas acetoxidans	Bond *et al.*, 2002
Rhodoferax ferrireducens	Chaudhuri and Lovley, 2003

Enzymatic Fuel Cells

Microbial based biofuel cells require the continuous fermentation of whole living cells performing numerous physiological processes, and thus dictate stringent working conditions. In order to overcome this constraint, the redox enzymes responsible for desired processes may be separated and purified from living organisms and applied as biocatalysts in biofuel cells rather then using the whole microbial cells. Enzymes have a complex structure comprising proteins. The electron-transferring unit of the enzyme, namely the apoenzyme and cofactor, is deeply buried inside its complex structure (Figure 2.3) Hence, redox enzymes lack direct electrical communication with electrodes due to the insulation of

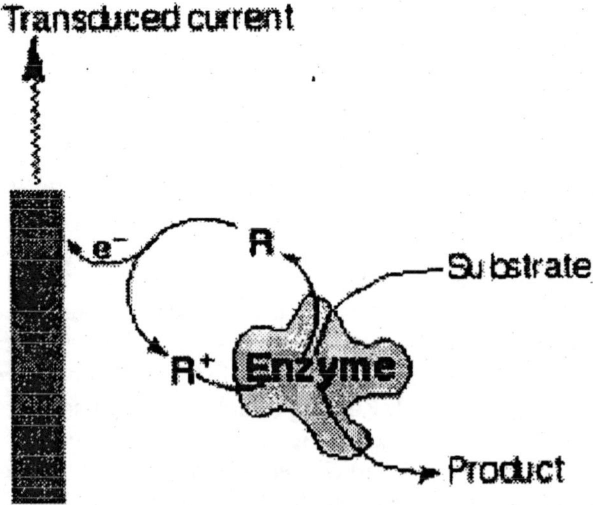

Figure 3.3: Redox-Enzymes and Electrode Supports (Shukla *et al.*, 2004)

the redox center from the conductive support by the protein matrices. Several methods have been applied to electrically contact redox enzymes and electrode supports (Aame *et al.*, 2004). Mediated bioelectrocatalysis used small redox molecules as electron-transfer mediators between the enzyme and the electrode. Mediated bioelectrocatalysis is useful for studying electron-transfer kinetics of the enzymes and is a key reaction in applying the enzymes to biological fuel cells. Table 3.3 shows the comparison of direct methanol fuel cell, microbial fuel cell and enzymatic fuel cell (Shukla *et al.*, 2004).

Table 3.3: Comparison of Direct Methanol Fuel Cell, Microbial Fuel Cell and Enzymatic Fuel Cell (Shukla *et al.*, 2004)

	DMFC	*Microbial Fuel Cell*	*Enzymatic Fuel Cell*
Fuel	3 per cent MetOH–H_2O	Glucose, fish meat	10 per cent MetOH–H_2O
Temperature (°C)	90	22–25	22–25
Output	0.5 V@ 300 mA/cm^2	0.5 V @ 0.21 mA/cm^2	0.5V @ 0.21 mA/cm^2
Electrode size	4" x 6" 154.8 cm2	3.5" x 5.5" 117.4 cm^2	2" × 3.5" 45.2 cm^2
Time (operation/hr)	1500	700	700
Catalyst	Pt and Pt-Rh	Bacteria	Methanol dehydrogenase

Design of Microbial Fuel Cells

Table 3.4 shows the summary of MFC components and the materials used to construct them (Logan *et al.*, 2006; Rabaey and Verstraete, 2005; Bullen *et al.*, 2006; Lovley, 2006). MFCs can be constructed in a manner that separates the anode and cathode into separate chambers (two chamber MFC) or in a manner where both are in one compartment (single chamber MFC). Figure 3.4 shows examples of one and two chambered MFCs. The primary difference between single and double chamber MFCs is the absence of an ioh'permeable membrane in the single chamber MFC. Both the anode and cathode are housed in the same compartment and are thus subjected to the same solutions and similar environments. The cathode is exposed to the solution in the chamber on one side and external air on the opposite side. In the double chamber MFC, different conditions (such as anaerobic, aerobic, etc.) can be maintained in each compartment. The rate of oxygen diffusion into the anode without an ion permeable membrane is 2.7 times higher than a double chamber design using a Nafion 117 membrane (Liu and Logan, 2004). The advantages of a single chamber MFC are reduced setup costs, due to the absence of expensive membranes and higher power output. A disadvantage to the single chamber is a decrease in Coulombic efficiency generally attributed to the diffusion of oxygen to the anode area. Facultative anaerobic bacteria will pass their electrons directly to oxygen rather than the surface of the electrode if oxygen is available, reducing the efficiency of the fuel cell.

Stacked Microbial Fuel Cell

A stacked MFC is shown in Figure 3.5E which was designed to investigate the performances of several MFCs when connected in series and in parallel (Aelterman *et al.*, 2006). Enhanced voltage or current output can be achieved by connecting several MFCs in series or in parallel. Coulombic efficiencies differed greatly in the two arrangements with the parallel connection giving about an efficiency six times higher when both the series were operated at the same volumetric flow rate. The parallel-connected stack has higher short circuit current than the series connected stack. This means that higher maximum bio-electrochemical reactirate is allowed in the connection of MFCs in parallel than

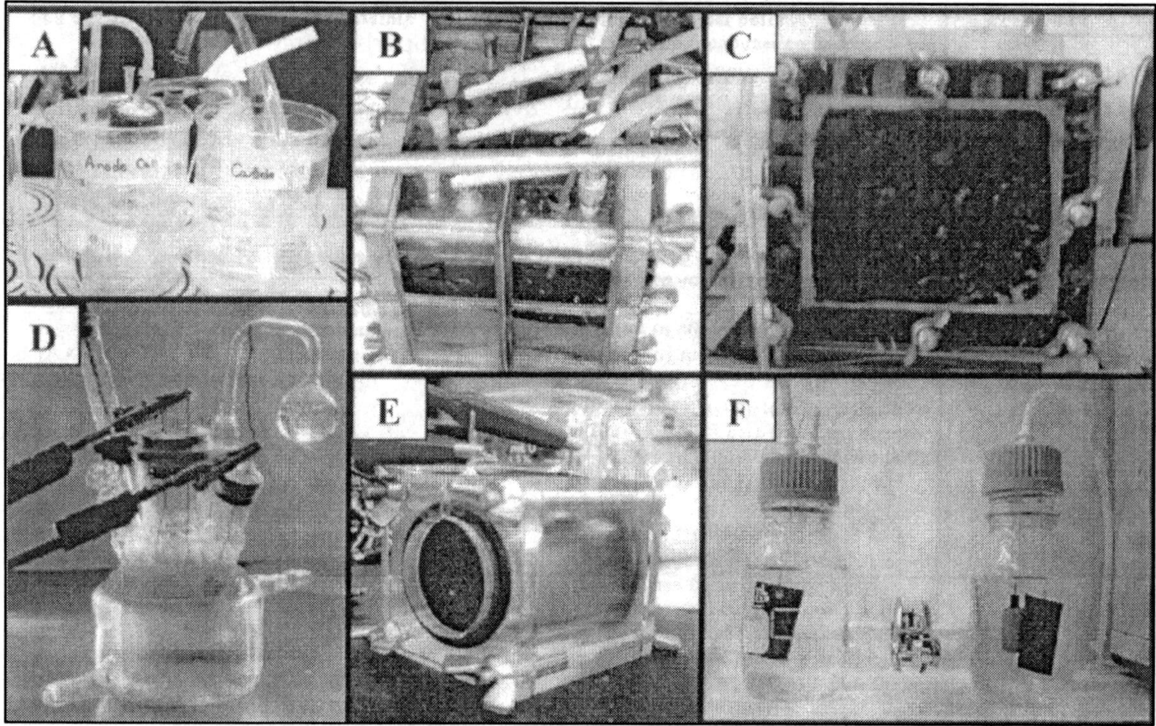

Figure 3.4: Different Types of MFCs Used in Studies (Logan *et al.*, 2006)

in series. Therefore to maximize Chemical Oxygen Demand (COD) removal, a parallel connection is preferred if the MFC units are not independently operated (Aelterman *et al.*, 2006; Zhuwei Du *et al.*, 2007).

Table 3.4: Basic Components of Microbial Fuel Cells (Zhuwei *et al.*, 2007)

Items	*Materials*	*Remarks*
Anode	Graphite, graphite felt, carbon paper, carbon-cloth, Pt, Pt black, reticulated vitreous carbon (RYC)	Necessary
Cathode	Graphite, graphite felt, carbon paper, carbon-cloth, Pt, Pt black, RVC	Necessary
Anodic chamber	Glass, polycarbonate, plexiglas	Necessary
Cathodic chamber	Glass, polycarbonate, plexiglas	Optional
Proton exchange system	Proton exchange membrane: Nafion, Ultrex, polyethylene, poly (styrene-co-divinylbenzene); salt bridge, porcelain septum, or solely electrolyte	Necessary
Electrode catalyst	Pt, Pt black, MnO_2, Fe^{3+}, polyaniline, electron mediator immobilized on anode	Optional

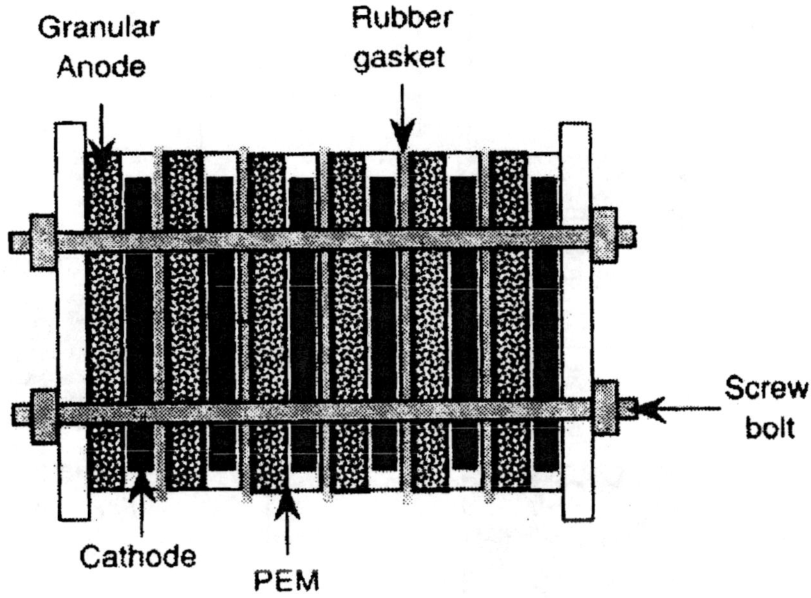

Figure 3.5: Schematic Representation of Stacked MFC (Aelterman *et al.*, 2006)

Membraneless MFC

The Schematic diagram of the membraneless MFC (ML-MFC) is shown in Figure 3.6 which was designed by Jang *et al.* (2004) working in continuous flow mode. A Plexiglass cylinder was partitioned into two sections by glass wool and glass bead layers. These two sections served as anodic and cathodic chambers, respectively. The disk-shaped graphite felt anode and cathode were placed at the bottom and the top of the reactor, respectively. Tartakovsky and Guiot (2006) designed another MFC with a rectangular container but no partition is achieved using glass wool and glass beads. The feed stream is supplied to the bottom of the anode and the effluent passes through the cathodic chamber and exits at the top continuously (Jang *et al.*, 2004; Moon *et al.*, 2005). There are no separate anolyte and catholyte. To ensure proper operation of the MFCs, DO gradient is achieved through the diffusion barriers between the anode and cathode. The electrodes were connected with platinum wire and the cathode compartment was aerated at various rates for the cathode reaction. In some experiments the graphite cathode was replaced with platinum–coated graphite and also NaCl or HCl were added to the bottom of the cathode.

Electrical Measurements

Power (W) was calculated using relation P = IV, where, I and V represents current (A) and voltage (V), respectively. Power density (mW/m^2) and current density (mA/m^2) were calculated by dividing the obtained power and current by the anode surface area (m^2). Power yield (W/kg COD) was obtained by dividing power with the corresponding substrate (COD) removal values. Volumetric power (mW/m^3) and current (mA/m^3) were calculated based on the anolyte volume. COD, pH and volatile fatty acids (VFA) were monitored in the anodic chamber of MFC during operation according to the standard methods. Overall, the cell current is controlled by the electrode size, Ion permeability and transport

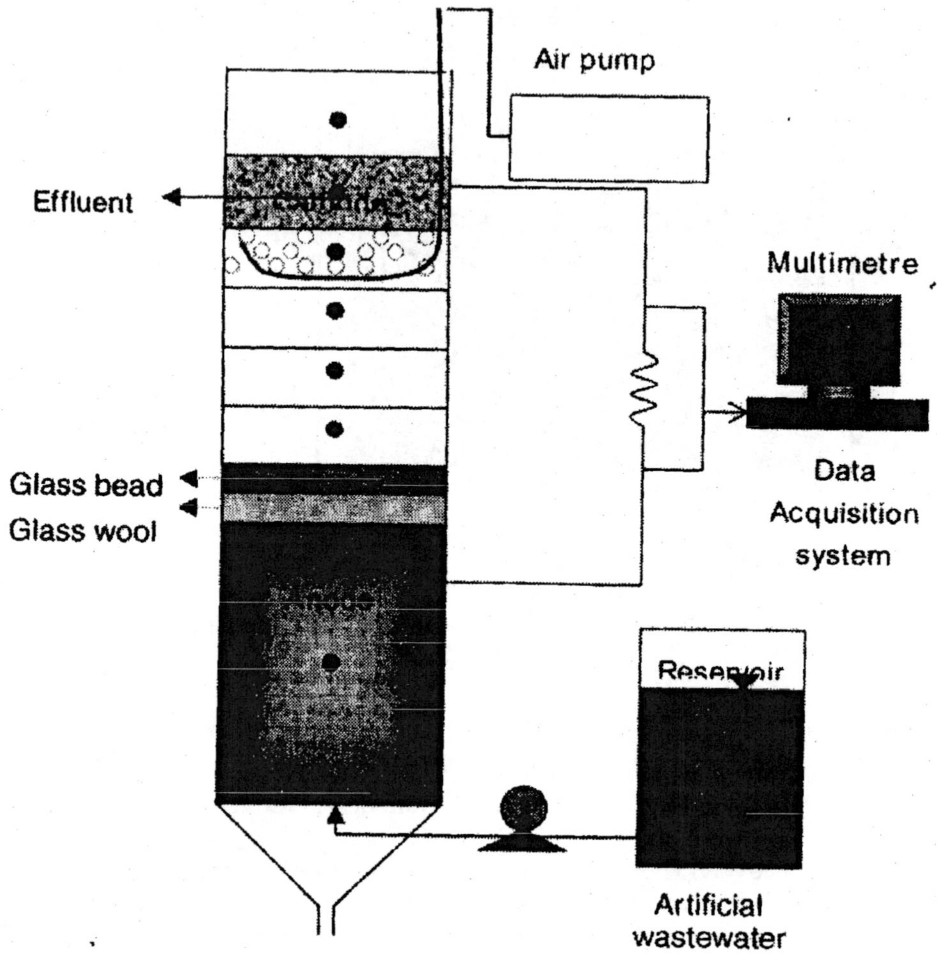

Figure 3.6: Schematic Representation of Membraneless-MFC (Jang *et al.*, 2004)

rates across the proton exchange membrane, which collectively influences the power of biofuel cell (Venkata Mohan S *et al.*, 2007)

Applications and Recent Developments

Power Generation from Wastewater

The application of BFC for wastewater treatment is very interesting due to energy recovery from waste as well as formation of less sludge, disposal of which is very expensive. Major part of energy, which is obtained from the oxidation of the organic contaminants, is converted to electricity and the remaining energy is used for microbial growth. Simultaneous electricity generation along with wastewater treatment provides the basis of a new type of treatment system. This system is referred as the "microbial POWER" (production of wastewater electricity reactor) system since bacteria is used in this wastewater system that results in power generation. Rabaey *et al.* (2004) found that serial enrichment of bacteria on the anode of a BFC resulted in increased power output and a change in the bacterial community.

Liu *et al.* (2004), demonstrated for the first time that electricity could be produced using ordinary domestic wastewater. In a large single chamber MFC, they produced up to 26 mW/m^2 from continuous treatment of domestic wastewater. Along with electricity generation, 80 per cent of the BOD was removed in the process. Removal of BOD was also proportional to detention time, with detention times of 12 hours typically used for this system and found that instead of pumping, passive air transfer to the cathode was sufficient for optimum performance of the system. Power generation was in proportion to wastewater strength (COD), and was linear over the range examined. Increasing the surface area of the electrode will increase power output per volume of reactor, and should decrease detention times needed for wastewater treatment. Table 3.5 shows various power densities achieved using different wastewaters (Logan *et al.*, 2005).

Table 3.5: Power Densities Achieved Using Different Wastewaters (Logan *et al.*, 2005)

Substrate	*Power (mW/m^2)*
Starch wastewater	20
Anaerobic sediments	16
Continuous–large SCMFC	26
Flat plate	76
Batch, small SCMFC (PEM)	28
Batch, small SCMFC (no PEM)	146
Animal wastewater	260

Further improvements have been made using domestic wastewater to increase power output. For example, when the small SCMFC was used to treat wastewater, a maximum power output of 146 m W/m^2 was achieved with domestic wastewater in fed-batch tests (Liu and Logan, 2004). The increase in the power output was attributed to the reduced internal resistance of the small SCMFC compared to the larger system (Logan *et al.*, 2005).

Biosensors

Toxic compounds inhibit metabolic activity of bacteria that results in lower electron transfer towards an electrode. Biosensors are constructed, by immobilizing bacteria onto an electrode and protected behind a membrane. If a toxic component diffuses through the membrane, this can be measured by the change in potential over the sensor. Such sensors could be extremely useful as indicators of toxicants in rivers, at the entrance of wastewater treatment plants, to detect pollution or illegal dumping, or to perform research on polluted sites (Meyer *et al.*, 2002; Chang *et al.*, 2004).

Electricity from Sediments

Based on the potential difference generated by bacteria between sediment and the aqueous phase above, MFCs can be used to generate electricity (Tender *et al.*, 2002). Two anode reactions appear to occur, *i.e.* oxidation of sulphide present in the sediment, which is formed through bacterial oxidation of organic carbon, and oxidation of organic matter by bacteria growing onto the anode. Inspite of the drawback in accessibility, the potential for energy generation from the seafloor is significant.

Bio-hydrogen

Instead of electricity generation, hydrogen production is achieved through MFC is with slight modification. In this, protons and electrons produced by the anodic reaction are combined at the cathode to form hydrogen. The required external potential for an MFC is theoretically 110m V, much lower than the 1210 m V required for direct electrolysis of water fit neutral pH because some energy comes from the biomass oxidation process in the anodic chamber. MFCs can potentially produce about 8–9 mol H_2/mol glucose compared to the typical 4 mol H_2/mol glucose achieved in conventional fermentation (Liu *et al.*, 2005c). In bio-hydrogen production using MFCs, oxygen is no longer needed in the cathodic chamber. This process has been referred to as a bio-electrochemically assisted microbial reactor (BEAMR). Thus, MFC efficiencies improve because oxygen leak to the anodic chamber is no longer an issue. To overcome the inherent low power feature of the MFC, hydrogen can be stored for later usage, thus providing a renewable hydrogen source. (Logan *et al.*, 2005).

Biomedical Applications

Biofuel cell can be applied not only to the biomass conversion, but also as energy source medical devices. Implantable devices operating on chemical batteries often require additional surgeries to replace their battery because of the depletion of chemical potential. However, microbial fuel cells have the ability of producing a continuous electricity source that could be utilized for a variety of purposes. Since humans and microorganisms already have a symbiotic relationship, there may be possibility of using MFC technology to solve the power source problem of implanted biomedical devices. MFCs can operate on a variety of substrates, and the glucose found in the bloodstream could produce electricity in the same way that it fuels the metabolic reactions in the cells of the body. At the same time, if the power output generated by a glucose-operated MFC could be tightly correlated to glucose concentration, which will provide a new long-term option of glucose sensing for millions of diabetic patients (Avinashshantaram *et al.*, 2005).

Conclusion

Power generation in BFC's requires maximum proton transport rates which will be achieved by reducing the internal resistance of the system and minimizing oxygen transport. Nafion membrane has been extensively used in many BFC studies, and it is clear that oxygen permeability through these membranes can reduce coulombic efficiency of the BFC. The use of mixed cultures may ,help to minimize the effects of oxygen diffusion into the anode chamber since it will scavenge any dissolved oxygen, maintaining anaerobic conditions in the anode chamber. A potential application of BFCs in the generation of electricity is from wastewater, includes domestic and high-starch content industrial wastewater. The microbial POWER technology is still in an early stage of development, but shows great promise as a new method to accomplish both wastewater treatment and electricity generation. A key factor in future implementation of the technology will be the cost to build and operate the system. Thus, as we develop costs to produce electricity we can compare costs for a combined wastewater treatment and power generation facility in contrast to that needed for separate facilities. It is clear that combining, wastewater treatment and power generation technologies will be both a challenging and rewarding engineering task.

References

Avinashshantaram, Halukbeyenal, Raajaraajan Angathevarveluchamy and Zbigniewlewandowsk, 2005. Wireless sensors powered by microbial fuel cells. *Environ. Sci. Technol.*, 39: 5037–5042.

Aarne Halme, Xia-Chang Zhang and Anja Ranta, 2000. Study of biological fuel cells, in the second annual advances in R&D the commercialization of small fuel cells and battery technologies for use in portable applications. April 26–28, New Orleans, USA.

Bond, D.R. and Lovely, D.R., 2003. Electricity production by *Geobacter sulfurreducens* attached to electrodes. *Appl. Environ. Microbiol.*, 69: 1548–1555.

Chaudhuri, S.K. and Lovely, D.R., 2003. Electricity generation by direct oxidation of glucose in mediatorless microbial fuel cells. *Nature Biotechnology*, 21: 1229–1232.

Holmes, D.E., Nicoll, J.S., Bond, D.R. and Lovely, D.R., 2004. Potential role of a novel psychrotolerant member of the *Geobacteraceae, Geopsychrobacter electrodiphilus* sp. in electricity production by the marine sediment fuel cell. *Appl. Environ. Microbiol.*, 70: 6023–6030.

Jang, J.K., Pham, T.H., Chang, L.S., Khang, K.H., Moon, H., Cho, K.S. and Kin, R.H., .2004. Construction and operation of a novel mediator and membraneless microbial fuel cells. *Process Biochemistry*, 39: 1007–1012.

Katz, E., Shipway, A.N. and Wilner, I., 2003.Biochemical fuel cells. In: *Handbook of Fuel Cells: Fundamental, Technology and Application*, (Ed.) W. Vielstich, A. Lamm and H.A. Gasteiger. John Wiley and Sons Ltd.

Kim, H.J., Park, H.S., Hyun, M.S., Chang, L.S., Kim, M. and Kim, B.H., 2002. A mediatorless microbial fuel cell using a metal reducing bacterium *Shewanella putrefaciens*. *Enzyme. Microb. Technol.*, 30: 145–152.

Korneal, Rabaey and Willy Verstraete, 2005. Microbial fuel cells: Novel biotechnology for energy generation. *Trends in Biotechnology*, 23: 291–298.

Leang, C., Coppi, M.V. and Lovely, D.R., 2003 .OmcB,a-c-type polyheme cytochrome, involved in Fe(III) reduction in Geobacter Sulfur reduces. *J. Bacterial*, 185: 2096–2103.

Liu, H., Ram Narayanan, R., Logan, B.E., 2004. *Environ. Sci. Technol.*, 38: 2281–2285.

Logan, B.E., 2005. Simultaneous wastewater treatment and biological electricity generation. *Water Science and Technology*, 52: 31–37.

Liu, H., Ramnarayanan, R. and Logan, B.E., 2004. Production of electricity during wastewater treatment using a single chamber microbial fuel cell. *Environ. Sci. Technol.*, 38: 2281–2285.

Logan, B.E. and Grot, S., 2005. A bioelectrochemically assisted microbial reactor (BEAMR) that generates hydrogen gas. Patent application 60/588,022.

Logan, B.E., Hamelers, B., Rozendal, R., Schrodel, V., Keller, J., Stefano Freguia. S., Peter Aelterman, P., Verstraete, W. and Rabaey, K., 2006. Microbial fuel cells: Methodology and technology. *Environ. Sci. and Technol* (In press).

Park, D.H. and Zeikus, J.G., 2000. Electricity generation in microbial fuel cells using neutral red as an electronophore. *Applied and Environmental Microbiology*, 66: 1292–1297.

Rabaey, K., Boon, N., Siciliano, S.D., Verhaege, M. and Verstreate, W., 2004. Biofuel cells select for microbial consortia that self mediate electron transfer. *Apply. Environ. Microbiol.* (In press).

Shukla, A.K., Suresh, P., Berchmans, S. and Rajendran, A., 2004. Biological fuel cells and their applications. *Current Science*, 87: 455–468.

Venkata Mohan, S., Saravanan, R., Veer Raghavulu, S., Mohanakrishna, G. and Sarma, P.N., 2007. Bio-electricity production from wastewater treatment in dual chambered microbial fuel cell (MFC) using selectively enriched mixed microflora: Effect of catholyte. *Bioresource Technology* (In press).

Zhuwei Du, Haoran Li and Tingyue Gu, 2007. A state of the art review on microbial fuel cells: A promising technology for wastewater treatment and bioenergy. *Biotechnology Advances*, 25: 464–482.

Chapter 4

Limnological Studies of Ooty Lake in Tamil Nadu

M. Manimegalai, S. Umavathi and S. Binukumari
P.G. and Research Department of Zoology,
Kongunadu Arts and Science College (Autonomous),
Coimbatore – 641 029, Tamil Nadu

ABSTRACT

Studies on the physico-chemical characteristics of the Ooty lake were carried out for a period of one year from December 2005 to November 2006. The study was performed to determine the pollution status of the pond and to implement the measures of pollution control. The parameters analyzed include the physical parameters such as colour, transparency, atmospheric temperature, water temperature, suspended solids, dissolved solids and total solids. The chemical parameters were evaluated such as pH, carbonate, bicarbonate, total alkalinity, DO, DCO_2, BOD, COD and nutrients such as calcium, magnesium, sulphate, phosphate, chloride and total hardness. The primary productivity of the Ooty lake was also analyzed.

Keywords: *Ooty lake, Physico-chemical characteristics, Primary productivity, Pollution.*

Introduction

Water is not only the most important essential constituent of all animals, plants and other organisms and it is also one of the most basic and important elements for human civilization. Our dependence on freshwater resources has accelerated in the last century due to rapid world population growth and economic development (Postel, 1992; Sharma and Sarang, 2004). The quality of life is closely associated with the quality of environment. With increasing industrialization and pollution growth water source available various purposes such as drinking, recreation, aquaculture, agriculture have been adulterated with industrial as well as human waste. Pollution water contains solids and dissolved organic compounds that impart an offensive odour and serve as an excellent medium for the growth and multiplications of microorganisms (Verma *et al.*, 2007).

As a result freshwater resources have deteriorated both in quantity and quality in many areas of the world. But most of our natural water bodies are gradually becoming, degraded to a. great extent due to rapid progress of industrialization and population explosion. All human civilizations have evolved in close proximity to natural sources of water, especially freshwater. An increasing tendency of these freshwater resources getting degraded due to dumping of industrial, chemical and agricultural effluents, domestic sewage and man made eutrophication etc., in an uncontrolled manner creates potential health hazards and pathogens present in domestic waste are responsible for industrial infections such as *Bacillus* dysentery, Typhoid fever, cholera and paratyphoid fever etc. renders them unsuitable for various purposes such as drinking irrigation, pisiculture and recreation (Jayaraman *et al.*, 2003). Drinking water is now-a-days becoming scares even though available in abundance due to its improper management (Subhadra *et al.*, 2003). The purpose of this study was to collect sufficient baseline water quality data to define current limnological conditions and to provide a basis for future water quality protection and monitoring.

Materials and Methods

The present work has been carried out on the water quality of Ooty lake. It is an artificial, freshwater lake, which was constructed for aesthetic reasons by John Sullivan in 1823. Ooty, the head quarters of Nilgiris biosphere reserve in the western ghats, is one of the two mega diversity density 'hotspots', located at about 2240m above MSL, is the highest place of Nilgiris (Venugopal, 1993). The Ooty lakes receives its water from the surface run-off in the following areas past of Larks Hills, St. Mary's Hill, Fern Hill and Elk Hill. Besides the above run-off, a perennial stream from upper part of the lake Kuddappamund Hill joins the lake.

About 2.6 km length of the channel receives almost all the polluting from the untreated daily waste product, unsewered waste water from the public, oil and grease from workshops and sewage from railway colony. Ooty also posses high ethnic value because of the large number of traditional human societies.

For the present study, water samples were collected for a period of twelve months from December 2005 to November 2006. Collections were made once in a month at the same time (12.00 pm) and same spot throughout the period of study. Surface water of the lake were collected in a sterilized polythene container completely to exclude the air space sealed tightly and brought to the laboratory within 24 hours and kept in the refrigerator for further analysis. The physical-parameters were analyzed by standard method of APHA (1998) and the primary productivity has been estimated by light and dark bottle technique following the method of Strickland and Parson (1978).

Results and Discussion

In the present study the variations in the physical, chemical parameters and the primary productivity were analyzed and shown in Tables 4.1–4.3.

In the present study it was observed that the colour of the water is pale green in colour throughout the period of the study. Colour in water may result because of the presence of natural metallic ions; manganese, planktons and weeds. Narain and Rajeev (2000) also observed the colour of the Kuwary river and it was well marked in dry season.

The transparency in the Ooty lake ranged from 58 to 79 cm. The minimum transparency was recorded in June and February while maximum in September after end of the rainy season. Usually lower transparency or higher turbidity values show high pollution load (Narain and Rajeev, 2000). The higher sediments during rainy season may increase the turbidity and decrease the transparency. According to Khalique and Afser (1995) turbidity and temperature correlate with each other. Our results also reveal the same condition.

In the present investigation the maximum temperature 20.3°C and minimum temperature 6.1°C were recorded (Table 4.1). Temperature is a vital parameter for growth of organisms and also plays an important role in the physico-chemical and physiological behavior of aquatic ecosystem; Atmospheric temperature was found to be higher than the water temperature except in the month on December and January. Fluctuations experienced in the air and water temperature may be due to the influence of season and location. Sreenivasan (1965) on Bhavani Sagar and Rani Sagar lake in Jodhpur and Sharma (1980) also noted similar relationship between air and water temperature. High temperature during summer was due to greater heating of water and insolent heat from sun, onset of monsoon result in fall in temperature water bodies because clouds acts as a barrier in the process of heating (Parvateesam and Mathur, 1992).

Table 4.1: Physical Properties of Ooty Lake for a Period of One Year from December 2005 to November 2006

Month	*Colour*	*Trans-parency (cm)*	*Atmos-pheric Temp. (°C)*	*Water Temp. (°C)*	*Suspended Solids (mg/l)*	*Dissolved Solids (mg/l)*	*Total Solids (mg/l)*
December 2005	Pale green	72	4.7	6.1	280.33±0.5774	160.33±0.5774	440.33±0.5774
January 2006	Pale green	76	4.9	8.2	290.33±0.5774	110.33±0.5774	400.33±0.5774
February	Pale green	62	11.5	11.0	320.33±0.5774	120.33±0.5774	440.33±0.5774
March	Pale green	78	17.6	11.4	313.33±5.7735	111.33±2.3094	421.67±2.8868
April	Pale green	75	17.6	14.1	361.67±2.8868	221.0±1.7321	573.33±11.547
May	Pale green	78	22.2	20.3	223.33±5.7735	136.67±11.547	440.0±17.321
June	Pale green	58	19.0	16.0	389.33±18.475	241.67±14.434	630.0±34.641
July	Pale green	76	18'.1	17.1	498.33±20.207	213.33±11.547	723.33±11.547
August	Pale green	71	19.5	16.2	590.0±17.321	193.33±5.7735	776.67±23.094
September	Pale green	79	17.8	16.0	413.33±11.547	275.0±8.6603	708.33±14.434
October	Pale green	71	14.0	6.5	388.33±2.8868	273.33±11.5470	630.0±17.3205
November	Pale green	69	6.8	6.1	353.33±5.7735	233.33±11.547	593.33±5.7735

Each value is the mean of four replications except colour, transparency, temperature.

The value of total dissolved solids (TDS) of the study area ranges from 110.33 to 275 mg/l and total suspended solids from 223.33 to 590 mg/l. The seasonal fluctuations shows that total dissolved solids are higher in monsoon months, they are favourable for biotic life and minimum in summer. The higher range during monsoon may be discharge of domestic effluents due to anthropogenic activities and leaching of surrounding rainwater. The TDS values are greater than ISI range of 500 mg/l. Similar observation of very high value of TDS was noticed in Thittagudi river water. Santhosh (1999) and Krishnakumar (2002) also reported similar findings in Ithikkara lake and Wellayani lakes respectively.

Table 4.2: Chemical Properties of Ooty Lake for a Period of One Year from December 2005 to November 2006

Month	*pH*	*Carbonate*	*Bicarbonate*	*Total Alkalinity*	*Dissolved Oxygen*	*Dissolved CO_2*	*BOD*	*COD*
December 2005	7.4	26.33±2.8868	30.667±0.57735	77.67±2.3094	2.9667±0.1155	5.2±0.1732	42.67±2.8868	231.67±14.434
January 2006	7.5	20.67±0.5774	32±1.7321	81.0±1.7321	3.77±0.2309	5.9±0.3464	52.67±0.2887	227.67±6.3509
February	7.9	19.67±0.5774	25.67±1.1547	88.0±1.7321	3.37±0.1155	4 3±0 3464	52.8±2.425	389.67±2.3094
March	7.3	20.67±1.1547	29.33±1.1547	138.0±6.9282	3.93±0.1155	6.13:f0.1155	37.2±0.6928	34333±2.8868
April	8.5	24.67±0.5774	28.67±1.1547	55.67±4.6188	2.6±0.1732	5.37±0.1155	49.53±3.06	260.0±8.6603
May	7.6	26.33±0.5774	31.33±1.1547	142.67±l0.97	3.03±0.1155	6±0.1732	45.87±0.9815	317.0±5.1962
June	7.3	27.67±0.5774	30.33±0.5773	44.33±2.8868	1.73±0.1155	4.23±0.2309	50.6±0.5196	295.33±4.0415
July	7.4	28.67±1.1547	30.67±0.5774	37.0±3.4641	1.5±0.1732	53±0.1732	52.23±0.6351	315.67±4.0415
August	8.1	29.33±0.5774	30.33±0.5773	133.33±l1.547	2.37±0.2309	4.17±0.115	60.0±1.7321	379.0±3.4641
September	8.3	30.0±1.7321	32.0±1.7320	43.33±5.7735	3.1±0.2309	4.4±0.3464	42.5±1.2124	321.7±2.8868
October	7.9	2430±0.5774	30.30±0.5774	37.30±1.1547	2.23±0.2309	3.9±0.0577	38.9±0.5196	301.7±2.8868
November	7.4	25.70±0.5774	31.30±1.1547	35.30±0.5774	3.0±0.1732	5.1±0.2309	41.03±0.4619	313±2.8868

Each value is the mean of four replications.

All values expressed in mg/l, except pH.

The reason may be sewage and sludge entering the river (Sharma and Kaur, 1998) and also may be due to contamination by leaching process through surface water during rainy season (Sing and Seemaraj, 1999). The high value of TDS influences the taste, hardness and corrosive property of water. High value of dissolved solids may aesthetically unsatisfactory for bathing and living (Ram Subramaniam *et al.*, 2004).

Table 4.3: Nutrient Content of Ooty Lake for a Period of One Year from December 2005 to November 2006

Months	*Calcium*	*Magnesium*	*Sulphates*	*Phosphates*	*Chloride*	*Total Hardness*
December 2005	123.0±1.7321	34.3±1.1547	28.7±1.1547	4.4±0.2309	74.0±0.866	156.3±1.1547
January 2006	124.7±4.0415	31.3±1.1547	34.0±1.7321	5.1±0.1155	63.9±2.5403	158.3±1.1547
February	113.0±1.7321	29.3±0.5774	31.3±1.1547	6.3±0.2309	66.9±1.6166	140.3±1.1547
March	124.3±1.1547	26.0±1. 7321	21.7±2.8868	5.2±0.2309	62.7±1.097	148.3±2.8868
April	138.3±2.8868	33.0±1.7321	29.3±1.1547	7.4±0.2887	85.7±2.3671	171.7±4.0414
May	11 0.0±1.7321	28.7±1.1547	36.7±2.3094	3.2±0.2309	81.3±0.5774	140.7±2.8868
June	99.7±0.5774	30.7±0.5774	35.3±1.1547	2.1±0.1732	72.0±1.7321	130.3±1.1547
July	106.3±2.3094	29.7±0.5774	20.7±1.1547	12.1±15.473	54.3±2.3094	136.0±1.7321
August	123.67±2.3094	31.3±1.1547	26.3±2.8868	7.4±0.0577	84.0±0.1155	154.7±4.0414
September	139.3±1.1547	26.7±1.1547	31. 7±1.1547	6.8±0.115	82.0±1.7321	170.7±8.0829
October	131.7±2.8868	34.7±2.3094	34.7±0.5774	6.4±0.2309	80.3±0.5774	214.0±3.4641
November	170.3±0.5774	29.7±1.1547	32.3±1.1547	5.1±0.1155	83.7±1.1547	199.7±1.1547

Each value is the mean of four replications.

All values expressed in mg/l.

Table 4.4: Primary Productivity of Ooty Lake for a Period of One Year from December 2005 to November 2006

Months	*Gross Primary Productivity*	*Respiratory Rate*	*Net Primary Productivity*
December 2005	1,6±0.0462	1.02±0.0 425	0.53±0.0577
January 2006	1.7±0.1501	1.01±0.0058	0.70±0.0866
February	1.3±0.1443	0.80±0.0866	0.53±0.5770
March	2.2±0.1732	1.38±0.0693	0.82±O.1039
April	2.6±0.0577	1.54±0.0346	1.02±0.0231
May	1.7±0.1559	1.01±0.0115	0.70±0.0866
June	1.6±0.0924	0.9±0.0289	0.61±0.0404
July	2.5±0.0577	0.5±0.0808	1.92±0.0924
August	2.5±0.1155	0.7±0.0577	1.85±0.1155
September	2.0±0.0058	0.4±0.0635	1.57±0.1732
October	3.0±0.0289	2.0±0.0115	1.13±0.1559
November	2.5±0.0560	0.5±0.707	1.90±0.0820

Each value is the mean of four replications.

All values are expressed in mg.c/m^2/hr

Table 4.5: Correlation Coefficient of Physico-chemical Parameters of Ooty Lake from December 2005 to November 2006

	NP	*TH*	*Cl*	*PO_4*	*SO_4*	*Mg*	*Ca*	*COD*	*BOD*	*DCo_2*	*DO_2*	*T.Alk*	*Bicar*	*Carbo*	*pH*	*Total*	*DS*	*SS*	*Tem. W*	*Tem. A*	*Trans*
NP	1.000																				
TH	–0.339	1.000																			
Cl	–0.114	0.106	1.000																		
PO.	0.295	–0.102	0.514	1.000																	
SO_4	–0.429	–0.011	–0.192	–0.065	1.000																
Mg	0.157	0.375	0.300	0.155	0.245	1.000															
Ca	–0.070	0.070	–0.646	–0.470	–0.017	0.131	1.000														
COD	–0.005	0.037	–0.239	0.057	–0.335	–0.137	0.593	1.000													
BOD	0.203	0.119	0.135	0.116	0.034	0.210	–0.228	0.015	1.000												
DCo,	–0.096	0.310	0.206	–0.293	–0.234	–0.356	–0.325	–0.309	–0.242	1.000											
DO,	–0.346	0.314	0.148	0.259	0.060	–0.451	–0.356	0.065	–0.286	0.504	1.000										
TA	–0.181	0.186	0.120	0.011	–0.101	–0.304	0.110	0.387	0.153	0.364	0.490	1.000									
Bicar	0.306	0.290	–0.348	–0.291	0.109	–0.154	0.116	–0.336	–0.344	0.431	0.124	–0.073	1.000								
Carbo	0.549	–0.102	–0.234	–0.137	–0.032	0.078	0.335	0.112	0.089	–0.301	–0.584*	–0.307	0.429	1.000							
pH	0.248	–0.251	0.510	0.803**	0.149	0.166	–0.470	*0.130	0.214	–0.351	–0.046	–0.129	–0.285	0.160	1.000						
T.Alk	–0.061	–0.633*	0.228	0.803*	0.223	0.048	–0.274	–0.249	0.318	–0.316	–0.409	–0.196	–0.493	–0.038	0.101	1.000					
DS	0.060	–0.560	0.250	–0.083	0.303	0.184	–0.241	–0.340	0.302	0.365	–0.457	–0.126	–0.282	0.127	0.159	0.939*	1.000				
SS	–0.164	–0.610*	0.234	0.123	0.149	–0.012	–0.263	–0.180	0.351	0.287	–0.384	–0.126	–0.632*	–0.137	0.099	0.967**	0.835**	1.000			
TemW	–0.251	–0.031	0.089	0.008	0.224	0.432	–0.351	–0.130	0.116	0.459	0.459	0.410	–0.102	–0.327	0.026	–0.109	–0.276	0.033	1.000		
Tem.A	–0.253	–0.024	0.502	0.188	0.143	–0.328	–0.584*	–0,342	0.151	0.477	0.470	0.543	–0.122	–0.395	0,368	0.098	0.043	0.186	0.789**	1.000	
Trans	0.120	0.221	–0.420	–0.316	0.252	–0.397	0.231	–0.045	–0.542	0.623	0.312	0.109	0.623*	–0.050	0.395	–0.734**	–0.754**	–0.696*	0.285	–0.001	1.000

*: For 5 per cent Significant.

**: 1 per cent Significant.

The ISI value for total suspended solid is 500 mg/l. The entire samples exceed this standard value, may be due to inorganic solids such as clay, silt and other soil constituents and organic material like plant fibers and biological solids are common constituents of surface water.

pH is also an important factor which controls the chemical charges, species composition and the metabolic activities of living organisms inhibiting the water body (Radhika *et al.*, 2004). The permissible limit of pH in drinking water is within 6.5 to 8.5. The pH range of 7 to 8 has been considering good for fish culture (Jhingram, 1977). During the present study the pH value ranged from 7.3 to 8.5 and this reveals that the pH of the water samples is associated in the desirable range and the samples are slightly alkaline nature. This may be due to high buffering of the system and also due to decomposition of organic matter and photosynthesis. The pH increases during the day time due to photosynthetic activity whereas it decreases during the night due to the respiratory action. Wetzel (1972) have observed the value of pH from 8.0 to 9 units in Indian waters. Similar result was observed in the present study. pH also reported to play an important role in the formation of algal bloom (King, 1970) and decreasing volume of water due to evaporation is often accompanied by progressive changes in pH (Adoni, 1975). Changes pH of a water body can indicate the presence of certain pollutants when it is continuously monitored. In general, the lake water is not fit for drinking purpose, the main contaminants being the municipal wastewater from the town. Alkalinity is measure of buffering capacity of water. The major portion of alkalinity in natural water is caused by hydroxide, carbonate and bicarbonate.

Alkalinity of the Ooty lake water is ranged from 37.0 mg/l to 142 mg/l. It was high during the summer and monsoon season. The acceptable limit for alkalinity is 200 mg/l. The observed values are well below the maximum value and hence it can be said that presence of alkalinity in such water is in tolerable range. The slight variation in total alkalinity may be due to the seasonal effect, planktonic population, bottom deposits, rocky bed and water current in the river (Jhingran, 1991). Alkalinity itself is not harmfull to human beings (Pandey *et al.*, 2002). Alkalinity also may be due to the contamination by leaching process through surface water during rainy season (Sing and Seema, 1999). Berman (1976) considered alkalinity as an important parameter in its view of its relationship with available carbondioxide for photosynthesis and its use as an indicator of productivity. Excess alkalinity gives a bitter taste to water (Guruprasad, 2003). An apparent positive correlation between total alkalinity and temperature was observed.

DO is one of the most important parameter in water quality standards. The DO is essential to maintain biotic community in water and to minimize the effect of waste discharge into fresh water body. Dissolved oxygen is obviously essential to the metabolism of all aquatic organisms that posses aerobic respiratory biochemistry (Wetzel, 1972). In the present study dissolved oxygen were found to low throughout the period and ranged from 1.5 mg/l to 3.93 mg/l. The low dissolved oxygen indicates high rate of biodegradation of organic matter. Amount of oxygen in lake is its purity indicator, also acts as an indicator of tropic status and magnitude of eutrophication in fresh water ecosystem. Trivedy and Goel (1980) also noticed that O_2 depletion might be due to accumulation of organic waste stabilization of water. The degradation of organic matter O_2 in the water may be used and it leads to toxic contamination. Minimum level of DO required for aquatic life production at different temperature is 5.8 to 6.8 mg/l. The overall effects make the water much less suitable for various needs particularly recreation, fish production and domestic uses. Such low level of DO greatly threatens the survival of aquatic organisms except anaerobic microorganisms (Chinmoy, 2000). Solubility of oxygen decreases with increase in temperature (Sabata and Nayar 1995). Similarly increase in dissolved oxygen is obviously related to the temperature of Ooty Lake. But dissolved oxygen shows positive correlation with COD.

The dissolved carbon dioxide concentration range from 3.9 to 6.13 mg/l having higher value in summer. This may be due to the release of carbon dioxide during microbial decomposition of organic matter. Some amount of carbon-dioxide is also released by the biotic components with the lake. Part of carbon-dioxide released gets converted into bicarbonates, which affects the pH slightly (Jain *et al.*, 1996). Dissolved carbon-dioxide shows negative correlation with BOD.

The maximum BOD values were recorded 60 mg/l in the month August and minimum value (37.2 mg/l) was recorded during the March may be due to retarded microbial activity, high organic decomposition and enhanced organic pollution load. This may lead to eutrophication and finally resulting in oxygen depletion, which affect the flora and fauna of the aquatic system, was reported by Yamuna and Balasubramaniam (2002).

Chemical oxygen demand (COD) is a measure of oxygen required for complete oxidation of organic matter by a strong oxidant. COD is a reliable parameter in judging the extent of organic pollution. Seasonal values of COD in water of Ooty lake ranged from 213.67 to 379.0 mg/l. During the course of study the values of COD were found to be higher than BOD values. The high COD values indicate that some degree of non-biodegradable oxygen demanding pollutants were present in the water. The COD was high during summer due to high rate of oxidation (Parvateesam, 1994).

Calcium is a common constituent of natural water and has important contribution to the hardness of water. It reduces the utility of the water for domestic use. Some $CaCO_3$ is desirable for domestic use of water because it provides a coating in the pipes, which protects them against corrosion. The maximum calcium content (170.3 mg/l) was recorded in the month November and the minimum (99.7) content was recorded in the month of June. Any value above 25 mg/l indicates calcium rich water. As per this definition, the water of Ooty lake is rich in calcium ions. Calcium shows positive correlation with magnesium.

The magnesium values were found to be ranging from 26 to 34.7 mg/l. The maximum value was observed in the month of March and minimum value in the month of October. Magnesium is often associated with calcium primarily due to its similar chemistry. The element is usually present in aquatic system in large amounts relative to plant needs (Goldman, 1972).

Minimum sulphate level of 20 mg/l was seen in the month of July and maximum 36.7 mg/l in the month of May. Most of the sulphate ions are probably derived from the solution of calcium and magnesium ions. Sulphate is a naturally occurring ion found almost in all types of water and its concentration occurs in wide ranges in nature (Loganayagi *et al.*, 2008).

The major natural sources of hardness in water are sedimentary rocks, sewage and runoff from soils (Moti Sharma, 2004). The total hardness was found to minimum of 131 mg/l in the month of June and maximum of 216 mg/l was found in the month of October (monsoon). As mentioned by Hynes (1979) increasing levels of hardness during post monsoon season tend to reduce the calcium content by precipitation of carbonates a phenomenon that has direct relevance with photosynthesis by phytoplankton. A low water level and flushing rate lead to increase in hardness during summer while increase in precipitation in August–September brings about dilution and resultant decline in hardness (Arvind Kumar, 1995). Minimum sulphate level of 20 mg/l was seen in the month of March and July and maximum of38 mg/l in the month of May. Sulphate shows positive correlation with hardness.

Phosphorous occurs in natural water almost as phosphate. The phosphate levels were found to be ranging from 2.1 to 7.4 mg/l. phosphate indicated the maximum during monsoon. Intensive algal growth in spring usually depletes lake phosphate to low levels. Zooplankton excretes approximately 10 per cent of their body phosphorous daily but there are considerable variations depending upon feeding rate, temperature, type of food, time of day and stages (Sharma and Sharma, 1992). The sources of phosphate were agricultural runoff from surrounding fields, domestic and industrial effluents, and use of detergents in lake area, urban runoff and faeces of birds, which enter the lake. Similar report was given by Das *et al.* (1997).

Chloride had been as a basic parameter for detecting pollution of water by sewage before development of bacteriological procedure (Sawyer, 1960). Generally unpolluted water contains low concentrations of chloride (Padmanabh and Santoshi, 2004). The concentration of chloride in the present study ranged from 54.3 to 85.7 mg/l. Adoni (1985) pointed out that high concentration of chloride is due to large quantity of organic matter in it.

The photosynthetic fixation of CO_2 and its quantitative measurement is considered a vital index of the productive potential of any aquatic system. Primary productivity of different lentic water bodies like ponds, lakes and reservoirs of India had been studied by Nasar and Dutta munshi (1975), Pandey *et al.* (1995), Gurusamy and Ramdass (2000), Mandal (2002) and Kumar *et al.*, (2006). Sharma and Sharma (1992) noted that phytoplankton encountered in the water body as well as they may be used as indicators of the water quality. In the present study the gross primary productivity ranged from 1.3 to 3.0 mg cm^{-2} hr^{-1} and net primary productivity ranges from 0.5 to 1.90 mg $cm^{-2}hr^{-1}$. A sustainable amount of the synthesized organic matter may be utilized during the respiratory activities of plankton leaving a small quantity of the net production comparatively at low levels. Sreenivasan (1976) recorded primary production rates between 2.2 and 9.8 mg $cm^{-2}hr^{-1}$ in the impoundments around the Nilgiris, 2.0 to 10.0 mg $cm^{-2}h^{-1}$ in Aliyar reservoir. Similar to the observations of Singh (1963), Harikrishnan and Azis abdul (2000) and Kumar *et al.* (2006), the present study revealed distinct seasonal variation in the GPP value, having winter and summer peaks. Solar radiations were perhaps the prime factor contributing the high values. NPP values showed the same seasonal and bimodal pattern of variations as shown by GPP values that is maxima attained during summer and minima in rainy months. The NPP had direct relationship with the present study.

The lake showed more productivity with more nutrients and it is progressing towards eutrophication. Awareness should be created among people to save water from pollution and it needs remedial measures. Since Ooty which is queen of hills and beauty of the lake should be retained.

Hence in the present study showed the heavy concentrations of all parameters that have been reported from the Ooty lake water might have been contributed to domestic sewage being drived into the lake water. On the basis of the observations made on physico-chemical features of the lake is regarded as of great economic, ecological and biological importance. The present lake would prove ideal, specially for the purpose of fish rearing and the pollution levels should be controlled. Siltation process is minimized and aquatic weeds population is minimized to positive approach.

References

Adoni, A.D., 1975. Studies on microbiology of Sagar lake. *Ph.D. Thesis*, Sagar Univ., Sagar, M.P., pp. 243.

Adoni, A.D., 1985. *Workbook on Limnology*. Pratibha Publishers, C–10, Gour Nagar, Sagar (M.P.), pp. 216.

APHA, 1998. *Standard Methods for Examination of Water and Wastewater*, 20th edn. Am. Pub. Hlth. Assoc. Washington, DC, p. 566.

Berman, T., 1976. Light penetration on lake Kinneret. *J. Hydrobiologia*, 49: 41–48.

Chatterjee, Chinmoy, 2000. Physico-chemical studies of water quality of the river Nunia at Asansol Industrial area. *Journal of Environment and Pollution,* 7(4): 266–261.

Das, J., Das, S.N. and Sahoo, R.K., 1997. Semi-diurnal variation of some physico-chemical parameters in the Mahanadi estuary, East Coast of India. *Indian J. Mar. Sci.,* 26: 323–326.

Dwivedi, Padmanabh and Sonar, Santoshi, 2004. Evaluation of physico-chemical and biological characteristics of water samples in water reservoirs around Rono Hills, Doimukh (Dist. Papum Pare) Arunachal Pradesh. *Poll. Res.,* 23(1): 101–104.

Gadi, Subhadra Devi, Barbuddhe, S.B., Hazel, D. and Dolly, C., 2003. Physico-chemical characteristics of Drinjing water at Velsao, Goa. *J. Ecotoxicol. Environ. Monit.,* 13(3): 203–209.

Goldman, 1972. The role of major nutrients in limiting the productivity of aquatic ecosystem. *Am. Sec. of Limnol. and Ouno.,* 1: 21–23.

Guruprasad, B., 2003. Physico-chemical and bacteriological of ground waters at Tedepalli Mandal of Guntur District Andhra Pradesh. *Nature, Environment and Pollution Technology,* 2(2): 173–178.

Gurusamy, K. and Ramadoss, V., 2000. Seasonal variation of chlorophyll and phytoplankton productivity in man made ponds. *J. Ecobiol.,* 12: 281–287.

Harikrishnan, K. and Aziz Abdul, P.K., 2002. Primary productivity studies in a freshwater temple tank in Kerala. *Indian J. Environ. and Ecoplan.,* 3: 127–130.

Hynes, H.B.N., 1979. *The Ecology of Running Waters.* Liver Pool University Press, Liver Pool.

Jain, S.M., Sharma, Meenakshi and Thakur, Ramesh, 1996. Seasonal variations in physico-chemical parameters of Halali Reservoir of Vidisha District, India. *J. Ecobiol.,* 8(3): 181–188.

Jayaraman, P.R., Gangadevi, T. and Vasudevan Nayar, T., 2003. Water quality studies on Karamanan River, Thiruvananthapuram District, South Kerala, India. *Poll. Res.,* 22(1): 89–100.

Jhingran, A.G., 1977. Optical appearance and Interpretation of annuli an scales of Gadusia Chapra (Ham). *J. Inland Fish. Soc., India,* 91: 138–153.

Jhingran, V.G., 1991. *Fish and fisheries of India,* 3rd edn. Hindustan Publishing Corporation (India), Delhi, p. 727.

Kalique, A. and Afser, M.R., 1995. Physico-chemical analysis of river Ganges at Farakka. *Biojournal,* 7(1–2): 101–105.

King, D.L., 1970. *J. Water Poll. Cont. Fed.,* 42: 2035–2051.

Krishnakumar, A., 2002. Environmental degradation of two river basins of South Kerala. *Ph.D. Thesis,* Department of Environmental Sciences, University of Kerala.

Kumar, Arvind, 1997. Comparative hydrological studies of tropical water bodies with special reference to sewage pollution in southern Bihar. *J. Ecobiol.,* 9(4): 225–262.

Kumar, Arvind, 1997. Studies on pollution in river Mayurakshi in south Bihar, India. *Journal of Environment and Pollution,* 2(1): 21–26.

Kumar, Ashok and Singh, N.K., 2006. Phytoplankonology of Pond at Deoghar, India. *J. Haematol and Ecotoxicol.,* 1: 61–66.

Loganayagi, A., Damodarkumar, S. and Murugesan, S., 2008. Quality of drinking water in and around Thiruvallur District, Tamil Nadu, 7(1): 133–138.

Mandal, O.P., 2002. Primary productivity in relationship to nutrient status of Kawar wetland, North Bihar. *Ph.D Thesis,* B.N. Mandal University, Madhepura, (Bihar).

Moti Sharma, R., 2004. Water quality of River Beas in Hamirpur Area of outer Himalayas. *Poll. Res.,* 23(1): 41–44.

Narian, S. and Chauchan, Rajeev, 2000. Water quality status of River complex Yamuna at Panchanada (Dist. Etawah, U.P., India). I. An integrated management approach. *Poll. Res.,* 19(3): 357–364.

Nasar, S.A.K. and Dutta Munshi, J.S., 1975. Studies on primary production in a freshwater pond. *Jap. J. Ecol.,* 25: 21–23.

Pandey, A.K., Dutta, S. and Sharma, K.C., 2002. Physico-chemical characteristics of Hamor pond of Kishangarh, Ajmer (Rajasthan). Indian J. Environ. and Ecoplan., 6: 167–169.

Pandey, K.N., Prakasan, V. and Sharma, U.P., 1995. Assessment of primary productivity of phytoplankton and macrophytes of Kawar lake wetland (Begusarai) Bihar. *J. Freshwater. Bio.,* 7: 237.

Parateesam, M. and Gupta, Sudha, 1994. Physico-chemical characteristics of lake receiving effluents from Textiles mills, Rajasthan. *Poll. Res.,* 13(4): 317–321.

Parateesam, M. and Mathur, M., 1992. Microbiological studies of Ansagar Lake, Ajmer, Rajasthan. *Indian J. Microbiol. Ecol.,* 2: 5–9.

Postel, S., 1992. *Last Oasis Facing Water Scarcity*. W.W. Norton and Company, Inc.

Radhika, C.G., Mini, I. and Gangadevi, T., 2004. Studies on abiotic parameters of a tropical freshwater lake vellayani Lake, Thiruvananthapuram District, Kerala. *Poll. Res.,* 23(1): 49–63.

Ramasubramaniam, V., Jeyaprakash, R., Ruby Mallika, D.A., Ramasubbu, R. and Mariappan, V., 2004. Analysis of physico-chemical characteristics of groundwater quality and quality index in and around Sivakasi town. *Indian J. Environ. and Ecoplan.,* 8(1): 171–176.

Sabata, B.C. and Nayar, M.P., 1995. *River Pollution in India: A Case Study of Ganga River*, APH Publishing Corp.

Sahu, B.K., Rao, R.J., Behera, S.K. and Pandit, R.K., 1995. Phytoplankton and primary production in the river Ganga from Rishikesh to Kanpur. *J. Ecobiol.,* 7(3): 219–224.

Santhosh, S., 1999. Studies on the geochemistry and productivity and productivity of Paruvur Canal system, Kollam district. *M.Phil. Thesis,* Department of Environmental Sciences, University of Kerala.

Sawyer, C.N., 1960. *Chemistry for Sanitary Engineers*. McGraw-Hill, New York, USA.

Sharma, B.K. and Kaur, H., 1998. *Environmental Chemistry*. Goel Publishing House, Meerut.

Sharma, L.L and Sarang, N., 2004. Physico-chemical limnology and productivity of Jaisamand lake, Udaipur (Rajasthan). *Poll. Res.,* 23(1): 87–92.

Sharma, L.L., 1960. Some Limnological aspests of Udaipur water in comparison to selected water of Rajasthan. *Ph.D. Thesis,* University of Udaipur.

Sharma, R. and Sharma, K.C., 1992. Diatoms of Anasagar lake of Aimer, Rajasthan, *Acta. Ecol.,* 14: 6–9.

Singh, B.H. and Seema Raj, 1999. Physico-chemical studies of Ganga River at Varanasi. *Journal of Environment and Pollution,* 6(1): 43–46.

Singh, N.K., 1963. Studies on density, productivity and species composition of phytoplanktonic in relation to abiotic spectrum of Ganes at Sahibganj. *J. Freshwater Biol.*, 5. 1–8.

Sivakumar, R., Mohanraj, P and Azeez, P.A., 2000. Physico-chemical sources of water sources of Ooty, South India. *Poll Res.*, 19(1): 143–146.

Sreenivasan, A., 1965. Limnology of tropical implements. III. Limnology and productivity of Amaravathy reservoir, Madras state (India). *Hydrobiologia*, 48(2): 117–123.

Sreenivasan, A., 1976. Limnological studies and primary production in temple pond ecosystems. *Hydrobiologia*, 48(2): 117–123.

Strickland, J.D.H. and Parson, T.R., 1968. A practical handbook of seawater analysis. *Bull. Fish. Res. Biol., Canada*, 167: 311.

Trivedy, R.K. and Goel, P.K., 1980. *Chemical and Biological Methods for Water Pollution Studies*. Environmental Publications, Karad.

Venugopal, D., 1993. *The Nilgiris*. Nilgiri Documentation Center, Ooty.

Verma, O.P., Nirmala Singh., Jain, P.A. and Choudhury, Onkar, 2007. Comparative study of organic pollution level from different water bodies in Allahabad city. *Indian J. Environ. and Ecoplan.*, 14(3): 705–708.

Vijayaragavan, S., 1971. Seasonal variation in primary productivity in three tropical ponds. *Hydrobiologia*, 38: 395– 408.

Wetzel, R.G., 1972. Productivity and nutrient relationship in Marl lakes of Northern India. *Verh. Int. ver. Limnol.*, 16: 321–332.

Chapter 5

Nutrient Dynamics and Phytoplankton Abundance in the Marine Environment of Kollam (Quilon), South West Coast of India

Wilma Cyril*[1], S.T. Shibu Joseph[2] and B.S. Sharma[3]
[1]Head, [2]Lecturer,
Post Graduate Department of Chemistry,
Fatima Mata National College, Kollam, Kerala
[3]Head, Department of Environmental Studies, School of Life Sciences,
Dr. Bhim Rao Ambedkar University, (Formerly Agra University),
Khandari Campus, Agra – 282 002
E-mail: sharmabswarup@rediffmail.com

ABSTRACT

Phytoplankton abundance was determined and nutrients measured in the Kollam Coast, South West Coast of India, at the sites Chavara, Neendakara, Tangasseri and Paravur. Marine water sampling was performed on a monthly basis during the period June 2003 to May 2004. The average concentrations of PO_4–P, NO_3–N and $[Si(OH)_4]$–Si in surface sea water varied in the range 1.07 µM to 1.66 µM, 3.49 µM to 4.79 µM and 3.18 to 5.06 µM respectively. In the bottom water, the values of PO_4–P, NO_3–N and $[Si(OH)_4]$–Si varied in the range 1.10 to 1.73 µM, 4.08 to 4.84 µM and 2.13 to 4.40 µM. Off the upwelling coast of Kollam, abundance of phytoplankton was noted, compared to other coastal water of India. Surface water phytoplankton density

* Corresponding author: E-mail: wilmaroy@rediffmail.com.

reached the highest level (114.80 × 10^2 numbers l^{-1}) during monsoon. During post-monsoon and pre-monsoon the phytoplankton counts were 100.44 × 10^2 and 29.76 × 10^2 respectively. Diatoms dominated during all seasons. The peak period of phytoplankton abundance coincided with low nutrient concentrations. Along with hydrographic and hydro-chemical factors anthropogenic factors contribute to the phytoplankton growth in the area.

Keywords: *Phytoplankton, Nutrient concentration, Kollam Coast, Arabian Sea.*

Introduction

The distribution and variation of nutrients in aquatic ecosystem are controlled by the physical, biological and chemical processes taking place in the pertinent environment. In the marine environment, biotic and abiotic environmental factors have an effect on phytoplankton succession and abundance (Polat and Piner, 2002). Besides, carbon, hydrogen and oxygen the major elements required by planktonic algae are nitrogen, phosphorus and silicon (Verlecar and Parulekar, 2001).

Nitrogen and phosphorus are the two major nutrients taken up by phytoplankton from the water for their healthy growth apart from siliceous organisms which utilises silicon. Nitrate is an essential nutrient for the growth of many photosynthetic autotrops and has been identified as the growth limiting nutrient. This parameter is of great importance in assessing the self-purification properties of water system and the nutrient balance in surface water (Rump and Krist, 1992). Phosphorus is one of the key elements of life. In almost all cases, the phosphorus is present as the orthophosphate group, but the low solubility of these inorganic compounds limits its availability as a nutrient. Silicate is an important nutrient to some marine organisms like diatoms and chrysophyceae.

Upwelling water imports large amounts of nutrients from bottom, thus enhancing biological productivity in surface sea water (Verlecar and Parulekar, 2001). Coastal areas of the Arabian sea are major areas of upwelling (Currie *et al.*, 1973). Prior work on phytoplankton studies from the Arabian sea are available (Hira *et al.*, 2001; Surekha *et al.*, 1996; Nair *et al.*, 1982). That the West Coast of India is a zone of upwelling has been reported by many investigators (Banse, 1959; Sarma *et al.*, 2000). Prior investigators have reported studies on phytoplankton along the South West Coast of India (Lierheimer and Banse, 2002; Madhupratap *et al.*, 1996; Qasim, 1977). But studies along the Coast of Kollam are limited. Moreover, if other factors like temperature and cloud cover are not limiting, the abundance of phytoplankton increases basic productivity (Balachandran *et al.*, 1997). Hence, the study pertaining to nutrient dynamics and phytoplankton abundance in the marine environment along the Coast of Kollam assumes importance.

Materials and Methods

The study was carried out in the Kollam, on the South West Coast of India located between 8°45′ and 9°28′N latitude and 76°28′ and 77°17′E longitude, having a coastline length of 37 km.

During the period June 2003 to May 2004 (the study period being divided into three seasons, June to September (monsoon) October to January (post-monsoon), February to May (pre-monsoon), marine water sampling was done from a mechanized canoe on a monthly basis during the first week of every month from eight sampling stations of the four sites of the Kollam Coast, Chavara nearshore S_1, Chavara offshore S_2, Neendakara nearshore S_3, Neendakata offshore S_4, Tangasseri nearshore S_5, Tangasseri offshore S_6, Paravur nearshore S_7 and Paravur offshore S_8 (Figure 5.1).

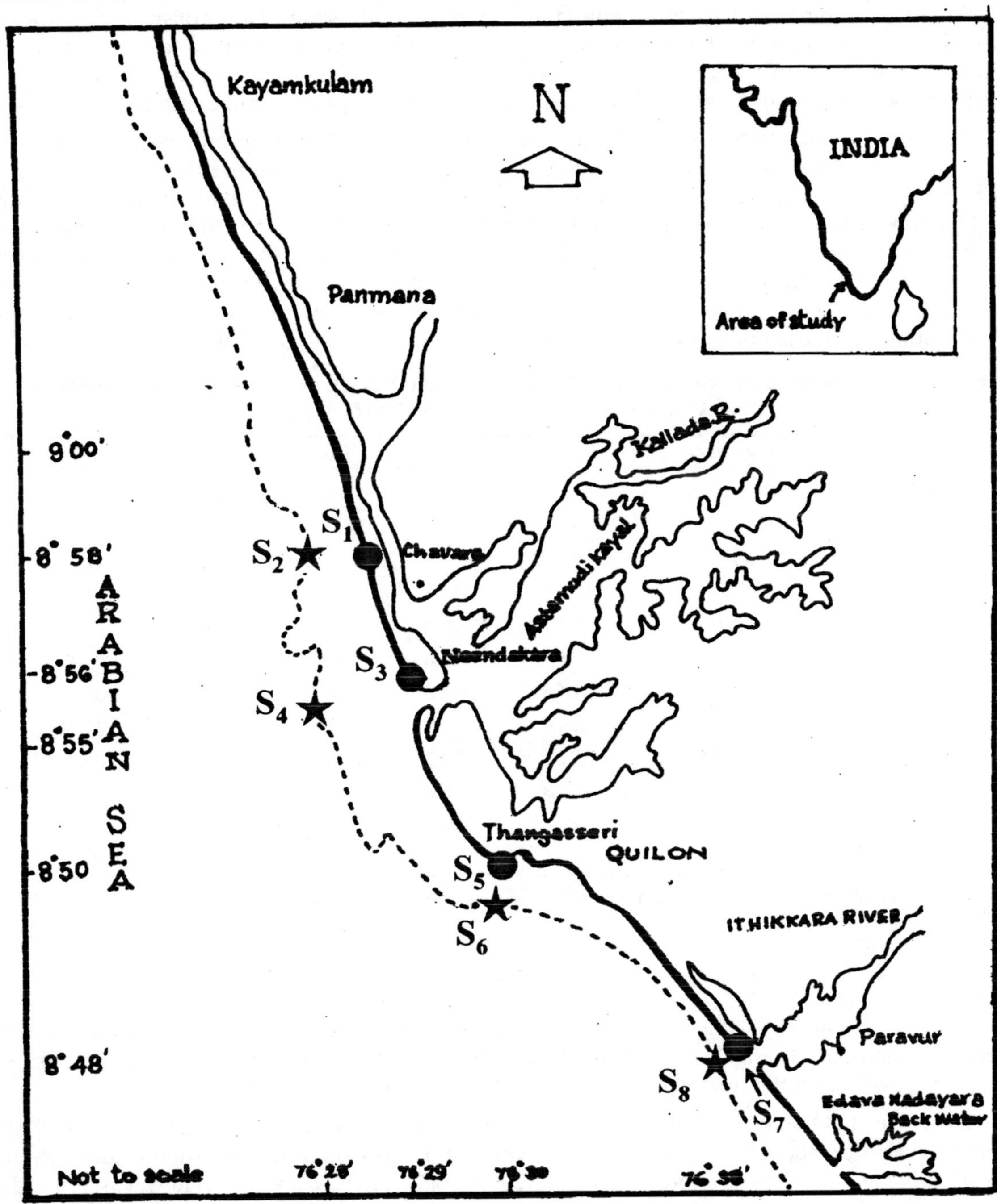

0 km zone of Kollam Coast ●

5 km Westward from the Kollam Coast ★

Figure 5.1: Map of Kollam Coast Showing the Sampling Stations

At each station phytoplankton samples were collected by filtering 50 litres of surface sea-water through plankton net of mesh size 20μ and preserved with 2 per cent neutral formalin solution. Density counts were made using Sedgwick-Rafter counting cell and are expressed as no/l and identification of different species carried out (Newell and Newell, 1979; Santhanam *et al.*, 1987).

Seawater for salinity and nutrient determination was sampled using non-metallic universal water sampler from Hydro-Bios, Germany. Salinity was estimated by Mohr-Knudson method and nutrients spectrophotometric ally (Strickland and Parsons, 1972; Parsons *et ai.*, 1984). Correlation coefficient between phytoplankton density and nutrient concentrations and standard deviation of mean phytoplankton count were determined with SPSS 8.0 Statistical Package Program. The surface water temperature, atmospheric temperature and relative humidity at each station were noted. The dates of sampling along with the average values of atmospheric temperature, water temperature, relative humidity and salinity of surface and bottom water are provided in Table 5.1.

Table 5.1: Average Value of Temperature (°C), Relative Humidity (Per cent) and Salinity (Per cent) in the Kollam Coast during June 2003 to May 2004

Season	*Date of Sampling*	*AT*	*WT*	*RH*	*SS*	*SB*
Monsoon	03.06.2003	23.4	21.5	90.5	25.5	27.34
	03.07.2004	24.5	22.4	89.1	25.4	27.4
	03.08.2003	26.5	24.3	90.5	28.8	29.9
	03.09.2003	24.1	23.9	91.1	33.1	34.9
Post-monsoon	02.10.2003	26.1	23.8	89.6	28.1	30.1
	02.11.2003	25.9	23.8	89.7	27.9	29-.8
	04.12.2003	24.9	22.8	90.5	33.9	35.1
	03.01.2004	247	22.1	90.5	35.1	36.9
Pre-monsoon	03.02.2004	24.1	22.8	88.7	35.2	36.7
	02.03.2004	24.3	25.9	89.7	35.1	36.8
	02.04.2004	29.1	27.1	80.7	35.1	36.8
	02.05.2004	26.4	36.9	81.1	27.4	29.2

AT: Atmospheric temperature; WT: Surface water temperature; RH: Relative Humidity; SS: Surface Water Salinity; SB: Bottom Water Salinity.

Results

The values of nutrient concentrations and phytoplankton count for the study period have been analyzed systematically and statistically in order, to find out differences between station and season. The average values for the study area have been used for statistical correlations.

The value of PO_4-P in the surface water varied from 0.81 μM at Paravur near shore S_7 to 1.29 μM at Neendakara near shore S_3 during the monsoon period. The bottom water recorded values ranging from 0.83 μM at Paravur near shore S_7 to 1.29 μM at Neendakara near shore S_3. The S_7 value was the lowest recorded in this study area during the study period. During post-monsoon, the values ranged from 0.98 μM at Paravur near shore S_7 to 1.71 μM at Tangasseri near shore S_5 in the surface water. The concentrations in the bottom water ranged from 1.1 μM at Paravur near shore S_7 to 1.81 μM Neendakara near shore S_3 and Tangasseri near shore S_5. During pre-monsoon, the PO_4-P concentration varied from 1.21 µM at Paravur offshore S_8 to 1.85 µM at Tangasseri near shore S_5 and Tangasseri offshore S_6 in the surface water. The bottom water recorded values ranging form 1.23 µM at Paravur near shore S_7 to 1.99 µM at Tangasseri near shore S_5. This S_5 value was the highest noted during the entire study period.

The values of NO_3-N in the surface water were in the range 3.4 µM at Chavara offshore S_2 to 4.81 µM at Neendakara near shore S_3 during the monsoon period. The S_2 value was the lowest recorded during the study period, where as in the bottom water it was 3.30 µM at Chavara near shore S_1 to 4.92 µM at Chavara offshore S_2. During post-monsoon the lowest value of NO_3-N 3.6 µM was recorded from the surface water at Tangasseri offshore S_6 and the highest value of 4.50 µM was recorded Neendakara near shore S_3. In the bottom water the values ranged from 3.5 µM at Tangasseri offshore S_6 to 4.72 µM at Paravur near shore S_7. During pre-monsoon the values ranged from 4.20 µM at Chavara offshore S_2 to.4.89 at Paravur hear shore S_7. In the bottom water the lowest value was 4.01 µM at Chavara near shore S_1 and the highest value was 4.96 at Paravur near shore S_7. This S_7 value was the highest recorded during the study period.

The value of silicate ranged from 3.4 µM at Paravur offshore S_8 to 5.5 µM at Neendakara near shore S_3 during the monsoon period. This S_3 value was the highest recorded during the study period. In the bottom water, the values ranged from 3.1 µM at three stations Neendakara offshore S_4. Tangasseri near shore S_5 and Tangasseri offshore S_6 to 4.8 µM at Paravur near shore S_7. During post-monsoon the values were in the range 2.02 µM at Chavara offshore S_2 to 4.30 µM three stations at Neendakara nearshore S_3. Tangasseri near shore S_5 and Tangasseri offshore S_6 in the surface water. In the bottom water the values ranged from 2.01 at Chavara offshore S_2 to 3.90 µM at Neendakara nearshore S_3. During pre-monsoon the values ranged from 2.40 µM at Chavara offshore S_2 to 4.70 µM at Chavara offshore S_3 in the surface water. In the bottom water the values ranged from 2.0 µM at Chavara offshore S_2 to 4.45 at Tangasseri near shore S_5. The S_2 value was the lowest recorded during the study period. The monthly variation of nutrient concentration at sites S_1 to S_8 are given in Figures 5.2a–h and the monthly variation of average nutrient concentration of the Kollam Coast is represented in Figure 5.2i.

During the present study, surface seawater phytoplankton abundance reached the highest level during monsoon. A secondary peak of abundance was noted in post-monsoon. The surface water phytoplankton density was 114.80×10^2, 100.44×10^2 and 29.76×10^2 cell nos/l during monsoon, post monsoon and pre-monsoon respectively. The overall phytoplankton count for Kollam Coast during the study period was 89.70×10^2 cell nos/l. The quantitative phytoplankton data represented as seasonal mean with standard deviation (SD) are given in Table 5.2. The percentage of diatoms, dinoflagellates and cyanophyta recorded in the Kollam Coast during monsoon, post-monsoon and pre-monsoon are shown in Figures 5.3a–c respectively. The most abundant phytoplankton along the Kollam Coast were *Asterionella, Chaetoceros, Rhizosolenia, Thalassionema, Trichodesmium, Cosinodiscus, Skeletonema, Peridinium, Thalassiothrix, Leptocylindrus, Eucampia* and *Plankotoniella* and the major phytoplankton species identified are:

Diatoms

Skeletonema costatum, Planktoniella sol, Ditylum brightwelli, Triceratium favus, Asterionella japonica, Coscinodiscus spp., *Biddulphia sinensis, B. mobiliensis, Rhizosolenia* spp., *Bacteriastrum varians, B. hyalinium, Chaetoceros lorenzianus, C. decipiens, C. curvisetus, C. affinis, Thalassionema nitzschioides, Thalassiothrix frauenfeldii, T. longisima, Nitzchia longissima, N. seriata, N. sejiata* and *Eucampia zoodiacus.*

Table 5.2: Overall Average Standard Deviation (SD) Phytoplankton Count (Cell Numbers/Litre) in the Kollam Coast

Phytoplankton Cell Nos/litre	*Monsoon*		*Post-monsoon*		*Pre-monsoon*		*Annual Average*	
	Mean	*SD*	*Mean*	*SD*	*Mean*	*SD*	*Mean*	*SD*
Diatoms								
Asterionella	8104.0	1604.9	6884.9	1195.3	36.1	28.7	5008.3	3791.6
Bacteriastrum	31.0	18.8	43.0	30.7	Nil	Nil	37.0	25.3
Biddulphia	13.0	6.2	14.0	4.3	93.8	146.3	44.4	93.8
Coscinodiscus	139.6	21.7	138.1	49.2	331.9	176.5	197.6	132.9
Chaetoceros	1762.9	542.3	1639.8	397.5	34.8	29.1	1304.5	830.9
Ditylum	65.1	50.6	61.0	43.6	50.4	43.6	58.8	44.5
Eucampia	84.3	19.4	79.4	7.4	222.3	147.4	147.8	121.8
Leptocylindrus	80.1	18.0	71.9	10.2	315.0	89.4	148.7	122.4
Nitzschia	92.6	33.4	Nil	Nil	Nil	Nil	92.6	33.4
Planktoniella	Nil	Nil	144.8	72.7	27.5	27.0	105.7	83.0
Rhizosolenia	816.8	187.8	278.9	119.7	278.0	135.3	457.9	296.4
Skeletonema	Nil	Nil	Nil	Nil	207.9	53.7	207.9	53.7
Thalassionema	Nil	Nil	Nil	Nil	372.6	59.8	372.6	59.8
Thalassiothrix	159.1	31.0	247.8	66.4	157.6	136.0	188.2	95.5
Triceratium	32.9	20.6	Nil	Nil	Nil	Nil	32.9	20.6
Dinoflagellates	–	–	–	–	–	–	–	–
Ceratium	28.8	17.3	192.3	55.7	36.4	37.7	85.8	86.0
Dinophysis	24.3	13.2	45.6	7.9	12.0	9.4	30.4	16.9
Peridinium	46.4	37.5	66.0	28.0	453.8	167.5	188.7	214.3
Prorocentrum	Nil	Nil	Nil	Nil	37.0	15.1	37.0	15.1
Cyanophyta	–	–	–	–	–	–	–	–
Trichodesmium	Nil	Nil	137.1	103.1	309.8	167.0	223.4	161.0

Dinoflagellates

Ceratiumfurca, Cfugus, Cbreva, Dinophysis caudata. D. miles, Prorocentrum micans, Peridinium depressum, P. oceanicum, P. claudicans and *P. pentagonium.*

Cyanophyta

The major cyanophyta identified was *Trichodesmium theibautii.*

Discussion

Earlier studies on phytoplankton abundance and hydrography show the South West Coast of India to be a region of upwelling and productivity. Kollam Coast is a populated area with hectic fishing activities. In this area, average values of phosphate, nitrate and silicate in surface sea water varied in the range 1.07 to 1.66 μM, 3.49 to 4.79 μM and 3.18 to, 5.06 μM respectively. In the bottom

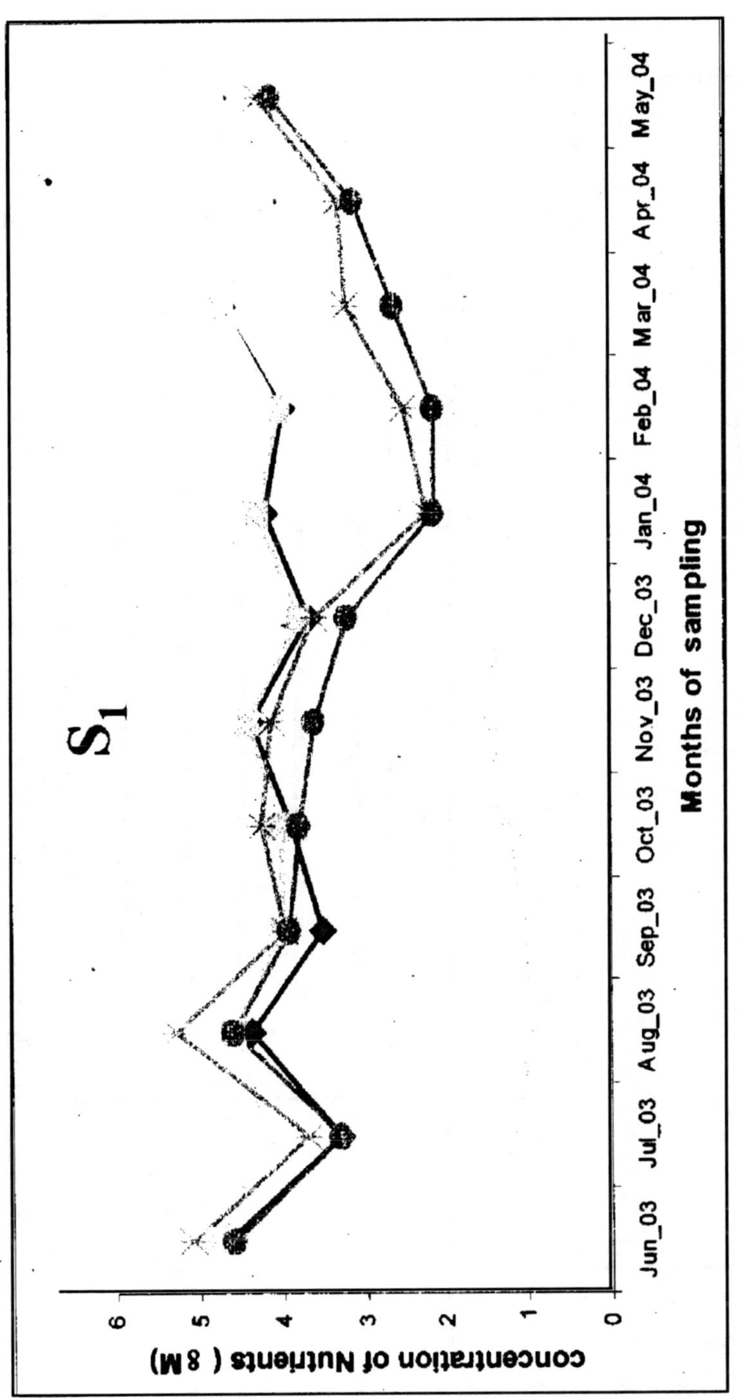

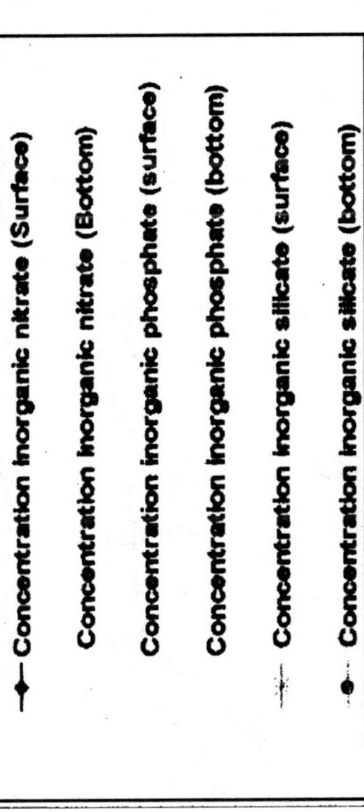

Figure 5.2a: Monthly Variation of Nutrient Concentration at Chavara Site S_1 during 2003–2004

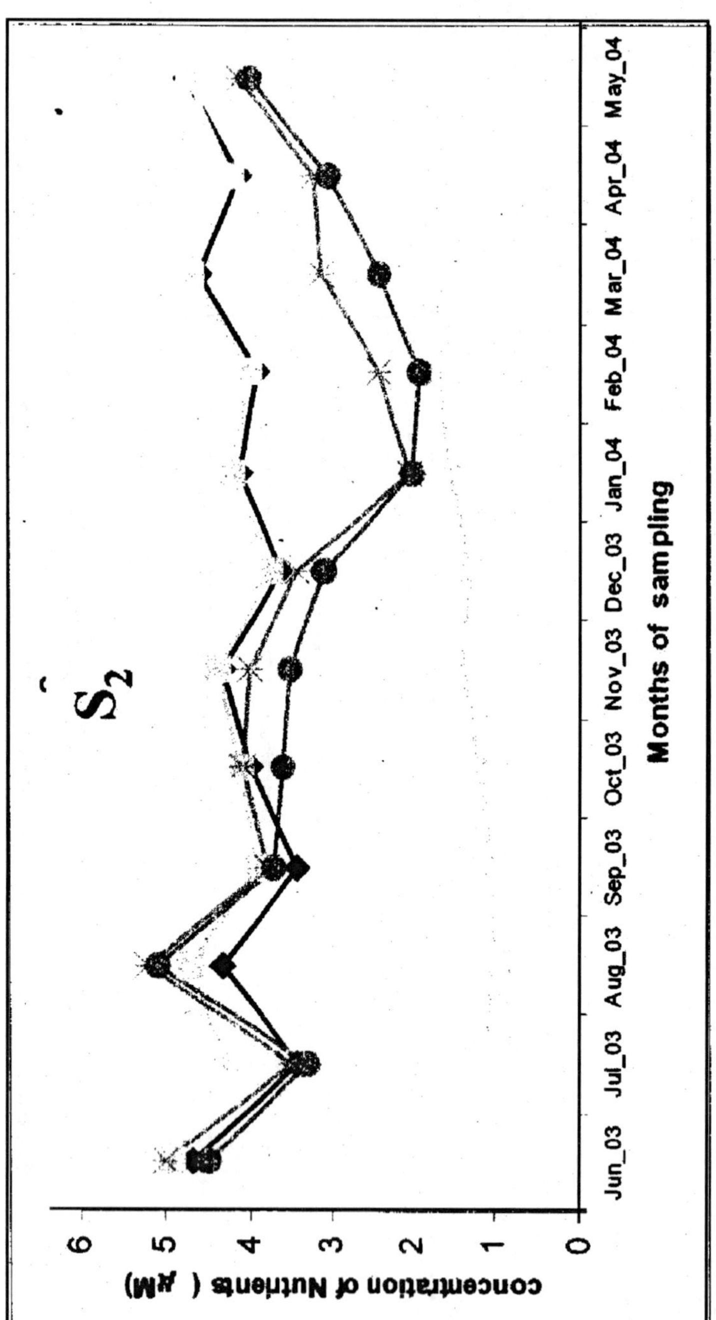

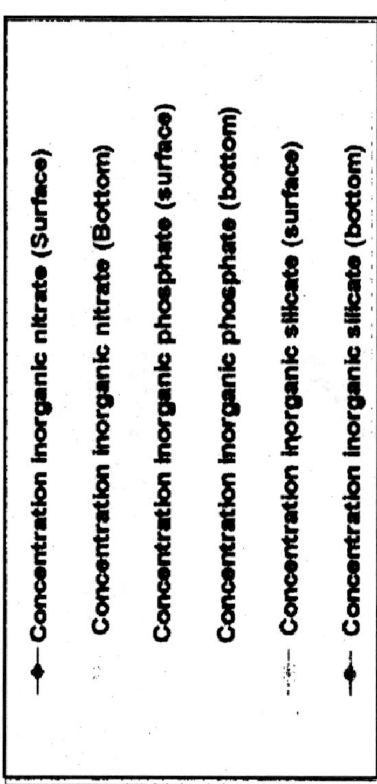

Figure 5.2b: Monthly Variation of Nutrient Concentration at Chavara Site S_2 during 2003–2004

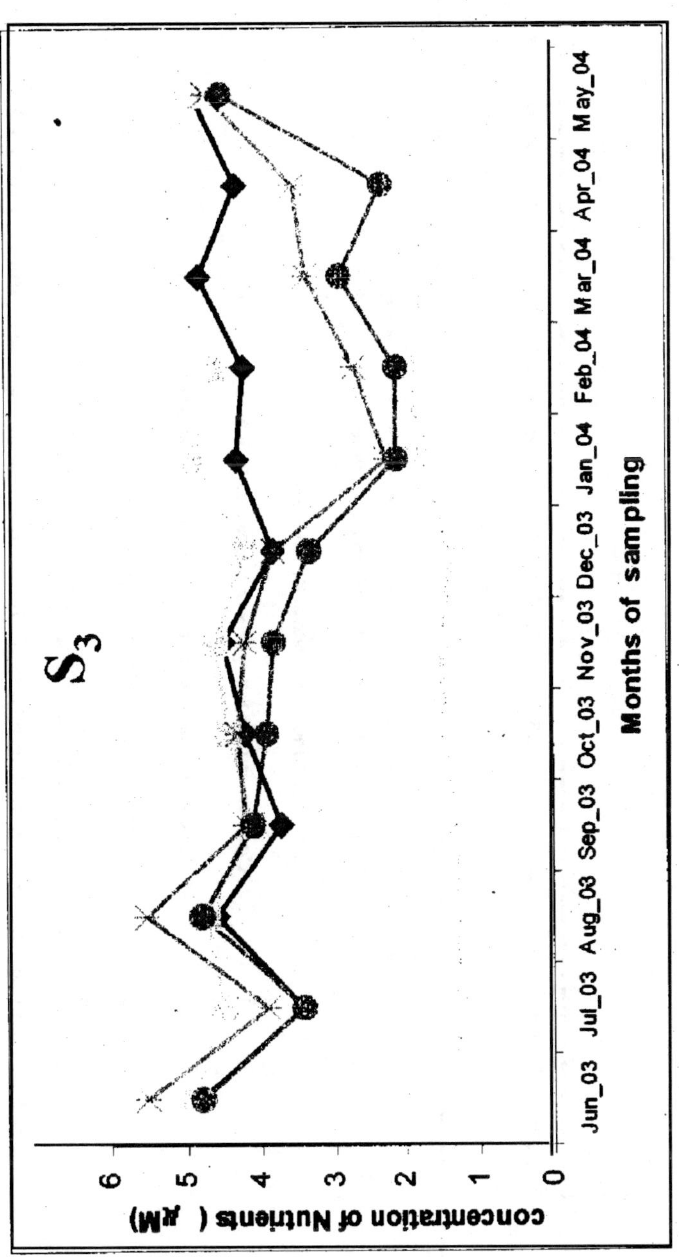

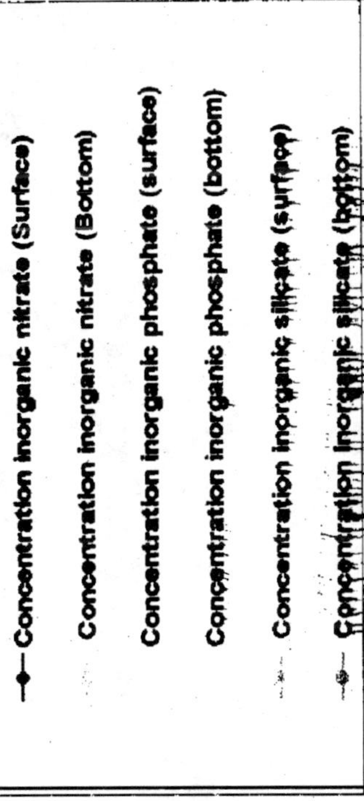

Figure 5.2c: Monthly Variation of Nutrient Concentration at Neendakara Site S_3 during 2003–2004

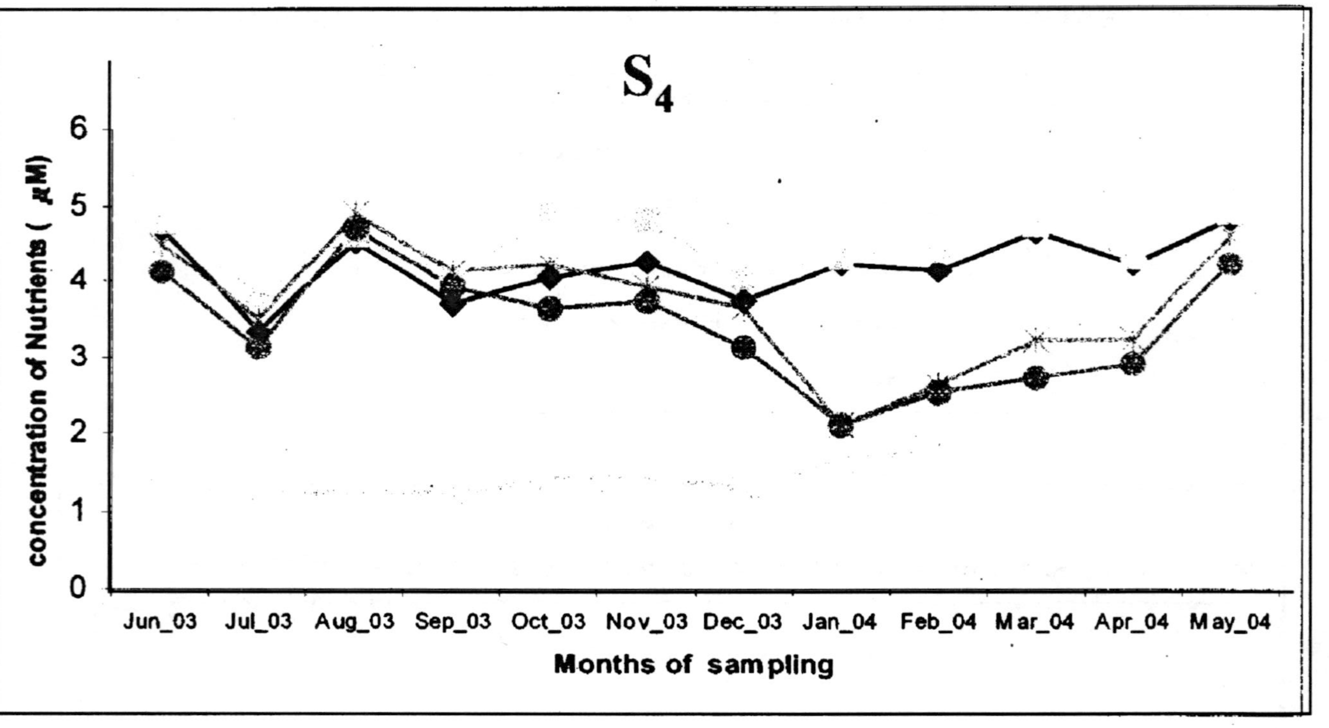

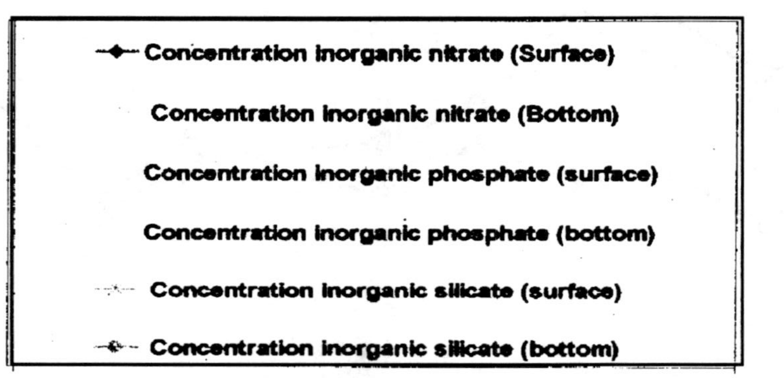

Figure 5.2d: Monthly Variation of Nutrient Concentration at Neendakara Site S_4 during 2003–2004

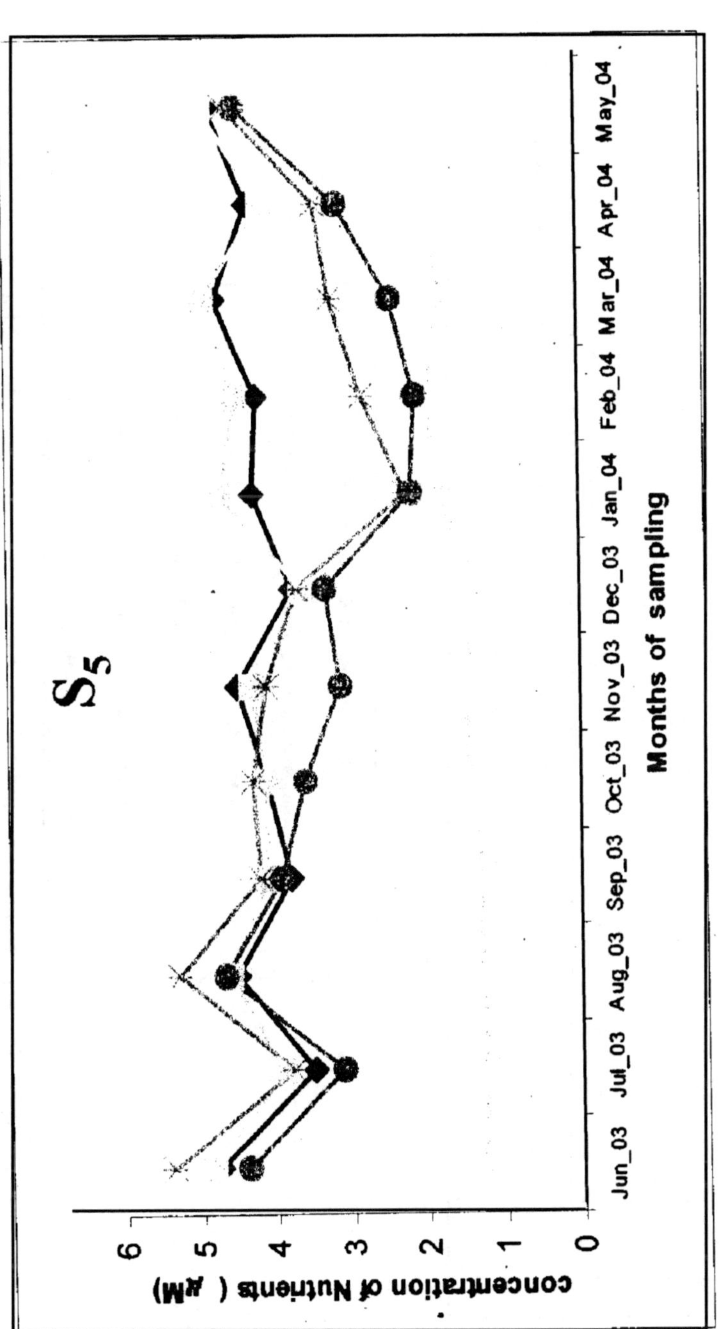

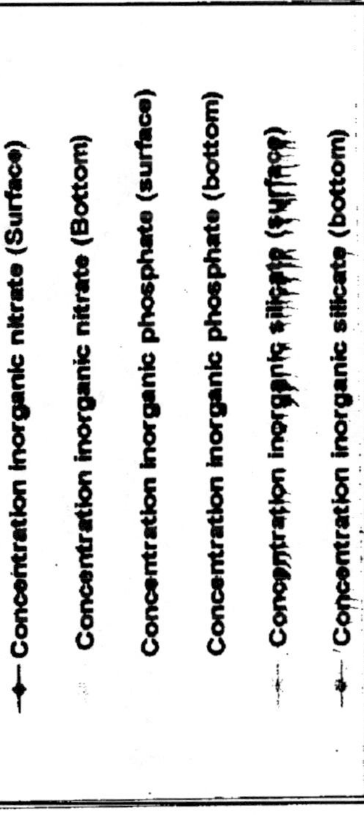

Figure 5.2e: Monthly Variation of Nutrient Concentration at Tangasseri Site S_5 during 2003–2004

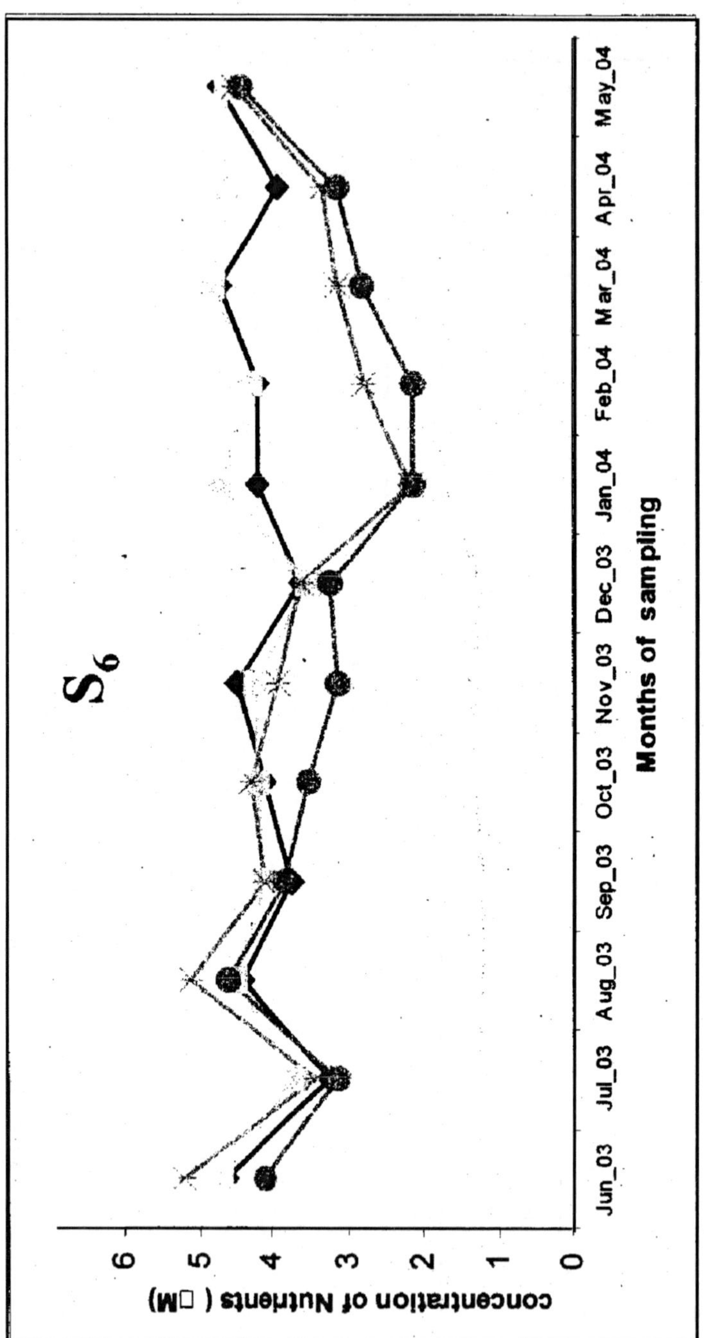

Figure 5.2f: Monthly Variation of Nutrient Concentration at Tangasseri Site S_6 during 2003–2004

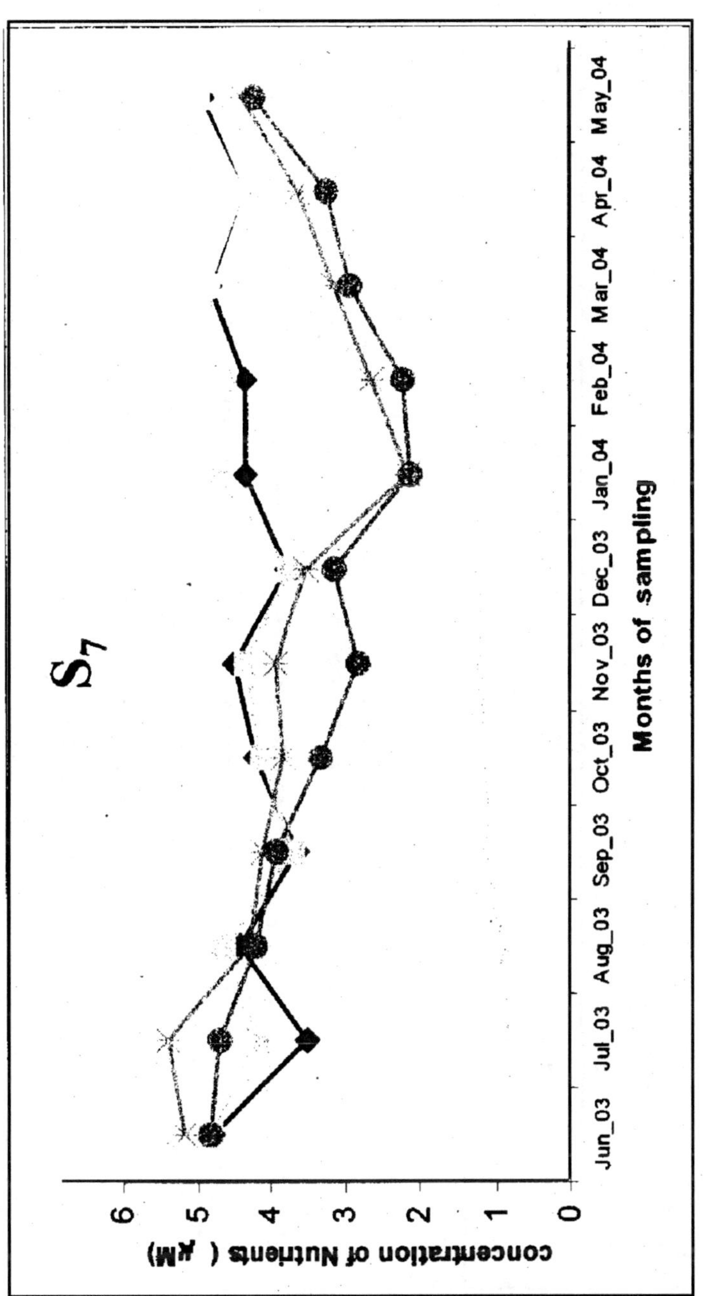

Figure 5.2g: Monthly Variation of Nutrient Concentration at Paravur Site S_7 during 2003–2004

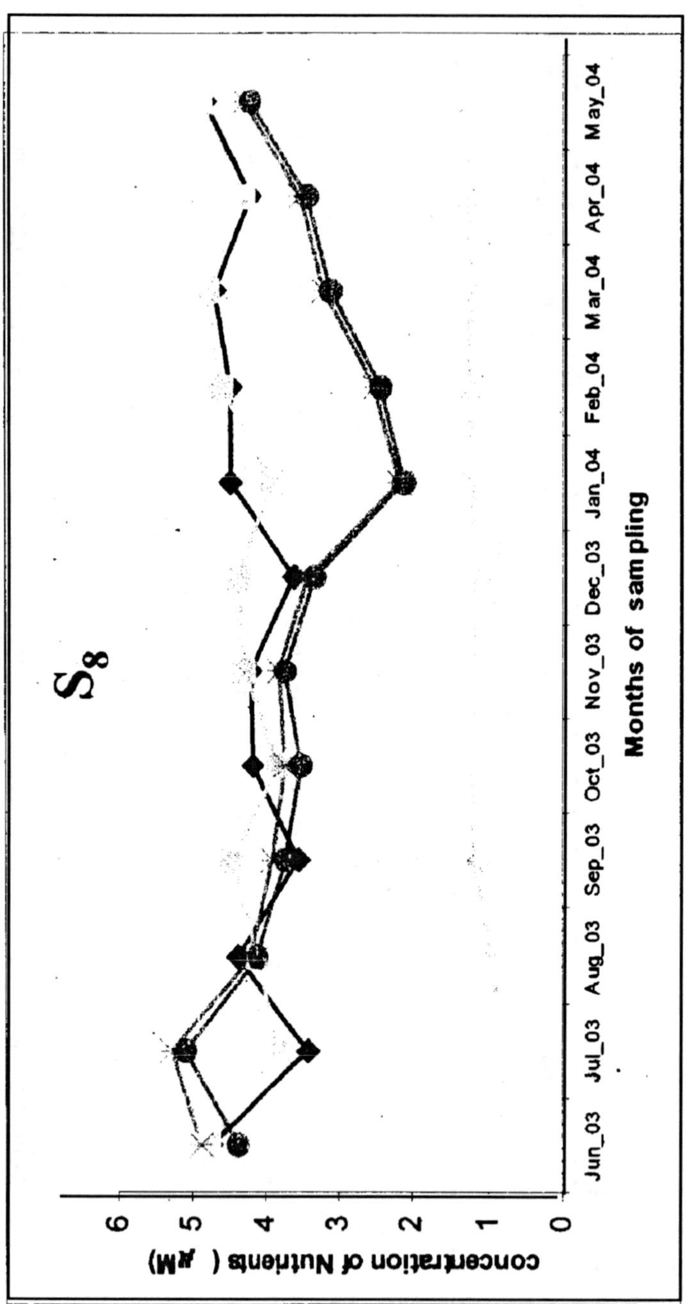

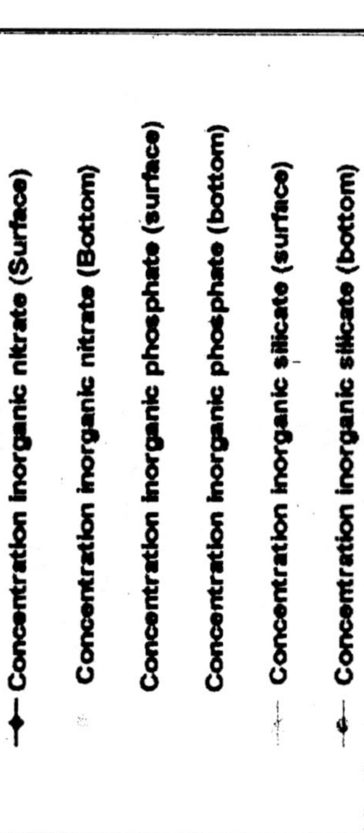

Figure 5.2h: Monthly Variation of Nutrient Concentration at Paravur Site S_8 during 2003–2004

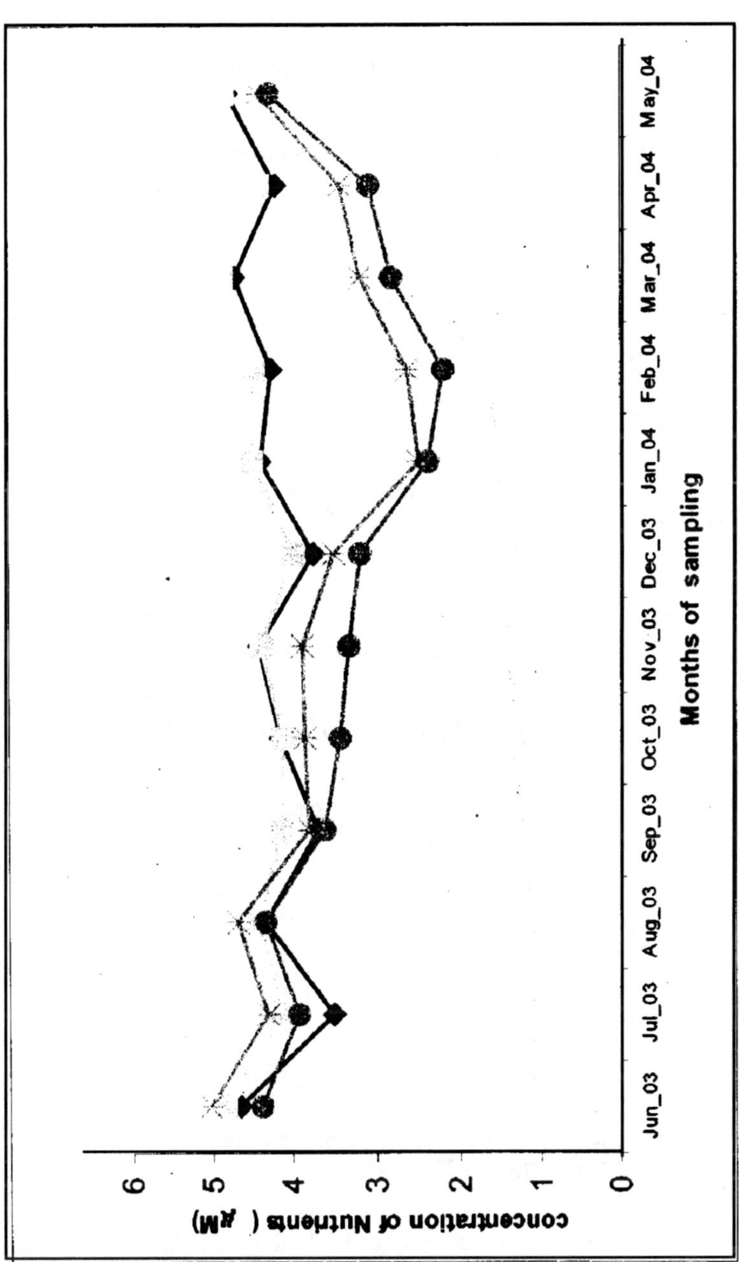

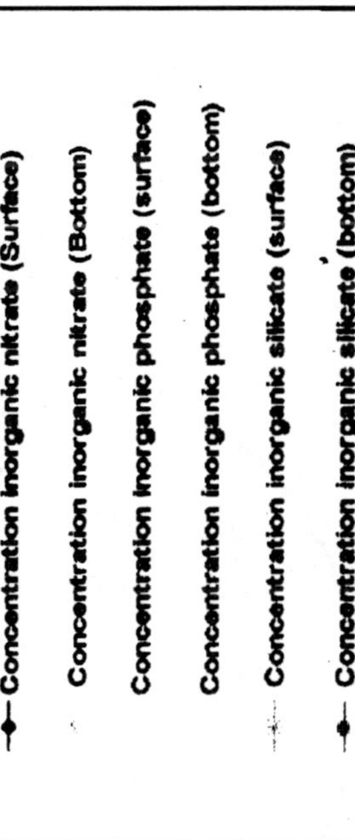

Figure 5.2i: Monthly Variation of Nutrient Concentration in the Kollam Coast during 2003–2004

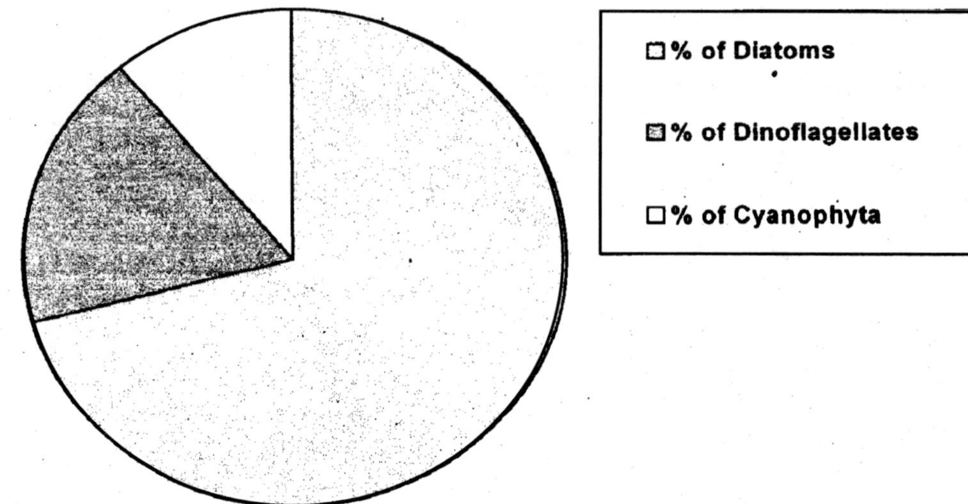

Figure 5.3a: Percentage Distribution of Phytoplankton during the Pre-monsoon in Kollam Coast during 2003–2004

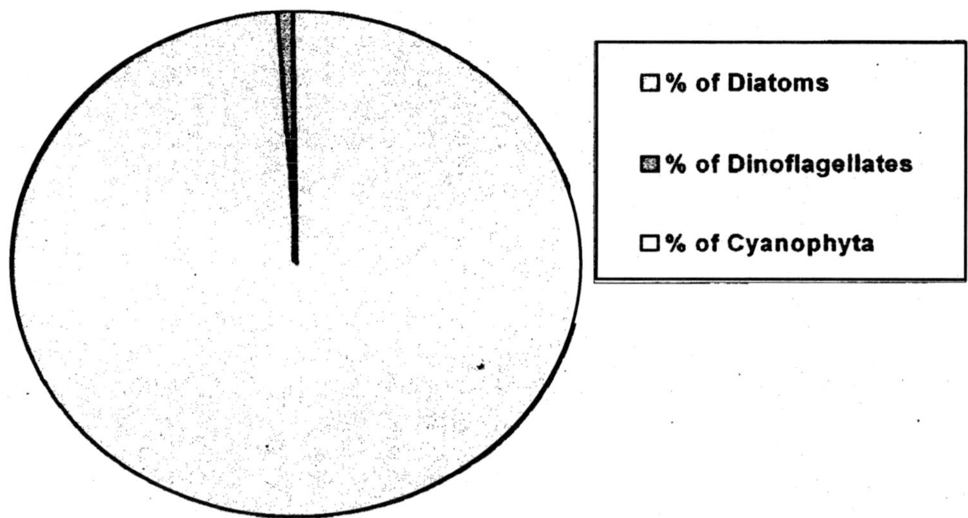

Figure 5.3b: Percentage Distribution of Phytoplankton during the Monsoon in Kollam Coast during 2003–2004

water, the values varied in the range. 1.10 to 1.73 µM, 4.08 to 4.84 µM and 2.13 to 4.40 µM. Nitrate is the limiting nutrient in the study area. During the present study, phytoplankton abundance decreased during pre-monsoon, whereas the NO_3-N level was relatively high. The average NO_3-N content during pre-monsoon period indicates higher values and did not exhibit any wide variation. The bottom water recorded higher concentration of NO_3-N than the surface water, which can be due to the continuous

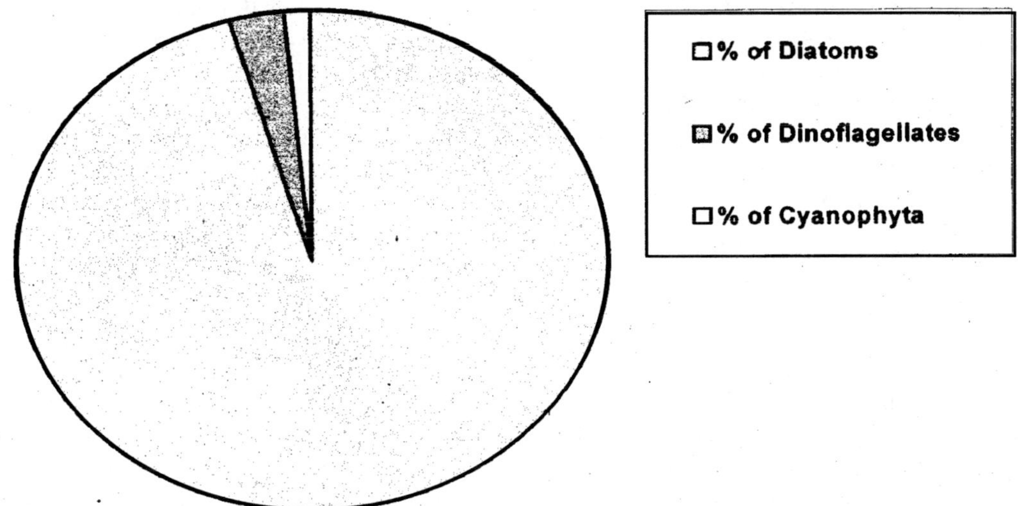

Figure 5.3c: Percentage Distribution of Phytoplankton during the Post-monsoon in Kollam Coast during 2003–2004

removal of phytoplanktons in the deep water from surface layers through sinking process as well as by grazing by zooplankton, thereby much of the nutrients absorbed from surface layer being regenerated and getting accumulated in the deeper and darker layers of water where photosynthesis does not occur.

Low phosphate during monsoon can be attributed to its removal by the influx of silt-laden freshwater from Kallada River, findings its way into the Ashtamudi estuary and therefore to the Arabia Sea at the 'Azhi' at Neendakara. The fresh water from the Ithikkara river, even though hindered during non monsoon period by closing of the 'Pozhi' at Paravur by a sand bar, overflowing took place during monsoon period. ('Azhi' in the vernacular Malayalam meaning permanent opening to sea and 'Pozhi' meaning temporary opening to sea). High values of phosphate during non-monsoon, period have earlier been reported from Indian coastal water (Sampath and Kannan, 1998) which is attributed to the release of phosphorus from mud and due to stirring action by strong winds and turbulence.

Average concentration of silicate in the non-monsoon period was nearly 50 per cent of the monsoon value. Surface run-off is the prime contributor of silica in the monsoon period. Silicate values showed wide variation between seasons, the highest mean value being 5.50 μM noted at the surface water at Neendakara near shore S_3 during monsoon and the lowest being 2.01 μM at the bottom of Chavara offshore S_2, during pre-monsoon which can be due to diatom dominancy.

The phytoplankton peak was observed during monsoon and a secondary peak during post-monsoon. Prior investigators have reported a bimodal peak of phytoplankton in the west coast of India (Chennubhotla, 1969; Subrahmanyan, 1959; Subrahmanyan and Sarma, 1960). Diatoms were dominant during all seasons. The dominancy of diatoms from Indian coastal water have been earlier reported (Raman and Prakash, 1989; Sarojini and Sarma, 2001). Statistical correlations show that differences were not significant among stations ($P > 0.05$). However, differences were significant among seasons. During all the seasons, PO_4-P was negatively correlated to phytoplankton cell number. During pre-monsoon the correction were $r = -0.985$; $P < 0.01$ and during post monsoon $r = -0.719$;

$P < 0.05$. The correlation for NO_3-N was not statistically significant. For $[Si(OH)_4]$-Si, the significant correlation were r = –0;891; $P < 0.01$ during post-monsoon and during monsoon r = –0.232; $P < 0.01$. Significant negative correlation between nutrient concentrations and phytoplankton has been reported from Goa in the west coast (Garg and Bhaskar, 2000) and this suggests that the concentration of nutrients play an important role in phytoplankton abundance. Relative higher concentration of nutrient concentration in the pre-monsoon period can be attributed to low phytoplankton abundance and effective regeneration process.

It is known that phytoplankton population are rapidly reduced by growing zooplankton and affected by physical process such as upwelling and turbulence, fronts and gyres, solar radiation and so forth. The results of the present study also show that the hydrographic and hydro-chemical factors play an important role in productivity along the Kollam Coast. The phenomenon of upwelling together with large influx of nutrient rich water into the sea increases productivity (Balachandran *et al.*, 1997). Higher values of phytoplankton density in the Kollam Coast, relative to many other coastal areas of India (Madhav and Kondatarao, 2004), are also attributed to land runoff following rains and the ret liquor from the Ashtamudi estuary and Edava Nadayara lake enriching the marine environment. Hence, along with hydrographic and hydrochemical factors, anthropogenic factors are also contributors to phytoplankton growth in the Kollam Coast.

Acknowledgement

We thank the University Grant Commission, New Delhi, India for financial assistance, the Faculty, Zoology and Botany of Fatima Mata National College, Kollam, Kerala and to Dr. J.S. Sharma (STA Fellow, Japan, Environment) Senior Chemist, ONGC, New Delhi for valuable suggestions. This is contribution FMNC-CD-09 from FMNC, Quilon.

Reference

Balachandran, V.K., Gopinathan, C.P., Pillai, V.K., Nandakumar, A. and Valsala, K.K., 1997. Chlorophyll profile of the euphotic zone in the Lakshadweep sea during the South-west monsoon. *Indian J. Fish.*, 44: 29–43.

Banse, K., 1959. On upwelling and bottom trawling off the south-west coast of India. *J. Mar. Biol. Assoc., India*, 1: 33.

Chennubhotla, V.S.K., 1969. Distribution of phytoplankton in the Arabian sea between Cape Comorinand Cochin. *Indian J. Fish.*, 16: 129–136.

Currie, R.I., Fischer, A.E. and Hargreaves, P.M., 1973. In: *The Biology of the Indian Ocean.* Springer-Verlag, Berlin, pp. 37.

Garg, A. and Bhaskar, P.V., 2000. Fluxes of diatoms in the Dona Paula Bay. *J. Plankton Res.*, 22: 2015–2038.

Hira, S., Nuzbai, K. and Niaz Rizvi, S.H., 2001. In: *The Indian Ocean: A Perspective*. Oxford and IBH Publishing Co. Pvt. Ltd., New Delhi, pp. 509.

Lierheimer, L.J. and Banse, K., 1960. Studies on the phytoplankton on the west coast of India. Part–III. Seasonal variations of the phytoplankters and environmental factors. *Indian J. Fish.*, 7: 307–336.

Madhav, G.V. and Kondalarao, B., 2004. Distribution of phytoplankton in the coastal waters of east coast of India. *Indian J. Mar. Sci.*, 33: 262–268.

Madhupratap, M., Gopala Krishnan, T.C., Haridas, P., Nair, K.K.C., Aravindakshan, P.N., Padmavathy, G. and Paul, S., 1996. Lack of seasonal and geographic variation in mesoplankton biomass in the Arabian Sea and its structure in the mixed layer. *Curr. Sci.*, 71: 863–868.

Nair, P.V.R., and Pillai, P.K. 1982. Productivity of Indian seas. *J. Mar. Biol. Ass. India,* 25: 41–50.

Newell, G.E. and Newell, R.C., 1979. *Marine Plankton: A Practical Guide*. Hutchinson, London.

Parsons, T.R.., Maita, Y. and Lalli, C.M., 1984. In: *A Manual of Chemical and Biological Methods for Seawater Analysis*. Pergamon Press, Oxford, pp. 173.

Polat, S. and Piner, M.P., 2002. Nutrients and phytoplankton in the Babadillimani Bight, North eastern Mediterranean Coast of Turkey. *Indian J. Mar. Sci.*, 31: 188–194.

Qasim, S.Z., 1977. Biological productivity of the Indian Ocean. *Indian J. Mar. Sci.*, 6: 122–137.

Raman, A.V. and Prakash, K.P., 1989. Phytoplankton in relation to pollution in Visakhapatnam harbour, east coast of India. *Indian J. Mar. Sci.*, 18: 33–36.

Rump, H.H. and Krist, H., 1992. In: *Laboratory Manual for the Examination of Water, Wastewater and Soil*, 2nd edn. VCH Publication, Weinheim, pp. 190.

Sampath, K.P. and Kannan, L., 1998. Seasonal variations in physico-chemical characteristics in the TRANQUEBAR–Nagapattinam Region, South-east coast of India. *Poll. Res.*, 17: 397–402.

Santhanam, R., Ramanathan, N., Venkataramanujan, K. and Jegatheesan, G., 1987. In: *Phytoplankton of the Indian Seas: An Aspect of Marine Botany*. Daya Publishing House, New Delhi.

Sarma, V.V.S.S., Dileep, K.M., Guans, M. and Madhupratap, M., 2000. Seasonal control on surface PCO in the Central and East Arabian Sea. *Proc. Indian Acad. Sci. (Earth Planet Sci.)*, 109: 471–479.

Sarojini, Y. and Sarma, N.S., 2001. Vertical distribution of phytoplankton around Andaman and Nicobar Islands, Bay of Bengal. *Indian J. Mar. Sci.*, 30: 65–69.

Strickland, J.D.H. and Parsons, T.R., 1972. In: *A Practical Handbook of Seawater Analysis*, 2nd edn. Bull. No. 167. Bull. Fish. Res. Bd., Can, Ottawa, pp. 310.

Subrahmanyan, R., 1959. Studies on the phytoplankton of the west coast of India. *Proc. Indian Acad. Sci. (Earth Planet. Sci.)*, 50: 113–252.

Subrahmanyan, R. and Sarma, A.H.V., 1960. Studies on the phytoplankton on the west coast of India. Part–III. Seasonal variations of the phytoplanktons and environmental factors. *Indian J. Fish.*, 7: 307–336.

Surekha, S. and Madhupratap, M., 1996. Seasonality and composition of phytoplankton in the Arabian Sea. *Curr. Sci.*, 71: 869–873.

Verlecar, X.N. and Parulekar, A.H., 2001. In: *The Indian Ocean: A Perspective* (Ed.) Rabin Sen Gupta and Ehrlich Desa. Oxford and IBH Publishing Co. Pvt. Ltd., New Delhi, pp. 400.

Chapter 6

Pre-impact Studies of the 2000MW Lower Subansiri Dam on Certain Hydrobiological Aspects of Down Stream of River Subansiri with Special Reference to Plankton and Fishes

Debojit Baruah[1], Lakhi Prasad Hazarika[2] and Sabita Borah[3]
[1]Department of Botany, [3]Department of Zoology, Lakhimpur Girls' College, Lakhimpur, Assam
[2]Department of Zoology, North Lakhimpur College, Lakhimpur, Assam

ABSTRACT

Pre-impact studies of the 2000MW lower Subansiri dam on certain hydrological aspects of down stream of river Subansiri *viz.*, physico-chemical parameters, phytoplankton, zooplankton and fresh water fish diversity have been carried out from January 2006 to December 2006.The water quality is suitable for planktonic and fish diversity. In the down stream of proposed dam, a total of 199 taxa have been recorded and identified which includes 48 species of algae, 23 zooplankton, 128 species of freshwater fishes belonging to 9 order, 27 families and 78 genera.

Keywords: *River dam, Down stream, Hydrobiology, Plankton and fish diversity, Pre-impact study.*

Introduction

Dam interrupts stream flow and generates hydrological changes along the integrated continuum of river ecosystem (Vannote *et al.*, 1980; Junk *et al.*, 1989). The impact of dam upon natural ecosystem and biodiversity has been one of the principal concerns raised by large dam (McCartney *et al.*, 2000). Over the course of last 10 years in particular, considerable investments have been made in the development of measures to alleviate these impact (Bergkamp *et al.*, 2000). Yet today wide spread concern remains that despite improvements in dam planning, design, construction and operation, they continue to result in significant negative impacts to a wide range of natural ecosystem and to the people that depends upon thus for their livelihood.

There is lots of information on post impact of dam on the ecology but in this aspect pre impact study is very limited. Study of this nature will help in future to assess the impact of dam on the various aspects of river ecology.

Materials and Methods

Study Area

River Subansiri, the largest tributary of river Brahmaputra, originates from western part of Mount Pararu (5059 km) in the Tibetan Himalaya. After flowing for 190 km. through Tibet, it enters India. It continues its journey through the Himalaya of India for 200 km and enters into the plains of Assam through a Gorge near Gerukamukh of Dhemaji District of Assam. Its total length is 520 km and drains a basin of 37000 sq. km.

For a systematic monitoring and study of each selected hydrological and other related parameters of the down stream river (areas between proposed dam and river Brahmaputra where the river Subansiri falls). The total study area is of about 150 km have been divided into four sectors namely:

1. Chowaldhowa Ghat,
2. Khabolo Ghat, and
3. Jamuguri Ghat.

Hydrobiological Parameters

Water samples were collected from January 2006 to December 2006, mead stream at a depth of one meter below the water surface in 2 liter bottles. Monsoon was provisionally reckoned as the period from July to October, winter from November to February and summer from March to June. The collected samples were analyzed in the spot and half of these brought to the laboratory for further analysis.

Water temperature by using mercury thermometer, transparency by Secchi disc, pH by pen type digital pH meter, conductivity by conductivity meter, current flow according to Trivedy *et al.* (1987), TDS, Total hardness, Calcium hardness, Chloride by APHA (1989), DO_2 by Winkler's modified method (APHA, 1989), Free CO_2, Carbonate and Bicarbonate alkalinity according to Welch (1952) and Silicate by Jhingran *et al.* (1969) were measured.

Planktonic Study

The planktons were collected on every 15 days of months at a fix time with the help of plankton net (no. 25). The planktons were collected by filtering 50 litre of water through the plankton net and preserved in 2.5 per cent formalin. The qualitative analysis of the plankton were done with the help of literature (Lackey. 1983; Nidham and Nidham, 1966, Edmonson, 1974).

Monsoon was provisionally reckoned as the period from July to October, winter from November to February and Summer from March to June.

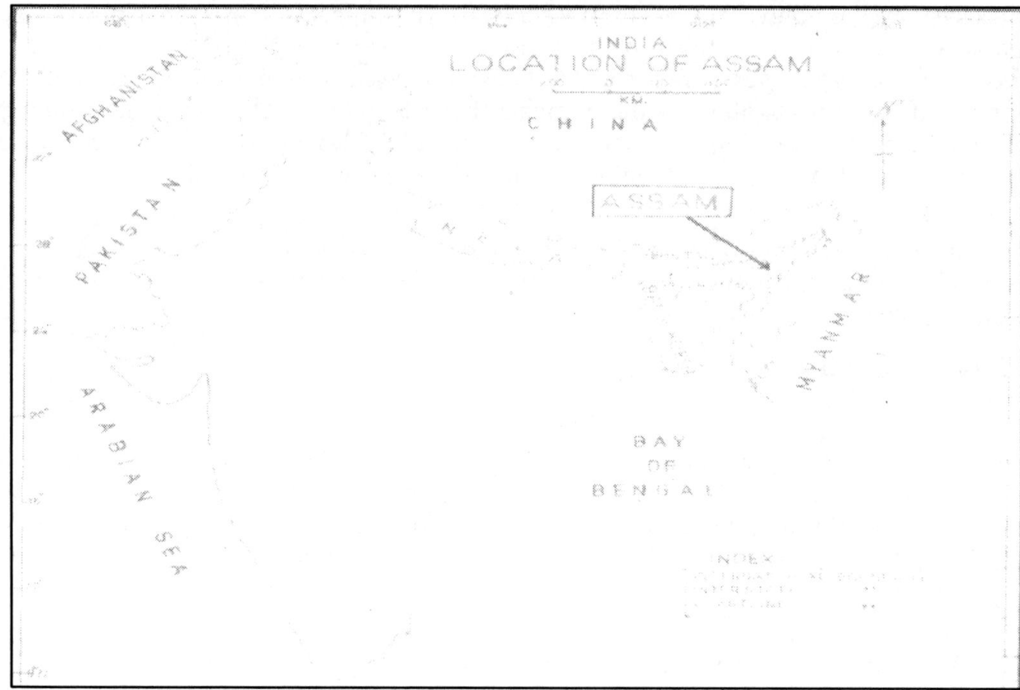

Figure 6.1a: Location Map of Assam and India

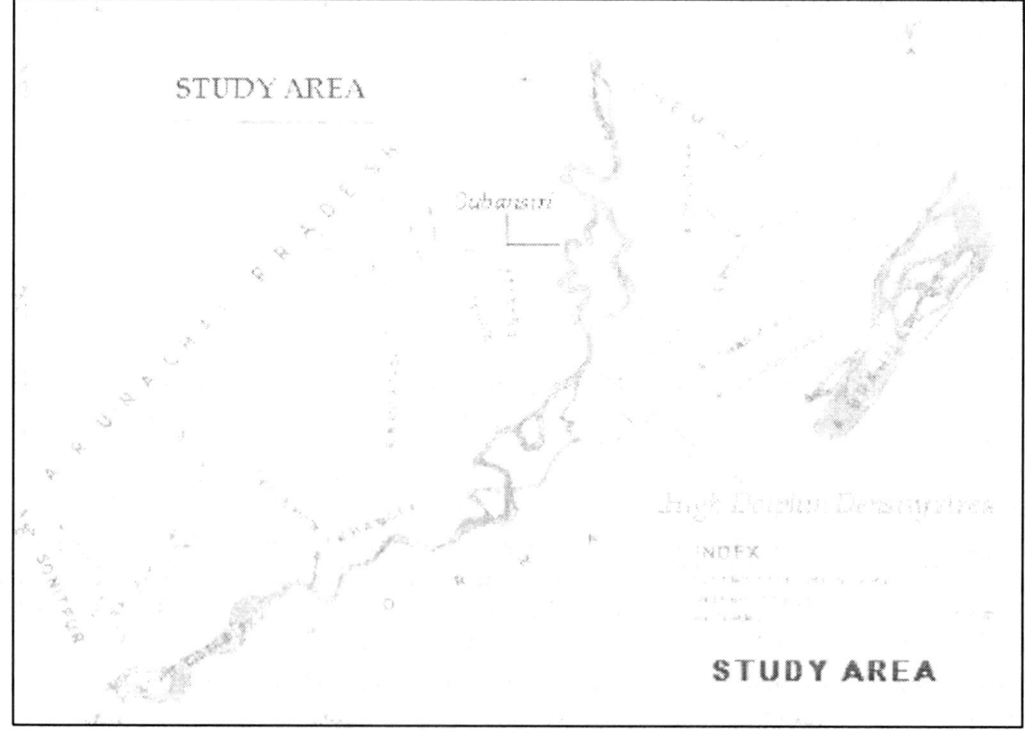

Figure 6.1b: Location Map of River Subansiri

Fish Diversity

Specimens of fishes were collected from the different landing centers of the four sectors. Fishes pertaining to the present study were also collected from surrounding lentic water bodies (Beels, wetlands, seasonal ponds). The collected fishes were preserved in 5 per cent formalin for taxonomic study. The specimen were identified according to the methods of Munshi and Srivastav (1988), Talwar and Jhingrang (1991), Biswas (1993) and Viswanath (2002).

One-Way ANOVA on all physico-chemical parameters of water quality among different sampling stations are carried out using ORIGIN 6.1 version.

Results

River Hydrology

The various physico-chemical and biological conditions of the water of downstream of river Subansiri presented in the Table 6.1(a), (b) and (c).

Table 6.1(a): Seasonal Variation in the Hydrological Parameters of the Down Stream (Station: Chowaldhowa Ghat)

Parameter	*Months*											
	Jan	*Feb*	*Mar*	*Apr*	*May*	*Jun*	*Jul*	*Aug*	*Sep*	*Oct*	*Nov*	*Dec*
Air Temp. (°C)	20.4	20.9	24.2	25.9	28.6	32.1	32.5	30.4	29.1	28.6	26.1	21.2
Water Temp. (°C)	14.3	15.1	20.4	23.6	26.00	29.8	31.2	27.7	26	24.2	21.6	19.7
Transparency (cm)	78.2	69.1	31.2	19.5	21.1	23.5	22.6	25.2	35.8	38.2	46.1	66.8
pH	7.7	7.2	7.1	7.1	7.2	7.1	7.1	7.1	7.1	7.4	7.6	7.3
TSS (ppm)	53	23	22	28	33	29	31	34	30	26	19	41
TDS (ppm)	193	146	141	]55	165	169	172	165	170	151	132	189
Total solids (ppm)	246	169	163	183	198	198	203	199	200	177	151	230
Conduct. (μ mho)	108.3	104.5	125.4	140.3	116.8	101.5	99.4	92.9	110.5	98.3	111.2	103.7
Alkalinity (mg/)l	119.1	112.6	109.2	102.8	108.1	115.3	118.1	102.8	95.9	87.6	80.7	86.5
Total hardness (mg/l)	61.8	60.3	77.4	75.2	80.1	87.2	74.8	66.9	62.5	61.2	60.8	64.1
Ca hardness (mg/l)	37.5	33.2	41.8	42.3	40.6	41.3	35.3	36.4	35.8	37.1	30.8	36.1
Chloride (mg/l)	9.8	8.3	7.5	6.6	5.4	4.1	4.3	9.2	7.2	7.3	8.6	9.8
Silicate (mg/l)	7.2	7.5	5.7	5.5	4.8	7.7	8.6	8.9	9.0	9.3	9.1	9.4
DO_2 (mg/l)	8.5	9.2	9.9	9.7	9.1	9.4	8.5	7.1	7.8	7.8	8.0	8.8
FCO_2 (mg/l)	3.1	3.4	5.1	7.1	6.3	5.9	7.1	5.1	9.4	3.6	3.3	4.6

Temperature

The air and water temperature of the three sectors ranged between 20.4°C (January) and 32.5°C (July) and 14.3° (January) respectively.

Table 6.1(b): Seasonal Variation in the Hydrological Parameters of the Down Stream (Station: Khabolo Ghat)

Parameter	Months											
	Jan	Feb	Mar	Apr	May	Jun	Jul	Aug	Sep	Oct	Nov	Dec
Air Temp. (°C)	21.3	20.9	23.2	26.0	28.3	33.2	32.9	31.5	28.1	29.5	25.2	21.9
Water Temp. (°C)	18.3	19.1	22.4	23.6	26	29.8	31.2	29.7	26	24.2	23.6	22.7
Transparency (cm)	80.2	71.2	30.9	27.8	29.1	22.5	25.7	27.9	39.3	44.2	54.1	71.8
pH	7.3	7.2	7.1	7.0	7.2	7.1	7.1	7.1	7.7	7.9	7.8	7.9
TSS (ppm)	43	20	19	18	29	26	31	34	55	46	33	31
TDS (ppm)	201	177	163	143	159	164	166	170	180	161	139	198
Total solids (ppm)	243	197	182	161	188	190	197	204	235	207	172	229
Conduct. (μ mho)	111.2	132.9	128.4	154.3	110.8	131.5	108.4	102.9	98.6	103.3	100.2	107.3
Alkalinity (mg/l)	113.2	109.6	106.5	112.3	98.9	105.9	114.1	96.8	75.8	77.9	74.7	71.5
Total hardness (mg/l)	56.4	55.1	63.9	66.6	71.7	79.7	69.2	54.8	50.3	53.8	50.9	66.1
Ca hardness (mg/l)	35.4	34.3	39.0	43.8	37.3	39.2	36.1	30.1	36.3	30.5	33.8	32.1
Chloride (mg/l)	7.0	8.8	7.9	6.4	6.0	4.5	4.0	3.9	7.9	6.6	6.3	7.0
Silicate (mg/)l	6.9	7.1	5.3	5.1	5.0	6.9	8.6	9.3	8.9	8.4	8.8	7.4
CO_2 (mg/l)	9.1	9.8	10.5	11.1	10.8	10.3	9.9	8.0	11.3	12.9	12.4	9.6
FCO_2 (mg/l)	3.9	4.7	5.0	7.9	7.0	6.4	8.3	4.4	3.4	5.2	4.9	6.4

Table 6.1(c): Seasonal Variation in the Hydrological Parameters of the Down Stream (Station: Khabolo Ghat)

Parameter	Months											
	Jan	Feb	Mar	Apr	May	Jun	Jul	Aug	Sep	Oct	Nov	Dec
Air Temp. (°C)	21.5	20.9	23.1	26.2	27.3	32.1	31.6	30.9	30.1	29.3	27.5	20.5
Water Temp. (°C)	16.1	17.6	20.9	24.5	27.5	29.9	30.4	28.3	25.7	23.2	21.3	22.7
Transparency (cm)	88.7	75.2	33.1	30.9	30.3	24.1	27.3	29.6	41.4	47.5	57.2	77.5
pH	7.5	7.3	7.4	7.3	7.2	7.0	7.1	7.2	7.5	7.7	7.9	7.1
TSS (ppm)	42	20	31	29	30	25	28	29	27	20	26	27
TDS (ppm)	190	149	145	150	173	175	177	178	188	157	140	185
Total solids (ppm)	232	169	176	179	203	200	205	207	215	175	166	212
Conduct. (μ mho)	101.4	130.6	125.3	160.3	104.8	143.5	124.4	98.9	90.4	115.3	97.2	1 11.5
Alkalinity (mg/l)	109.2	107.3	99.5	101.6	1 00.4	98.5	102.8	93.9	69.8	70.6	72.3	70.4
Total hardness (mg/l)	60.7	54.5	70.4	69.1	78.8	83.6	70.4	60.5	55.4	60.6	54.7	60.4
Ca hardness (mg/l)	30.6	30.1	40.0	41.5	30.9	29.2	30.4	29.1	26.7	28.3	30.3	28.1
Chloride (mg/l)	6.3	7.5	7.0	5.3	6.1	3.3	5.7	3.0	6.9	5.4	5.1	6.0
Silicate (mg/l)	6.0	6.1	4.9	5.0	5.5	5.9	7.3	9.0	8.1	7.2	7.8	7.1
CO_2 (mg/l)	9.1	8.9	1 1.0	11.3	12.1	11.9	9.7	10.6	12.6	12.3	13.1	10.2
FCO_2 (mg/l)	3.0	2.9	4.5	5.3	5.7	5.8	7.2	3.4	4.1	3.8	3.3	5.4

Transparency

Transparency values ranged seasonally between 21.1 cm in May and 88.7 cm in March in Jamuguri Ghat and Chowaldhowa Ghat respectively, the highest being in the water flow was slow and carried relatively less amount of silt. It was recorded to be the lowest during the monsoon months as the water was highly turbid.

pH

The water was recorded to be relatively alkaline throughout the year (between 7.1–7.9). The highest recorded during the post monsoon months (September and October) in all the sectors.

TDS

The total dissolved solids (TDS) values of all the sectors fluctuated seasonally between 132 and 201 ppm. The lowest was recorded in May (Chowaldhowa Ghat) and highest in July (Jamuguri ghat).

TSS

The total suspended solids (TSS) ranges between 19 and 52 ppm.

Conductivity

The highest conductivity (160.3 µ mho) was recorded in the month of April at Jamuguri Ghat and lowest in the month of September at Jamuguri Ghat.

Alkalinity

The total alkalinity of the water ranged between 70.4 and 119 mg/lit. The maximum and minimum contents were recorded in Chowaldhowa Ghat in January and Jamuguri Ghat in December respectively.

Calcium Hardness

In all the sectors it ranges from 26.7 to 43.8 mg/lit annually.

Chloride

The chloride content was relatively low in all the sectors with a range of 3.0 mg/lit (December) to 9.8 mg/lit (February).

Silicate

The silicate content ranged between 5.1 and 9.4 mg/lit, the maximum recorded in the post monsoon period.

Plankotonic Diversity

A total of 48 phytoplanktonic species have been recorded (Table 6.2a). The phytoplankton is composed of three classes of algae namely Myxophyceae (Blue green algae, 08 species), Chlorophyceae (Green algae, 17 species) and Bacillariophyceae (Diatom, 23 species) Bacillariophyceae showed dominance over other two algae. Similarly a total of 23 different species of zooplankton belongs to four different categories namely Protoza (07 species), Rotifera (06 species), Cladocera (05 species) and Copipodes (05 species) are recorded and identified (Table 6.2b).

Fish Diversity

A total of 128 fish species belongs to 09 order, 27 families and 78 genera were recorded in the study area (Table 6.3). Observation reveals that near Chowaldhowa Ghat *Tor putitora, Tor tor,*

Acrossocheilus hexagonelips, Labeo pungusia, Labeo dyocheilus are available and all these fishes are either endangered (Tor species) or threatened. The migratory fish *Hilsa hilsa,* culture fishes namely *Cirrhinus mrigala, Ctenophyryngodon idellus* and *Cyprinus carpio* are found to be present in the river. Lots of fishes having ornamental values namely *Botia dario, Oreichchys* species, *Botia rostata, Macrognatus pancalus, Macroganthus oral, Chela cachius, Psilorhynchus balitora, Barillius barna, Carra kempi, Chaca chaca, Chanda nama, Closogobius giurius, Danio dongrila, Erecthetis pusilas, Channa stewarti, Chaca chaca, Xenetedon concilla,* members of colisa (05 species), *Tetradon cutcutia, Semiplotus semiplotus.* etc. are available in the down stream basin.

Table 6.2(a): Phytoplankton Composition in the Down Stream of River Subansiri

Sl.No.	*Name of the Species*	*Sl.No.*	*Name of the Species*
	MYXOPHYCEAE		**CHLOROPHYCEAE**
1.	*Anabaena fertilissima*	25.	*Zygnema khannae*
2.	*Anabaena vaginicoda*		**BACILLARIOPHYCEAE**
3.	*Gloeocapse compacta*	26.	*Asterionella gracillima*
4.	*Lyngloya allorgii*	27.	*Calonesia sp.*
5.	*Lyngloya limnetica*	28.	*Coconeis sp.*
6.	*Nostoc vaginicola*	29.	*Cylotella ocellata*
7.	*Oscillatoria limnetica*	30.	*Cymbella affinis*
8.	*Oscillatoria homogeneae*	31.	*Cymbella kalbei*
	CHLOROPHYCEAE	32.	*Denticula elegans*
9.	*Ankistrodesmus convolutes*	33.	*Diatoma elogatum*
10.	*Bulbochaeta varinis*	34.	*Diploneis elliptica*
11.	*Chaetohora elegans*	35.	*Epithemia sorex*
12.	*Chlorella vulgaris*	36.	*Eunotia pectinalis*
13.	*Cladophora glomerata*	37.	*Fragilaria capucina*
14.	*Closterium turgidium*	38.	*Frustulia rhomboids*
15.	*Clolestrum reticulatum*	39.	*Gomphonema montanum*
16.	*Cosmarium granatum*	40.	*Gyrosigma spencerii*
17.	*Microspora quadrata*	41.	*Melosira varians*
18.	*Mougeotia calcarea*	42.	*Navicula* spp.
19.	*Oedogonium gracilius*	43.	*Nizchita thermalis*
20.	*Pediastrum tetras*	44.	*Pinnularia undulate*
21.	*Spirogyra pratensis*	45.	*Raphidonema* sp.
22.	*Spirogyra singularies*	46.	*Stauroneis acuta*
23.	*Tetradesmus smithii*	47.	*Surirella ovalis*
24.	*Ulothrix zonata*	48.	*Syndera pulchella*

Discussion

Physico-chemical parameters *i.e.* hydrological study of down stream of river Subansiri indicates its pollution free environment. All the parameters are suitable for growth and development of planktons and fishes.

Table 6.2(b): Zooplankton Composition in the Down Stream of River Subansiri

Sl.No.	*Name of the Species*	*Sl.No.*	*Name of the Species*
	PROTOZOA		**PROTOZOA**
1	*Arecella vulgaris*	13.	*Monystyla lunaris*
2	*Arecella discoides*		**CLADOCERA**
3	*Amoeba adiosa*	14.	*Alona species*
4	*Centropyxis ecornis*	15.	*Daphnia carinata*
5	*Euglena species*	16.	*Diaphanosoma exisum*
6	*Paramoceium caudatum*	17.	*Diaphanosoma sarsi*
7	*Vorticella campanula*	18.	*Moina brachiata*
	ROTIFERA		**COPEPODA**
8.	*Bruchionus calcyflorus*	19.	*Cyclops vicinus*
9.	*Bruchionus anagularies*	20.	*Heliodiaptomus exisum*
10.	*Filinita longiseta*	21.	*Mesocyclopes bankart*
11.	*Keratella tropica*	22.	*Nauplius larva*
12.	*Lecane luna*	23.	*Phyllodiaptomus* species

Table 6.3: Freshwater Fishes of the Subansiri Floodplain

Sl.No.	*Order*	*Family*	*Scientific Name*	*Habitat and Status*
1.	Osteoglossiformes	Notopteridae	*Chitala chitala*	R, W; ++
2.			*Notopterus notopterus*	R, W; ++
3.	Anguilliformes	Anguillidae	*Anguilla bengalensis*	R, W; R
4.	Clupeiformes		*Gudusia chapra*	R; +++
5.			*Hilsa hilsa*	M
6.			*Gonilosa manmina*	R, W; +++
7.		Engraullidae	*Setipinna phasa*	R, W; +
8.	Cypriniformes	Cyprinidae	*Aspidoparie jaya*	R, W; +++
9.			*Aspidoparie morar*	R, W; +++
10.			*Amblypharyngodon mola*	R, W; +++
11.			*Barilius barna*	R, W; ++
12.			*Barilius bendelisis*	R; +
13.			*Barilius tileo*	R; +
14.			*Barilius vagra*	R; +
15.			*Barilius barila*	R; +++
16.			*Bengana elanga*	R, W; En
17.			*Brachydanio acuticephala*	R, W; +
18.			*Chela cachius*	R, W; +++
19.			*Chela laubuca*	R, W; +++

Contd...

Table 6.3–Contd...

Sl.No.	Order	Family	Scientific Name	Habitat and Status
20.			*Oanio dingala*	R, W; +++
21.			*Danio regina*	R, W; +++
22.			*Esomus danricus*	R, W; +++
23.			*Raiamas bola*	R; +
24.			*Rasbora daniconius*	R, W; ++
25.			*Rosbora rosbora*	R; +
26.			*Salmophasia phulo*	R, W; +++
27.			*Salmophasia bacalia*	R, W; +++
28.			*Bangana dero*	R, W; ++
29.			*Catla catla*	R, W; +
30.			*Chagunius chagunio*	R; En
31.			*Cirihinus mrigala*	Cui
32.			*Cirihinus reba*	Cui
33.			*Ctenopharyngodon idellus*	Cui
34.			*Cyprinus carpio*	Cui
35.			*Labeo bata*	R, W; +++
36.			*Labeo calbasu*	R, W; +++
37.			*Labio gonius*	R, W; +++
38.			*Labio nandina*	R; +++
39.			*Labeo rohita*	R, W; ++
40.			*Labeo pangusia*	R; +
41.			*Neolissochilus hexasticus*	R; +
42.			*Neolissochilus hexagonoleps*	R; +
43.			*Oreichthys cosuatis*	R, W; ++
44.			*Osteobrama cotio*	R, W; ++
45.			*Puntius conchonius*	R, W; +++
46.			*Puntius guganio*	R, W; +++
47.			*Puntius phutunio*	R, W; ++
48.			*Puntius sarana sarana*	R, W; +
49.			*Puntius saphorea*	R, W; +++
50.			*Protopuntius clavatus*	R, W; ++
51.			*Semiplotus semiplotus*	R; +
52.			*Tor putitora*	R; En
53.			*Tor tor*	R; En
54.			*Tor progensus*	R; ++
55.			*Schizothorax richardsoneei*	R; +

Contd...

Table 6.3–Contd...

Sl.No.	Order	Family	Scientific Name	Habitat and Status
56.			*Schizothoraichthys progastus*	R; +
57.			*Crossocheilus laticus*	R; ++
58.			*Garra annandalei*	R; +
59.			*Garra kempi*	R; +
60.			*Garra lissarhynchus*	R; +
61.			*Garra rupecula*	R; +
62.			*Gara nusta*	R; +
63.		*Psilorhynchusynchidae*	*Psilorhynchus balitora*	R; +
64.			*Psilorhynchus homaloptera*	R; +
65.		*Balitoridae*	*Aborichthys kempi*	R; H +
66.			*Acanthocobitis botia*	R; H +
67.			*Acanthococobitis paronaceous*	R; H +
68.			*Mesonoemacheilus reticulofasciatus*	R; H +
69.			*Nemacheilus corica*	R; H +
70.			*Nenoemacheilus assamensis*	R; H+
71.			*Schistura devdevi*	R; H +
72.			*Schistura rupesola inglisi*	R; H +
73.		*Cobitidae*	*Botia dario*	R; +
74.			*Botia rostrata*	R; En
75.			*Botia histironica*	R; En
76.			*Acantophthalamus pangia*	R, W; +++
77.			*Lepidocephalus guntea*	R, W; +++
78.	*Scturiformes*	*Bagridae*	*Aorichthys aor*	R; +
79.			*Aorichthys senghola*	R; +
80.			*Mystus bleekeri*	R; +
81.			*Mystus montanus*	R; +
82.			*Mystus tengara*	R, W; +++
83.			*Mystus vittatus*	R; +
84.		*Siluridae*	*Ompok bimaculatus*	R, W; En
85.			*Ompok pabda*	R, W; +
86.			*Ompok pabo*	R, W; +
87.			*Wallago attu*	R, W; +++
88.		*Schilbeidae*	*Ailia coila*	R; ++
89.			*Clupisoma gaura*	R; +
90.			*Eutropiichthys vacha*	R; +
91.		*Sisordiae*	*Bagarius bagarius*	R; +

Contd...

Table 6.3–Contd...

Sl.No.	Order	Family	Scientific Name	Habitat and Status
92.			*Bagarius yarrellii*	R; +
93.			*Conta conta*	R; +
94.			*Erethistes pussilus*	R; H +
95.			*Gagata gagata*	R; H +
96.			*Gagata cenia*	R; H +
97.			*Hara hara*	R; H +
98.			*Gangra viridiscens*	R; H +
99.			*Pseudochenis sulcatus*	R; H +
100.			*Sisor rhabdophorus*	R; En
101.		*Claridae*	*Clarias batrachus*	W; +
102.		*Heteropneustidae*	*Heteropneustes fossilis*	W; +++
103.		*Chacidae*	*Chaca chaca*	R; +
104.		*Olyridae*	*Olyra langicuda*	R, W; +
105.	*Beloniformes*	*Belonidae*	*Xenentodon cancila*	R, W; +++
106.	*Synbranchiformes*	*Symbranchidae*	*Monopterus cuchia*	W; ++
107.			*Monopterus albus*	W; ++
108.			*Macrognathus oral*	R, W / +++
109.			*Macroganthus pancalus*	R, W; +++
110.			*Mastacembalus armatus*	R; +
111.		*Chaudhuridae*	*Chaudhuria khajuriae*	R; +
112.	*Perciformes*	*Chandidae*	*Chanda nama*	R, W; +++
113.			*Parambassis ranga*	R, W; +++
114.		*Nandidae*	*Nandus nandus*	RW; +++
115.			*Badis assmensis*	W; ++
116.			*Badis badis*	W; ++
117.		*Gobidae*	*Glossogbius giuris*	R, W; ++
118.		*Anabantidae*	*Anabas testudineus*	W; ++
119.		*Belontidae*	*Colisa fasciatus*	W; +
120.			*Colisa sofa*	W; ++
121.		*Channidae*	*Channa barca*	W; En
122.			*Channa marulius*	W; ++
123.			*Channa orientais*	W; ++
124.			*Channa punctatus*	W; +++
125.			*Channa belheri*	R; En
126.			*Channa stwartii*	W; +
127.			*Channa striatus*	W; +
128.	*Tetradontiformes*	*Tetradontidae*	*Tetradon cutcutia*	R, W; +++

R: River; W: Wetland; +: Rare; ++: Normal; +++: Abundant; En: Endangered; M: Migratory;
Cl: Fishes released from culture fisheries.

Table 6.4: One-Way ANOVA on Test Values of Various Parameters Among Different Sampling Stations

Parameters	Station	Mean	Variance	N	F value	P value	Comment
Air Temp.	1	26.74167	19.29174	12			
	2	26.74167	21.96629	12	F = 0.01186,	p=0.99821	NS
	3	26.66667	18.9497	12			
Water Temp.	1	22.23	21.65455	12			
	2	22.87	24.93295	12	F = 0.21964	P = 0.88225	NS
	3	21.25	27.45	12			
Transparency	1	39.77	430.065	12			
	2	43.72	423.523	12	F = 0.23441	P = 0.87191	NS
	3	46.90	504.225	12			
pH	1	7.25	0.04455	12			
	2	7.36	0.12242	12	F = 0.7153	P = 0.54814	NS
	3	7.40	0.10447	12			
TSS	1	30.75	84.022	12			
	2	30.41	77.901	12	F = 0.51173	P = 0.67629	NS
	3	27.83	32.151	12			
TDS	1	162.333	335.151	12			
	2	168.416	350.628	12	F = 0.2359	P = 0.87086	NS
	3	167.250	321.840	12			
Total Solids	1	193.08	728.083	12			
	2	200.66	644.060	12			NS
	3	195.08	440.810	12	F = 0.36031	P = 0.78194	
Conductivity	1	1 09.40	171.418	12			
	2	115.81	292.266	12	F = 0.63203	P = 0.59828	NS
	3	116.96	430.753	12			
Alkalinity	1	103.22	168.336	12			
	2	97.766	228.686	12	F = 1.28999	P = 0.28969	NS
	3	91.358	246.537	12			
Total Hardness	1	69.358	83.802	12			
	2	61.275	85.022	12	F = 2.2203	p = 0.09913	NS
	3	64.925	90.838	12			
Calcium Hardness	1	37.35	12.704	12			
	2	35.65	15.348	12	F = 4.988	P = 0.00459	NS
	3	31.26	21.206	12			
Chloride	1	7.341	3.833	12			
	2	6.358	2.435	12	F = 2.23884	p = 0.09703	NS
	3	5.633	1.872	12			

Contd...

Table 6.4–Contd...

Parameters	*Station*	*Mean*	*Variance*	*N*	*F value*	*P value*	*Comment*
Silicate	1	7.725	2.625	12			
	2	7.308	2.382	12	F = 1.0187	P = 0.39352	NS
	3	6.6583	1.660	12			
Dissolved	1	8.65	0.73364	12			
Oxygen	2	10.64	1.43114	12	F = 5.44892	P = 0.00283	NS
	3	9.941	3.56174	12			
Free	1	5.333	3.686	12			
Carbon-dioxide	2	5.458	2.488	12	F = 1.15841	P = 0.33636	NS
	3	4.55	1.788	12			

At the 0.05 level, the means are NOT significantly different.

Observation reveals that both the phytoplankton and zooplankton numbers are fairly good in the study area, which plays vital role in the aquatic food chain of the river ecosystem. Phytoplankton production was directly related to the dissolved oxygen and inversely related to the free carbon dioxide. Due to low photosynthesis carbon dioxide was easily detectable throughout the year and the growth of phytoplankton did not affect the concentration of free carbon dioxide. This is in agreement

Plate 6.1: Some Ornamental and Endangered Fishes of Downstream Floodplain of the Subansiri River

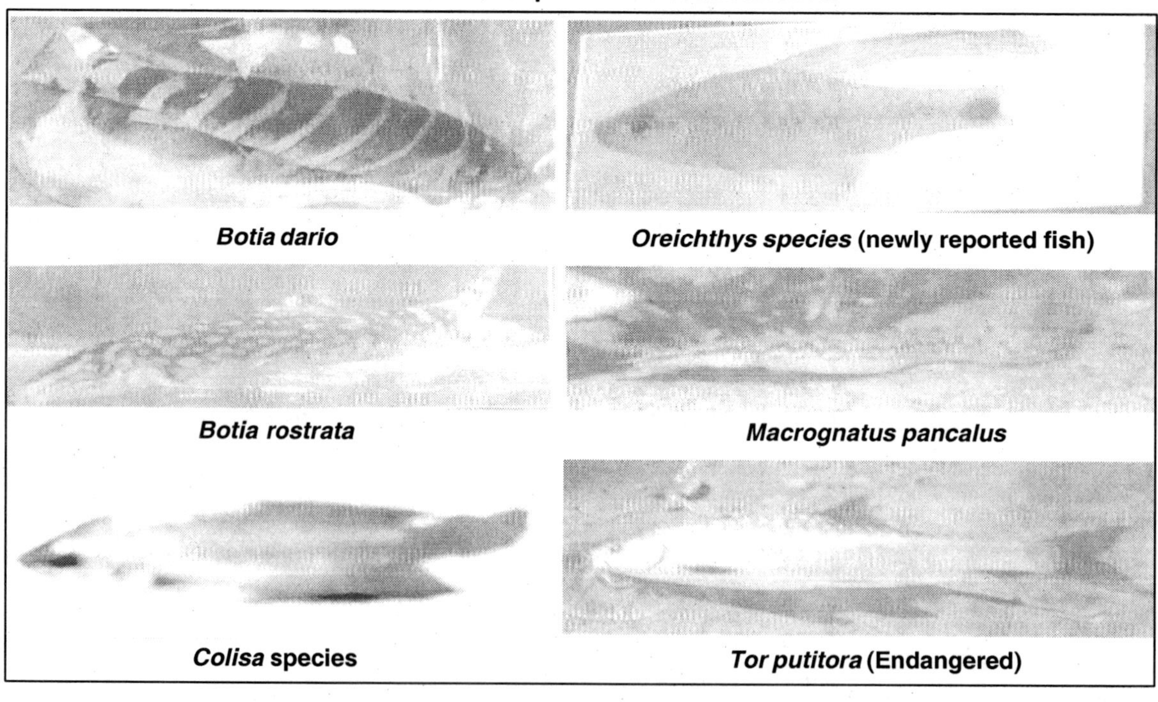

Botia dario | ***Oreichthys species*** **(newly reported fish)**

Botia rostrata | ***Macrognatus pancalus***

Colisa **species** | ***Tor putitora*** **(Endangered)**

Contd...

Plate 6.1–Contd...

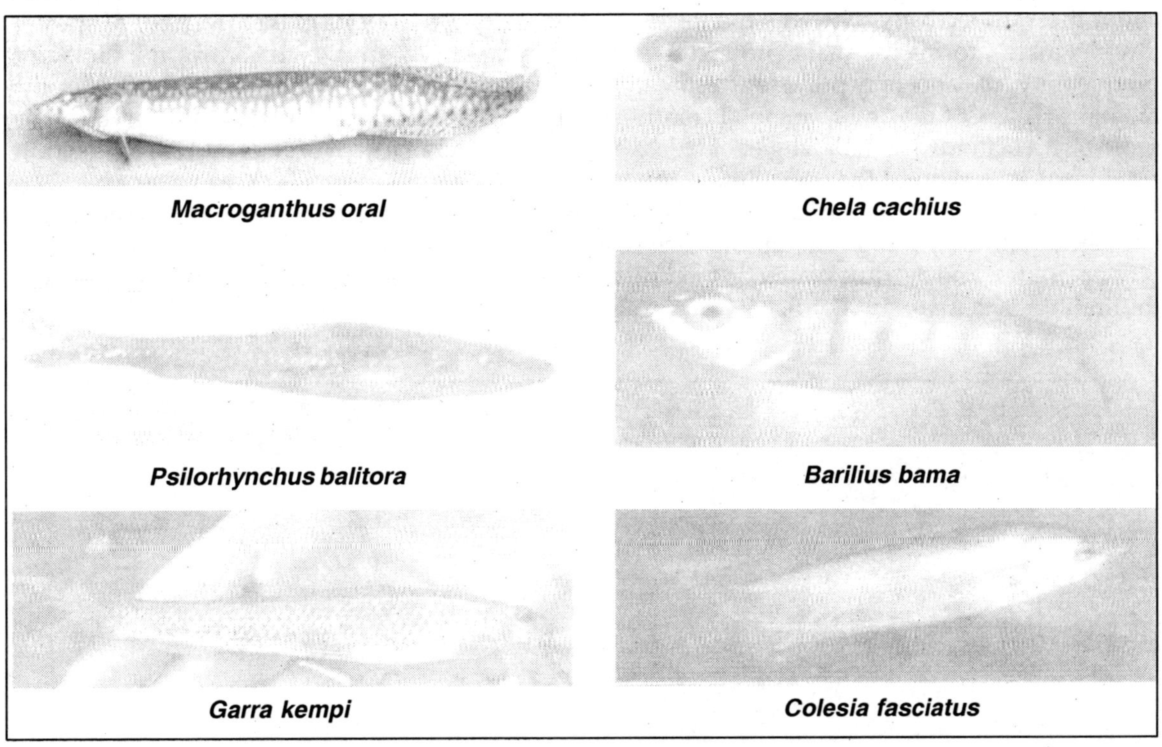

Macroganthus oral ***Chela cachius***

Psilorhynchus balitora ***Barilius bama***

Garra kempi ***Colesia fasciatus***

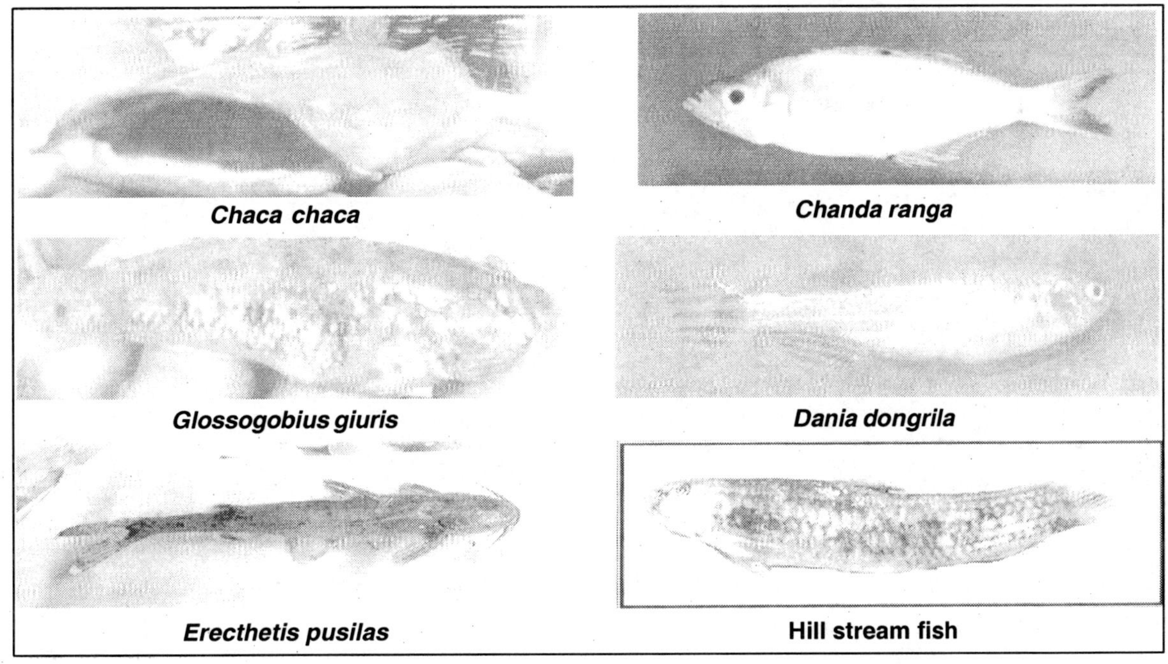

Chaca chaca ***Chanda ranga***

Glossogobius giuris ***Dania dongrila***

Erecthetis pusilas **Hill stream fish**

with the earlier findings of Biswas *et al.*, 1996. Among the phytoplankton and zooplankton diatoms and rotifers respectively showed superiority over other groups in terms of their number of species. Phytoplankton forms the main producers of an aquatic ecosystem, which control the biological productivity. They not only provide an estimation of standing crop but represent more comprehensive biological index of the environmental conditions. Phytoplankton, which include blue green algae, green algae, diatoms, desmids, euglenoides etc. are important among aquatic flora and are ecologically significant as they form the basic link in the food chain of aquatic animals (Kortnicker and Shon, 1971), serves as ecological indicators (Puri *et al.*, 1964) and act as secondary host for a number of fish parasite (Hoffman, 1967). The abundance of rotifera may be attributed to its dependence on phytoplankton and detritus matter as food (Pandey *et al.*, 2004). This findings is in accordance with the findings of Pennak (1944), Michel (1968), Saha *et al.* (1971), Sarwar and Parveen (1995). The rotifer

Blasting at Dam Site | **Habitat Degradation at Dam Site**

Felling of Trees (Illegal) | **Damping of Muck to the River (Illegal)**

Plate 6.2: Habitat Degradation at Dam Site

communities were reported to have some association with water quality, and any variation in suspended solids, dissolved solids, organic matter etc. in the water should immediately affect their distribution (Holland *et al.*, 1983).

Still the river is undisturbed habitat of endangered fish species *viz. Tor putitora, Tor tor,* threaten and rare fishes *viz. Acrossocheilus hexagonelips, Labeo pungusia, Labeo dyocheilus, Chitlala chitala, Cirrhinuh mrigala, Ctenophyryngodon idellus* and *Cyprinus carpio* are not indigenous species but somehow they are being released from the culture fisheries (particularly during the flood) in to the river. Some air breathing fishes like *Channa barca, Heteropneustes fossilis, Clarias batrachus, Anabas testudineus* etc. are not directly found in the river but they are abundant in the nearby lentic water bodies (Swamps, wetland, seasonal ponds etc.) of the Subansiri Drainage system. The floodplain ecosystems have jeopardized the flora and fauna of these habitat (Biswas *el al.*, 1998). The river floodplain includes river channels, oxbows, relict channels, floodplain pools and seasonally inundated areas of the terrestrial environment. The seasonal inundation of the floodplain that is associated with monsoon rains determines the magnitudes and extent of land water interactions and allochthonous inputs in the river (Welcomme, 1979; Junk *et al.*, 1989; Sedell *et al.*, 1989). The wetlands or flood plain lakes of the river have the potentialities of well developed fishery and in fact, most fish production in the large river is derived from floodplain habitat (Welcome, 1979; Junk *et al.*, 1989). The migratory fishes *Hilsa hilsa* is migrated through the river Brahmaputra and reached the Subansiri River.

Plate 6.3: Damping of Muck in the River (Illegal)

Conclusion

The physico-chemical properties of water influence the species composition and productivity of any aquatic ecosystem. In this study status of water quality is a good indicator of pollution free environment. Likewise the river ecosystem is very rich in planktonic and fish diversity.

The NHPC violating all the rules by continuing blasting operation for road construction are being accompanied by extensive muck and debris disposed in the river which already badly affected the breeding ground of Mahasheer, the endangered fish (*Tor tor* and *Tor putitora*). Dumped the excavated material in to the river, felling of trees for construction of residential colony and for the purpose of fuel wood for the labours of construction site etc. will changes down stream ecology of the river (Plate 6.2 and 6.3). If the project work is continuing in this manner serious ecological problems in the downstream will come very shortly.

Acknowledgement

Financial assistance provided by UGC (NERO) and ASTEC, Assam in the form of Minor Research Project is gratefully acknowledged. Authors are thankful to Mr. Bhaskar Bora, Green Heritage of Lakhimpur for photographs.

References

Bergcamp, G., McCartney, M.P., Dugan, P., McNelly, J. and Acreman, M.C., 2000. *Dam Ecosystem Function and Environmental Restoration WCD Thematic Review: Env. Issue*.

Biswas, S.P., 1993. *Manual of Methods in Fish Biology*. South Asian Publishers Pvt. Ltd., New Delhi.

Biswas, S.P., Baruah, A., Baruah, D., Mohan, R.S.L. and Kotoky, P., 1996. Phytoplanktonic composition and their productivity in the upper stretches of river Brahmaputra. *Bull. Life Sciences*, 6: 53–58.

Biswas, S.P., Baruah, A., Baruah, D. and Baruah, S., 1998. Ecology of chars or river island of the Brahmaputra with special reference to fisheries. *FTR*, DoEN Project No. 14/38/93–MAB/RE.

Edmonson, W.T., 1974. *A Manual on Methods for Measuring Production in Aquatic Environment*. Academic Press, New York.

Hoffman, G.L., 1967. *Parasites of North American Fishes*, University of California Press, 486 pp.

Holland, C.E., Brayan, C.F. and Neman, Jr., J.P., 1983. Water quality and the rotifer populations in the Achafalalaya river basin, Laousiana. *Hydrobiologia*, 70: 103–117.

Jhingran, V.G., Natarajan, A.V., Banerjee, S.M. and David, A., 1969: *Methodology on Reservoir Fisheries Investigation in India*. Bull. No. 12 CIFRI Barrackpore, India, 475 pp.

Junk, W.J., Dayley, P.B. and Sparks, R.E., 1989. The flood pulse concept in a river flood plain system. In: *Proceeding of the International Large Dam Symposium*, (Ed.) D.P. Dodge. Canadian Special Publication of Fisheries and Aquatic Science; 106; Ottawa, pp. 110–127.

Kortnicker, L.S. and Shon, I.G., 1971. Viability of Ostracod Shell, Ecology and palecologic implementations. *Bull. Centre. Rech. Pan.*, 5: 125–135.

Lackey, J.B., 1983. The manipulation and counting of river plankton and changes in some Organism due to formalin preservation. Publ. Health Deptt., 53: 2080.

McCartney, M.P., Sullivan, C. and Acreman, M.C., 2000. *Ecosystem Impact of Large Dam WCD Thematic Review*, II–I: 1–80.

Michael, R.G., 1968. Studies on the zooplankton of a tropical fish pond. *Hydrobiology*, 32(1–2): 47–60.

Munshi, J.S.D. and Srivastava, M.P., 1988. *Natural Histories of Fishes and Systematic of Freshwater Fishes of India*. Narendra Publ. House, New Delhi.

Needham, J.G. and Needham, P.R., 1966. *A Guide to Study of Freshwater Biology*. Holden Day Inc., San Francisco, pp. 180.

Pandey, B.N., Hussain, H., Jha, A.K. and Shyamanand, 2004. Seasonal fluctuation in zooplankton community in relation to certain physico-chemical parameters of river Ramjan of Kissangang, Bihar. *J. Nat. Env. Pollut. Tech*., 3(3): 325–330.

Pennak, R.W., 1944. Diurnal movements of zooplankton organism in some Colorado Mountain Lakes. *Ecology*, 25: 387–403.

Puri, H.S., Bonaduce, G. and Malloy, J., 1964. Ecology of the Gulf of Neples. Publ. Staz. Zool. Napoli, 33, Suppi: 87–99.

Sarwar, S.G. and Parveen, A., 1995. Community structure and population density of zooplankton in two interconnected lakes of Srinagar. Kashmir. *Pollution Research*, 15(1): 35–38.

Saha, G.N., Sehgalki, Mitra, E. and Nandy, A.C., 1971. Studies on the seasonal and diurnal variation in physico-chemical and biological conditions of a perennial freshwater pond. *J. Inland. Soc. India*, 3: 79–102.

Sedell, J.R., Richey, J.E. and Swanson, F.G., 1989: The river continuum concept: A basis for the expected ecosystem behaviour of very large rivers. Canadian Spl. Publ. in *Fisheries and Aquatic Sciences*, 106: 49–55.

Talwar, P.K. and Jhingran, A.G., 1991. *Inland Fisheries of India and Adjustant Countries*. Oxford and IBH Publ. Com., New Delhi, Vols. 1 and 2.

Trivedy, R.K., Goel, P.K. and Trisal, C.L., 1987. *Practical Methods in Ecology and Environmental Sciences*. Enviro Media Publication, Karad (India), pp. 123–126.

Vennote, R.L., Minshall, G.M., Cummins, K.W., Sedell, J.R. and Cushing, C.E., 1980. The river Continuums concept. *Can. J. Fisheries and Aq. Sc.*, 37: 130–137.

Viswanath, W., 2002. *Fishes of North East India: A Field Guide to Species Identification*. NATP, NBFGR, Lucknow.

Welch, P.S., 1952. *Limnology*. McGraw-Hill Book Company Inc., New York.

Welcomme, R.L., 1979. *Fisheries Ecology of Floodplain Rivers*. Longman, London, 317 pp.

Chapter 7

Impact of Paper Mill Effluent on Agroecosystem: A Review

P. Dhevagi, C. Udayasoorian and G. Oblisami
Department of Environmental Science,
Tamil Nadu Agricultural University, Coimbatore – 3

Introduction

Water requirement to meet agriculture, domestic, industrial and other demands indicate the need for regeneration of wastewater. Industrial wastewater is being used for irrigation in dry areas. Pulp and paper industry ranks third in terms of fresh water usage. Pulp and paper industry ranks third in terms of fresh water usage. Water usage worked out to 400 m^3 per tonne of paper produced and the wastewater quantity let out was about 330 m^3 per tonne of paper produced per day (Udayasoorian *et al.,* 1999a). Nearly 75 to 95 per cent of fresh water used in the paper and pulp mill was discharged as effluent containing organic and inorganic pollutants and colouring materials. These compounds may enhance the growth of crop plants or retard the growth.

The pulp and paper industry wastewater contain compounds originating from the raw materials of the pulp and the chemicals added during the manufacturing processes. In the kraft process, wood chips are cooked under pressure with a mixture of hot caustic soda and sodium sulphide to promote cleavage of various ether bonds in the lignin. The inorganic chemicals are recovered from the spent black liquor and the pulp washing liquor in the recovery of boiler, where the organic residue is combusted to generate steam for the process. In this approximately 55 per cent of the total weight of the wood is dissolved in the pulping liquor which contains degradation products of its constituents *i.e.* lignin, polysaccharides and wood extractive (Rydholm, 1965). A part of sodium sulphide reacts with lignin and carbohydrates in wood to form malodorous sulphur compounds such as hydrogen sulphide, methyl mercaptan and dimethyl sulphide during the cooking of wood chips in kraft pulping process.

The other gaseous emissions include oxides of sulphur and nitrogen (Subramanyam and Sundaresan, 1982). The first stage in the manufacture of pulp is wood debarking, usually wet drum debarking. In this process, water is used to thaw the frozen logs, remove the bark and wash the logs after debarking. The contribution of wet debarking water to a final chemical process, total effluent is about 5 per cent as BOD (or) COD and 16 per cent as suspended solid (Virkola and Honkanen, 1985).

The main wastewater sources in the pulp and paper manufacturing are shown in Figure 7.1.

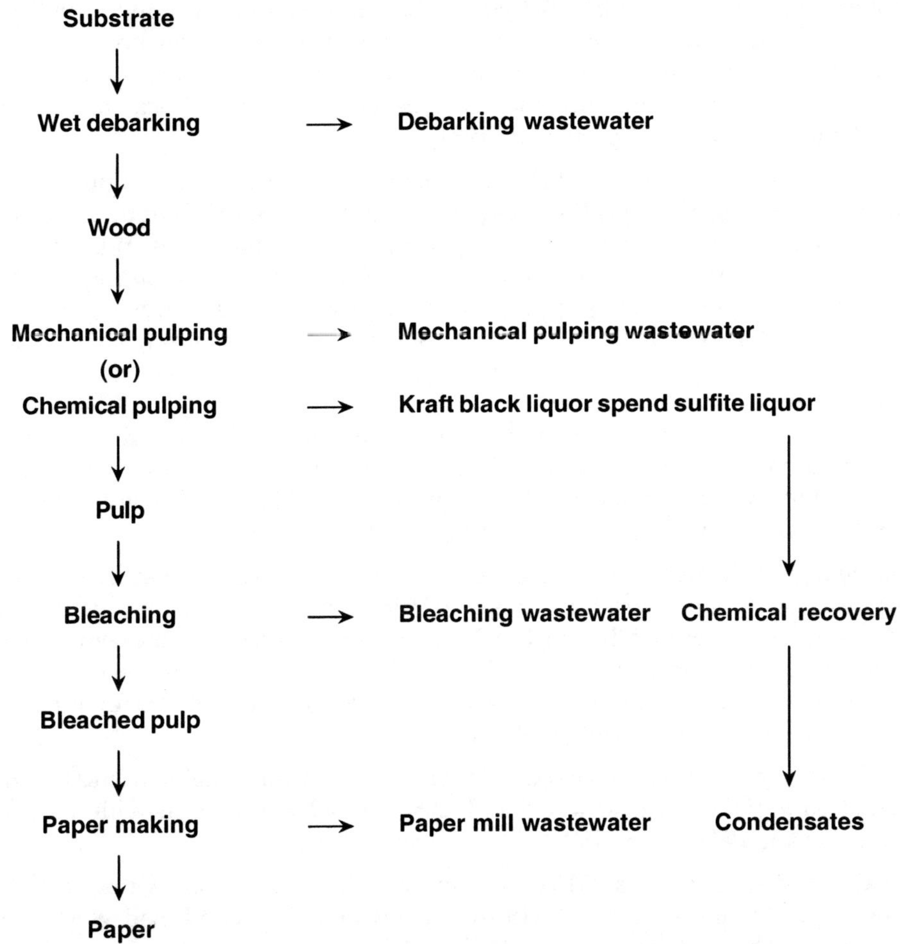

Figure 7.1: The Main Wastewater Sources in the Pulp and Paper Industry (Sierra-Alvarez, 1990)

Effluent Characteristics

The estimated total annual world capacity of pulp production in 1992 was around 261 million tonnes and that of paper and paper board production to 196 million tonnes (FAO, 1988). The quantity of fresh water used in the small paper mill for a production of 20 tonnes of paper is about 8000 m^3 per day.

Pollutants that generally arise from the industry include wood sugars, cellulose fibre, lignin and other spent chemicals which impart Biochemical Oxygen Demand (BOD), Chemical Oxygen Demand (COD), colour etc. Effluent streams in the industry are black liquor, white liquor, machine wastewater and sanitary wastes (Sharma and Bandhyopathyay, 1991). They also observed that wood sugars represent 70 per cent of the BOD and lignin to the extent of 30 per cent of the original weight of the raw material. Pollutants from the paper industry include suspended solids, foam, colour due to lignin compounds, dissolved organic salts, sodium, toxic compounds like chlorinated lignin, resin acids, fatty acids, and phenolic derivatives (Sahaya Alfred, 1997 and Sundaramoorthy and Kunjithapatham 2000). Discharge of untreated wastewater will therefore create serious water pollution problems resulting in the deterioration of water quality and toxicity to aquatic life (Larson *et al.*, 1988). Most of these wastes are therefore applied to the land, since soil is believed to have a capacity for receiving and decomposing wastes and pollutants (Young *et al.*, 1981).

The characteristic dark brown colour of the effluents is due to formation of lignin degradation products during the processing of lignoceliulosics (Sundman *et al.*, 1981). Colour due to lignin is one of the difficult problems to solve in the pulp and paper industry (Ramanathan and Parabrahmam, 1982). The brown colour in debarking water is due to oxidation products of tanning agents (Niirenen and Vaeaetaenen, 1983). The colour due to lignin compounds persist in water body for long distances (Subrahamanyam *et al.*, 1984). Nearly 80 to 90 per cent of the colour originates from bleaching. The presence of colouring matter in pulp mill wastewater had a very serious influence on the aesthetic sense of the consumer and also enhances the cost of wastewater treatment (Upadhyaya and Singh, 1991). Colour of the wastewater is due to the presence of lignin and lignin derivatives which are not easily biodegradable and hence account for high BOD to COD ratio. Colour load in the Wastewater vary depending upon the raw material used and is mainly attributable to lignin degradation products formed during pulping and bleaching operations (Mittar *et al.*, 1992).

After discharge from pulp and paper mills, the concentration of oxygen dissolved in the water receiving the wastes, falls to a low level as a result of break down of various organic materials. Rajannan and Oblisami (1979) reported that no dissolved oxygen was present in the combined effluent from paper and pulp industry. Presence of reduced levels (0.9 to 1.3 mg l^{-1}) of dissolved oxygen in the combined effluent was reported by Somashekar *et al.* (1984). Verma *et al.* (1988) also reported the paper mill effluent having dissolved oxygen content of 6.08±0.40.

The combined wastewater from a paper and pulp mi II had pH value ranging from 7.0 to 9.4. Dave (1982) reported that pulp mill brown stock had a pH 8.6. The pH was alkaline with high electrical conductivity (Oblisami and Palanisami, 1991).

The BOD, COD and suspended solid content in paper machine wastewater were 153, 393 and 327 kg respectively for the volume of 216 m^3 wastewater. The combined wastewater had BOD : nitrogen : phosphorus in the proportion of 100 : 2 : 0.7 which indicated that the wastewater is deficient in nitrogen and phosphorus for aerobic treatment (Sandand *et al.*, 1995). COD of brownstock was 1440 mg l^{-1}. The COD of approximately 1000 mg l^{-1} of various debarking effluents, consists of tannins (30–55 per cent), monomeric phenols (10–20 per cent), simple carbohydrates (30–40 per cent) and resin compounds (5 per cent) as reviewed by Field *et al.* (1988).

The effluents containing rich organic carbon content was reported by Billings and DeHass (1984) and Dhevagi *et al.* (2000). The primary toxic compounds in pulping effluents were found to be chlorinated lignin compounds and resin acids (Dhevagi *et al.*, 2000). According to Udayasoorian *et al.* (1999b) considerable amounts of toxic substances were discharged from paper and pulp industry and among those pentachlorophenol was very common. The predominant toxicant was the resin acids and isopimaric substances together accounted for 60–90 per cent of overall toxicity. Total phenol was in extremely higher (917100 mg l^{-1}) concentration. Pushpavalli (1990) observed the presence of traces of toxic elements in the effluent.

The paper and pulp industry had an increase in salinity with low sodium hazard and residual sodium carbonate was absent (Narashimha Rao and Narashimha Rao, 1992). The higher concentrations of calcium, magnesium, sodium, chloride, sulphate and bicarbonate were reported by number of authors (Reddy *et al.*, 1981 and Gomathi and Oblisami, 1992).

Fifty per cent of the particulate phosphorus in both kraft mill and paper mill effluents were potentially available for algae (Priha, 1994). The effluent having carbonate, bicarbonate, sulphate and chloride ions were 0.4, 4.8–5.2, 3.8–4.4 and 6.8–9.2 meq l^{-1} and these values were within the safe limit for successful crop growth. The effluent from paper industry has intense brownish colour, alkaline, rich in inorganic and organic sodium compounds, dissolved inorganic salts, chlorinated lignin, fatty acids, resin acids, BOD and COD (Gilliespie, 1985).

Manikya Reddy and Venkateswarlu (1985) analysed the concentration of heavy metals in the paper effluent and found that it is in the order of Zn > Cu > Pb > Ni > Co > Mn. Wastewater from bleaching process contained chlorinated non-bleach pulp products like hemicellulose, monosaccharide, resins and substance extracted from the fibre (Bharathi and Krishnamoorthy, 1988). Combined effluent from all the sampling points were low in nutrient content with considerable amounts of chloride, sulphate and polyphenols.

Many wastewaters from the forest industry contain toxic substances. An important fraction of toxicity is derived from organic compounds like wood resin (Walden *et al.*, 1986), tannins and chlorophenol (Temmink *et al.*, 1989) and these compounds are toxic to aquatic organisms.

Higher population of bacteria, actinomycetes and fungi were recorded in the effluents of paper industry. Nitrogen fixing *Azotobacter* and *Klebsiella* were abundant and nitrogen fixing *Bacillus* was also present. Of 129 isolates of *Klebsiella* from pulp mill wastewater, 32 per cent possessed, nitrogen fixing ability and it was increased by the addition of 0.5 to 3 per cent glucose. In wastewater most of the yeasts found were *Hansenula oxpichia.* Extensive documentation supported the presence of bacteria belonging to the family Enterobacteriaceae in effluent from the paper manufacturing industry (Caplenas *et al.*, 1981).

Vaeaetanean and Niemelae (1983) studied the factors regulating the density of bacteria in process water of a paper mill and reported the variations in the number of to.tal colonies of *Klebsiella pneumoniae, Acenetobacter* spp. and *Pseudomonas* in process water. Datta Roy and Malty (1984) recorded significant numbers of *Salmonella* present in the paper mill effluent which is discharged into the Hoogly river. Dhevagi *et al.* (2000) reported that the nutrient present in the effluent enhanced the microbial population. Neilson and Allard (1985) recorded reduction of Enterobacteriaceae population in the paper mill process water. Chuang (1985) reported the presence of nitrogen fixing bacteria *Azotobacter* and studied the effects of nutrients on waste treatment.

Krueger and Sheikh (1987) reported a new selective medium for isolating *Pseudomonas* spp. from wastewater. Despite the unfavourable nature of wastewater from paper mill on the growth and multiplication of algae, 26 species belonging to four genera were recorded. Blue green algae were abundant followed by diatoms. Rivers receiving paper and pulp industry effluents had the algae belonging to members of Bacillariophyceae and members of *Oscillatoria, Arthrospira* and *Spirulina* (Dhevagi and Oblisami 2000). Dhevagi *et al.* (2003) reported that the microbial load of the effluent was higher for bacteria and considerable numbers of actinomycetes, fungi, *Azotobacter* and yeasts were also present.

Effect of Effluent on Soil Properties

Land application of wastewater is a preferred alternative for its disposal since soil is believed to have a capacity for receiving and decomposing wastes and pollutants, where organic materials are stabilized through the activity of microbial flora in the soil. The removal of different constituents is accomplished by physical, chemical and microbial interaction with the soil matrix and a cover crop (or) plant uptake. In the paper industry, the treatment of wastewater generated during paper production produces several kinds of residues which constitute a disposal problem. The application of wastewater from the pulp and paper industry leads to the deterioration of soil physical, chemical and biological properties.

Effect of Effluent on Soil Physico-chemical Properties

Paper mill wastewater had appreciable concentration of carbonate and bicarbonate alkalinity and exhibited a tendency to precipitate calcium in the soil as $CaCO_3$ and thus increased the resultant proportion of sodium to calcium and magnesium and sodium adsorption ratio (SAR) of the soil solution vis-à-vis ESP. The treated effluent altered the pH of the soils. Significant relationship between SAR of the water and resultant ESP of soil at equilibrium were also reported by Juwarkar and Bhalkar (1984) in Amlai, Akola region soils.

Igbounamba (1972) and Patra and Behera (1974) reported that due to treated effluent irrigation N, P and K values increased. The toxicity of high sodium could be reduced by decreasing the SAR of the wastewater which would be augmented by increasing the proportion of Ca^{+2} and Mg^{+2} to Na^{+} in the wastewater by the addition of Ca and Mg salts. The concentration of dissolved salts in irrigation water having an EC of 2000 to 3000 mhos cm^{-1} has been successfully used for crop irrigation on soils containing 20 to 30 per cent clay with good drainage.

Dhevagi *et al.* (2006) reported that treated effluent altered the pH of the soils. This property of effluent could be utilized to reclaim acidic soils and nutrient status of the soil was increased due to effluent irrigation. They also reported that the per cent pore space of the soils got reduced due to irrigation with the effluents and the water holding capacity of black soil increased due to the effluent; whereas it decreased in red soil. There was a build-up in the content of Na in soils due to continuous irrigation with effluents, which could be overcome by using either gypsum or mixing hypochlorite effluents.

Reddy *et al.* (1981) reported that continuous irrigation with pulp and paper mill effluent resulted in increased soluble salts, organic matter content. The retention of moisture necessary for growing sugarcane in sandy soil under effluent irrigation was attributed to the constant humification and mineralization of organic matter and also no adverse effect on the soil characteristics due to adsorption of bases including sodium in soils treated with paper industry effluent.

Irrigation with undiluted paper and pulp mill effluent led to increased soil pH, electrical conductivity, organic carbon and available nutrients (Juwarkar and Subrahmanyam, 1987 and Kannan and Oblisami, 1990b). They observed that the effluents of paper mill altered the physico-chemical characteristics of the soils. The pH value in case of brown soil increased from 7.7 to 8.6 after treatment with effluents. In the case of red soil the pH value change was from 8.0 to 8.7. The concentration of N, P, K and Ca increased in all the soils after irrigation with effluents. The per cent organic carbon and exchangeable sodium percentage in the treated soil were significantly higher as compared to the control and water hording capacity of the soil also substantially decreased after effluent treatment.

Sandana (1995) observed that no appreciable decrease in the rate of infiltration in soils treated with effluent. They also observed that there was greater accumulation of salts during rabi season compared to kharif. They also conducted studies with anaerobically treated pulp mill wastewater from a 250 TPD paper mill in Maharashtra and reported that the effluent had an EC of 2225 to 2600 µmhos cm^{-1} which on irrigation increased the soil EC. The exchangeable sodium per cent in effluent irrigated soil rose to 40 after 3 years of irrigation. According to Notter and Red (1985) land treatment is feasible for lignin removal which would also reduce wastewater treatment costs. Due to continuous irrigation with paper factory effluent the EC, soluble salts, organic matter content and exchangeable sodium percentage of the soils increased (Dhevagi 1996 *et al.*).

Juwarker *et al.* (1987) observed that mixing of pulp mill wastewater with chlorination and hypochlorite wastewater in 7 : 1, 4 : 1 and 3 : 1 proportions reduced the SAR from 18.8 to 11.1, 6.6 and 5.3 respectively without any appreciable change in total dissolve solids. They also reported the relationship between the SAR of the wastewater and resultant ESP of soil. Increase in the wastewater SAR increased the soil ESP and the increase was highly significant. Stabilization of hydraulic conductivity was noticed by Hansbrix (1987) in wetland soils irrigated with mechanically treated wastewater which got purified by microbial degradation.

According to Palaniswami (1989) the higher amount of available nitrogen, phosphorus, potassium, zinc, copper, organic carbon, calcium carbonate, pH, EC, CEC, exchangeable cations, exchangeable sodium and potassium per cent was found in soils irrigated with effluent for over three years.

Dhevagi *et al.* (2006) reported that pH, EC, organic carbon, CEC, exchangeable cations, available P,K, micro-nutrients and enzymes like amylase, invertase, cellulase, dehydrogenase and phosphatase activities were more in the 15 year irrigated plots than in the other irrigated plots. For 15 years the available micronutrient level did not reach the toxic levels. The pH of the soil increased slightly (7.6 to 7.8). The increase in nitrogen, phosphorus and potassium content of soil was not significant due to irrigation with combined effluents. Palaniswami and Sree Ramulu (1994) observed that even after 15 years of continuous irrigation with effluent of pH 8.28 similar trend was observed in EC, available micronutrients and enzyme activities in the surface soil and in subsurface soil. They also reported that the available nitrogen content was more in second and third year effluent irrigated soil than 15 year irrigated soil. Udayasoorian *et al.*, 1999a and 1999b also observed the same results.

Effect of Effluent on Soil Microbial Properties

Paper mill is one among the industries which releases large quantities of effluents. Land application of wastewater is a common practice and this leads to addition of enormous quantity of pollutants, where organic materials are stabilised through the activity of microflora in the soil. This may affect the soil microflora. Spermosphere microflora is also influenced by effluent irrigation. Spermosphere microflora play an important role in better establishment of the plants in soil.

Higher population of bacteria, fungi and actinomycetes were recorded in the effluent irrigated soil (Dhevagi *et al.*, 2006). The effluent irrigation was not deleterious to nodulation by either native rhizobia (or) inoculated rhizobia. The enrichment in the nutrient content of the soil due to the effluent irrigation also enhanced the microbial population. The increase in organic matter content of the soil due to effluent irrigation accelerated the microbial activity which in turn accelerated the substrate decomposition with the release of carbon dioxide. Effluent irrigation over a period of 15 years had maximum population of bacteria, actinomycetes, fungi, *Rhizobium, Azotobacter* and yeast. The combined effluent irrigation resulted in an increase in the rhizosphere microflora of sugarcane and the activities of cellulase and invertase were greater in submerged soil than in garden land soils to which effluent had been applied.

Udayasoorian *et al.*, 1999a reported that paper factory effluent irrigation resulted in an increase in soil amylase, invertase, cellulase, dehydrogenase and phosphatase activity. Chauhan and Kaur (1991) observed that diluted effluent found to increase the microbial population and fertility of soil. Kannapiran (1995) observed that more microbial activity in the rhizosphere with 50 and 75 per cent effluent concentration. Sandana (1995) reported that the population of spermosphere microbial load was lower at higher concentration of effluent whereas, higher counts were recorded at lower effluents concentration. She also observed that marked differences were exhibited in non rhizosphere and rhtzosphere microflora of tomato by the effluent irrigation along with amendment.

Effect of Effluent on Plants

Water requirement to meet agricultural, domestic, industrial and other demand indicate the need for regeneration of wastewater. Industrial effluents are being used for irrigation in dry areas; these effluents not only contain nutrients that enhance the growth of crop plants but also have toxic materials. So it is essential that the implications of the use of industrial effluents in the crop field and their effect should be assessed before they are recommended for use in irrigation. The results obtained after the germination and growth study showed that raw effluent (as such) affected the germination and vigour index of maize, groundnut, soybean and blackgram. Growth behaviour of the crops showed increasing trend towards increase in dilution of the effluent. The increase in growth with increase in effluent dilution might be due to decrease in inhibitory compounds present in the effluent (Dhevagi *et al.*, 2003).

Irrigation of paper mill effluent had shown encouraging results on the growth of oats and orchard grass. Prasad *et al.* (1977) and Subrahmanyam *et al.* (1984) reported that kraft mill wastewater can be used to grow crops like paddy, barley, wheat maize and kenaf on coarse textured soils (Dhevagi *et al.*, 2006). Reduction in root length, shoot length, dry matter production and grain yield of cumbu and sorghum was observed due to irrigation of different concentration of the effluent excepting 25 per cent effluent. The vigour index of rice was significantly poor in undiluted effluent when compared to water control; it increased with the decrease in the concentration of the effluents.

Udayasoorian *et al.* (1999b) reported that the diluted effluent showed favourable effect on seedling growth of maize, cotton and paddy. There was an inhibition in germination of rice seeds with increasing concentration of card board factory effluents and high presoaking period. Early seedlings growth and dry biomass were found to be better in 25, per cent effluent concentration with continuous application (Dixit *et al.*, 1986). Ketkar (1989) observed that discharge of treated effluent from paper mill for irrigating 300 to 500 acres land for growing rice, wheat, sugarcane, onion, groundnut, hybrid napier, subabul and *Eucalyptus* since 1972, dose not caused any build up of salts for the last 16 years. Oblisami and Palanisami (1991) reported that the combined effluent irrigation from paper factory did not affect the germination of maize and ragi, however there was no increase in the vigour index of maize, ragi, groundnut and cotton. Misra and Behera (1991) showed that the percentage of germination, water imbibing capacity, growth, pigment, carbohydrates and protein content showed a decreasing trend with increase in effluent that an concentration and time. Verma and Poornima (1992) reported appreciable increase in anomalies and pollen sterility with increase in concentration of paper factory effluent was observed on irrigation to barley crop. Irrigation with the combined raw effluent from paper and pulp mill reduced the germination and vigour indices of maize, wheat, ragi, and cotton (Kannan and Oblisami, 1992). Narashimha Rao and Narashimha Rao (1992) recommended the use of effluent water for rice and cotton without affecting the growth. Sandana (1995) observed better seed germination, shoot length, root length, and vigour index of rice seedlings at 50 to 25 per cent effluent concentrations comparable with that of tap water and with increase in effluent concentration (75 and 100 per cent), the above parameters decreased.

The vigour index of blackgram and tomato was significantly poor in undiluted effluent when compared to water control; it increased with the decrease in the concentrations of the effluents (Dhevagi *et al.*, 2000). The paper factory effluent reduced the germination, shoot length, root growth, chlorophyll content, protein content, DNA, RNA, free amino acids and soluble sugar of *Phaseolus aureus* (Rout, 1990). Kannan and Oblisami (1992) observed that the irrigation with the combined raw effluent from paper and pulp mill reduced the germination and vigour of cowpea and horsegram. Sandana (1995) observed better seed germination, shoot length, root length and vigour index of lab at 50 and 25 per cent effluent concentrations. Spulink (1949) reported that the use of sulphite mill wastewater had no adverse effect on germination of sunflower seeds.

Sahaya Alfred, 1999 showed a significant reduction in germination, root and shoot length of sunflower at higher concentration of effluent. He reported that combined raw effluent from paper and pulpmill reducing the germination and vigour indices of groundnut and sunflower. Fazeli *et al.* (1991) reported that accumulation of toxic metals like copper, lead, zinc, nickel, cadmium and cobalt in coconut water, roots and leaves above the safety limit when effluent irrigation was given to coconut plantation.

Irrigation with treated effluent from paper factory stimulated the carbohydrates, protein ascorbic acid and organic acid metabolism of methi (*Trigonella foenumgraecum*) and increased uptake of N, P, K and Na by the plant was also observed (Mane and Shitole, 1989).

Srivastava (1991) reported that chloralkali plant effluent was highly deleterious for germination and early growth performance of seeds of *Raphanus sativa* and *Allium cepa* as compared to paper mill effluent. Narashimha Rao and Narashimha Rao (1992) observed spots of dead tissues and curling of leaves on the tobacco and chilli leaves in fields due to absorption of chlorine from the soil irrigated with the effluent water.

Sandaria and Oblisami (1999b) reported the effect of paper factory effluent on tomato. They observed the better seed germination, shoot length, root length and vigour index of tomato seedlings at 50 and 25 per cent effluent concentrations comparable with that of tap water and with increase in effluent concentration (75 and 100 per cent), the above parameters decreased.

Hansen *et al.* (1980) suggested that kraft mill wastewater could be safely used for populus crops. The use of paper factory effluent for irrigating sugarcane was reported by several researchers *viz.*, Udayasoorian (1999b), Pushpavalli (1990) and Oblisami and Palanisami (1991). An increase in jaggery yield was observed when sugarcane was grown with paper and pulp mill wastewater (Subrahmanyam *et al.*, 1984). The irrigation with effluent is superior than fresh water with respect to plant characteristics and yield of sugarcane.

Calcium along with organic matter present in the effluent in appreciable quantity play an important role in formation of stable aggregates. Organic matter constantly subjected to humification and mineralization makes the soil as store house for moisture necessary for sugarcane especially in sandy soils (Reddy *et al.*, 1981). The sugarcane varieties raised in 15 years effluent irrigated soil had no reduction in quality of juice but cane and sugar yields were reduced in the CV COC 671 (Oblisami and Palanisami, 1991).

Neelay and Dhondizal (1985) reported that effluent from pulp and paper mill was used for irrigating *Eucalyptus canaldulensis, Pongamia pinnata, Acacia auriculiformis, Leucaena leucocephala* and *Dendrocalmus strictus* and increased the height and collar diameter of these tree species. Gomathi and Oblisami (1992) also reported the use of pulp mill waste for irrigation by mixing with water for the crops like neem, pungam and tamarind without any inhibitory effect.

Kannapiran (1995) observed that better seed germination and growth of seedling at higher concentration of effluent (75 and 100 per cent) for *Acacia leucophloea, Albizzia lebbek* and *Leucaena leucocephala;* whereas in *Cassia hybrida* seed germination and growth of the seedlings were affected with increased effluent concentrations.

Wastewater emanating from paper and pulp industry has been tried as a source of irrigation for agricultural crops and the study revealed that the 50 per cent effluent concentration can be utilised for maize, groundnut, blackgram, greengram without reduction in growth. For castor and gingelly, effluent can be utilised with appropriate dilution before using for irrigation (Dhevagi, 1996 and Dhevagi *et al.*, 2000).

Combined Use of Effluents and Amendments

Combined use of industrial effluents along with amendments might provide a soil with enough nutrients and with better physical and microbiological environment, thus improving the soil fertility. The adverse effects of effluent from paper factory would be alleviated by resorting to the application of N, P, and K along with organic and inorganic amendments such as pressmud, FY M and gypsum. Application of 100 per cent fertilizer plus 20 tons of gypsum along with effluent irrigation recorded better growth of the crop, higher dry matter production and nutrient uptake of sugarcane. The gypsum application was found to increase the availability of micronutrients in the soil. The sugarcane crop raised in pressmud added soils had higher values of N, K, crude protein, amino acid, chlorophyll and carotenoid content. The 100 per cent effluent irrigation plus 20 tons gypsum per ha treatment recorded maximum germination, shoot length, dry weight of sugarcane setting and increase in content and uptake of N, P_2O_5, K_2O, Mn, Zn and also the uptake of Fe, and Co (Oblisami and Palanisami, 1991). Better growth of plants *viz.*, brinjal, onion, sunflower, banana and pulpwood such as *Eucalyptus* were observed while paper mill sludge was applied along with effluent irrigation (Veena *et al.*, 1992).

Sandana (1995) showed that different organic amendments along with effluent irrigation to tomato revealed considerable decrease in bulk density of the soil, slight increase in pH, EC and organic carbon content. Number of fruiting clusters, fruit weight, fruit yield was highest in the treatment receiving bioearth, pressmud and FYM along with effluent irrigation. Sahaya Alfred 1997, Kalaiselvi *et al.*, 2000 and Malathi, 2001 also observed the same results. The treated paper and pulp mill effluent could be safely used for irrigation in light textured soils along with amendments.

Recycling and reuse of industrial wastewater in agriculture is not only helpful for conserving the water for irrigation, also the plant nutrients. The treated effluent did not cause any adverse effect on physico-chemical and microbial population dynamics under stress conditions of treated effluent irrigation. So it is essential that the implications of the use of industrial effluents in the crop field and their effect should be assessed before they are recommended for use in irrigation. If the wastewater from paper and pulp industry could successfully be used for irrigation, it is possible to prevent river water pollution and also to augment already scarce irrigation water resources.

References

Bharathi, S.G. and Krishnamoorthy, S.R., 1988. Effect of paper mill effluent on the distribution of algal flora of river Kali around Dandeli, Karnataka. *Proc. Ind. Sci. Cong.*, 3: 19–20.

Billings, R.M. and De Hass, G.G., 1984. Pollution control in the pulp and paper industry. In: *Industrial Pollution Control Handbook.* (Ed.) H.F. Lund. McGraw Hill Book Co., New York, p. 18.1–18.28.

Caplenas, M.R., Kanarek, M.S. and Dufour, A.P., 1981. Source and extent of *Klebsiella pneumoniae* in the paper industry. *Appl. Environ. Microbial.*, 42: 779–785.

Chauhan, U.K. and Kaur, P., 1991. Impact of pulp and paper mill effluent on soil microflora. In: *Proc. Ind. Sci. Congr. Ass., Indore,* 3(8): 24–25.

Chuang, Y.T., 1985. Studies on waste treatment with nitrogen fixing bacteria. Isolation of *Azotobacter* and effects of nutrients on waste treatment. *Rep. Taiwan Sug. Res. Inst.*, 107: 41–50.

Datta Roy, S. and Malty, B.R., 1984. Pollution load of paper mill effluents discharged into the Hoogly river at Kalyani, West Bengal. *Env. Eco.*, 1: 29–34.

Dhevagi, P., Rajannan, G. and Oblisami, G., 2000. Effect of paper mill effluent on spermosphere microflora. *J. Ecobiol.*, 12: 149–152.

Dhevagi, P. and Oblisami, G., 2000. Effect of paper mill effluent on germination of Agricultural crops. *J. Ecobiol..*, 12: 243–249.

Dhevagi, P., Maheswari, M., Ilamurugu, K. and Oblisami, G., 2003. Influence of paper mill effluent on soil microflora. *J. Ecobiol.*, 16: 33–36.

Dhevagi, P., Maheswari, M. and Oblisami, G., 2006. Effect of paper mill effluent on soil microflora of maize. *Indian J. Env. Sci.*, 1: 33–36.

Dixit, A.M., Lailman, S. and Srivastava, S.K., 1986. Effect of cardboard factory effluent on seed germination and early seedling growth of rice seeds. *Seeds Res.*, 14: 66–71.

FAO, 1988. *Pulp and Paper Capacities, 1987–1992*. Food and Agriculture Organization of the United Nations, Rome.

Fazeli, M.S., Sathyanarayanan, S. and Satish, P.N., 1991. Effect of paper mill effluent on accumulation of heavy metals in coconut trees near Nanjangud, Mysore Distt., Karnataka. *Environ. Biol. Water Sci.*, 17: 47–50.

Field, J.A., Leyendeckers, M.J.H., Alvarez, S.R., Lettinga, G. and Habets, L.H.A., 1988. The methanogenic toxicity of bark tannins and the anaerobic biodegradability of water soluble bark matter. *Wat. Sci. Tech.*, 20: 324–329.

Gilliespie, N.J., 1985. Pulp and paper effluent management. *J. Water Pollut. Cont. Fed.*, 57: 587–596.

Gomathi, V. and Oblisami, G., 1992. Effect of pulp and paper mill effluent on germination of tree crops. *Indian J. Environ. Hlth.*, 34: 326–328.

Hansbrix, 1987. Treatments of wastewater in rhizosphere of wetland plants: The root zone method. *Wat. Sic. Tech.*, 19: 107–118.

Hansen, E.A., Dawson, M.P. and Tolsted, N.P., 1980. Irrigation of intensively cultured plantation with paper mill effluent. *TAPPI.*, 63: 139–144.

Igbounamba, O., 1972. Exchangeable reserve potassium and other cation reserves in agodi and ado soils. *Soil Sci.*, 113: 394–409.

Juwarkar, A.S. and Bhalkar, D.V., 1984. Effect of pulp and paper mill wastewater on crop growth and soil properties. *Ph.D. Thesis*, Punjabrao Krishi Vidyapeeth, Akola.

Juwarkar, A.S., Bhalkar, D.V. and Subramanyam, P.V.R., 1987. Effect of pulp and paper mill wastewater on soil properties. *Indian J. Environ. Hlth.*, 29: 313–321.

Juwarkar, A.S. and Subrahmanyam, P.V.R., 1987. Impact of pulp and paper mill wastewater on crop and soil. *Water Sci. Technol.*, 19: 693–700.

Kalaichelvi, K., Rajeaswari, K., Vijayakumari and Manian, S., 2000. Effect of paper mill factory effluent on the growth parameters of *Vigna mungo* (L.). In: *Proc. National Level Conference on Eco-friendly Technologies for Sustainable Development*, held at PSG CAS, Cbe 14 March 27th to 29th, p. 8.

Kannan, K. and Oblisami, G., 1990a. Influence of paper mill effluent irrigation on soil enzyme activities. *Soil Biol. Biochem.*, 22: 923–926.

Kannan, K. and Oblisami, G., 1990b. Influence of irrigation with pulp and paper mill effluent on soil chemical and microbiological properties. *Biol. Fertility Soils*, 10: 197–201.

Kannan, K. and Oblisami, G., 1992. Effect of raw and treated paper mill effluent irrigation on vigour indices of certain crop plants. *Madras Agr.ic. J.*, 79: 18–21.

Rannapiran, S., 1995. Studies on the effect of solid and liquid wastes from paper and pulp industry on forest species. *M.Sc. (Env. Sciences)*, Tamil Nadu Agril. Univ., Coimbatore.

Ketkar, C.M., 1989. Pollution control of pulp and paper mills effluent and sludge by recycling through land disposal for various productive uses. *J. Intn. Public. Hlth. Engrs., India*, 3: 1–6.

Krishnamoorthy, Ch., 1971. Relation of soil test with crop response. In: *Proc. International Symposium on Soil Fertility Evaluation*, New Delhi, 1: 755–760.

Krueger, C.L. and Sheikh, W., 1987. A new selective medium for isolating *Pseudomonas* spp. from water. *Appl. Environ. Microbiol.*, 53: 895–897.

Larson, A.T., Andersson, A., Forlw, L. and Hardig, J., 1988. Physiological disturbances in fish exposed to bleached kraft mill effluents. *Wat. Sci. Tech.*, 20: 67–76.

Malathi, G., 2001. Impact of treated pulp and paper mill effluent on vegetable soil ecosystem. *M.Sc.(Ag.) Thesis*, Tamil Nadu Agricultural University, Coimbatore.

Mane, T.T. and Shitole, M.G., 1989. Productivity of methi (*Trigonella foenumgraecum* L.) on physico-chemical properties of soil under pulp and paper mill treated effluent irrigation. *Proc. Ind. Sci. Cong.*, 3: 144.

Manikya Reddy, P. and Venkateswarlu, B., 1985. Ecological studies in the paper mill effluents and their impact on the river Tungabhadra: heavy metals and algae. *Proc. Indian Acad. Sci.*, 95: 139–146.

Misra, R.N. and Behera, P.K., 1991. The effect of paper industry effluent on growth pigments, CH_2O, proteins of rice seedlings. *Environ. Poll.*, 72: 159–197.

Mittar, D., Khanna, P.K. Marwaha, S.S. and Kennedy, J.F., 1992. Bio-bleaching of pulp and paper mill effluents by *Phanerochaete chrysosporium*. *J. Chem. Tech. Biotechnol.*, 53: 81–92.

Narashimha Rao, P. and Narashimha Rao, Y., 1992. Quality of effluent water discharged from paper board industry and its effect on alluvial soil and crops. *Indian J. Agric. Sci.*, 62: 9–12.

Neelay, V.R. and Dhondizal, L.P., 1985. Observation on the possibility of using industrial effluent water for raising forest plantation. *J. Trop. Forestry,* 2: 131–139.

Neilson, A.K. and Allard, A.S., 1985. Acetylene reduction (N_2 fixation) by Enterobacteriaceae isolated from industrial wastewater and biological treatment systems. *Appl. Microbiol. Biotechnol.*, 23: 67–74.

Niirenen, M. and Vaeaetaenen, P., 1983. Treatment of pulp wood in a mill. In: *Manufacture of Pulp,* (Eds.). N.E. Virkota, Turuku, Oy furum Sanomat, 5: 125–258.

Notter, W.L. and Red, T., 1985. Treatment of wastewater by application to forest land. *Tappi.*, 68: 114–117.

Oblisami, G. and Palanisami, A., 1991. Studies on the effect of paper and sugar factory on soil microflora and agricultural cropping system report. Tamil Nadu Agricultural University, Coimbatore, pp. 297–305.

Palaniswami, G., 1989. Studies on the effect of continuous irrigation with paper factory effluent on soil properties and on sugarcane nursery and main crop and development of techniques for clarification of effluent *M.Sc. (Ag.) Thesis,* TNAU, Coimbatore, p. 146.

Palaniswami, C. and Sree Ramulu, U.S., 1994. Effects of continuous irrigation with paper factory effluent on soil properties. *J. Indian Soc. of Soil Sci.,* 42: 139–140.

Patra, S.K. and Behera, S., 1974. Effects of paper mill sludge on mineralization of soil nitrogen. *Agra Univ. J. Res.,* 4: 111–113.

Prasad, M., Juwarkar, A.S. and Subrahmanyam, P.V.R., 1977. Utilization of pulp mill wastewater for crop irrigation with reference to wheat and paddy. In: *Proc. Seminar on Industrial Waste,* Calcutta, India, pp. 31–47.

Priha, M., 1994. Bioavailability of pulp and paper mill effluent phosphorus. *Wat. Sci. Tech.,* 29: 93–103.

Pushpavalli, R., 1990. Studies on the characterization of pulp and paper mill effluent and its effect on soil profile characteristics and on germination, yield and juice quality of sugarcane (Var. CO 63004 and COC 671) *M.Sc. (Ag.) Thesis,* Tamil Nadu Agricultural University, Coimbatore, p. 165.

Rajannan, G. and Oblisami, G., 1979. Effect of paper factory effluents on the soil and crop plants. *Indian J. Environ. Hlth.*, 21: 120–130.

Ramanathan, N.L. and Parabrahmam, M., 1982. Environmental issues pertaining to non wood pulp and paper industry in India. In: *Proc. International Seminar on Management of Environmental Problems in the Pulp and Paper Industry,* (Ed.) M.N. Ramanathan. Delhi Productivity Council, New Delhi, pp. 21–26.

Reddy, G.B., Cheng, C.N. and Dunn, S.J., 1983. Survival of *Rhizobium japonicum* in soil. *Soil Biol. Biochem.,* 15: 343–345.

Reddy, M.R., Jivendra, S. and Jain, S.C., 1981. Paper mills effluent for Sugarcane irrigation. *IAWPC Technol. Annu.*, 8: 129–146.

Rout, G.R., 1990. Studies on physiotoxicity of paper mill effluent on *Phaseolus aureus. Andhra Pradesh Agri. J.*, 37: 24–27.

Rydholm, S.A., 1965. *Pulping Processes.* Interscience Publishers, New York.

Sahaya Alfred, R., 1997. Impact of paper and pulp factory effluent on soil, water and plant ecosystem. *Ph.D. Thesis*, Tamil Nadu Agricultural University, Coimbatore.

Sandana, K.M.C., 1995. Studies on the effect of liquid and solid wastes from industries and the growth of certain crops. *M.Sc. Thesis*, Tamil Nadu Agricultural University, Coimbatore.

Sandana, K.M.C. and Oblisami, G., 1996. Effect of paper factory effluent on soil characteristics and growth of tomato. In: *Abst. National Conference on Environmental Biotechnology Utilization of Industrial Wastes: Potentials and Limitations*, held at Bharadhidasan University, Trichy, p. 65.

Sharma, R.R.S. and Bandhyopadhyay, M., 1991. Treatment of pulp and paper mill effluent by up-flow anaerobic filter. *Indian J. Environ. Hlth.*, 33: 456–463.

Sierra-Alvarez, R., 1990. The role of natural wood constituents on the anaerobic treatability of forest industry wastewater. *Ph.D. Thesis*. Wageningon Agril. Univ., Wageningen, The Netherlands.

Somashekar, R.K., Gowda, M.T.G., Shettigar, S.L.N. and Srinath, K.P., 1984. Effect of industrial effluents on crop plants. *Indian J. Environ. Hlth.*, 26: 136–146.

Srivatsava, R.K., 1991. Effect of paper mill effluent on seed germination and early growth performance of radish and onion. *J. Eco. Toxicol. Environ. Monit.*, 1: 13–18.

Subrahmanyam, P.V.R., Juwarkar, A.S. and Sundaresan, B.B., 1984. Utilization of pulp and paper mill wastewater for crop irrigation. In: *Proc. Asian Chemical Conference on Priorities in Chemistry in Development of Asia*, Priochem Asia, Kulalumpur, Malaysia, pp. 26–31.

Subrahmanyam, P.V.R. and Sundarasen, B.B., 1982. Magnitude of pollution from pulp and paper industry in India. In: *Proc. International Seminar on Management of Environmental Problems in the Pulp and Paper Industry*, (Ed.) M.N. Ramanathan. Delhi Productivity Council, New Delhi, pp. 50–65.

Sundman, G., Kirk, T.K. and Chang, H.M., 1981. Fungal decolorisation of kraft bleach plant effluents: Fate of the chromophoric material. *Tappi*, 64: 145–148.

Sundaramoorthy, P. and Kunjithapatham, J., 2000. Effect of paper mill effluent on seed germination and seedling growth of six varieties of groundnut (*Arachis hypogea*). *J. Ecotoxicol. Environ. Monit.*, 10: 53–57.

Temmink, J.H.M., Field, J.A., Van Haastrecht, J.C. and Merckelbach, R.C.M., 1989. Acute and subacute toxicity of bark tannins to carp (*Cyprinus carpio* L). *Wat. Res.*, 23: 341–344.

Udayasoorian, C., Dhevagi, P. and Ramasami, P.P., 1999a. Case study on the utilization of paper and pulp mill effluent irrigation for field crops. In: *Proceedings of the Workshop on Bioremediation of Polluted Habitats*, p. 71–73.

Udayasoorian, C., Sagaya Alfred, C. and Ramasami, P.P., 1999b. Long term effect of bagasse based paper mill effluent irrigation on soil and groundwater. In: 2nd *International Conference on Contaminants in Soil Environment in the Australia–Pacific Region*, p. 80 –81.

Upathyaya, J.S. and Singh, B., 1991. Decolourization of effluent from pulp and paper industry. *Indian J. Environ. Hlth.*, 33: 350–356.

Vaeaetaenen, P. and Niemelae, S.I., 1983. Factors regulating the density of bacteria in process water of a paper mill. *J. Appl. Bacteriol.*, 54: 367–371.

Veena, V., Sahu, A.K. and Patel, K., 1992. Pulp and paper mill sludge in agricultural soil amendment and effluent water for irrigation. Sharanpur. *IPPTA*, 4: 112–121.

Verma, A.M., Dudani, V.K., Kumari, B. and Kargupta, A.N., 1988. Algal population in paper mill wastewater. *Indian J. Environ. Hlth.*, 30: 388–390.

Verma, A.K. and Poornima, S., 1992. Evaluation of the genotoxicity of a paper industry effluent on barley. *J. Cytol. Genet.*, 27: 73–78.

Virkola, N. and Honkanen, K., 1985. Wastewater characteristics. *Wat. Sci. Tech.*, 17: 1–28.

Walden, C.C., Mc Leay, D.J. and Mc Kague, A.B., 1986. Cellulose production processes. In: *The Handbook of Environmental Chemistry, Vol. 3, Part D: Anthropogenic Compounds*, (Ed.) O. Hutzinger. Springer-Verlag, Berlin, pp. 1–34.

Young, J.C., Mc Dermott, G.N. and Jenkins, D., 1981. Alterations in the BOD procedures for the 15th edition of Standard methods for the examination of water and wastewater. *J. Water Pollut. Control Fed.*, 53: 1253–1259.

Chapter 8

PCP Degradation by *Pseudomonas aeruginosa* and *Pseudomonas putida*: A Comparative Study

T. Byragi Reddy[1], K. Swapna Vahini[2] and M. Sumanth Kumar[3]
[1]Associate Professor, [3]Research Scholar,
Department of Environmental Sciences, Andhra University
[2]Assistant Professor, Department of Environmental Studies,
GITAM Institute of Science, GITAM University

ABSTRACT

Chlorophenolic compounds are generated from a number of industrial manufacturing processes including pulp and paper. These compounds are found to be toxic and recalcitrant and hence their discharge into the environment must be regulated. Pentachlorophenol (PCP), a chlorophenolic compound has been used as a wood preservative, insecticide and herbicide. As it is a human carcinogen and toxic to numerous aquatic organisms, PCP is considered as priority pollutant by various regulating agencies. Because of its toxicity, there is need for removal of PCP from natural environment and bioremediation is a useful alternative to conventional clean-up method.

The aim of the present work was to study the substrate utilization and influence of pH of the medium on PCP degradation by two commercially available bacterial strains, *Pseudomonas aeruginosa* (MTCC 4996) and *Pseudomonas putida* (MTCC 1194). The degradation of PCP was established by measuring the PCP aromatic ring clearage using UV-VIS spectrophotometer at 320 nm. Both the bacterial strains could degrade PCP within 64 hrs.

PCP was degraded rapidly at pH value 6 to 8 for *P. aeruginosa*, whereas in *P. putida* rapid degradation was only seen at pH 6 and 7. PCP degradation at pH 5 was significantly low for both the bacterial strains and PCP was not utilized within 72hrs. These results show the broad range of pH for PCP degradation for *P. aeruginosa* over *P. putida*.

Keywords: *Pentachlorophenol, Bioremediation, UV-VIS spectrophotometer.*

Introduction

Phenolic compounds are ubiquitous in aqueous streams of industrial processes including petroleum refining, coking and coal conversions, chemical plants, foundries and pulp and-paper plants (Van Schie and Young, 1998; Aitken *et al.*, 1994; Daniel Ryan *et al.*, 2007). The presence of these compounds in drinking and irrigation water represents a health and environmental hazard (Alberti *et al.*, 1981).

As compared to phenol, chlorophenolic compounds are more persistent in the environment. Toxicity and bioaccumulative potential of chlorophenols increases with the degree of chlorination (R.C. Loehr *et al.*, 1988). Pentachlorophenol (PCP), a achlorophenolic compound has been used as a wood preservative, insecticide and herbicide and was introduced into the environment by waste streams of several industrial operations. A large amount of PCP is used by the wood preserving industry (Hoos 1978; Cirelli 1987) and PCP contamination in soil represents a serious problem to the environment that surrounds several wood treatment plants and sawmills (Rao, 1978). PCP is considered a priority pollutant by various regulatory agencies. This polychlorinated aromatic compound is toxic to numerous aquatic organisms at a concentration of as low as 0.5 mg (Borthwiek and Sehimmel *et al.*, 1978) and adversely affects flora and fauna (Chu and Kirsch, 1973). It is also suspected to be a human carcinogen (Me Allister *et al.*, 1996).

Although conventional methods for dephenolization do exist, they are often expensive, deliver incomplete purification, can form hazardous by products and are often applicable to only a limited concentration range. The use of micro organisms for bioremediation of PCP-contaminated sites may prove to be a viable alternative to conventional clean-up methods. Biological treatment of chlorophenols attracts more attention than physical and chemical methods, because a variety of micro-organisms are known to utilize chlorophenols as their sole carbon or energy source and because the reaction products are Cl^- ions, CO_2 and biomass. PCP-degrading bacteria include species of *Arthrobacter* (Stanlake and Finn. 1982). *Flavobacterium* (Saber and Crawford, 1985) *Pseudomonas* (Radehaus and Schmidt, 1992) *Rhodococcus* (Apajalathi and Salkinoja-Salonen, 1986). *Pseudomonas* sps. are used effectively for the bioremediation of phenolic compounds (Erika Aljandra Wolski *et al.*, 2006).

The objective of the present study was to evaluate and compare the capacity of two bacterial *Pseudomonas* strains to degrade PCP at various cultural parameters.

Materials and Methods

Strain Preservation

The commercial slant cultures of *Pseudomonas aeruginosa* (MTCC 4996) and *Pseudomonas putida* (MTCC 1194) were obtained from IM Tech, Chandigarh. The cultures of both the bacterial strains were maintained on nutrient agar slants.

Chemicals and Reagents

Pentachlorophenol (PCP 99 per cent pure) was obtained from Sigma Chemical Co. (St. Louis MO 63178, USA). All other chemicals used were of the highest purity commercially available.

Culture Conditions and Media

Minimal Salts Media (MS) consisted of (in g.l^{-1}): $NaNO_3$–2.25; K_2PO_4–2.83, KH_2PO_4–0.507, $MgSO_4.7H_2O$-0.45, $FeSO_4$–0.135 and $CaCl_2$–0.45. pH was adjusted to 7. PCP was first converted to its sodium salt by dissolving it in 0.2N NaOH and then autoclaved. This stock solution of PCP (5000 mg.l^{-1}) was used to prepare the MS media supplemented with 25 mg.l^{-1} of PCP as the sole source of carbon.

Flask Culture

All experiments were conducted in 250ml ehrlenmeyer flasks containing 45 ml autoclaved MS medium. These were inoculated with 1 ml of the inoculam (after 24 hrs growth on nutrient medium) and incubated at 30°C.

PCP Degradation

PCP degradation was established by measuring the PCP aromatic ring cleaving using a UV–VIS Spectrophotometer (Systronic UV–VIS Spectrophotometer 108) at A320 nm and compared with the standard curve.

Influence of pH of the Medium on PCP Degradation

P. aeruginosa and *P. putida* cells were grown in 50 ml MS medium with 25 mg/l of PCP at different pH values (5, 6, 7 and 8). Samples were analyzed at 8 hr interval for 72 hrs. The cells were pelleted by centrifugation at 10,000 rpm for 5 mins. PCP degradation was assessed by measuring A320nm of the supernatant. Simultaneously, controls were kept which include cells in MS without PCP added and MS medium supplemented with PCP without cells.

Results and Discussion

The degradation capacities of the two bacterial strains were tested at substrate concentration 25 mg/l and at pH 7. It was found that in *P. aeruginosa* PCP concentration began to decrease rapidly after 16 hrs and almost completely degraded PCP in 56 hrs (Figure 8.1) whereas in *P. putida*, there was a progressive decrease in PCP concentration and degradation was complete after 72 hrs (Figure 8.1).

The change in the degradation capacities of the two bacterial species was tested at pH 5, 6, 7 and 8. PCP was completely degraded at pH 6, 7 and 8 by *P. aeruginosa* but at pH 5, complete degradation of phenol within 72 hrs was not observed (Figure 8.2). In *P. putida*, PCP was completely degraded within 48–56 hrs at pH 6 and 7 but was not degraded completely within 72 hrs for pH 5 and 8 (Figure 8.3). In both the bacterial species, the degradation was taking place quickly at pH 6. This was in similar with the fact given by Erika Alejandra *et al.* (2006) that the enzyme for PCP degradation has their optimum pH at 6.3. So at pH either higher or lower than 6.3, the rate of degradation was comparatively low. In the present study, *P. aeruginosa* could degrade PCP at a wide range of pH from 5 to 8 when compared to *P. putida*, which had a narrow range as compared to *P. aeruginosa*. The acidification of natural environments may enhance the toxicity of PCP (Rutgers *et al.*, 1998). Therefore it very important to find strains that degrade PCP at low pH. In the present study, the strain *P. aeruginosa* was found to be a better bacterial strain.

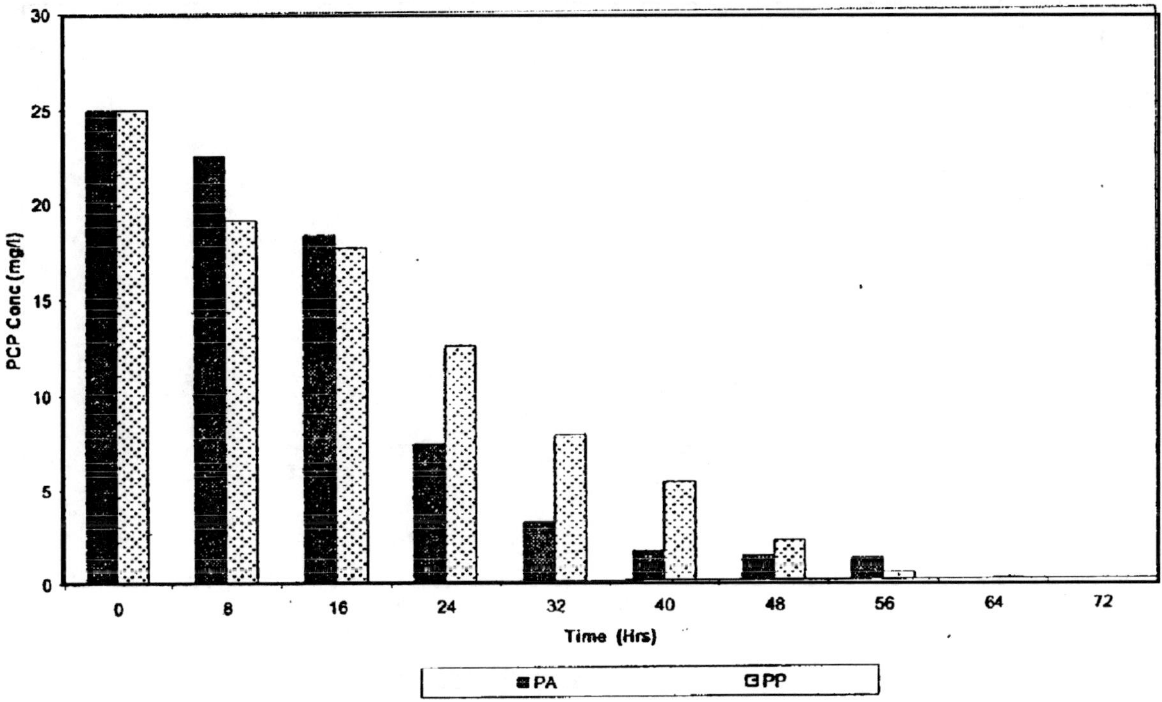

Figure 8.1: A Comparison of Biodegradation Capability of PCP by *Pseudomonas aeruginosa* and *Pseudomonas putida*

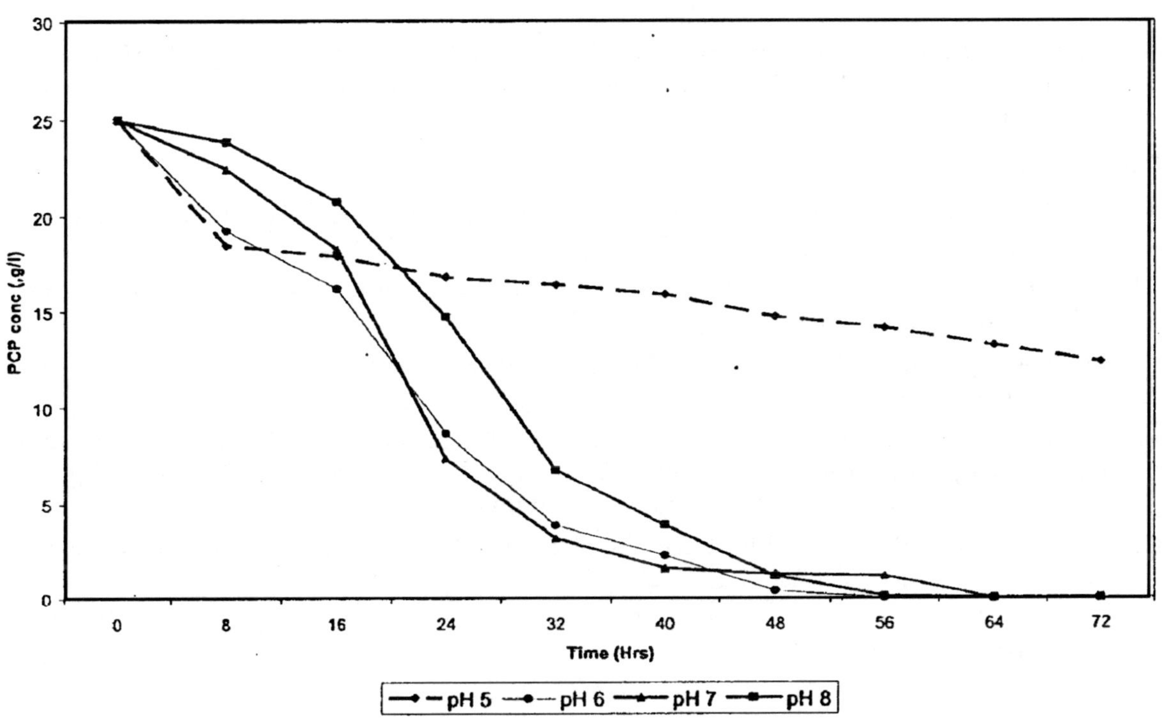

Figure 8.2: PCP Biodegradation by *P. aeruginosa* at Different pH Values

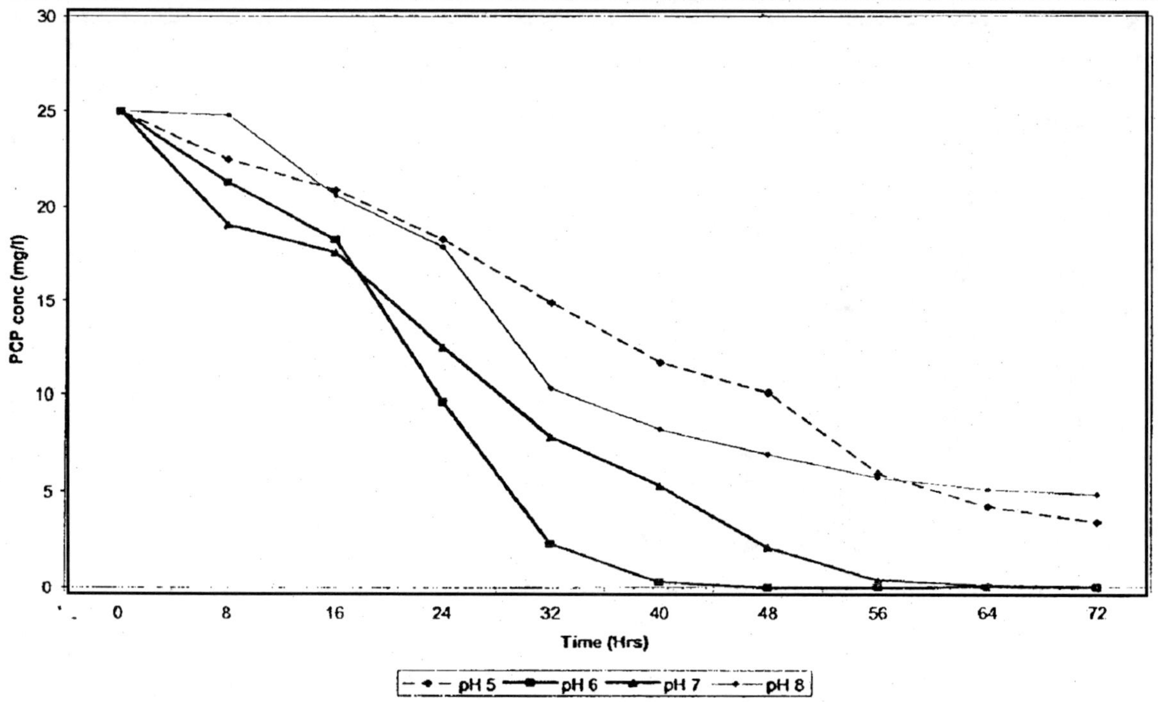

Figure 8.3: PCP Biodegradation by *P. putida* at Different pH Values

Conclusion

Based on the utilization of substrate concentration and tolerance to the range of pH values *P. aeruginosa* was having a better adaptability than *P. putida.* However more research is necessary to check with other parameters to derive to a conclusion.

References

Aitken, M.D., Massey, L.J., Chen, T. and Heck, P.E., 1994. Characterization of reaction products from the catalized oxidation of phenolic pollutants. *Water Res.,* 28: 1879–1889.

Alberti, B. and Klibanov, A., 1981. Enzymatic removal of dissolved aromatics from industrial aqueous effluents. *Biotechnol. Bioeng. of Symp.,* 2: 373–379.

Jha and Salkinoja-Salonen M.S., 1986. Degradation of PCP by *Rhodococcus chlorophenolicus. Appl. Microbiol. Biotechnol.,* 25: 62–67.

Borthwick, P.W. and Schimmel, S.C., 1978. Toxicity of pentachlorophenol and related compounds to early life stages of selected estuarine animals. In: *Pentachlorophenol: Chemistry. Pharmacology and Environmental Toxicology,* (Ed.) K. Ranga Rao. Plenum Press, New York.

Chu, J.P. and Kirch, E.J., 1972. Utilization of halophenols by a pentachlorophenol metabolizing bacterium. *Dev. Ind. Microbiol.,* 14: 264–273.

Cirelli, D.P., 1978. Patterns of PCP usage in USA: An overview. In: *Pentachlorophenol,* (Ed.) K.R. Rao. Plenum Press, New York, USA.

Daniel, R., Leukes, W. and Burton, S., 2007. Improving the bioremediation of phenolic wastewaters by *Tremetes versicolor*. *Biores. Tech.*, 98(3): 579–587.

Erika Alijandra Wolski, Silvia Elina Murialdo and Jorfe Frolian Gonzalez, 2006. Effect of phenols and inoculum size of pentachlorophenol degradation by *pseudomonas* species. *Water SA*, 32(1).

Hoos, R., 1978. Patterns of pentachlorophenol usage in Canada: An overview. In: *Pentachlorophenol*, (Ed.) K.R. Rao. Plenum Press, New York, USA.

Mc Allister, Ka. Leeh and Trevors Jt., 1996 Microbial degaradation of pentachlorophenol. *Biodegradation*, 7: 1–10.

Loehr, R.C. and Krishnamoorthy, R., 1988. Terrestrial bioaccumulation potential of phenolic compounds. *HG. Waste and Haz. Madl.*, 5: 109–119.

Radehaus, R.M. and Schmidt, S.K., 1999. Characterization of a novel *Pseudomonas* sp. that mineralizes high concentrations of pentachlorophenol. *Appl. Environ. Micobiol.*, 58: 2879–2885.

Rao, K.R., 1978. Chemistry, pharmacology and environmental toxicology. In: *Pentachlorophenol*, (Ed.) K.R. Rao. Plenum Press, New York, USA.

Rutgers, M., Van Bommel, S., Breure, A.M., Van Andel J.G. and Duetz, W.A., 1998. Environmental toxicology and effect of pH on toxicity and biodegradation of PCP by *Sphingomonas* sps. strain P5 in nutrient culture. *Chem.*, 17: 792–797.

Saber, D.I. and Crawford, R.I., 1985. Isolation and characterization of *Flavobacterium* strains that degrade pentachlorophenol. *Appl. Environ. Microbiol.*, 50: 1512–1518.

Stanlake, G.J. and Finn, R.K., 1982. Isolation and characterization of a pentachlorophenol-degrading bacterium. *AppL. Environ Microbiol.*, 44: 1412–1427.

Van Schie, G. and Young, P., 1998. Isolation and characterisation of phenolic degrading denitrifying bacteria. *Appl. Environ. Microbiol.*, 64: 2432–2438.

Chapter 9

Study of Spoilage Causing Pathogenic Microorganisms in Indian Cheese (*Paneer*) Sold in Allahabad City

Om Prakash Verma*, Ankit Kumar, Nirmala Singh, Ajay Kumar Singh, Purushottum Kumar and Birendra Kumar Singh
College of Biotechnology and Allied Sciences, Allahabad Agricultural Institute-Deemed University, Allahabad – 211 007

ABSTRACT

40 *Paneer* samples were collected from different shops in Allahabad. Shops were divided into hygienic and unhygienic shops, which were further sub divided into high traffic area and low traffic area shops. In addition Coliform, *Staphylococcus aureus* and *Enterococcus faecalis* were studied. On the basis of differential counts, almost all the *Paneer* samples were found to be unsatisfactory in quality. 100 per cent samples showed presence of pathogenic organism comprising of Coliforms (75.0 per cent), *Staphylococcus aureus* (62.5 per cent), *Enterococcus faecalis* (65.0 per cent) and Yeasts and Molds (100.0 per cent). Differential counts and incidence of spoilage causing organism were higher in shops located in high traffic area.

Keywords: *Paneer, Total Viable Count (TVC), Yeast and molds count (Y and M), Staphylococcus aureus, Enterococcus faecalis.*

---

* F.I.S.E.C. (Fellow of International Society for Ecological Communications).

Introduction

Paneer is, an important indigenous milk product, prepared from Cow's or Skimmed milk by removing whey. The *Paneer* should be granular and should have fat not less than 15 per cent. *Paneer* contains fat (25–35 per cent), protein (15–20 per cent), lactose (2.0–2.5 per cent), ash (0.3–0.4 per cent) and water (up to 70 per cent) in the form of moisture (Yadav *et al.*, 1993). A good quality *Paneer* should not have very high microbial count as well as any undesirable microorganisms in it. But contamination of *Paneer* occurs during the manufacturing and subsequent handling. Various kinds of microorganisms gain entry during different manufacturing steps of *Paneer* and also during its storage. Here it is worth to mention that conditions under which they are sold are far from satisfactory, which results in further contamination of *Paneer* and deterioration of the product, causing odour defects, acidity, proteolysis, lipolysis and moldy texture etc. In some cases due to unavailability of deep freezers, chances and level of contamination also increases. The microbiological profile (Frazier and Westhoff, 1995) of *Paneer* includes acid producing bacteria, chromogenic, bacteria, lipolytic bacteria, aerobic spore forming bacteria, psychrotrophic bacteria, thermophilic bacteria and pathogenic strains of bacteria (Ghodekar *et al.*, 1974; Varadraj and Nambudaripad, 1983; Kumar and Sinha, 1989; Vaishnavi *et al.*, 2001). Besides, due to conditions prevailing at the time of *Paneer* production, subsequent handling and storage, yeast and molds frequently contaminate the products (Ghodekar *et al.*, 1980).

Since, *Paneer* is one of the ready-to-eat milk products, which consumed as such by the consumers, and contamination of *Paneer* may cause serious public health hazards, hence it is extremely necessary to ensure good quality of the *Paneer*. Therefore the present study was undertaken.

Materials and Methods

Collection of Samples

The *Paneer* samples were collected and transported to laboratory in 100 ml. sterilized, wide mouthed sample bottle. In the laboratory the samples were kept at 4.0±0.1°C, till they were analyzed microbiologically.

Preparation of Sample

For the study, the samples were prepared as the method described by Kumar and Sinha (1989). From a flask containing 99 ml sterilized Ringer solution, 10 ml was transferred into sterilized mortar and 1 gram of *Paneer* sample was crushed with the help of sterilized pestle, to form a homogenous mixture. The contents were then transferred in to the flask. It gave the dilution of 1 : 100. The sample was further diluted to 1 : 1000 by transferring 1 ml from the flask (1 : 100 dilutions) to a test tube containing 9 ml of sterilized Ringer solution using sterile 1 ml pipette. By this method, the sample was diluted up to 1 : 1 0000000 (1×10^{-7}).

The dilutions thus prepared were used for various microbiological tests.

Evaluation of Microbiological Quality of *Paneer*

Following parameters were chosen to evaluate the microbiological quality of *Paneer* samples:

1. Total Viable Count (TVC)
2. Yeast and Molds Count (Y and M)
3. Coliform bacteria
4. Pathogenic bacteria *viz. Staphylococcus aureus* and *Enterococcus faecalis*.

Total Viable Count (TVC)

Total Viable Count in *Paneer* samples were estimated by Serial Dilution Pour Plate Technique using 10^{-7} dilution and Nutrient Agar media, in duplicates, as the procedure described in APHA manual (1966). The plates were incubated at 37.0±0.5°C for 24–48 hours. After incubation, numbers of bacteria per gram were counted using the dilution factor.

Yeast and Molds Count (Y and M)

To observe yeast and molds count in *Paneer,* diluted samples (10^{-4}) were inoculated on Sabouraud's Dextrose Agar media, following incubation period of 4–5 days at 25.0±0.5°C. After incubation colonies were counted using dilution factor

Detection of Coliform Bacteria in Samples

To detect the presence of Coliform bacteria in the samples, Presumptive Coliform Test (PCT) was performed by using MacConkey's Bile Salt Broth (Rajmany *et al.,* 1989). After incubation at 37.0±0.5°C, for 24–48 hours, change in the colour of media from purple to yellow showed acid production and gas in Durham's tube gave the presumptive evidence of faecal contamination in the sample.

Detection of Pathogenic Bacteria in Samples

Two pathogenic bacteria *viz., Staphylococcus aureus* and *Enterococcus faecalis* were selected for the study.

To isolate the *Staphylococcus aureus* from the sample, Manitol Salt Agar media was used. The diluted sample (10^{-5}) was inoculated in the media using Serial Dilution Pour Plate Technique, following incubation at 37.0±0.5°C for 24–48 hours. The pinpoint colonies with yellow zone exhibit the presence of *Staphylococci.*

For *Enterococcus faecalis,* Sodium Azide Broth was used (Aneja, 2003). 1 ml sample from the original solution was transferred in sterilized Sodium Azide Broth tubes. After incubation of 24–48 hours at 37.0±0.5°C, positive samples (tubes) showed change in colour from purple to yellow, which indicate presumptive evidence of *Enterococcus faecalis* in the sample.

Results and Discussion

Paneer samples analyzed microbiologically showed Total Viable Count (TVC) ranged from 500–18000 × 10^5 per gram with an average of 4041 × 10^5 per gram while Yeast and Molds Count from 0.5–15 × 10^5 with an average of 5.17 × 10^{-1} per gram (Table 9.1, Figure 9.1a and b). *Paneer* samples showed a very high incidence (100.0 per cent) of spoilage causing and pathogenic microorganisms (Cappuccino and Sherman, 2004). Of the 40 samples analyzed 30 (75.0 per cent) showed presence of Coliform bacteria, 26 (65.0 per cent) showed presence of *Enterococcus faecalis,* 25 (62.5 per cent) showed presence of *Staphylococcus aureus* and all 40 (100 per cent) showed presence of Yeast and Molds (Table 9.2, Figure 9.2).

Table 9.1: Microbiological Quality of *Paneer*

Sl.No.	Microbial Parameter	Range (Counts/gm)		Average (Counts/gm)* 10^5	Grade*
		Min* 10^5	Max* 10^5		
1.	TVC	500	18000	4041	Unsatisfactory
2.	Y and M	0.5	15	5.17	Unsatisfactory

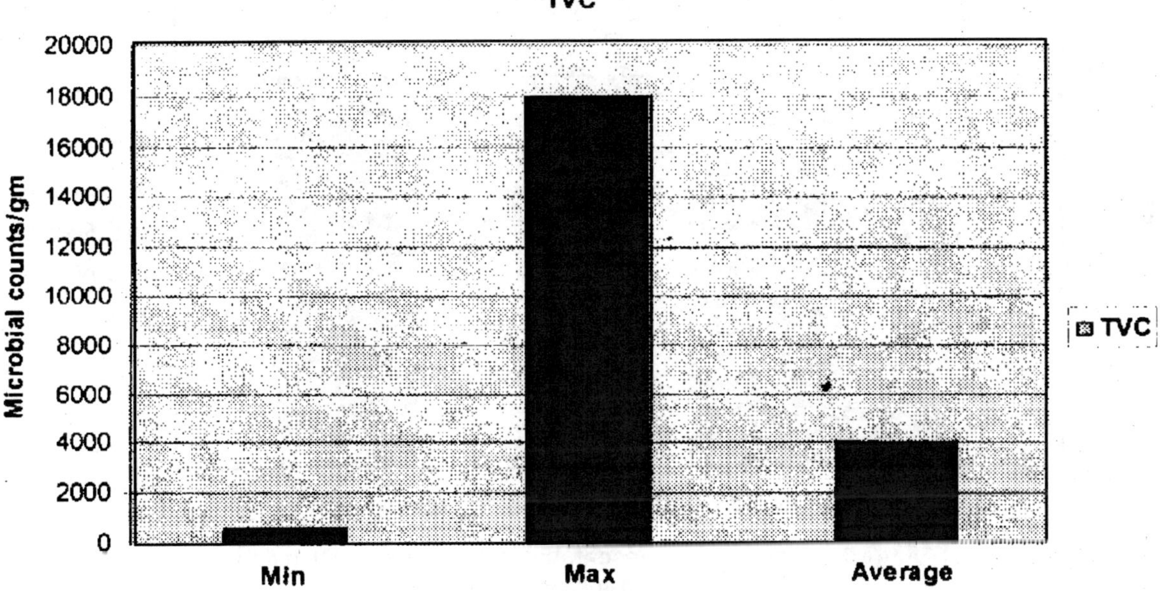

Figure 9.1a: Graphical Representation of Total Viable Count in *Paneer*

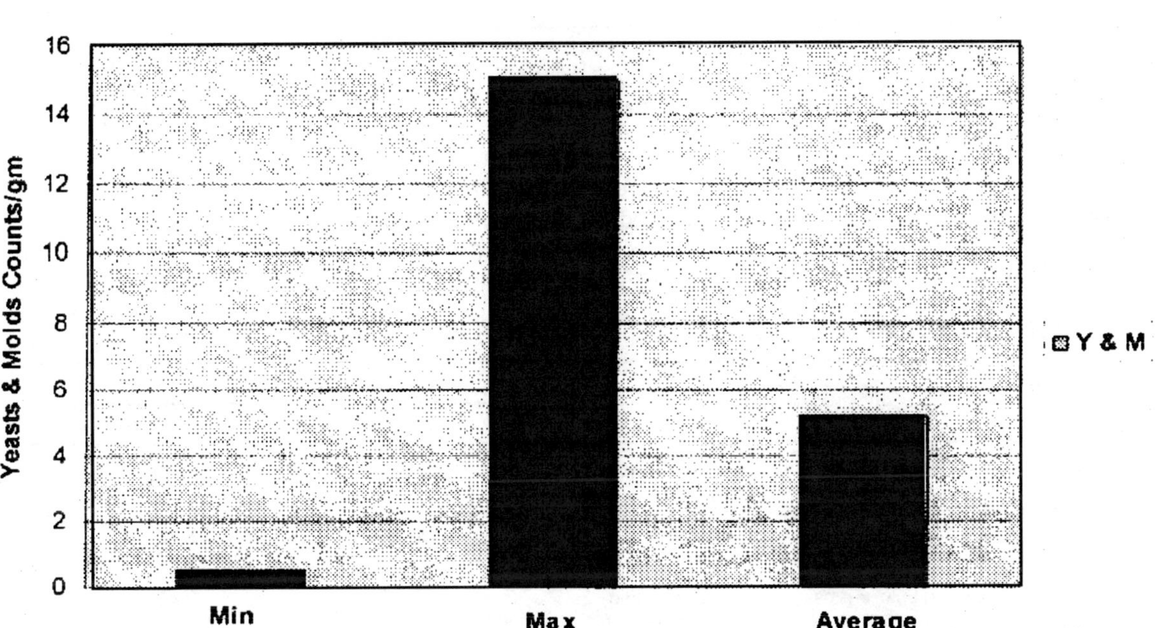

Figure 9.1b: Graphical Representation of Yeast and Molds Count in *Paneer*

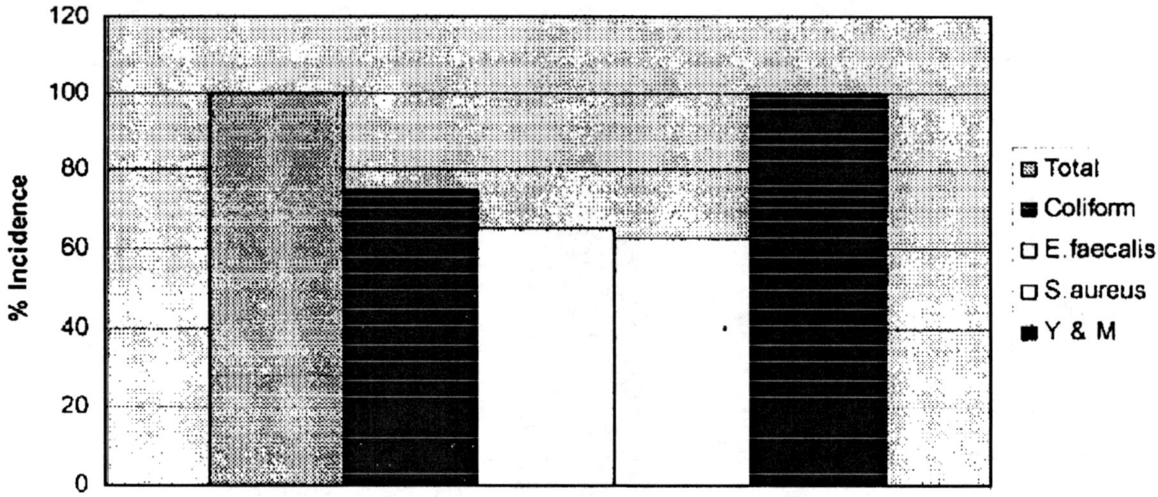

Figure 9.2: Graphical Representation of Spoilage Causing and Pathogenic Microorganisms in *Paneer*

Table 9.2: Incidence of Spoilage Causing Pathogenic Microorganisms in *Paneer*

Sl.No.	*Total No. of Samples*	*No. of Samples Positive for Organism*	*Pathogenic and Spoilage Causing Organisms*			
1.	40	40 (100.0 per cent)	30 (75.0 per cent)	26 (65.0 per cent)	25 (62.5 per cent)	40 (100.0 per cent)

On the basis of sanitary conditions, shops were grouped into two categories *viz.*, Hygienic shops (n=20) and unhygienic shops (n=20), which were further divided into two subgroups on the basis of traffic level of the area where they were situated *i.e.*, shops of hygienic conditions situated at low traffic area ($n_1 = 12$) and at high traffic area ($n_2 = 8$) and shops of unhygienic conditions situated at low traffic area ($n_3 = 10$) and high traffic area ($n_4 = 10$) (Table 9.3, Figure 9.3a and b).

Table 9.3: Comparison of Microbiological Quality of *Paneer* on the Basis of their Sources

Sl.No.	*Source*		*Mean Differential Count/gm in* Paneer	
			TVC 10^5*	*Y and M* 10^5*
1.	Hygienic shops (n = 20)	Low traffic shops (n_1 = 20)	1862.50	1.875
		High traffic shops (n_2 = 8)	5990.00	8.75
2.	Unhygienic shops (n = 20)	Low traffic shops (n_3 = 10)	2320.00	2.47
		High traffic shops (n_4 = 10)	7025.00	8.95

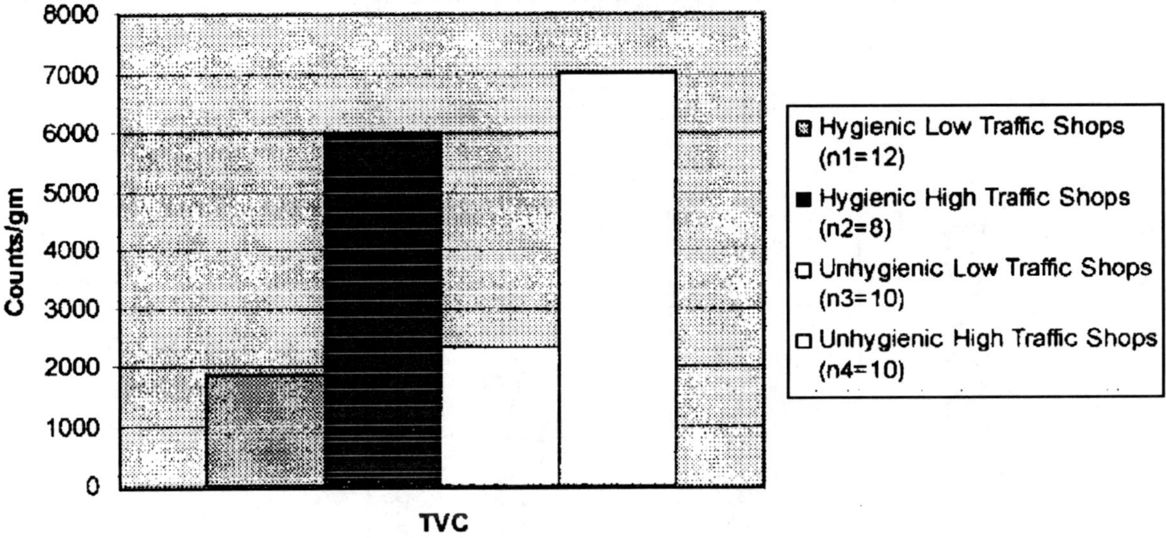

Figure 9.3a: Graphical Representation of Total Viable Count of *Paneer* Samples on the Basis of their Sources

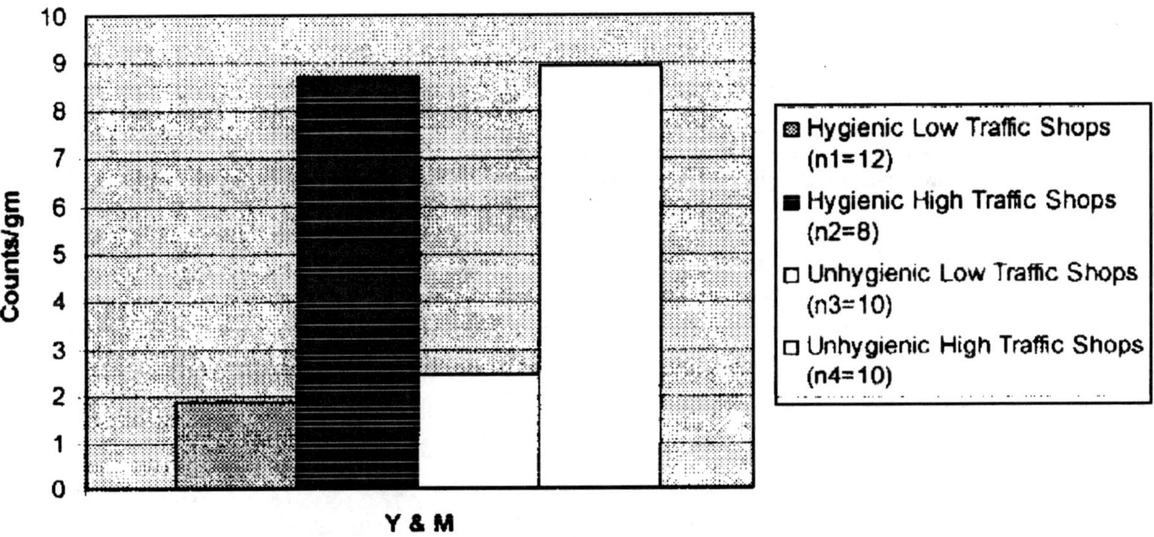

Figure 9.3b: Graphical Representation of Yeast and Molds Count of *Paneer* Samples on the Basis of their Sources

In general, the impact of sanitary condition on microbial count of *Paneer* was found to be non significant with the traffic but the traffic may show a significant effect on microbial quality of sample. The shops situated at high traffic and unhygienic sanitary conditions were found to have higher microbial counts as compared to shops situated at low traffic and hygienic conditions, with significant difference in TVC and Yeast and Molds count. The results observed in the study for Total Viable Count were showing quite high number of Total Viable Count in comparison to previous report *i.e.* 2021×10^5 per gram (Rajorhia *et al.*, 1984).

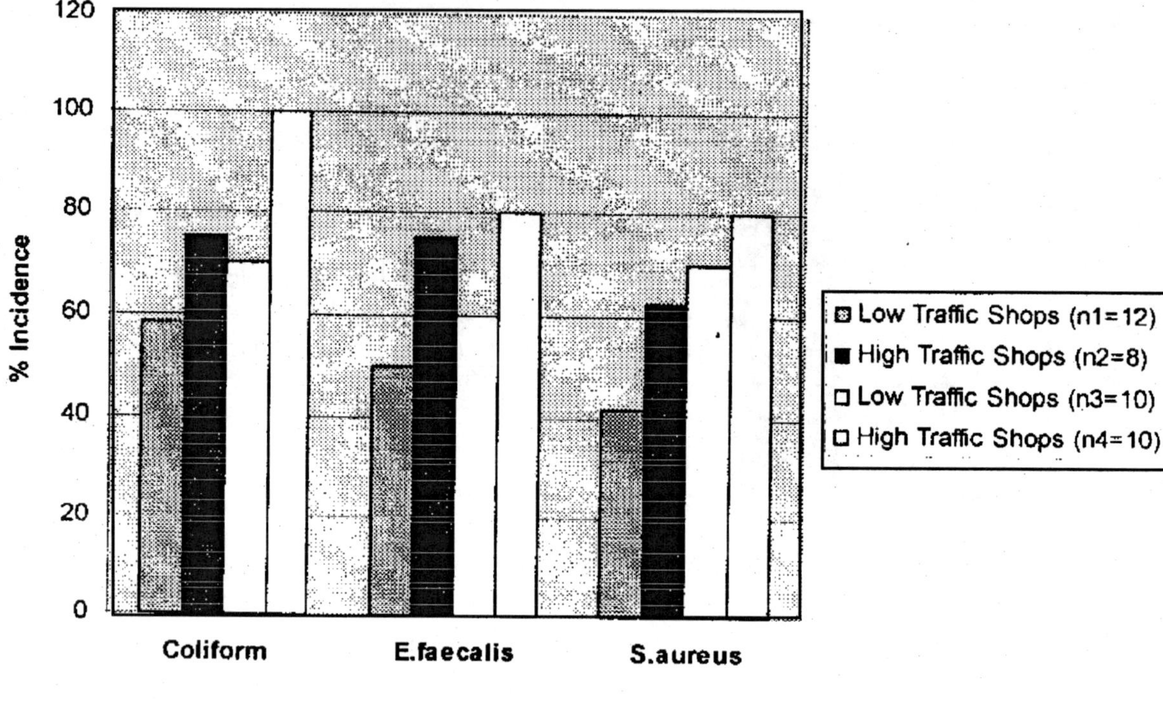

Figure 9.4: Graphical Representation of Incidence of Spoilage Causing and Pathogenic Microorganism in *Paneer* Samples on the Basis of their Sources

Table 9.4: Incidence of Spoilage Causing and Pathogenic Microorganisms in *Paneer* on the Basis of their Sources

Sl.No.	*Source*		*Microorganisms*		
			Coliforms	*E. faecalis*	*S. aureus*
1.	Hygienic shops (n = 20)	Low traffic shops (n_1 = 20)	58.3 per cent (n = 7)	50.0 per cent (n = 6)	41.7 per cent (n = 5)
		High traffic shops (n_2 = 8)	75.0 per cent (n = 6)	75.0 per cent (n = 6)	62.5 per cent (n = 5)
2.	Unhygienic shops (n = 20)	Low traffic shops (n_3 = 10)	70.0 per cent (n = 7)	60.0 per cent (n = 6)	70.0 per cent (n = 7)
		High traffic shops (n_4 = 10)	100.0 per cent (n = 10)	80.0 per cent (n = 8)	80.0 per cent (n = 8)

The counts obtained from low traffic area were showing low Total Viable Count in comparison to shops situated in high traffic area. The result was found statistically significant. The most possible reason would be the difference in environmental conditions of the surroundings due to which in high traffic area, chances of contamination in food products is more. Same thing was found with Yeast and

Molds count where again the results were statistically significant. High traffic, unhygienic storage condition and improper handling would be responsible for high microbiological counts.

The incidence of pathogenic bacteria was 100 per cent, which shows a high risk at public health hazard. When the results were compared to standards the overall microbial quality of *Paneer* was found to be unsatisfactory. The major factors for the unsatisfactory quality of *Paneer* are high traffic, unhygienic storage conditions, optimum temperature for growth of microorganism and the favourable environment of *Paneer* (moisture, nutrient content etc.). When *Paneer* is stored in unhygienic conditions, moisture and nutrients provide favourable conditions for microbial growth, which again increases through flies and unclean utensils.

As *Paneer* is one of the ready to eat milk product, which is being consume directly by the people, there is an immediate need to standardize the techniques, which are being used for the production and storage of the *Paneer* especially in temperate regions of India, like Allahabad. It is advisable to the vendors to use sanitized utensils during the production, storage and selling and also use deep freezers while storing of the *Paneer.*

As traffic level is playing very significant role it is also advisable to vendors and shop owners to take necessary steps to discourage the entry of microorganism through traffic pollution and apply some additional precautional measures such as close glass doors, freezer chambers, freezer counters, fly catchers, air conditioners and sanitized conditions in and around the shops, and also consumers are advised to prefer to buy food products from hygienic shops located at low traffic areas.

Acknowledgement

Authors gratefully acknowledge to Prof. (Dr.) M. Imtiyaz, Dean, College of Biotechnology and Allied Sciences, Allahabad Agricultural Institute-Deemed University, Allahabad, for helping and providing laboratory facilities. Authors also thank all the staff members and co-workers for their valuable suggestions.

References

Aneja, K.R., 2003. *Experiments in Microbiology, Plant Pathology and Biotechnology,* 4th edn. New Age International (P) Ltd., New Delhi.

Cappuccino, J.G. and Sherman, N., 2004. *Microbiology: A Laboratory Manual,* 4th edn. Pearson Education Pte. Ltd., Singapore.

Frazier, W.C. and Westhoff, D.C., 1995. *Food Microbiology,* 4th edn. Tata McGraw Hill Publishing Co. Ltd., New Delhi.

Ghodekar, D.R., Dudani, A.T. and Rangnathan, B., 1974. Microbiological quality of Indian Milk Products. *J. Milk Food Tech.,* 37(30): 119–122

Ghodekar, D.R., Dudani, A.T. and Rangnathan, B., 1980. Yeast and molds in indigenous milk products. *Indian J. Dairy Sci.,* 33(2): 255–259.

Kumar, V. and Sinha, R.N., 1989. Incidence of coliforms in indigenous milk products. *Indian J. Dairy Sci.,* 42(3): 579–580.

Rajmany, L., Garg, S.K. and Yadav, R.K., 1989. Incidence of *Staphylococcus* in milk and milk products. *Indian J. Dairy, Sci.,* 42(1): 136–138.

Varadraj, M.C. and Nambuderipad, V.K.N., 1983. *Staphylococcus aureus* in association with a few common contaminants in a dairy. *Indian J. Dairy, Sci.*, 34(2): 200–202.

Vaishnavi, C., Singh, S., Grover, R. and Singh, K., 2001. Bacteriological study of Indian cheese sold in Chandigarh. *Indian J. Medical Sci.*, 19(4): 224–226.

Yadav, Y.S., Grover, S. and Batish, V.K., 1993. *A Comprehensive Dairy Microbiology*. B.V. Gupta Publishers, Metropolitan Book Co. Pvt. Ltd., New Delhi.

Chapter 10

Comparative Analysis of Variability in Rice (*Oryza sativa* L.) Under Different Land Situations

Monalisa Manna, Hossain Ali Mondal* and Biman Gopal Sasmal

Department of Genetics, Bidhan Chandra Krishi Viswavidyalaya, Mohanpur – 741 252, West Bengal, India

ABSTRACT

The present investigation was undertaken with a view to establish a basis of selection indices of rice genotypes belonging to rain fed medium land, shallow deep (30 cm) and low land (70 cm) situations. Genetics variability, heritability (broad sense) and GA (at 5 per cent selection intensity) of 9 morphoeconomic and 7 grain traits of each land situation were estimated following RBD during wet season. The varieties of three ecogroups exhibited different estimations and combinations of variation. heritability and GA values. High estimates of heritability were recorded in majority of characters of medium land varieties followed by shallow deep and low land varieties. Shallow deep varieties had higher estimations of GA for more number of characters than the other two groups. The results indicated different selection criteria for combination of morphoeconomic and grain characters suitable for different land situations. Emphasis should be given on FLA, spikelets/panicle, filled grains/panicle, grain yield/plant and embryo weight for medium land; plant height, FLA, high density grains/panicle, grain breadth, kernel weight and grain weight for shallow deep and high density grains/panicle, grain length, kernel length and breadth for low land situation during selection of genotypes.

Keywords: *Selection indices, Rice genotypes, Different land situations.*

---

* Correspndong author presently at Bose Institute, Kolkata, India. E-mail: hossain_bic@rediffmail.com

Introduction

Estimation of genetic variability, heritability and genetic advance is a prerequisite to select superior genotypes from the germplasms. Yield is a complex character being governed by a large number of cumulative, duplicate and dominant genes, and highly influenced by environment. This necessitates through knowledge of variability owing to genetic factors, actual genetic variation heritable in the progeny, and the genetic advance that can be achieved through selection. The aim of the present study was to estimate the magnitude of genotypic and phenotypic variations, heritability and genetic advance so as to help in developing a basis to select superior genotypes from among the rice genotypes belonging to three land situation *viz.* medium, shallow and low land situations.

Materials and Methods

All the 45 varieties were grouped according to their agro ecological adaptability. Group-l included 15 rainfed medium land rice genotypes (IR-8, IR-30, IR-64, Jaya, Ratna, T-N-1, CR-44-1, Vijaya. Pusa-2-21, Satika, Dular, Kunti, NC-1626, Latisail, Laldhan). Group-II was represented by 15 shallow deep (up to 30 cm) types (Aditya, Bipasa, Pankaj, Marichbuti, Swarnamasuri, Neijersail, Mugibalam, Lankagar, Saktigar, Suakalma, Kalamkati, Nunaus, Kalonunia, Patnai, Gobindabhog) and group-III had 15 low land (up to 70 cm) cultivars (Dinesh, Jaladhi-1, Jitendra, Matta, Sabita, Nagra, Daharnagra, Dudkalma, Champakushi, Hatipinjra, Rasapinjar, NC-678, NC-1281, Kumargore, Hansraj). The genetic parameters of different characters were analysed in each group separately to know the influence of adaptation to different agro-ecological conditions.

The experiment was laid out in a RBD with three replications during wet season of 2000 at RRSS, Chakdah, BCKV. 21 to 22 days old seedlings were transplanted as one seedling per hill. All the genotypes were grown in 1.5 m long rows keeping 15 cm × 10 cm spacing. There were 5 rows/variety/replication with 16 plants in each row. Recommended intercultural practices were followed. 10 random plants/variety/replication were taken for recording observations on 9 morphoeconomic characters *viz.*, plant height (cm), effective tillers/plant, panicle length (cm), individual panicle weight (g), flag leaf area (FLA, cm^{-2}) spikelets/panicle, filled grains/panicle, high density grains/panicle (HDG), grain yield/plant (g), and 9 grain traits *viz.* grain length, breadth, thickness; kernel length, breadth, thickness (mm); 1000-kernel weight (g), 100-embryo weight (mg) and 1000-grain weight (g), HDG are those which remain submerged in a solution having specific gravity > 1.20.

The coefficients of variation and heritability parameters (broad sense) were estimated (Burton, 1952) and the expected genetic advance (at 5 per cent selection intensity) was also calculated (Johnson *et al.*, 1955).

Results and Discussion

The experiment result (Table 10.1) showed the significant differences among the genotypes in respect of all the characters in different land situations. Group-III exhibited high mean value in most of the characters. The range of different traits was higher in Group-I followed by Group-II and Group-III.

Pattern of Variability

The coefficient of variability (CV) indicated that tillers/plant, panicle weight, FLA, HDG/panicle, grain yield/plant and embryo weight varied substantially in three ecogroups, and the magnitude of HDG/panicle was higher in Group-III while for other five characters it was higher in Group-I. Plant height, spikelets/panicle, filled grain/panicle, grain breadth and kernel breadth showed appreciable differences in three groups, otherwise, the extent of variation for remaining traits was similar. The lowest variation was recorded in group-I and group-II for grain length and group-II for plant height.

Table 10.1

Characters	*Range*	*Mean*	*F' value*	*Coefficient of Variability (per cent)*
Plant Height (cm)				
Group–I	78.87–110.47	98.33	4.32**	6.83
Group–II	107.07–159.00	138.56	29.72**	3.67
Group–III	150.43–183.50	166.9	4.92**	4.62
Tillers/plant				
Group–I	6.67–12.17	9.49	3.49**	13.41
Group–II	7.40–11.90	9.85	4.27**	11.26
Group–III	6.50–11.33	9.1	3.57**	11.37
Panicle length (cm)				
Group–I	16.53–29.27	25.82	13.77**	5.27
Group–II	19.67–28.53	24.48	4.60**	6.93
Group–III	22.87–28.73	25.69	3.95**	6.29
Panicle weight (g)				
Group–I	1.22–3.79	2.76	6.29**	22.17
Group–II	1.19–3.76	2.93	10.28**	16.19
Group–III	1.66–4.8	4.15	6.96**	13.46
Flag leaf area (cm^2)				
Group–I	9.42–32.91	18.79	12.87**	15.09
Group–II	13.74–45.07	24.38	23.04**	9.58
Group–III	15.60–46.22	29.26	14.44**	11.8
Spikelets/panicle				
Group–I	56.33–226.33	128.9	22.76**	9.94
Group–II	66.66–203.00	138.14	22.89**	9.79
Group–III	113.00–180.67	144.22	12.84**	14.62
Filled grains/panicle				
Group–I	42.10–115.07	80.77	8.62**	10.08
Group–II	62.40–92.13	73.8	5.14**	7.91
Group–III	53.27–79.80	71.03	5.84**	7.83
HDG/panicle				
Group–I	15.93–52.67	35.23	14.10**	10.79
Group–II	20.67–56.83	32.35	11.49**	10.36
Group–III	23.10–52.33	36.38	8.12**	15.09

Contd...

Table 10.1–Contd...

Characters	Range	Mean	F' value	Coefficient of Variability (per cent)
Grain yield/plant (g)				
Group–I	15.40–43.97	26.86	9.55**	15.64
Group–II	17.46–40.40	25.23	7.40**	12.65
Group–III	12.32–38.90	28.3	8.67**	12.32
Grain length (mm)				
Group–I	6.18–10.16	8.45	14.88**	4.72
Group–II	6.22–10.47	8.41	21.33**	4.09
Group–III	6.77–10.50	9.11	20.59**	4.53
Grain breadth (mm)				
Group–I	2.26–3.10	2.66	2.33**	11.28
Group–II	1.87–3.45	2.71	7.8**	8.25
Group–III	2.23–3.64	2.92	9.5**	6.85
Kernel length (mm)				
Group–I	4.73–7.13	6.24	12.62**	5.78
Group–II	4.36–8.64	6.36	20.27**	5.21
Group–III	4.56–9.59	6.92	17.22**	6.93
100–kernel weight (g)				
Group–I	9.42–25.05	18.58	12.62**	7.74
Group–II	9.37–24.25	18.25	36.77**	7.49
Group–III	13.63–30.84	21.89	19.93**	7.23
100–embryo weight (mg)				
Group–I	11.43–44.33	28.69	11.07**	12.9
Group–II	14.33–54.00	35.12	7.64**	10.44
Group–III	23.33–41.33	35.71	12.53**	11.03
100–grain weight (g)				
Group–I	11.21–28.63	22.61	10.07**	6.33
Group–II	9.63–30.48	22.19	55.51**	5.99
Group–III	18.16–36.48	27.24	45.76**	5.58

Group–I: Medium land; Group–II: Shallow deep; Group–III: Low land.

Significant at 1 per cent level.

Heritability and Genetic advance for 16 characters in Rice.

Coefficient of variation of group-II and group-III was nearly same in magnitude for tillers/plant, filled grains/panicle, filled grains/panicle and yield/plant and the same pattern was also recorded in Group-I and Group-II for spikelets and HDG per panicle. The results also revealed that CV was maximum in Group-I for majority of the characters followed by Group-III and Group-II. Further. majority of the morphoeconomic traits of three ecogroups demonstrated higher CV, while grain traits had only in a few cases. Again, group-II and Group-III exhibited poor CV in majority of grain characters. The results suggested that grain characters of these two groups were not much influenced by ecosystem (Chauhan *et al.*, 1993; 1995).

Coefficient of Variability

In all cases, phenotypic coefficient of variability (PCV) was higher than genotypic coefficient of variability (GCV). This indicated that the characters under study had reacted with the environment to some extent.

Environmental functions were high for the characters, panicle length, panicle weight, FLA, spikelets/panicle, filled grains/panicle, grain yield/plant and embryo weight of group-I; plant height, HDG/panicle, grain breadth, kernel weight and grain weight of group-II and tillers/plant, grains length, kernel length and kernel breadth of group-III. However, tillers/plant and grain length of group-I; plant height, grain length, kernel length, breadth, and grain weight of group-II and grain weight and kernel weight of group-III showed little difference between PCV and GCV. It pointed out that these characters were not much influenced by the environment. Therefore, selection based on phenotypic values of such stable attributes is expected to be better (Chauhan, *et al.*, 1993; Rao *et al.*, 1996; Gupta *et al.*, 1999).

Heritability and Genetic Advance

Out of 16 characters high heritability in broad sense was estimated in 13, 12 and 9 characters in group-I, group-II and group-III respectively. It was also noted all the 7 grain traits and only two morphoeconomic traits (FLA and HDG/panicle) gave higher magnitude of heritability in three groups. Highest heritability was recorded in the varieties of group-III for HDG/panicle, grain length, kernel length, grain breadth and kernel breadth, and in group-II for FLA, embryo weight, kernel and grain weight. The genetic advance (GA) at 5 per cent selection intensity was highest for embryo weight in three groups, and it was maximum in group-I Majority ofthe morphoeconomic traits and only three grain traits (kernel weight, embryo weight and grain weight) demonstrated high GA in three groups. Varieties of group-II showed highest GA in seven characters followed by group-I. GA expressed, as percentage of population mean was high in group-I for spikelets/panicle, filled grains/panicle, yield/plant, kernel and embryo weight followed by group-II, and group-III, and was also high in group-II for plant height, FLA and grain weight followed by group-l and group-II.

Higher values of heritability (broad sense) are advantageous when improvement is sought through phenotypic selection. Johnson *et al.* (1955) suggested that heritability estimates along with genetic advance would be more useful in predicting the grain under phenotypic selection than heritability estimate alone. Ramanujam and Tirumalachar (1967) advocated reliability of heritability accompanied with genetic advance as the broad sense heritability estimates include both additive and epistatic gene effects. The result of the present study exhibited high heritability coupled with high GA in three groups for FLA, HDG/panicle, weight of kernel, embryo and grain. These five common traits of three ecogroups should be considered for effective selection. Further, spikelets and filled grains per panicle of group-I and group-II should also be taken for consideration during phenotypic selection as they showed high heritability and high GA (Angrish *et al.*, 1992; Chaubey *et al.*, 1994; Reddy *et al.*, 1997; Sarvanan *et al.*, 1997).

Low heritability accompanied with high GA was recorded in plant height of group-I, in grain yield/plant of group-II and in plant height, spikelets and filled grains per panicle and grain yield/

plant of group-III. It revealed that additive gene effects governed these characters. The low heritability was being exhibited due to high environmental effects (Pandey *et al.*, 1992; Selvarani, 1997).

Grain length and breadth, kernel length and breadth of three ecogroups demonstrated non-additive gene effects as they had high heritability with low GA. High heritability was due to favorable influence of environment rather than genotype. Varieties of three ecogroups showed both poor heritability and poor GA for tillers/plant and panicle weight. This pattern was also noted in group-II and group-III for panicle length. This indicated that these characters were highly influenced by the environmental effect and hence selection should be ineffective.

The result also revealed that FLA, spikelets/panicle, filled grains/panicle, yield/plant and embryo weight of group-I; plant height, FLA, HDG/panicle, grain breadth, kernel and grain weight of group-II and grain length, kernel length and breadth, and HDG/panicle of group-III showed high GCV, heritability as well as GA. According to Panse (1957) this association would indicate that additive gene effect is probably more important in the inheritance of these traits under three ecogroups. The above result clearly suggested that varieties of three ecogroups showed different estimates and combinations of variations, heritability and GA values. Therefore, emphasis should be given on ecogroups rather than random selection during breeding work.

References

Angrish, R. and Panwar, D.V.S., 1992. High density grain variability in some traditional rice cultures. *Oryza*, 29: 59–60.

Burton, G.W., 1952. Quantitative inheritance in grasses. In: *Proc. 6th Int. Grassland Cong.*, 1: 227–283.

Chaubey, R.K. and Sing, R.P., 1994. Genetic variability, correlation and path analysis of yield components in rice. *Madras Agric. J.*, 81(9): 468–470.

Chauhan, J.S., Chauhan, V.S. and Lodh, S.B., 1995. Comparative analysis of variability and correlations between quality components in traditional rainfed upland and low land rice. *Indian J. Genet.*, 55(1): 6–12.

Chauhan, J.S., Chauhan, V.S and Vrikar, M., 1993. Genetic variation and character association in rainfed upland rice. *Oryza*, 30: 116–119.

Chauhan, V.S., Chauhan, J.S. and Tandon, J.P., 1993. Genetic analysis of grain number, grain weight and grain yield in rice (*O. sativa* L.). *Indian J. Genet.*, 53(3): 261–263.

Gupta, A., Sharma, R.K., Mani, V.P and Chauhan, V.S., 1999. Variability and association analysis for grain yield and its components in hill rice. *J. Hill Res.*, 12(2): 99–101.

Johnson, H.W., Robinson, H.F. and Comstock, R.E., 1955. Genetic and environmental heritability in soybean. *Agron. J.*, 47: 314–318.

Pandey, M.P., Seshu, O.V. and Akbar, M., 1992. Genetic variation and association of embryo size to rice seed and seedling vigour. *Indian J. Genet.*, 52: 310–320

Panse, V.G., 1957. Genetics of quantitative characters in rotation to plant breeding. *Indian J. Genet.*, 17: 318–328.

Ramanujam, S. and Tirumalachar, D.K., 1967. Genetic variability of certain characters in red piper (*Capsicum annum*). *Mysore J. Agric. Sci.*, 1: 30–36.

Rao, T.P., Gomathinayagam, P. and Soundrapandian, G., 1996. Genetic variability and character association studies in semi-dry rice. *Madras Agric. J.*, 83(3): 185–188.

Reddy, J.N., Pani, D. and Roy, J.K., 1997. Variability and character association in low land rice. *Indian Agric.*, 41(3): 159–165.

Saravanan, R. and Senthil, N., 1997. Genotypic and phenotypic variation, heritability and GA in some important traits in rice. *Madras Agric. J.*, 84(5): 276–277.

Selvarani, M. and Rangaswamy, P., 1997. Variability studies for physiological traits-I. Yield components in rice. *Madras Agric. J.*, 84(1): 44–45.

Chapter 11

Alterations in Major Nutrient Molecules of Albino Mice Under Monocrotophos Stress

P. Madhaveelatha[1], K. Sankaraiyah[1], R. Jayakumar[2] and K. Jayantha Rao[2]
[1]Department of Zoology, Dr. S.R.K. Government Arts College, Yanam – 533 464, UT of Puducherry
[2]Department of Zoology, S.V. University, Tirupati – 517 502

ABSTRACT

Major nutrient molecules such as carbohydrates, proteins, free amino acids and lipids were estimated in brain, liver, kidney and intestine tissues of albino mice exposed to single and multiple doses of monocrotophos. There was a decrease in carbohydrates, proteins and lipids concentration and raise in free amino acids in all the tissues after exposure. The change in the major nutrient molecules content showed dose dependency and this might be an adjustment of body's metabolic activities to over come the stress condition.

Keywords: *Carbohydrates, Proteins, Free amino acids, Lipids and Monocrotophos.*

Introduction

Since pesticides impose stress condition in exposed animals, they may induce change in the major nutrient molecules such as carbohydrates, proteins amino acids and total lipids level. Elumalai, *et al.* (1999) reported decreased free sugar levels in rats exposed to monocrotophos, Sharma, *et al.*

(2000) noticed gluconeogenesis and increased energy demand in rats. Sublethal exposure of monocrotophos and fenvalerate mixture induced a decrease in total carbohydrates level in *Labeo rohita* (Tilak, *et al.*, 2001). Monocrotophos treated albino rats showed decreased protein content and increased free amino acid levels (Elumalai, *et al.*, 1999).

In the present investigation, an attempt was made to assess the toxic effect of sub-lethal concentration of monocrotophos on major nutrient molecules in different tissues of albino mice after exposure to single and multiple doses.

Materials and Methods

The animals were obtained from Indian Institute of Science, Bangalore and mice colony was maintained in the laboratory at 27±2°C with 12 h light and 12 h darkness. Mice were fed on standard diet and water supplied *ad libitum*. As the acute LD_{50} value of monocrotophos through intra peritoneal injection was 11.2 mg/kg body weight (Madhaveelatha *et al.*, 2007) sub lethal concentration (1/10th of LD_{50}) was administered for the present study. Healthy animals weighing 30±2 gm were divided into 3 groups. To the first set of animals single dose was administered on first day and to the second group four multiple doses were administered on alternate days (*i.e.*, on 1st day, 3rd day, 5th day and 7th day). The animals of third set served as corresponding controls. The control and treated animals were sacrificed on the 9th day by cervical dislocation and the tissues were isolated and processed immediately for study. Carrol *et al.* (1956), Lowry *et al.* (1951), More and Stein (1954) and Folch *et al.* (1957) methods were used to estimate carbohydrates, proteins, tree amino acids, and total lipids.

Results and Discussion

The current investigation implies great dose dependent decline in carbohydrates, proteins, lipids and raise in free amino acids of all tissues in experimental animals.

The significant decline in carbohydrates in all tissues may be due to higher energy demand under monocrotophos stress and their decreased rate of gluconeogenesis (Sharma, *et al.*, 2000) as a consequence of toxic stress. Carbohydrates that are convertible (or related) to glycogen might be trapped extensively to meet the energy demand under insecticide stress (Sivaprasada Rao, 1980). The decrease may also be due to stimulation of hormones that accelerates carbohydrate utilization or inhibition of other factors, which contribute to the carbohydrate synthesis under monocrotophos intoxication. The pesticides in general cause either anoxia or hypoxia in non-target organisms, with a damage in respiratory system (Sangeeta *et al.*, 2002). Since energy synthesis from glycolysis in anaerobic conditions is very less, lots of carbohydrate molecules might be utilized leading to their depletion. Massicotte *et al.* (2005) efforts suggest that mitochondria are important early target for OP compounds, and exposure to them results in diminution of ATP production and in this condition more carbohydrates might be consumed to meet the energy demand. Exertion of Rajamannar and Manohar (1998); Elumalai *et al.* (1999); Sharma *et al.* (2000) and Tilak *et al.* (2001) with Mcp on different tissues of different animals divulge the same trend.

A significant shrink of total proteins and boost in total free amino acids was observed in all the tissues of experimental mice. The results designate an adjustment in protein metabolism under Mcp stress. Since energy production was reduced in hypoxic conditions created by the pesticide, proteins might be utilized to combat the crisis or their synthesis might be reduced leading to low protein levels. The decrease in protein may be due to their deceased synthesis of break down into amino acids, leading to increased amino acid levels in different tissues. Elumalai *et al.* (1999) observed similar change in muscle and kidney of Mcp treated albino rats. Experiments of several workers (Richardson,

1981; Zaid *et al.*, 1990; Rajamannar and Manohar, 1998; Das and Mukherzee, 2000) with different pesticides on different animals correlate with the present annotations. The rise in free amino acids level should lead to increased synthesis of proteins or protein turnover, but the decrease in protein indicates animal's metabolic adaptation to use the tree amino acid pool as a feasible source of energy, to meet the energy requirement under stress conditions.

An increase in the tree amino acid content in the tissues indicates metabolic adaptation to overcome the effect of monocrotophos toxicity. It appears that free amino acid pool may form a possible source of energy to meet the energy requirement. The elevation in amino acids may be due to active transmination and decreased utilization of amino acids from other sources like glucose and fatty acids. The animals under xenobiotic stress may elevate their amino acid levels to maintain osmoregulation (Sambasiva Rao, 1984) or the damage of tissue caused a raise in amino acids (Dikshith, *et al.*, 1975; Dubale and Shah, 1979; Jayantha Rao *et al.*, 1983). The predominant changes in liver may be due to its active role in metabolizing the xenobiotic. The present observations are in accordance with the findings of Zaid *et al.* (1990), Rajamannar and Manohar (1998), Elumalai *et al.* (1999), Das and Mukherzee (2000).

A significant decline in total lipid content in the present cram suggests lipolytic action of monocrotophos. Bhatia and Venkatasubramanian (1972), Peter *et al.* (1973) advocated that pesticides affect the lipids. Besides carbohydrates, lipids might have been channeled to meet the metabolic demand for extra energy that is needed to mitigate the toxic stress. As protein synthesis was reduced in stress condition the reduction in fatty acid synthesizing enzymes might resulted in low lipid levels in the tissue of experimental animals. The high energy demand, low carbohydrate levels and reduced energy production might resulted an increase in lyses of reserves fats to meet the energy requirement leading to decrease in lipid content of different tissues of experimental animals. Reduced cholesterol levels, triglycerides, phospholipids and increased lipid peroxidation might also cause the decrease in total lipid content of experimental animals. Investigations of Sivaprasada and Rao (1981 in the parallel field rope the present analysis.

References

Bhatia, S.C. and Subramanian, T.A Venkata, 1972. Mechanism of dieldrin induced fat accumulation in rat liver. *J. Ag. Food. Chem.*, 20: 993–99.

Carrol, M.V., Langley, R.W. and Row, J.M., 1956. Glycogen determination in liver and muscle by use of anthrone reagent. *J. Biol. Chem.*, 22: 583–593.

Das, B.K. and Mukherjee, S.C., 2000. Chronic toxic effects of quinolphos on some biochemical parameters in *Labeo rohita* (ham). *Toxicol. Lett.*, 119(1–3): 11–18.

Dikshith, T.S.S., Behari, J.R., Datta, K.K., and Mathur, A.K., 1975. Effect of diazinan in male rats, histopathological and biochemical studies. *Environ. Physiol. Biochem.*, 5: 293–299.

Dubale, M.S. and Shah, P., 1979. Histopathological lesions induced by malathion in the liver of *Channa punctatus*. *Ind. J. Exp. Biol.*, 17(7): 693–697.

Elumalai, M., Jayakumar, R. and Balasubramanian, M.P., 1999. Impact of Monocrotophos on protein and carbohydrate metabolism in different tissues of albino rat. *Cytobios*, 98(389): 131–136.

Folch, J., Lees, M. and Sloane G.M. Slanley, 1957. A simple method for the isolation and purification of total lipids from animal tissues. *J. Biol. Chem.*, 226: 497–509.

Jayantha Rao, K., Madhu, Ch. and Murthy, V.S.R., 1983. Histopathology of malathion effect on gills of freshwater teleost *Tilapia mossambica* (Peters.). *J. Environ. Biol.*, 4(1): 19–23.

Lowry, O.H., Rosebrough, N.J., Farr, A.L and Randall, R.J., 1951. Protein measurement with Folin-phenol reagent. *J. Biol. Chem.*, 193: 265–275.

Madhaveelatha, P., John Sushma, N., Sankaraiyah, K. and K. Jayantha Rao, 2006. Effect of monocrotophos on some enzymes of albino mice. *J. Natcon*, 18(2): 397–401.

Massicotte, C., Knight, K., Van der Schyf, C.J., Jortner, B.S. and Ehrich, M., 2005. Effects of organophosphorus compounds on ATP production and mitochondrial integrity in cultured cells. *Neurotox Res.*, 7(3): 203–217.

Moores and Stein, W.H., 1954. A modified ninhydrin reagent for the photometric determination of amino acids and related compounds. *J. Biol. Chem.*, 221: 907.

Rajamannar, K. and Manohar, L., 1998. Sublethal toxicity of certain pesticides on carbohydrates, proteins and amino acids in *Labeo rohita. J. Ecobiol.*, 10(3): 185–191.

Richardson, R.J., 1981. Toxicology of the nervous system. In: *Toxicology Principles and Practice*, (Ed.) A.L. Reeves. Wiley Interscience Pub., Toranto.

Sambasiva Rao, K.B.S, 1984. Toxic impact of carboryl-1 and phenthoate on some metabolic aspects of a murrel [*Channa punctatus* (Block)]: A synergistic study. *Ph.D. Thesis*, S.V. University, Tirupati, India.

Sangeeta, Handa, S.M. and Mittal, P.K., 2002. SEM study on cytotoxic effect of monocrotophos (MCP) on lungs of rat. *Indian J. Exp. Biol.*, 40(7): 835–838.

Sharma, Neeta Raj, Kushwah, Ameeta, and Kushwah, B.S., 2000. Effect of dietary protein against monocrotophos toxicity. *J. Environ. Sci. Health, B.* 2004; 39(5–6): 805–818.

Sivaprasada Rao, K., 1980. Studies on some aspects of metabolic changes with emphasis on carbohydrate utilization of cell free systems of the freshwater teleost *Tilapia mosambica* (peters) subject to methylparathion exposure. *Ph.D. Thesis*, S.V. University, Tirupati, India.

Sivaprasad, K.S.R and Rao, K.V.R., 1981. Lipid derivative in the tissue of the freshwater teleost *Tilapia mosambica* (Peters.): Effect of methyl parathion. *Ind. Nat. Sci. Acad.*, B.47.

Tilak, K.S., Veeraiah, K., Lakshmi and Vijaya, 2001. Biochemical changes induced in freshwater fish *Labeo rohita* (Hamilton), exposed to pesticide mixture. *Asian Journal of Microbiology, Biotechnology and Environmental Sciences*, 3(4): 315–319.

Zaid, S.A., Singh, S., Palni, L. and Singh, V.S., 1990. Biochemical alterations in the levels of DNA, RNA and protein in discrete areas of rat brain following nuvacron toxicity. *J. Pak Med. Assoc.*, 40(11): 261–263.

Chapter 12

Non-keratinizing Carcinoma in the Sputum of Occupational Workers (Stone Crusher Mill Workers) as an Indicator of Health Hazard

Puspanjali Parida
P.G. Department of Zoology, North Orissa University, Baripada – 757 003

ABSTRACT

Sputum samples were cytologically analyzed from 30 persons who are not exposed to stone dust containing silica as stone crusher control group (SCCG) against 120 workers of stone crusher mill exposed to stone dust containing silica as stone crusher experimental group (SCEG). Cytological analysis of sputum showed that SCCG and most of SCEG did not exhibit any abnormal cytological features. However, only 3 (2.5 per cent) persons from SCEG showed non-keratinizing carcinoma in their sputum including stone dust (pollution) hazards in work environment.

Keywords: *Sputum, Stone crusher, Non-keratinizing carcinoma, Silica.*

Introduction

The global incidence of lung cancer is increasing at a rate of 0.5 per cent per year and as a consequence, lung cancer is the leading cause of cancer mortality in most countries (Kurihara *et al.*, 1989; Magrath and Litvak, 1993). In many occupations, a large number of workers work in a polluted dusty environment. The deposition of this dust or respirable particles in respiratory tract is a function of its aerodynomic diameter of a sphere of unit density having the same terminal setting velocity as the particle itself (Jones, 1993). According to Steenland Stayner (1997), silica is toxic to pulmonary macrophages which engulf silica particles, a variety of chemotactic and toxic substances released from lysed macrophages result in collagen production which causes fibrosis or silicosis. In addition to silicosis, other diseases contributed by silica are bronchitis, tuberculosis, scleroderma, rheumatoid arthritis, renal diseases and emphysema (Shi *et al.*, 1998).

For building and road construction, usually granite stone containing silica are used as chief raw materials, processed in the stone crusher mills. During operation of the mill stone dusts containing silica are liberated. The present work in designed to know whether the occupational workers of stone crushers mills (starting from Bhubaneswar on both sides of National Highway Number 5 up to Jankia in Khurda district of Orissa) exposed to silica are prone to respiratory malignancy or not.

Materials and Methods

The study population is divided in to 2 groups: (*i*) Control group (SSSG) and (*ii*) Experimental group (SCEG). Control group consisted of 30 persons (15 women and 15 Men with age between 12–50 years and all were non-smokers) who are not living in close vicinity of stone crusher mills but far away about 100 km distance from stone crusher mills with almost parallel socio-economic status as that of workers in stone crusher mill were taken as stone crusher control group (SCCG) (Table 12.1).

Table 12.1: Sputum Samples Cytologically Screened from the Individuals not Exposed to Crystalline Silica (SCCG) Showing their Sex and Age Group

No. of Individuals (n = 30)	*Male*	*Female*	*Age Group (in Years)*
3	2	1	12–22
12	6	6	23–33
8	3	5	34–44
7	4	3	45–50

Experimental group consisted of 120 workers (n = 120) including 61 women belonging to age group 12–50 years in 12 difference stone crusher mills, working about a period of 5-20 years constitute as stone crusher experimental group (SCEG) (Table 12.2).

Table 12.2: Sputum Samples Cytological Screened from the Workers of Stone Crusher Mills (SCEG) Showing their Sex, Age Group and Occupational Exposure Period to Crystalline Silica

No. of Workers (n = 120)	*Male*	*Female*	*Age Group (in years)*	*Exposure Periods (in years)*
30	12	18	12–22	5
54	29	25	23–33	10
28	13	15	34–44	15
08	5	3	45–50	20

Each person (both SCCG and SCEG) is requested to answer certain question in a specific format of data sheet which covers following questions: (*i*) Medical questionnaire: covering each person as to his/her personal background especially respiratory conditions, smoking habits, cough, phlegm, shoreness of breath, sneezing and chest tightness; (*ii*) occupational questionnaire: include history of employment in current and past occupational activities; (*iii*) Life style questionnaire: including smoking, alcohol drinking, diet, parity, marital status and economic status.

In the early morning after leaving the bed each person (both control and experimental group) is requested to brush their teeth and gargle with saline solution. Sputum was collected in screw capped, properly labeled, wide mouthed bottle, half filled with 70 per cent ethanol. The bottle was shaken briskly for a few second to disperse the mucus threads and cells for proper fixation.

As soon as possible a thin layer of sputum was drawn on a clean, dried and labeled glass slide and was fixed in 95 per cent ethanol and stained with Papanicolau's staining procedure (Papanicolau, 1954) with little modification the stained slides were observed under research microscope.

Mineralogical and Chemical Analysis of Stone

For stone crusher mill, the stones are obtained from the Eastern Ghat groups of rock and are used in the mills by blasting the rocks (quarries) with the help of dynamite (explosive), there are about a dozen of rock-quarries. The stone from different rock quarries were obtained for mineralogical (thin section) and chemical analysis (SiO_2 percentage only) to know constituent of the stone and percentage of silica in the stone.

When studied under petrological microscope (mineralogical analysis) the stone gives a granular texture and are composed of orthoclase, quartz, hypersthene, plagioclase, perthites, and magnetite. The SiO_2 percentage (chemical analysis) in these stone is 53.44 per cent, due to the above features, the stone is identified to be "hypersthene granite" or "intermediate charnokite".

Results and Discussion

In control group (SCCG), sputum from 30 individuals were analyzed, 21 (70 per cent) individual exhibited sqamous cells were not seen in sputum of rest 9 (30 per cent) individuals. So also ciliated columnar cells, goblet cells, pulmonary histiocytes, inflammatory, cells and different contaminates were seen from 15 (50 per cent), 3 (10 per cent), 6 (20 per cent), 6 (20 per cent), 3 (10 per cent) individuals respectively Table 12.3.

In experimental group (SCEG), sputum from 120 workers were analyzed. Most of the samples showed similar cytological features as the above control group (SCEG). Besides that squamous metaplasis, 3 (2.5 per cent) basal cell hyperplasisa, crystals were seen in sputum samples 8 (6.7 per cent), 10 (8.3 per cent), 76 (63.3 per cent) workers respectively (Table 12.4). However, only 3 (2.5 per cent) male workers from SCEG showed non-keratinizing Adenocarcinoma.

Table 12.3: Different Cytological Features in Sputum Samples of 30 Individuals in Control Group (SCCG)

	Total No. of Individuals	*Female*	*Male*
Squamous cell	21 (70 per cent)	12 (40 per cent)	9 (30 per cent)
Ciliated columnar cell	15 (50 per cent)	8 (26.6 per cent)	7 (23.4 per cent)
Goblet cell	3 (10 per cent)	3 (10 per cent)	0 (0 per cent)
Pulmonary histiocytes	6 (20 per cent)	1 (3.3 per cent)	5 (16.7 per cent)
Inflammatory cells	6 (20 per cent)	2 (6.7 per cent)	4 (13.3 per cent)
Contaminates	3 (10 per cent)	3 (10 per cent)	0 (0 per cent)

Table 12.4: Number of Individuals Showing Different Characteristics of Cells, Acellular Structures and Contaminates in Sputum of 120 Workers in Experimental Group (SCEG)

	No of Workers	*Female*	*Male*
Squamous mataplasia	8 (6.7 per cent)	3 (2.5 per cent)	5 (4.2 per cent)
Basal cell hyperplasia	10 (8.3 per cent)	3 (2.5 per cent)	7 (5.8 per cent)
Squamous cell Carcinoma	4 (3.3 per cent)	0 (0.0 per cent)	4 (3.3 per cent)
Parasites, bacteria, fungi	32 (26.7 per cent)	18 (15 per cent)	14 (11.7 per cent)
Other contaminates crystals	76 (63.3 per cent)	41 (34.2 per cent)	35 (29.1 per cent)

Table 12.5: Data Obtained from Questionnaire Regarding Life Style Pattern

	SCCG (n = 30)	*SCEG (n = 120)*		
		No of Workers	*Female*	*Male*
Habit	Alcohol	15 (12.5 per cent)	0 (0 per cent)	15 (12.5 per cent)
	Smoking	49 (40.8 per cent)	8 (6.7 per cent)	41 (34.1 per cent)
	Both alcohol and smoking	8 (6.7 per cent)	0 (0 per cent)	8 (6.7 per cent)
Bronchial problem	During work	21 (17.4 per cent)	9 (7.5 per cent)	12 (1 0 per cent)
	After work	31 (25.8 per cent)	8 (67 per cent)	23 (19.1 per cent)
	Periodic medical check-up	0 (0 per cent)	0 (0 per cent)	0 (0 per cent)
	Use of mask while working	87 (72.5 per cent)	57 (47.5 per cent)	30 (25 per cent)

Chemical analysis showed silicon dioxide percentage in stone obtained from the Eastern Ghat group of rocks for crusher mill is 53.44 per cent, the data obtained through questionnaire from workers relating to habit, bronchial problem, periodic medical check-up, use of mask while working are summarised in Table 12.5. It is also found that in case of SCEG, 21 (17.4 per cent) individuals and bronchial problem during work. While 32 (25.8 per cent) had such after work. All individuals of SCCG are non-smokers and not addicted to alcohol while 15 (12.5 per cent), 49 (40.8 per cent), and 8 (6.7 per cent) workers of SCEG are alcoholics, smokers and both alcoholics and smokers respectively.

The above finding showed that sputum samples of all individuals from SCCG and majority of workers (96.7 per cent) from SCEG showed normal cytological features while male workers showed 3 (25 per cent) non-keratinizing carcinoma, which ultimately leads to long cancer in due course of time may be due to percentage of silica in stone, exposure period, age group, smoking habit and other factors.

There are no effective screening tests or early detection procedures for cancer, and treatment is generally not effective, even when they are detected in early stages. Primary prevention is therefore the only strategy to reduce mortality from cancer by reducing the occupational exposure through use of mask and education of workers as well as employers regarding the health hazards to stone dust and governmental legislation to ban stone crusher mills in close vicinity of human habitats.

References

Jones, A.D., 1993. *Respirable Industrial Fibers and Treatment*. Orient Paper Backs, New Delhi.

Kurihara, M., Aok, K. and Hisamichi, S., 1989. *Cancer Mortality Statistics in the World, 1950–1985*. Japan, Nagoya, University Press, Nagoya.

Magrath, I. and Litvack, J., 1993. Cancer in developing countries: opportunity and challenge. *JNCI*, 85: 862.

Papanicolaou, G.N., 1954. *Atlas of Exfoliative Cytology.* Cambridge Harvard Univ. Press, pp. 54–55

She, X., Castranova, V., Halliwell, B. and Vallyathan, V., 1998. Reactive oxygen species and silica induced carcinogenesis. *J. Taxicol. and Env. Hlth.*, 1: 181–197.

Steenland, K. and Stayner, L., 1997. Silica asbestos man-made mineral fibers and cancer causes and control, 8: 491–503.

Chapter 13

Correlation and Path Analysis of Some Economic Traits in Potato (*Solanum tuberosum* L.)

K.G. Santosh, H.R. Bhoomika, N.S. Mavarkar and S.L. Biradar

ABSTRACT

The studies on phenotypic and genotypic correlations and path analysis were conducted in a collection of 42 genotypes. Tuber yield exhibited positive and significant association with all the characters under study. Average tuber weight had negative correlation with most of the characters. Rest of the pairs showed positive correlation both at phenotypic and genotypic levels. Path coefficient analysis indicated that total tuber weight per plant had very high positive direct effect on tuber yield, leaf area index and average tuber weight had a negative influence on yield in general. It is suggested that maximum weightage should be given to number of tubers per plant and total tuber weight per plant for improvement of potato by selection.

Keywords: *Potato, Correlation, Path coefficient analysis.*

Introduction

In potato (*Solanum tuberosum* L.), the tuber yield is a complex polygenic trait and is the product of contributions and interactions of different component characters (Luthra, 2001). Therefore, it is imperative to know the contributory role of component characters on yield while carrying out the selection experiments. The understanding of correlation between yield and its components is necessary to develop a breeding strategy for desired improvement in tuber yield. But correlation study is simply a measure of degree of association between yield and its attributes and does not give the actual dependence of yield on correlated characters. Path analysis is an effective method to determine the direct and indirect association between the yield components and permits to examine the specific forces acting to produce a given correlated effect, thus, the study of character association and path analysis together can give a clear picture about the nature of the associations between yield and its attributes which is essential for formulating selection programme in early stages of selection experiments. Hence, the present study was undertaken to generate information on nature of association between yield and different yield attributes and to workout selection criterion for selecting high yielding genotypes, in potato.

Materials and Methods

Forty two genotypes were grown in randomized block design with three replications during Kharif 2003–04 at the Agricultural Research Station, Madenur, Hassan. In each replication, the plot size was 3 × 2.4 m, which accommodated 60 plants. The standard manurial and cultural practices were followed through out the crop season.

The crop was harvested at 90 days after planting and data were recorded on five randomly selected plants per replication for six characters *viz.*, per cent emergence, leaf area index (LAI), number of tubers per plant, average tuber weight (g), total tuber weight per plant (g) and total tuber yield per hectare (t). The mean values of the above six characters were used for the correlation studies. The correlations were worked out according to the method adopted by Al-Jibouri *et al.* (2000). Genetic correlation coefficients were tested after transformation and T test. The path co-efficient analysis was done as suggested by Dewey and Lu (1959) taking total tuber yield/ha (t) as depending trait.

Results and Discussion

Phenotypic and genotypic correlations among six characters were worked out and are presented in Table 13.1. In the present investigation, except in a few cases, the genotypic correlations were higher than phenotypic correlations, indicating little influence of environment and the presence of inherent association between various characters. Higher magnitude of genotypic correlation was observed earlier also by Srivastava and Srivastava (1976).

Table 13.1: Phenotypic (P) and Genotypic (G) Correlations Between Pairs of Characters

Characters	*P and G*	*Leaf Area Index*	*No. of Tubers/Plant*	*Average Tuber Weight (g)*	*Total Tuber Weight/Plant (g)*	*Total Tuber Yield/ha (t)*
Per cent emergence	P	0.053	0.093	–0.026	0.057	0.328*
	G	0.091	0.490**	–0.078	0.026	0.524**
Leaf area index	P		0.137	0.337*	0.357*	0.353*
	G		0.232	0.547**	0.603**	0.426**
No. of tubers/plant	P			–0.383*	0.353*	0.367*
	G			–0.342*	0.472**	0.726**
Average tuber weight (g)	P				0.604**	0.347*
	G				0.599**	0.400**
Total tuber weight/plant (g)	P					0.674**
	G					0.843**

**: Significant at 1 per cent level; *: Significant at 5 per cent level.

Tuber yield was positively and significantly correlated both at phenotypic and genotypic levels with per cent emergence, leaf area index (LAI), number of tubers per plant, average tuber weight (g), total tuber weight per plant (g). These findings are in conformity with those of Patel *et al.* (1973), Gaur *et al.* (1978) and Pandita *et al.* (1980).

Number of tubers per plant showed positive and significant association with the total tuber weight per plant however, it was negatively and significantly correlated with average tuber weight. Gaur *et al.* (1978) suggested that to increase yield, it is important to break the negative genetic linkage between tuber number and average tuber weight. The character leaf area index showed positive and significant correlation with that of average tuber weight and total tuber weight per plant. These are in agreement with the findings of Pandey and Gupta (1995) in TPS families and Gaur *et al.* (1978), in clones.

Path coefficient values based on genotypic correlation are given in Table 13.2. Only per cent emergence, number of tubers per plant and total tuber weight per plant had a positive direct effect on the yield. However, leaf area index and average tuber weight had negative direct influence even though their correlation coefficients with yield were positive and higher in magnitude than the direct effects. This indicates that correlation was due to the direct and indirect factors, which have to be considered simultaneously for selection (Singh and Chaudhary, 1977). Therefore, direct selection through leaf area index and average tuber weight will not help in improvement of tuber yield. These findings are in agreement with the findings of Desai and Jaimini (1998).

Table 13.2: Genotypic Path Analysis Among Five Important Traits in Potato

Characters	*1*	*2*	*3*	*4*	*5*	*r*
1	0.165	–0.024	0.076	0.005	0.303	0.524**
2	0.015	–0.269	0.036	–0.034	0.678	0.426**
3	0.081	–0.062	0.155	0.021	0.531	0.726**
4	–0.013	–0.147	–0.053	–0.062	0.674	0.400**
5	0.044	–0.162	0.073	–0.037	0.925	0.843**

**: Significant at 1 per cent level R = 0.2332

r: Genotypic correlation values.

1: Per cent emergence at 30 Days after planting

2: Leaf Area Index (LAI)

3: Number of tubers per plant

4: Average tuber weight (g)

5: Total tuber weight per plant (g)

The maximum positive direct effect on the yield was exerted by total tuber weight per plant both at phenotypic and genotypic levels followed by per cent emergence. This indicates the existence of the true relationship between this trait and the total tuber yield and direct selection through this trait will be effective in the improvement of tuber yield. The number of tubers per plant had the lowest positive direct effect followed by the per cent emergence at genotypic level. However, the magnitude of their correlation were higher than the direct effects which again calls for the simultaneous selection.

Conclusion

With the above results it is evident that, the trait total tuber yield/ha is positively and highly correlated with characters like per cent emergence, number of tubers per plant and total tuber weight per plant at both phenotypic and genotypic levels. These traits also exhibited direct, positive effect on the yield at genetic level. Hence, it can be concluded that for getting higher tuber yields in potato, further selection can be based upon these traits *viz.*, per cent emergence, number of tubers per plant and total tuber weight per plant.

References

Al-Jibouri, H.A., Miller, P.A. and Robinson, H.F., 1958. Genotypic and environmental variances in upland cotton crosses of inter-specific origin. *Agron. J.*, 50: 633–636.

Dewey, D.R and Lu, K.H., 1959. A correlation and path coefficient analysis of components of crested-wheat grass seed production. *Agron. J.*, 51: 515–518.

Gaur, P.C., Gupta, P.K., and Kishore, H., 1978. Genetic components of tuber yield and quality characters in Potato. *J. Indian Potato Assoc.*, 5: 32–37.

Luthra, S.K., 2001. Heritability, genetic advance and character association in potato. *J. Indian Potato Assoc.*, 28: 1–3.

Pandita, M.L., Sidhu, A.S. and Arora, S.K., 1980. Variability and interrelationships between yield and yield components in potato. *Haryana J. Agric. Res.*, 10: 424–426.

Pandey, S.K. and Gupta, P.K., 1995. Genetic divergence in some Indian and exotic potato varieties and advanced hybrids. *J. Indian Potato Assoc.*, 22: 38–45.

Patel, O.P.L., Patel, R.M., Mondloi, K.K. and Sikka, L.C., 1973. Genetic variability and correlation studies in potato. *Haryana J. Hort. Res.*, 2: 75.

Singh, R.K. and Chaudhary, B.D., 1977. *Biometrical Methods in Quantitative Genetic Analysis*. Kalyani Publishers, New Delhi.

Srivastava, V.K. and Srivastava, L.S., 1976. Genetic parameters, correlation coefficients and path coefficient analysis in bitter gourd (*Momordica charantia* L.). *Indian J. Hort.*, 33: 66–70.

Chapter 14

Genetic Variability Studies in Potato (*Solanum tuberosum* L.)

K.G. Santosh, K.M. Indiresh, H.R. Bhoomika, S.L. Biradar and N.S. Mavarkar

ABSTRACT

Genetic variability, heritability and genetic advance for 25 various morphological and quality characters were studied in 42 potato genotypes. The genetic variability for tuber yield and other related characters was found to be higher than that for tuber quality characters. The estimates of heritability ranged from 7.40 to 98.70 per cent. In general these estimates for the tuber yield components and quality characters were high. The expected genetic advance was relatively higher for total tuber yield and total marketable yield, than for other characters.

Keywords: *Potato, Genetic variability, Heritability, Genetic advance.*

Introduction

Genetic variability in the breeding material is important for the improvement of a crop plant. Such information. helps in locating suitable parental lines for a breeding programme and also indicates the extent of genetic advance expected by selection, in the subsequent generation. Since, the estimates of genetic variability vary with the population, it was considered necessary to study it in the material available. The present study reports on the genetic variability for 25 morphological and tuber quality traits in 42 potato genotypes.

Materials and Methods

Forty two genotypes were grown in randomized block design with three replications during *Kharif* 2003–04 at Agricultural Research Station, Madenur, Hassan, UAS, Bangalore. In each replication, the plot size was 3 × 2.4 m, which accommodated 60 plants. The observations were recorded on five plants in each replication and their means were calculated. The recommended production practices were followed through out the crop season.

Data on different characters were analyzed on mean values based for experimental design. Analysis of variance for the experimental design was carried out as per the procedure of Cochran and Cox (1959). The genotypic and phenotypic co efficient of variation present in a particular character was estimated by the formula suggested by Singh and Chaudhary (1977). The heritability and genetic advance were computed by using the methodology of Johnson *et al.* (1955).

Results and Discusion

The analysis of variance (Table 14.1) revealed significant differences among genotypes for all the characters studied. Variation was significant for almost all the characters, except for number of branches per plant, total non-marketable yield/ha and number of eyes per tuber.

Table 14.1: Mean Sum of Squares for Yield and Other Ancillary Characters of Potato Genotypes

Character/Source	*Mean Sum of Squares*	
	Genotypes	*Error*
	(df= 41)	*(df= 82)*
Per cent emergence	442.58**	101.17
No. of leaves (30 DAP)	126.95**	72.70
No. of leaves (45 DAP)	173.43**	106.30
Plant height (cm) (45 DAP)	147.57**	19.81
Plant height (cm) (60 DAP)	114.44**	39.87
No. of branches/plant (60 DAP)	0.75	0.40
Leaf area (cm^2) at harvest	3383422.80**	23660.29
Leaf area index (LAI)	2.35**	0.02
No.of tubers/plant	3.88**	1.32
Average tuber weight (g)	1455.01 **	496.81
Average tuber weight/plant (g)	22559.03**	8255.03
Total marketable yield /plot (kg)	49.97**	94.00
Total non-marketable yield/plot (kg)	0.54	0.44
Total tuber yield/plot (kg)	55.97**	6.76
Total tuber yield/ha (t)	107.96**	13.03
Total marketable yield/ha (t)	96.53**	13.07
No. of eyes/tuber	90.50	85.86
Per cent dry matter	30.85**	0.19
Reducing sugars (per cent)	0.17**	0.00

Contd...

Table 14.1–Contd...

Character/Source	Mean Sum of Squares	
	Genotypes	*Error*
	(df= 41)	*(df= 82)*
Non-reducing sugars (per cent)	0.10**	0.00
Total sugars (per cent)	0.48**	0.01
Total starch content (per cent)	6.31 **	0.07
Chlorophyll '*a*' content (mg/g)	0.0001 **	0.00
Chlorophyll '*b*' content (mg/g)	0.0031 **	0.00
Total chlorophyll content (mg/g)	0.0029**	0.00

DAP: Days after planting

**: Significant at 1 per cent level.

*: Significant at 5 per cent level.

The Genotypic co-efficient of variation (GCV per cent) for majority of traits was quite close to the estimates of phenotypic co-efficient of variation (PCV per cent) indicating negligible environmental role and that the genotypic performance appeared to be well adaptive to the fullest phenotypic expression of the traits. High GCV as well as PCV percentage was observed for leaf area at harvest, leaf area index, number of tubers per plant, average tuber weight, average tuber weight per plant and total tuber yield per hectare suggesting greater variability exists for these characters among the varieties due to the genetic causes which are less affected by environment and hence could be improved through selection (Table 14.2).

Heritability estimates for plant height, leaf area, leaf area index (LAI), total marketable yield, total tuber yield, sugar content, dry matter, starch content, total chlorophyll content and early blight incidence were high ranging from 67.90 to 98.90 per cent. These findings are in a consonance with the results reported by Dayal *et al.* (1972), Gaur *et al.* (1978), Birhman and Kaul (1989) and Chaudhary and Sharma (1984).

The attributes like per cent emergence, number of tubers per plant, average tuber weight, late blight incidence and phoma incidence exhibited moderate heritability estimates, meaning that these characters are moderately influenced by environment. These results were in conformity with the earlier results obtained by Gaur *et al.* (1978) and Sood and Singh (1979). Low heritability estimate was observed for number of leaves per plant and number of eyes per tuber, thus limiting the scope of selection for these traits.

Heritability estimates along with genetic gain (genetic advance as per cent mean) is more useful than heritability alone in predicting the resultant effect of selecting the best individuals (Johnson *et al.*, 1955). Genetic advance is the measure of improvement that can be achieved by practicing selection in a population. High heritability with low genetic advance indicates importance of non additive gene action, while high heritability with high genetic advance indicates additive gene effects.

In the present study high genetic advance over mean coupled with high heritability was observed in characters like leaf area, leaf area index (LAI), total tuber yield, total marketable tuber yield, dry matter, sugar content and early blight incidence. However, the estimates were moderate for number of

Table 14.2: Variability, Heritability and Genetic Advance for Growth, Yield and Quality Parameters in 42 Genotypes of potato

Sl.No.	*Character/Source*	*Mean*	*Range*	*GV*	*PV*	*GCV (per cent)*	*PCV (per cent)*	*h^2 (per cent)*	*GAM (per cent)*
1.	Per cent emergence	83.78	43.88–98.44	113.80	214.98	12.73	17.50	52.90	19.09
2.	No. of leaves (30 OAP)	58.00	39.27–70.13	18.08	90.78	7.33	16.43	19.90	6.74
3.	No. of leaves (45 OAP)	77.14	54.87–89.87	32.38	128.68	6.13	14.71	17.40	5.26
4.	Plant height (em) (45 OAP)	48.49	30.87–62.93	42.59	62.40	13.46	16.29	68.30	22.91
5.	Plant height (em) (60 OAP)	68.15	53.87–84.20	24.86	64.72	7.32	11.80	38.40	9.33
6.	No. of branches/plant (60 OAP)	2.52	1.47–3.47	0.17	0.51	13.54	28.47	22.60	13.12
7.	Leaf area (cm^2) at harvest	4230.02	1876.88–5991.01	1119927.50	1143587.79	25.02	25.28	97.90	51.01
8.	Leaf area index (LAI)	3.53	1.56–4.99	0.78	0.79	25.01	25.27	97.90	51.06
9.	No. of tubers/plant	4.95	3.07–7.93	0.85	2.17	18.65	29.75	39.30	24.02
10.	Average tuber weight (g)	74.99	42.65–152.17	319.40	816.21	23.83	38.09	39.10	30.71
11.	Average tuber weight/plant (g)	340.08	171.33–559.67	4768.00	13023.03	39.14	47.49	36.60	25.30
12.	Total marketable yield /plot (kg)	9.69	1.83–18.25	14.40	21.19	20.30	33.56	67.10	65.67
13.	Total non-marketable yield/plot (kg)	1.08	0.33–2.17	0.03	0.47	17.21	63.48	7.40	9.26
14.	Total tuber yield/plot (kg)	10.77	2.17–20.08	16.40	23.16	37.62	44.71	70.80	65.21
15.	Total tuber yield/ha (t)	14.95	3.01–27.89	31.64	44.68	37.62	44.70	70.80	65.20
16.	Total marketable yield/ha (t)	13.46	2.55–25.35	27.82	40.90	39.16	47.48	68.00	66.52
17.	No. of eyes/tuber	11.87	6.60–14.27	1.55	4.94	10.48	18.73	31.38	12.10
18.	Per cent dry matter	17.90	11.82–24.45	10.22	10.41	17.86	18.02	98.20	36.48
19.	Reducing sugars (per cent)	1.24	0.73–1.90	0.55	0.58	18.90	19.46	94.30	37.96
20.	Non-reducing sugars (per cent)	0.91	. 0.46–1.31	0.03	0.04	20.14	20.95	92.40	39.72
21.	Total sugars (per cent)	2.14	1.19–2.76	0.16	0.16	18.48	18.83	96.30	37.30
22.	Total starch content (per cent)	20.61	17.26–25.12	2.08	2.15	7.00	7.11	96.90	14.21
23.	Chlorophyll 'a' content (mg/g)	0.68	0.67–0.69	0.0000032	0.0000035	0.83	0.88	89.60	1.46
24.	Chlorophyll 'b' content (mg/g)	0.56	0.51–0.62	0.0001	0.0001	5.73	5.76	98.70	3.63
25.	Total chlorophyll content (mg/g)	1.25	1.19–1.31	0.000096	0.000099	2.49	2.53	96.40	4.81

GV: Genotypic variance; PV: Phenotypic variance; GCV: Genotypic coefficient of variation; PCV: Phenotypic coefficient of variation; h^2: Heritability; GAM: Genetic advance over mean.

tubers per plant and per cent emergence. Therefore, the higher heritability coupled with moderate to high genetic advance values observed for these traits in the present findings suggest the existing variability among the genotypes with respect to the traits is mainly due to additive type of genes (Panse, 1957). This confirms that these traits would be improved through simple selection methods. Similar findings have been reported by Gaur and Rana (1982) for tuber dry matter content, Desai and J aimini (1997) for tuber number and further supported by the findings of Dayal *et al.* (1972), Pandita and Sidhu (1979), Chaudhary (1985) and Pandita *et al.* (1980).

Higher heritability estimates were accompanied by lower genetic advance (as per cent of the mean) for chlorophyll content. This suggests that selection may not be useful for the improvement of this trait because of the narrow range of phenotypic variation among the genotypes in respect of this character. These findings are in contrary with the earlier findings of Sidhu and Pandita (1979) and Hosamani (1990) who have noticed low genetic advance for dry matter and starch content in potato.

Low heritability accompanied by very low genetic advance has been observed for number of leaves and number of eyes per tuber. Similar report has also been made by Hosamani (1990).

The salient feature of this study was that the wide range of variability observed for majority of the traits such as average tuber weight, dry matter and reducing sugars, which calls attention and consideration towards these traits during selection process for improvement of tuber yield.

References

Birhman, R.K. and Kaul, M.H., 1989. Variability components and genetic parameters in andigena Potatoes. *J. of Genetics and Plant Breeding,* 108: 209–219.

Chaudhary, S.K. and Sharma, S.K., 1984. Genetic variability for yield and its components in potato breeding material. *Indian J. Agric. Sci.,* 54: 941–942.

Chaudhary, S.K., 1985. Variability for some quantitative characters in potato (*Solanum tuberosum* L.). *Crop Improvement,* 12: 197–199.

Cochran, W.G. and Cox, G.M., 1959. *Experimental Designs* 2nd *edn.* John Wiley and Sons, New York, pp. 450–457.

Dayal, T.R., Upadhya, M.D., Malhotra, V.P. and Mehra, K.L., 1972. Heritability and correlation in yield and other qualitative characters in Potato. *Indian J. Agric. Res.,* 42: 464–466.

Desai, N.C. and Jaimini, S.K., 1997. Correlation and path analysis in potato. *J. Indian Potato Assoc.,* 25: 25–29.

Gaur, P.C., Kishore, H. and Gupta, P.K., 1978. Morphological and genetic variability for some quantitative characters in Potato. *J. Indian Potato Assoc.,* 5: 70–75.

Gaur, P.C. and Rana, M,S., 1982. Variation for tuber dry matter in andigena potatoes. *J. Indian Potato Assoc.,* 9: 88–90.

Hosamani, R.M, 1990. Genetic analysis in potato (*Solanum tuberosum* L.) for rainfed conditions on black soils. *M.Sc. Thesis,* University of Agricultural Sciences, College of Agriculture, Dharwad, p. 42.

Johnson, H.W., Robinson, H.F. and Comstock, R.E., 1955. Estimates of genotypic and environmental variances in soy bean. *Agron. J.,* 47: 477– 483.

Pandita, M.L. and Sidhu, A.S., 1979. Stability parameters and genotype environment interaction in Potato. *J. Indian Potato Assoc.,* 6: 52–60.

Pandita, M.L., Sidhu, A.S. and Arora, S.K., 1980. Variability and interrelationships between yield and yield components in potato. *Haryana J. Agri. Res.*, 10: 424–426.

Panse, V.G., 1957. Genetics of quantitative characters in relation to plant breeding. *Indian J. Genetics and Plant Breeding.*, 17: 318–328.

Sidhu, A.S. and Pandita, M.L., 1979. Genetic variability and correlation studies on Potato. *J. Indian Potato Assoc.*, 6: 103–108.

Singh, R.K. and Chaudhary, B.D., 1977. *Biometrical Methods in Quantitative Genetic Analysis*. Kalyani Publishers, New Delhi.

Sood and Singh, P., 1979. Rate of tuberization in different varieties/hybrids of potato. *J. Indian Potato Assoc.*, 6: 24–29.

Chapter 15

Effect of Different Packing Materials on the Shelf Life and Quality of Acid Lime (*Citrus aurantifolia* Swingla.) at Room Temperature

V. Bali Reddy[1], G. Bindu Madhavi[2], V. Chenga Reddy[3], D. Venkata Reddy[1] and B. Srinu[4]

[1]Research Associate, [2]Scientist (Plant Pathology), [3]Senior Scientist (Plant Breeding), Regional Agricultural Research Station (ANGRAU), Lam, Guntur – 522 034, A.P.

[4]Post Graduate Student in Allahabad Agricultural Institute-Deemed University, Allahabad, U.P.

ABSTRACT

An experiment was carried out entitled "Effect of different packaging materials on the shelf life and quality of Acid lime (*Citrus aurentifolia* Swingla) at room temperature" in the Horticulture Laboratory of Allahabad Agricultural Institute-Deemed University, Allahabad during the year 2006. Acid lime fruits of uniform size were stored under ten different treatments and found that packing acid lime fruits with Low Density Polyethene (LDPE) was most effective in preventing the physiological loss in weight, reducing the specific gravity, percentage of juice, diameter of the fruit. Minimum spoilage was observed in fruits packed with LDPE film. LDPE packing was effective in minimizing the rate of increase in TSS, acidity, ascorbic acid and pH of the juice. Rest of the treatments, packing with High Density Polythene (HDPE), polyvinyl chloride (PVC) also had significant effects on storage and also one or other characters studied.

Keywords: *Acid lime, Packing material, Storage, Shelf life, LDPE, HDPE*

Introduction

Acid lime occupies a place of considerable importance in citrus industry world wide. The use of poly-films or polythene bags as a packing material is widely practiced in some Western countries which benefit the grower, retailer and consumer. Studies have shown that pre packing of fresh produce in polythene bags helps in extending the shelf life of fruits, both at room temperature and also under cold storage. Recently plastic films have been made a significant penetration in packing of multifarious food items, along with fruits due to their multiple advantageous properties relating to transparency, moisture proof ness, water resistance, gas barrier, heat, chemical and cold resistance and light interception. Therefore the experiment was conducted to develop an economical and acceptable method of packing material in order to avoid enormous wastage and to improve the marketability of acid lime fruits.

Materials and Methods

The experiment was laid out in a CRD with three replications having ten treatments *viz.*, check (T_0), high density ploythene (T_1), low density polythene (T_2), polyvinyl chloride (T_3), bamboo basket with paper (T_4), bamboo basket with rice straw (T_5), wooden boxes with paper (T_6), wooden boxes with rice straw (T_7), cartoon boxes with paper (T_8) and cartoon boxes with rice straw (T_9). Freshly harvested fruits of uniform size, shape, colour and maturity showing slightly yellowing colour were selected for the study. Three hundred sixty (360) fruits of uniform maturity, similar in size and colour, unblemished, free from diseases and pests were harvested from the tree growing in the orchard. Three hundred and sixty fruits were divided into 10 equal lots and each lot consist of 36 fruits. The fruits were washed and allowed to dry under electric fan and then put in different packaging materials and stored at room temperature. One of these lots were put as check, 36 fruits were divided in to equal lots again and each lot contain 12 fruits were kept for storage of 10, 20 and 30 days, respectively.

Observation of fruits for various parameters were recorded at 10 days interval of storage. The physiological loss in weight was determined by weighing the fruits in each treatment before and after 10 days and the difference in weight was expressed in gram of the original fresh weight of the fruit. Specific gravity was determined by water displacement methods. Juice was extracted by juice extractor and percentage was calculated. Diameter of fruits was measured with Vernier caliperse and expressed in cm. pH of juice was determined with the help of an electronic pH meter. Total soluble solids from fruits were measured with the help of hand refractometer. For determination of titratable acidity and ascorbic acid content the titrimetric method as described by Rangana (1979) was used. Spoilage of fruits were judged on the basis of visual observation and expressed as percentage over total number of fruits. Initial observations on various parameters of fruits were recorded (Table 15.1).

Results and Discussion

Physical Characters

Fruit Weight Loss

The physiological loss in weight of fruits (Table 15.2) during storage may be attributed to the accumulation of CO_2, restricted intake of O_2 from atmosphere, low level of relative humidity of the air in the storage space, and the lower rate of respiration of fruits. It was found that significant losses occurred in all the packaging materials except with LDPE stored at room temperature, with the progress of storage period, the check treatment fruits showed sign of loss in weight and shriveling. This result is in conformity with the findings of previous workers Cohen *et al.* (1978) in Eureka lemon, Sozzi *et al.* (1983) in lemons, Paul (1984) in limes and Singh *et al.* (1984) in malta fruits.

Table 15.1: Initial Readings of Different Characters of Acid Lime Before Treatment for Storage

Sl.No.	*Characters*	*Unit*	*Value*
A.	Physical characters		
1.	Physiological loss in weight	Gram	50.75
2.	Specific gravity of fruits	–	1.61
3.	Percentage of juice content of fruits	Per cent	67.6
4.	Diameter of fruits	cm	4.8 × 4.4
B	Chemical constituents		
1.	Total soluble solids	Brix	7.15
2.	Titrable acidity (citric acid)	Per cent	7.10
3.	pH of juice	–	2.60
4.	Ascorbic acid (vitmin C)	mg/100 ml	55.95

Specific Gravity

All the treatment have very significant effect on specific gravity of fruit during storage period at room temperature. The specific gravity was found 1.61 of fresh fruit (Table 15.2). Maximum specific gravity was recorded in LDPE storage fruits. The specific gravity of the stored fruits decreased correspondingly as the period of storage progressed.

Percentage of Juice

In fresh condition the juice content of fruit was 67.6 per cent and the maximum fruit juice content was recorded in fruits packed with LDPE, result being 68 per cent, 66 per cent and 61.82 per cent at 10, 20 and 30 days of age respectively, where as minimum juice content was found in check treatment. These findings are in conformity with the report of Gaspar *et al.* (1997) that mature green guavas was wrapped in Pro LDPE, stored at 80°C for 20 weeks, LDPE was more effective than PVC. LDPE delayed the development of peal colour and loss of firmness by up to two days and maintained fruit quality with out of significant loss of quality.

Diameter of Fruits

It was noticed that LDPE stored fruits retained the size of fruits up to a remarkable extent at room temperature. This result was in conformity with the finding of Gasper *et al.* (1997) in Guava. Fruits wrapped with LDPE delayed the peel colour and loss of firmness by up to 2 days.

Chemical Constituents

Percentage of Total Soluble Solids

A steady rise in TSS was noted up to the end of the study with all packaging materials (Table 15.3). The lowest TSS was noticed under low density polyethylene (LDPE) 6.79 (0 Brix) store. Increase in TSS may be due to hydrolysis of polysaccharides and concentration of juice as a result of dehydration. These findings were in conformity with the work of Kohli *et al.* (1966) in lime and. Dalal *et al.* (1962) observed decreased vitamin 'C' and concentration of juice and TSS in coorg mandarin stored at room temperature.

Table 15.2: Effect of Different Packing Materials on the Shelf Life of Acid Lime at Room Temperature

Treatment	Physiological Loss in Weight (gm)			Specific Gravity			Percentage of Juice			Diameter of Fruits		
	Storage Period (Days)			Storage Period (Days)			Storage Period (Days)			Storage Period (Days)		
	10	20	30	10	20	30	10	20	30	10	20	30
T_0	15.85	25.01	37.89	1.21	0.87	0.66	58.97	40.77	20.36	4.0 × 3.9	3.5 × 3.0	2.5 × 2.4
T_1	6.90	7.57	9.50	1.35	1.30	1.22	62.26	58.44	53.83	4.5 × 4.1	4.4 × 4.0	4.3 × 3.9
T_2	5.90	6.90	8.70	1.40	1.35	1.26	68.00	66.00	61.82	4.5 × 4.1	4.2 × 3.9	4.1 × 3.7
T_3	6.00	7.20	9.80	1.28	1.24	1.18	60.00	55.00	51.00	45x 4.4	4.2 × 4.1	3.6x2.7
T_4	12.32	20.08	24.01	1.25	1.03	0.83	59.18	51.00	32.60	4.1 × 3.7	3.4 × 2.6	4.0x 3.1
T_5	13.46	21.09	25.40	1.21	0.70	0.62	51.11	44.44	30.60	4.1 × 3.6	4.0 × 3.1	3.4 × 2.6
T_6	10.05	18.00	23.01	1.20	1.04	0.77	55.76	46.15	29.00	4.5 × 4.1	4.0 × 3.8	3.6 × 3.6
T_7	13.35	19.45	22.58	1.25	0.98	0.78	54.54	50.00	31.00	3.8 × 3.6	3.8 × 3.43	3.6 × 3.2
T_8	14.05	20.00	23.65	1.16	1.00	0.87	52.08	46.00	30.25	4.2 × 3.8	4.1 × 3.7	3.8 × 2.8
T_9	11.55	18.83	24.83	1.22	0.97	0.84	57.14	48.21	28.57	4.2 × 4.1	4.1 × 3.9	3.8 × 3.6
S.ED+	3.25	0.65	0.81	0.07	0.04	0.02	5.522	0.69	0.79	1.73	0.36	0.58
C.D ($p = 0.05$)		1.35	1.69		0.07	0.04		1.43	1.66		0.74	1.21

Table 15.3: Effect of Different Packing Materials on the Quality Parameters of Acid Lime at Room Temperature

Treatment	TSS (Brix)			Total Titrable Acidity (as citric acidity (per cent))			pH of Juice Content (mg/100 g)			Ascorbic Acid			Per cent of Spoilage of Fruits		
	Storage Period (Days)			Storage Period (Days)			Storage Period (Days)			Storage Period (Days)					
	10	20	30	10	20	30	10	20	30	10	20	30	10	20	30
T_0	7.60	7.98	9.50	6.52	6.14	4.54	3.00	3.04	3.06	49.25	48.04	47.00	15.25	63.00	100.00
T_1	7.16	7.24	7.31	6.78	6.65	6.14	2.60	2.67	2.75	52.16	52.00	51.40	0.00	2.77	11.11
T_2	6.70	6.77	6.79	6.91	6.72	6.45	2.56	2.62	2.70	53.35	52.85	52.30	0.00	0.00	5.55
T_3	7.30	7.38	7.45	6.72	6.59	6.08	2.58	2.69	2.83	52.00	51.50	51.00	0.00	6.00	19.44
T_4	7.36	7.41	8.27	6.20	6.00	5.63	2.58	2.95	3.04	51.04	50.83	49.60	0.00	14.55	36.11
T_5	7.40	7.46	8.61	6.14	5.92	5.50	2.87	2.93	3.03	49.98	48.85	47.78	0.00	12.65	30.55
T_6	7.47	7.53	8.06	6.33	6.08	5.50	2.67	2.74	2.98	50.80	50.60	49.20	0.00	16.04	47.22
T_7	7.46	7.51	7.73	6.46	6.20	5.44	2.69	2.72	2.83	49.98	48.88	47.79	0.00	15.55	41.66
T_8	7.44	7.48	7.86	6.40	6.08	5.31	2.68	2.72	2.81	50.85	50.40	49.10	0.00	19.55	58.33
T_9	7.40	7.46	8.59	6.01	5.76	5.24	2.74	2.80	2.94	49.97	48.90	47.76	0.00	16.66	52.77
S.Ed.±	0.28	0.07	0.14	0.32	0.05	0.06	0.16	0.02	0.02	1.19	0.35	0.35	054	0.88	0.98
C.D ($p = 0.05$)		0.15	0.29		0.09	0.12		0.04	0.04		0.74	0.73		1.74	1.97

Total Titrable Acidity

The retention of acidity was maximum in the fruits stored in LDPE being 6.91 per cent, 6.72 per cent and 6.45 per cent at 10, 20 and 30 days of storage, whereas the minimum acidity being recorded with check fruits 6.52 per cent, 6.14 per cent and 4.54 per cent with 10, 20, 30 days, respectively. The result of the present study corroborate the reports of Kohli *et al.* (1966) in lime. High level of acid retention seems to be associated with the lower rate of respiration, since it forms a necessary respiratory substrate for this catabolic process in fruits.

Table 15.4: Effect of Different Packing Materials Combinations on Economy (Rs.) of Acid Lime at Room Temperature

Treatment	*Common Expenditure (Rs.)*	*Extra Expenditure (Rs.)*	*Total Cost (Rs.)*	*Quantity of Fruits in Each Treatment*	*No. of Spoiled Fruits*	*Gross income (Rs.) (Rs.)*	*Net Income*
T_0	18.00		18.00	36	36	0.00	18.00
T_1	18.00	6.00	24.00	36	4	64.00	40.00
T_2	18.00	3.60	21.60	36	2	68.00	46.40
T_3	18.00	2.00	20.00	36	7	58.00	38.00
T_4	18.00	24.44	42.44	36	13	46.00	3.56
T_5	18.00	24.05	42.05	36	11	50.00	7.95
T_6	18.00	18.44	36.44	36	17	38.00	1.56
T_7	18.00	18.05	36.05	36	15	42.00	5.95
T_8	18.00	9.44	27.44	36	21	30.00	2.56
T_9	18.00	9.05	27.05	36	19	34.00	6.95

Input cost:

1: Purchasing cost of fruits @ 0.50/fruit = Rs. 180.00

2: News paper @ Rs. 4/kg

3: Rice straw @ Rs. 0.50/kg

4: Cost of HDPE bag @ Rs. 30/gm= Rs. 6 (20gm)

5: Cost of LDPE bag @ Rs. 24/100 gm = Rs. 3.60 (15gm)

6: Cost of PVC bags @ Rs. 20/100 gm = Rs. 2.00 (10gm)

7: Cost of bamboo baskets @ Rs. 8/box = Rs. 8 × 6 = 48.00

8: Cost of wooden box @ Rs. 6/box = Rs. 6 × 6 = 36.00

9: Cost of carton box @ Rs. 3/box = Rs 3 × 6 = 18.00

pH of Juice

The lowest pH value was recorded with low density polyethylene (LDPE) stored fruits being 2.56, 2.62 and 2.70 at 10,20 and 30 days of storage, respectively. The pH value of juice increased probably due to hydrolysis of polysaccharides. It may be attributed to reduced acidity in fruits. This observation is in accordance with the findings of Neeraj *et al.* (2002) in kinnow mandarin and Kohli *et al.* (1966) in kagjilime.

Ascorbic Acid

Ascorbic acid content of the fruits decreased significantly with the increase of the storage period. The maximum ascorbic acid was recorded with low density polyethylene (LDPE) stored fruits. The decrease in ascorbic acid under all the treatments may be attributed to the process of oxidation at which the change occurs is largely conditioned by pH. Similar findings were also reported by Kohli *et al.* (1966) in lime, Ojha (1987) in anoia, Angadi (1988) in coorg mandarin and Paul *et al* (2004) in guava fruits.

Spoilage of Fruits

The highest percentage of spoilage (100 per cent) was observed under check treatment fruits. Minimum spoilage of 5.55 per cent was recorded in LDPE packed fruits. This finding is in conformity with the findings of Sadasivam *et al.* (1972), Cohen *et al.* (1978) in Eureka lemon, Adsule *et al.* (1993), Singh *et al.* (1998) in oranges and Rane *et al.* (2002) in kinnow mandarin fruits.

Economics

Economics of different packaging materials at room temperature was studied and found that the maximum net return was obtained from LDPE (Rs. 46.40) followed by HDPE (Rs.40.00) and PVC (Rs. 38.00).This observation is in corroboration with the report of Maini *et al.* (1993) in citrus.

Summary and Conclusion

It is found that low density polyethylene (LDPE) bags are effective in reducing the physiological loss in weight, Specific gravity and juice content of fruits. There was significant effect of LDPE bags retaining the total titrable acidity in terms of citric acid. The lowest pH value and maximum ascorbic acid retention were recorded with low density polyethylene (LDPE). Minimum spoilage and maximum net return was obtained from LDPE packed fruits followed by HDPE and PVC packed fruits.

References

Adsule, P.G. and Tandon, D.K., 1983. An assessment of LDPE bags for enhancing shelf life of guava. *Indian Food Packer*, 37(3): 82.

Angadi, S.G., 1988. Effect of temperature on storage of Coorg mandarin. *M.Sc. Thesis*, University of Agricultural Science, Bangalore.

Cohen, E., Lurie, S., Shapira, B., Ben Yehoshva, S., Shalom, Y. and Rosenberger, I., 1990. Prolonged storage and lemons using individual seal-packaging. *J. Am Soc. Hortic. Sci.*, Alexandrin, Va, The Society, 115(2): 251–255.

Cohen, E., Shiffmann and Nodal, M., 1978. Storage capability at different temperatures of lemon grown in Israel. *Scientia Horticulture*, 9(3): 251–257.

Dalal, V.B., Subramanyam, H. and Srivastava, H.C., 1962. Studies on the effect of repeated wax coating on the storage behaviour of Coog oranges. *Food Sci.*, 11: 240–241.

Gaspar, J.W., Couto, F.A.A., Salomao, L.C.C., Finger, F.L., Cardoso, A.A and Donadio, L.C., 1997. Effect of low temperature and plastic films on postharvest life of guava. *Acta Horticulture*, 452: 107–114.

Kohil, R.R. and Bhambota, J.R., 1996. Storage of oranges. *Jr. Hort.*, 22(2): 667–674.

Maini, S.S., Lal, B.B. and Anand, J.C., 1993. Fruits packaging. *Advances in Hort.*, 4: 1776–1794.

Neeraj, Joon, M.S. and Bhatia, S.K., 2003. Use of plastic in fruit packaging: A review. *Haryana Journal of Horticultural Sciences*, 32(1/2): 1–7.

Ojha, C.M., 1987. Physico-chemical changes during growth and development and storage of *aonla* fruits. *M.Sc. Thesis*, N.D.V.A.T., Faizabad.

Pal, R.K., Ahmad, M.S., Roy and Singh, Manoj, 2004. Influence of storage environment, surface coating and individual shrink wrapping on quality assurances of guava. *Plant Foods for Human Nutrition*, 59(2): 67–72.

Paul, Dominane de Saint, 1984. Storage of limes. *Fruits*, 41(9): 570–582.

Rana, G.S., Sharama, R.K., Garg, S.P. and Bhatia, S.K., 2002. Effect of different storage conditions on shelf-life of Kinnow. *Research on Crops*, 202, 3:2: 403–405; 3 ref.

Sadasivam, R., Muthuswamy, S., Sunderraj, S. and Yasudeva, V., 1972. Storage studies of sathugudi (sweet orange). *South Indian Horti.*, 20(1/4): 37–40.

Singh, I.S. and Modhi, G., 1984. *Postharvest Handling of Fruits*. H.A.U., Hissar, p. 123–151.

Singh, R.V., Tewari, J.D. and Chauhan, B.B.S., 1998. Effect of different containers and packaging materials on the transportation and storage behavior of orange. *Progress Horticulture*, 25(3/4): 202–205.

Sozzi, A., Testoni, A., Grassi, M. and Lanza, F., 1983. Storage of lemons: A comparison between types of wrapping at different temperatures. *Hort. Abst.*, 56(2): 34–36.

Chapter 16

Seasonal Periodicity and Pollution Level of a LNMU Campus Pond, Darbhanga

Ashwini Kumar[1], Onkar Mishra[2] and S.B. Shashi[3]

[1]Research Scholar, Department of Zoology, LNMU, Darbhanga – 846 008

[2]Unveirsity Department of Zoology, L.N.M.U., Darbhanga – 846 008

[3]Department of Zoology, R.B. Jalan Bela College, Bela, Darbhanga – 846 005

ABSTRACT

The paper deals with the study of physico-chemical characteristics of a Mithila University pond of District Darbhanga situated just near Manokamna mandir from February 2007 to January 2008. The water was analyzed for various physico-chemical parameters *e.g.*, AT, WT, TR, pH, TDS, DO_2 etc. The paper also includes the population density of planktons in this pond during the studied period.

Keywords: *University Pond, Water Quality, Planktons, Physico-chemical parameters.*

Introduction

Darbhanga is one of the boon city of Mithilanchal, which is rich in ponds, fish culture, makhana culture and Singhara culture basically. Among the various ponds there are some ponds in LNMU, Darbhanga campus. It is also a fact that rain and flood effects the ponds of the whole Mithilanchal region, therefore, this pond has been selected for the study. So for as literature are concerned, numerous workers have investigated the different pollution levels of several ponds and rivers (Siddique and Ahmad, 1992; Dwivedi and Pandey, 2001; Brautigam, 1999; Khare, 1999 and 2002; Jha *et al.*, 2004; Shashi and Kumar, 2007).

Materials and Methods

Random grab samples were collected in multiples from the pre-selected sites of the studied pond seasonally using propylene water samples prescribed by APHA *et al.* (1999). Population density of phytoplankton and were made by standard methods prescribed by APHA (1999).

Results and Discussion

The result of plankton population and physicochemical profiles are shown in Tables 16.1 and 16.2. During study it was found higher temperature in summer season, lower in winter season and medium in rainy season. The range was between 17.2–33.6 (January–May). It has been observed that an indirect effect to toxicity intensifying deoxygenating mid finally increasing the biomagnifications, that is why the dissolved oxygen depletion and plankton community intensify their span in summer. Eutrophic waters are characterised by cyanobacterial algal blooms (Khare 1999, 2002 and Jha *et al.*, 2004). In the present studied pond of LNMU, Darbhanga Campus *Oscillatoria* sp., *Nostoc, Rivularia aquatica* including nine more common phytoplanktons were found, whereas eleven zooplanktons of phylum *Protozoa, Rotifera* and *Arthropoda* were observed. The nature and health of aquatic communities is an expression of water (APHA, 1985). Water pollution manifests through changes such as physical, chemical and biological pond water of the studied pond was observed alkaline throughout the study period (pH 7.1–8.8 *i.e.* September–June). The alkalnity was lower in the month of September, higher in summer and winter seasons. The whole observation showed that the studied LNMU Campus pond did not showed not trend of normal pond. However the eutrophication trends with normal and developing as polluted pond has been studied by some workers (Jha *et al.*, 2006; Siddique, 1992; Dwiwedi and Pandey, 2001).

Table 16.1: Physico-chemical Properties of Pond Water of LNMU Campus (February 2007–January 2008)

Seasons →	*Summer Season*			*Rainy Season*					*Winter Season*			
Months →	*Feb*	*Mar*	*Apr*	*May*	*Jun*	*Jul*	*Aug*	*Sept*	*Oct*	*Nov*	*Dec*	*Jan*
Parameters												
AT (°C)	21.9	24.2	28.5	33.6	32.5	28.4	26.2	25.8	23.5	21.0	18.3	17.2
WT (°C)	18.5	23.0	26.2	28.2	29.5	25.7	24.0	24.2	21.5	21.1	20.5	16.2
TR (°C)	25.0	27.2	30.6	36.6	32.3	28.2	22.8	14.2	17.9	20.3	19.9	21.5
pH	8.7	8.3	8.3	8.7	8.8	8.1	7.3	7.1	7.4	7.7	7.8	7.8
TDS (ppm)	485	460	600	670	835	995	700	625	400	415	570	555
DO_2 (ppm)	11.7	8.9	5.9	3.8	6.2	7,0	7.9	8.2	9.5	10.1	12.5	14.2
FCO_2 (ppm)	7.04	5.62	4.30	4.10	4.70	6.75	6.10	4.8	3.76	4.14	2.86	6.0
Ca (ppm)	34.5	36.0	40.1	46.5	45.2	42.5	30.3	26.6	26.8	25.5	28.0	32.2
Mg (ppm)	12.4	14.5	20.6	20.5	24.0	17.6	6.4	3.8	7.5	9.2	10.8	10.5
Cl (ppm)	42.0	50.2	64.2	65.6	70.4	78.6	76.5	60.2	55.5	52.2	46.1	43.0

AT: Atmospheric Temperature; WT: Water Temperature; TR: Transparency; TDS: Total Dissolved Solids; DO_2: Dissolved Oxygen; FCO_2: Free Carbon dioxide; Ca: Calcium; Mg: Magnesium; C: Chorine.

Table 16.2: Population Density of Planktons in a LNMU Campus Pond (2007–2008)

Sl.No.	*Phytoplanktons*	*Class*	*Zooplanktons*	*Phylum*
1.	*Chlorella*	Chlorophyceae	*Euglina* Sp.	Protozoa
2.	*Spirogyra* sp.	Chlorophyceae	*Cyclops* sp.	Arthropoda
3.	*Eudomia* sp.	Chlorophyceae	*Moina* sp.	Arthropoda
4.	*Closterium* sp.	Chlorophyceae	*Nauplius* larvae	Arthropoda
5.	*Pediastrum tetras*	Chlorophyceae	*Amoeba*	Protozoa
6.	*Pinnularia* sp.	Bacillariophyceae	*Paramecium aurelia*	Protozoa
7.	*Garnphonema* sp.	Bacillariophyceae	*Lecane* sp.	Rotifera
8.	*Diatoma* sp.	Bacillariophyceae	*Bosmania* sp.	Arthropoda
9.	*Nostoc*	Cyanophyceae	*Brachionus* sp.	Rotifera
10.	*Oscillatoriasp*	Cyanophyceae	*Monostyla* sp.	Arthropoda
11.	*Rivularia aquatica*	Cyanophyceae	*Verticella companula*	Arthropoda

Acknowledgement

The authors wish to thanks Prof. N.K. Dubey, Head, University Department of Zoology, L.N. Mithila University, Darbhanga.

References

APHA, AWWA, WPCF, 1999. *Standard Method for the Examination of Water and Wastewater,* 20th Edition. American Public Health Association, Washington D.C.

Brautigam, A., 1999. The freshwater diversity crisis. *World Conservation,* 30(2): 4–5

Dwiwedi, H.K. and Pandey, G.C., 2001. Seasonal dynamics of cyanobacterial toxin producing algal species water ponds. *Aquacult.*, 2(2): 141–146.

Jha, V., Shasbi, S.B. and Singh, T., 2004. Inland water resources of North Bihar: An environmental appraisal. In: *Proceedings of the 3rd National Seminar on Inland Water Resources and Environment,* University of Kerala, Thiruvanantpuram, 2–4 Feb., p. 63–66.

Jha, V., Shashi, S.B. and Singh, T., 2006. Management strategies in Makhana based integrated aquaculture. *Fishing Chimes,* 26(3): 16–19.

Khare, P.K., 1999. Phytoplankton as indicator of water quality and pollution status of Jagat Sagar pond, Chhatarpur (M.P.). *Geobios News Reports,* 18: 107–110

Khare, P.K., 2002. Assessment of organic pollution and water quality of Satri tank, Chhatarpur (M.P.). *Ind. J. Environ. Ecoplan.*, 6(1): 85–88.

Shashi S.B. and Kumar, S., 2007. Seasonal periodicity of toxic cyanobacteria in a fish culture pond with reference to its controlling factors. In: *International Symposium on Recent Advances in Contemporary Biology: Environmental Issue and Sustainable Development,* 29–30 September, S.K.M. University, Dumka (Jharkhand), p. 63.

Siddique, E.N. and Ahmad, M.S., 1992. Green algae of Darbhanga. *Bio. J.*, 4(1 and 2): 149–152.

Chapter 17

Variations in the Primary Productivity of Singanallur Pond in Coimbatore, Tamil Nadu

S. Umavathi and K. Logankumar
PG and Research Department of Zoology,
Kongunadu Arts and Science College, Coimbatore – 641 029, Tamil Nadu

ABSTRACT

The primary productivity of the Singanallur pond water was studied during the month of July 2006 to December 2006. The primary productivity of the ecosystem is essential in food chain studies. The sustainable yield of a fishery can be estimated by using only measurements of primary production. In addition to long-term effects, changes in photosynthetic rates may show the almost instantaneous effect of a nutrient, toxicant or physical changes in the environment. By primary productivity autotrophic organisms transform inorganic carbon and nutrients into living matter (Tilzer, 1989).

Keywords: *Pond ecosystem, Gross primary productivity, Respiratory rate, Net Primary productivity.*

Introduction

Water is priceless gift of nature and it is essential for living organisms. Most of our water source is polluted with untreated or partially treated waste from industry domestic sewage and fertilizer/ pesticide run off from the agricultural fields. The quality and quantity of available water is decreasing and aggravating the already serious situation. When water quality deteriorates, water loses its economic value (Gitanjali and Kumeresan, 2006). Pond are small bodies of water in which the littoral zone is relatively large and limnetic and profundal regions are small or absent. Pond may be found in most regions of adequate rainfall in which the abiotic environment includes all those physiological and chemical aspects, which exert an influence on living organisms.

In an aquatic body, primary productivity gives information relating to the amount of energy available to support bioactivity of the system (Vollenweider, 1969). Estimation of primary productivity of the aquatic systems are adversely affected by anthropogenic activities. Primary production is an important biological phenomenon in the aquatic environment in which phytoplanktons act as primary producers, their physical activities are greatly controlled by physicochemical characters of the water body (Sahu *et al.*, 1995; Arvind Kumar, 1997). Generally plankton are considered of an index of fertility of the water column (Fraser, 1962). Several studies are available relating to the primary productivity of different ponds, lakes and reservoirs in different parts of India (Munawar, 1966, 1970; Datta and chaudhury, 1984; Ahmad, 1990; Prasad, 1990; Basheer *et al.*, 1996).

The plankton constitutes the basic food. source of any aquatic system, which supports fish and other aquatic animals. Odum (1960) stated that fish production in a pond has direct relation to its primary productivity and suggested that harvesting should be made depending on the primary productivity. Pandey *et al.* (1994) and Sharma and Sharma (1992) noted phytoplankton encountered in the wastewater. Virtually all the dynamic features of rivers such as colour, clarity, trophic state, water taste, zooplankton and fish production depends upon the large degree of the phytoplankton (Boney, 1975). Productivity of lake depends on the presence of plankton biomass. Environment of nutrients and dissolved matter in the water bodies affect diversity of plankton and also physico-chemical properties of water.

Planktonic organisms have a short life cycle with a high metabolic activity, which facilitates them to respond to any pollution stress quickly and significantly compared to benthic and nektonic organisms (Perkins, 1976). The Planktonic population of lakes and reservoir of India are scanty and a few works were made by Chakrabarthy *et al.* (1959) in river Jammu, Sreenivasan *et al.* (1964) in Bhavanisagar, Mitra *et al.* (1986) in Sagar islands, Shoba *et al.* (2006) in Kadinamkulam lake and Umavathi *et al.* (2007) in Sulur pond, Coimbatore, Tamil Nadu. In Kerala, studies on the plankton population were done by Pillai *et al.* (1975) in Vembanadu lake, Silas and Pillai (1973) in Cochin backwater and Nair *et al.* (1981) in Asthumudi estuary. Palmar (1969) also stressed that the phytoplankton composition as well as the individual algal species were the indicators of the organic pollution of the water bodies.

Materials and Methods

The Singanallur pond is located 0.5 km. south west of Singanallur village in Coimbatore district. It is situated at an altitude of 450 m from the mean sea level. The pond receives water from Noyyal river and Sanganoor river during rainy season. The total depth of the pond is 13.95 feet and the area is about 1.143 sq.km. The water storage of the pond is about 14.80 million cubic feet. The pond water is mainly used for Agriculture, about 1318 acres of land have irrigation facility in the villages Nanjundapuram, Singanallur, Ondipudur of Coimbatore district. The pond receives flood water during rainy seasons from the catchment areas of western ghats. The pond also gets sewage water throughout the year from the Coimbatore city through the Sanganur pallam.

For the present study 2 different sampling stations were fixed in the pond to study the primary productivity of the pond. The water samples were collected for a period of six months from July 2006 to December 2006. Samples were collected regularly from the pond at 12 noon and this was uniformly maintained throughout the study period. Samples collected were brought to the laboratory and kept in the refrigerator for further analysis. Primary productivity has been estimated by light and dark bottle technique following the method of Strickland and Parson, 1978; Vollenwider, 1969 and APHA, 1984.

Results and Discussion

The primary productive value of the Singanallur pond is presented in the Table 17.1. Planktons qualitative and quantitative estimate provides good indices of water quality and the capacity of water to sustain-heterotrpic communities. Zooplanktons usually act as primary consumers and constitute an important link between phytoplanktons (primary producers) and the consumers like fishes in aquatic food chain (Sawane *et al.*, 2006). The values of Gross Primary Productivity (GPP), Respiratory Rate (RR) and Net Primary Productivity (NPP) were fluctuated between 2.44–3.40, 0.69–0.92 and 1.45–2.65 mgC/L/hr respectively. The high productivity was noted in the month of July have been attributed to the relatively low salinity when the photosynthesis would be maximum and also due to the eurihaline nature of the phytoplankton crop prevailing at the time. Fogg (1982) has suggested that nutrient concentration and availability of light are the major parameters, which govern phytoplankton production. Primary productivity increased with light penetration and these variations in the values were directly or indirectly influenced by the environmental factors. Increase in temperature enhances the release of nutrients from sediments through bacterial decomposition (Thilaga *et al.*, 2004).

Table 17.1: Variations in the Primary Productivity of Singanallur Pond (2006)

Month	*Gross Primary Productivity*	*Respiratory Rate*	*Net Primary Productivity*
July	3.16±3.152	0.81±0.014	2.3 7±0.009
August	3.02±0.045	0.69±0.006	2.31±0.007
September	2.82±0.045	0.88±0.045	1.92±0.045
October	3.40±0.071	0.74±0.006	2.65±0.009
November	3.14±0.055	0.71±0.011	2.37±0.009
December	2.44±0.055	0.92±0.070	1.45±0.027

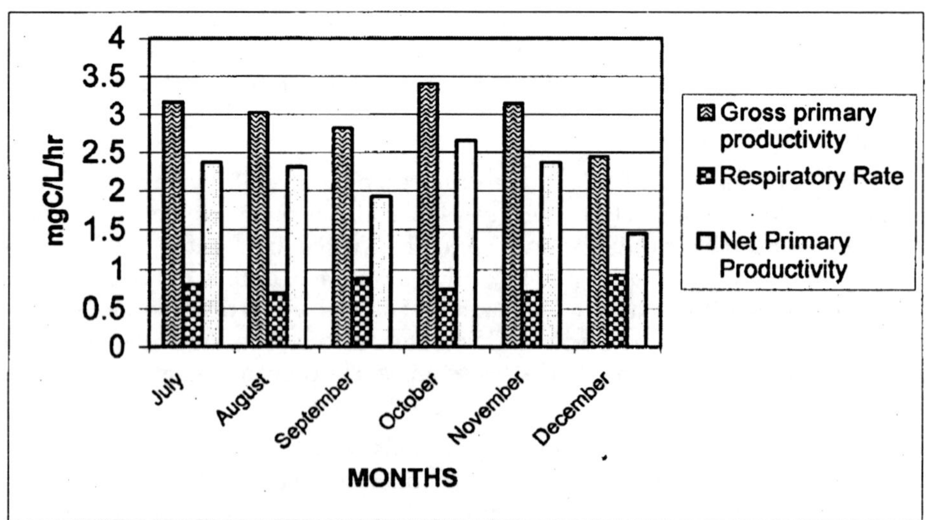

Figure 17.1: Variations in the Primary Productivity of Singanallur Pond (2006)

The significant direct relationship of temperature with productivity was established by several authors (Vijayaragavan, 1971; Arvola, 1983; Eloranta and Salimar, 1984; Verma and Mohandy, 1995). Nair *et al.* (1981) have mentioned that the average rates of primary production and magnitude of variations in the tropical brackish water bodies are high. Probably this is due to high fertility, with regards to the availability of nutrients and large standing crop of mixed phytoplankton derived from marine, estuarine and freshwater masses suspended in a widely fluctuated ambient medium (Shoba *et al.*, 2006).

Based on the present observation, it is believed that the multiple points of pollution in the pond, the quality of water is adversely affected and leads to high fluctuations in the primary productivity. The. primary productivity is increased by artificial means which is better for fish production.

References

Ahmad, S.S., 1990. Primary productivity in a tropical pond at Darbhanga. *J. Naccon.*, 2(1): 1–6.

APHA, 1984. *Standard Methods for the Examination of Water and Wastewater, 18th edn.* APHA, AWWA, WFE, Washington.

Arvind Kumar, 1997. Comparative hydrological studies of tropical water bodies with special reference to sewage pollution in south Bihar. *J. Ecobiol.*, 9(4): 255–262.

Basheer, V.S., Khan, A.A. and Alam, A. 1996. Seasonal variations in the primary productivity of a pond receiving sewage effluents. *J. Inland Fish. Soc., India,* 28(1): 76–82.

Boney, A.D., 1975. *Phytoplankton Institute of Biology Study*, 52, Crane, Russakio, New York, pp. 116–125.

Chackrabarthy, R.D., Ray, P. and Singh, S.B., 1959. A qualitative study of the plankton and physico-chmical conditions of the River Yamuna at Allahabad in 1954–1955. *Indian J. Fish.*, 6(1): 186–203.

Datta, N.C and Choudhury, S., 1984. Primary productivity and its correlation with physico-chemical parameters of a fresh water sewage fed impoundment near Kolkota. *Environ. Ecol.*, 2: 65–68.

Eloranta, P.V. and Salimar, R., 1984. Phytoplankton in eutropic cooling water plant. *Hydrobiologia,* 118: 267–274.

Fogg, G.E., 1982. Phytoplankton primary production. In: *Fundamentals of Aquatic Ecosystem*. Blackwell Scientific Publications, Oxford.

Fraser, J., 1962. *Nature Adrift: The Story of Marine Plankton Fouls*. G.T. and Co. Ltd., London.

Gitanjali, G. and Kumaresan, A., 2006. Impact of tourist influx on the courtallam water quality index. *Indian J. Environ and Ecoplan.*, 10(3): 825–830.

Mitra, A. Detk and Chaudhury, A., 1986. Studies on phytoplankton and productivity around Sagar island and Prentice island, Sunderbans. In: *Proc. Nat. Sem. Estuarine Management,* Trivandrum, 1987, p. 327–329.

Munawar, M., 1966. Limnological studies in certain polluted and unpolluted freshwater ponds of Hyderabad, India. *Ph.D. Thesis,* Osmania University, Hyderabad, India.

Munawar, M., 1970. Limnological studies in freshwater ponds of Hyderabad, India. *J. The Biotope. Hydrobiologia,* 35: 127–162.

Nair, N.B. and Abdul Aziz, P.K., 1981. Hydrology Ashtamudi estury: A tropical back water system in Kerala. In: *Proc. Nat. Sem. Estuarine Management,* Trivandrum, 1987, p. 268–280.

Odum, E.P., 1960. The role of tidal marshes in estuarine production. *N.Y. State Conservation and Information Leaflet*, L. 60.4 pp.

Palmer, C.M., 1969. A composite rating or algae tolerating organic pollution. *J. Phycol.*, 5: 78–82.

Pandey, B.N., Jha, A.K. and Das, P.K.L., 1994. Hydrobiological study of a swamp at Purnia, Bihar, in relation to its phytoplankton fauna. *J. Ecobiol.*, 6(1): 13–16.

Perkins, E.J., 1976. *The Biology of Estuaries and Coastal Waves*. Academic Press, London, p. 25–37.

Pillai, P.P., Qassium, S.Z. and Nair, A.K.K., 1973. Copepod components of zooplankton in a tropical estuary. *Indian J. Mar. Sci.*, 2: 38–46.

Prasad, D.Y., 1990. Primary productivity and energy flow in upper lake, Bhopal. *Indian J. Env. Hlth.*, 23: 132–138.

Sahu, R.K., Rao, R.J., Behera, S.K. and Pandit, 1995. Phytoplankton and primary production in the river Ganga from Rishikesh to Kanpur. *J. Ecobiol.*, 7(3): 219–224.

Sawane, A.P., Puranik, P.G. and Lonkar, A.N., 2006. Preliminary study on the seasonal distribution of plankton in Irai river at Irai dam site, District Chandrapur, Maharashtra. *Indian J. Environ and Ecoplan.*, 12(1): 207–212.

Sharma, R. and Sharma, K.C., 1992. Diatoms of Anusagar lake of Ajmer, Rajasthan. *Acta Ecol.*, 14: 6–9.

Shoba, V., Anuradha Rammohan., Santhosh, S. and Chandra Thara, J., 2006. Phytoplankton productivity assessment of Kadinamkulam lake with special reference to the retting activity, Thiruvananthapuram South Kerela. *Indian J. Environ. and Ecoplan.*, 12(3): 511–518.

Sreenivasan, A., Raj, R.S. and Antonym, K.J., 1964. Limnological studies of tropical impoundments 2 hydrological features and plankton of Bhavanisagar reservoir Madras state, from 1961–1962. In: *Proc. Indian Acad. Sci.*, 59B(2): 53–71.

Strickland, J.D.H. and Parson, T.R., 1968. A practical handbook of sea water analysis. *Bull. Fish. Res. Biol., Canada*, 167: 311.

Thilaga, A., Subhashini, S., Logankumar, K. and Shobana, S., 2004. Effect of sewage water on the primary productivity of the Ooty lake. *Him. J. Env. Zool.*, 18(2): 125–127.

Tilzer, M.M., 1989. The productivity of phytoplanktons and its control by resource availability. In: *Phytotalk*, (Ed.) H.D. Kumar. Rastogi and Company, India, 1: 1–40.

Umavathi, S., Logankumar, K., Subhashini, S. and Logaswamy, S., 2006. Studies on the primary productivity of Sulur pond, Coimbatore, Tamil Nadu. *Nature Environment and Pollution Technology*, 6(3): 491–494.

Verma and Mohandy, 1995. Phytoplankton of Malyanta pond of Laxmisagar and its correlation with physico-chemical parameters. *Poll. Res.*, 14: 243–253.

Vijaya Raghavan, S., 1971. Seasonal variation in primary productivity in three tropical ponds. *Hydorbiologia*, 38: 395–408.

Vollen Weider, R.A., 1969. *A Manual of Methods for Measuring Primary Production in Aquatic Environment.* IBP Hand Book No. 2. Blackwell Scientific Publication.

Chapter 18

Effect of Water Activity on the Germination of Certain Fungi Isolated from Working Environments in Imphal, Manipur

A. Premila Devi[1], S. Mukta Singh[2*] and N. Irabanta Singh[1]

[1]Aerobiology, Microbiology and Plant Pathology Laboratory, Department of Life Sciences, Manipur University

[2]Department of Botany, D.M. College of Science, Imphal

ABSTRACT

The present chapter deals with the effect of different water activity grades on the germination of eleven air borne fungal spores *viz. Asperxillus niger, A. echinulatus, Alternaria tenuis, Rhizopus* sp., *Mucor* sp.. *Cladosporium* sp., *Fusarium* sp., *Chaetomium* sp., *Penicillium* sp., *Curvularia lunata* and *Torula* sp. isolated from working environments in Imphal. The study revealed limits of a_w for germination ranges from 0.80 to 1.00. The test fungi also exhibited specific lower limits of water activity (a_w) for germination. Species of *Mucor, Fusarium, Rhizopus, Chaetomium* and *Curvularia* have a lower limit of a_w for germination at 0.90 whereas *Cladosporium* and *Penicillium* recorded at 0.85. The lowest water activity was exhibited by *A. echinulatus* at 0.70 a_w. The reduction of a_w below certain value led to the lengthening of lag phase. At 0.90 a_w all the test fungal species could germinate with a higher lag phase whereas at 0.85 a_w only six fungal species germinate. Thus, the number of germinating spores progressively reduces with lower water activity grade.

***Keywords**: Water activity, Germination, Working environment.*

---

* Corresponding Author.

Introduction

Microorganisms like fungi, actinomycetes and bacteria are the main agents for degrading biological materials. Each organism has a specific maximal, optimal and minimal water activity (a_w) for germination and, development. If a_w is reduced below the optimal level, there is a lengthening of the lag phase of growth and decrease in the rate of growth. The decrease in the amount of cell substance synthesized changes and is variable with the organism and also with solute employed to reduce a_w. The a_w induces a relationship between moisture in foods and other microorganisms and their activities to grow on them (Scott, 1957). Moreover, the problem faced by food and fibre suppliers are aerobiological in nature (Edmonds, 1979). Therefore, an attempt was made to find out the effect of a_w on the germination of certain fungi isolated from some working environments in Imphal, Manipur.

Materials and Methods

Air borne fungi of certain working environments *viz.*, rice mill, hostel kitchen and bakery located in different localities of Imphal, Manipur were isolated by employing gravity petridish culture technique. About 20ml sterile Czapek's agar medium was poured aseptically into 100mm diameter borosil glass petridishes and allowed to cool down at room temperature (25±2°C). The petridishes were exposed for half an hour at one metre above the ground level. The exposed plates were then incubated in a BOD incubator at 25±2°C for a week. The colonies developed in the petridishes were identified and purified by further subculturing. The cultures were maintained on Czapek's agar medium.

The water activity (a_w) is

$$a_w = \frac{n_2}{n_1 + n_2} \times \frac{p}{p_0}$$

where,

n_1: No. of moles of solute

n_2: No. of moles of solvent

p: Partial presence of water vapour under test condition

p_0: The saturation vapour pressure of water under the same condition.

Water activity is also numerically equal to the equilibrium relative humidity (ERH) expressed in decimal. Further, a_w is a fundamental property of aqueous solutions.

Thus, a_w (food) = ERH (air)/100

where, ERH = Equilibrium relative humidity of air

Different water activity (a_w) levels *viz.* 0.55, 0.60, 0.75, 0.80, 0.85 and 0.90 were prepared by dissolving required amount of analytical grade chemicals in distilled water. The chemicals taken were calcium nitrate, sodium bromide, cobalt chloride, sodium chloride, ammonium sulphate, potassium chloride and barium chloride. Moreover, the remaining a_w grades of 0.70, 0.95 and 1.00 were prepared with Glycerol (60 gm/100 ml distilled water), Sulphuric acid (7.8 ml/100 ml distilled water) and water respectively (Winston and Bates, 1960; Michael, 1984). About 1/3rd volume of tall, big and clean 1000 ml capacity glass specimen jars were filled up with these solutions. Again, a loopful of spores of fungi up with the help of a sterile inoculating needle and inoculated in the cavity of a cavity glass slides having a little amount of liquid Czapek's medium and distilled water. The inoculated cavity slides were incubated inside each specimen jar having different grades of a_w. The cavity slides were held inside the jars with the help of a slide holder. The caps of the specimen jars were replaced tightly and sealed with wax. The jars were kept in the laboratory at room temperature (25±2°C) in the laboratory. Four replications were also taken for each treatment. The germination of the inoculated spores were examined at an interval of 24 hours and recorded for 165 days. The data thus recorded were analysed and plotted in order to determine the relationship between a_w and latent period of germination of the spores.

Results and Discussion

Altogether, eleven fungal genera *viz., Aspergillus niger, A. echinulatus, Mucor* species, *Alternaria tenuis, Cladosporium* sp., *Fusarium,* sp., *Penicllium* sp., *Torula* sp., *Rhizopus* sp., *Chaetomium* sp. and *Curvularia lunata,* isolated from the air of different working environments of Imphal, were tested for spore germination in ten different grades of a_w. Tables 18.1 and 18.2 revealed suitable limits of water activity for growth at room temperature. It varies from 0.80 to 0.95 for *A. niger,* 0.70 to 0.95 for *A. echinulatus,* 0.80 to 0.95 for *A. tenius,* 0.90 to 1.00 for *Mucor* sp., 0.85 to 0.95 for *Cladosporium* sp., 0.90 to 1.00 for *Fusarium* sp., 0.85 to 1.00 for *Penicillium* sp., 0.80 to 1.00 for *Torula* sp., 0.90 to 1.00 for *Rhizopus* sp., 0.90 to 1.00 for *Chaetomium* sp. and 0.90 to 1.00 for *Curvularia lunata* respectively. Such differential response of the test fungi may be due to their different physiological and metabolic requirements of the test fungi. Similar findings were also recorded by other workers (Christian, 1965; Garner, 1966; Tomkins, 1929 and Scott, 1957).

Table 18.1: Germination (Per cent) of Selected Fungi in Media in Different a_w

Sl.No.	*Fungal Types*	a_w									
		1.00	*0.95*	*0.90*	*0.85*	*0.80*	*0.75*	*0.70*	*0.65*	*0.60*	*0.55*
1.	*Aspergillus niger*	–	1.3	1.4	2.0	3.0	–	–	–	–	–
2.	*Aspergillus echinulatus*	–	1.8	1.9	2.0	2.2	2.6	3.5	–	–	–
3.	*Alternaria tenuis*	–	1.0	1.5	2.0	2.8	–	–	–	–	–
4.	*Mucor*	1.5	1.8	3.5	–	–	–	–	–	–	
5.	*Cladosporium*	–	0.8	1.7	3.3	–	–	–	–	–	–
6.	*Fusarium*	0.9	1.4	3.0		–	–	–	–	–	–
7.	*Penicillium*	0.8	1.1	1.3	3.2	–	–	–	–	–	–
8.	*Torula*	0.9	1.3	1.5	1.8	3.0	–	–	–	–	–
9.	*Rhizopus*	1.5	1.8	3.4	–	–	–	–	–	–	–
10.	*Chaetomium*	1.4	1.7	3.2	–	–	–	–	–	–	–
11.	*Curvularia*	1.5	1.8	3.0	–	–	–	–	–	–	–

Further, the test fungi showed specific lower limits of a_w for germination. During the present investigation, *Mucor* sp., *Fusarium* sp., *Rhizopus* sp., *Clwetomium* sp. and *Curvularia lunata* indicated a lower limit of a_w at 0.90 whereas *Cladosporium* and *Penicilliwn* have a lower limit of a_w at 0.85. However, the lowest grade of 0.70 a_w was indicated by *A. echinulatus.* These findings thus clearly indicated a close correlation between the fungal spore types and a_w of the environment in which they may germinate and recycle the spore production chain. Such findings are in conformity with the findings of Chayton (1942) and Scott (1957).

Table 18.2: Germination (Per cent) of Selected Fungi in Water in Different a_w

Sl.No.	*Fungal Types*	*a_w*									
		1.00	*0.95*	*0.90*	*0.85*	*0.80*	*0.75*	*0.70*	*0.65*	*0.60*	*0.55*
1.	*Aspergillus niger*	–	1.6	1.8	2.2	3.3	–	–	–	–	–
2.	*Aspergillus echinulatus*	–	2.0	2.2	2.5	2.7	2.8	3.6	–	–	–
3.	*Alternaria tenuis*	–	1.4	1.7	2.1	2.9	–	–	–	–	–
4.	*Mucor*	*1.8*	2.0	3.6	–	–	–	–	–	–	
5.	*Cladosporium*	–	1.1	1.9	3.4	–	–	–	–	–	–
6.	*Fusarium*	1.4	1.8	3.2		–	–	–	–	–	–
7.	*Penicillium*	1.3	1.9	2.1	3.4	–	–	–	–	–	–
8.	*Torula*	1.4	1.7	1.9	2.3	3.2	–	–	–	–	–
9.	*Rhizopus*	1.9	2.3	3.6	–	–	–	–	–	–	–
10.	*Chaetomium*	1.8	2.4	3.6	–	–	–	–	–	–	–
11.	*Curvularia*	2.0	2.5	3.4	–	–	–	–	–	–	–

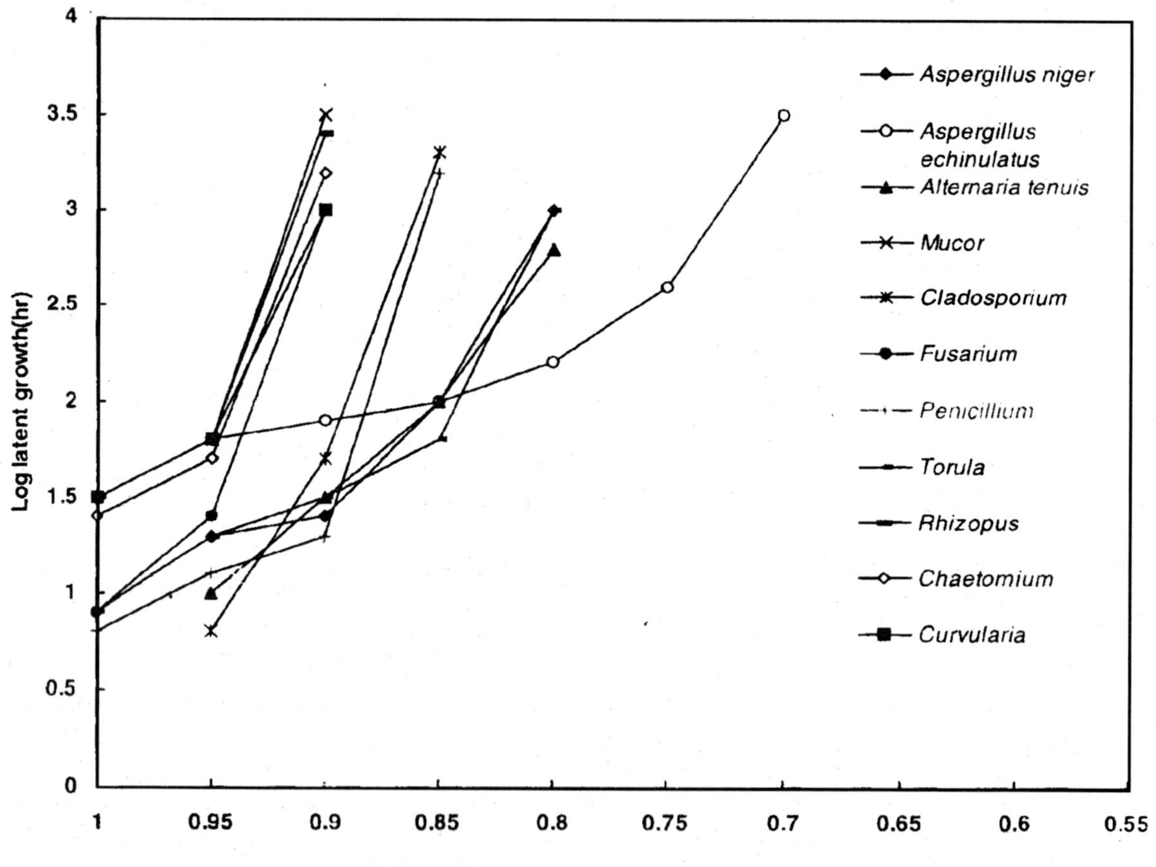

Figure 18.1: Correlation Between Water Activity (a_w) and Latent Growth Period for Germination of Certain Indoor Airborne Fungal Spores in Culture Media

Again, Figures 18.1 and 18.2 indicated the reduction in a_w below certain value for a particular spore led to an increase in lag period (phase) and finally checked the growth. Beyond this limit the fungal spores completely fail to germinate even upto a period of four months. In Czapek's medium, the lag phase was less than 1.5 log value of a_w above 0.95, slightly greater at 0.95 and 30 hours to four months at a_w of 0.90 to 0.95 respectively. However, in distilled water, higher number of lag period hours was required for germination of all the spore types tested with corresponding a_w ranging from 0.80 to 1.00. Germination was observed in the range of 63 to 100 percent both in Czapek's medium and distilled water during latent hour in log range from 0.8 to 1.5 which corresponds to 6.30 to 31.00 hours. All the test spore types germinate freely at 0.95 a_w both in Czapek's medium and distilled water with latent period of log ranging from 1.1 to 2.5 and 0.8 to 1.8 respectively. At 0.90 a_w all fungal spores

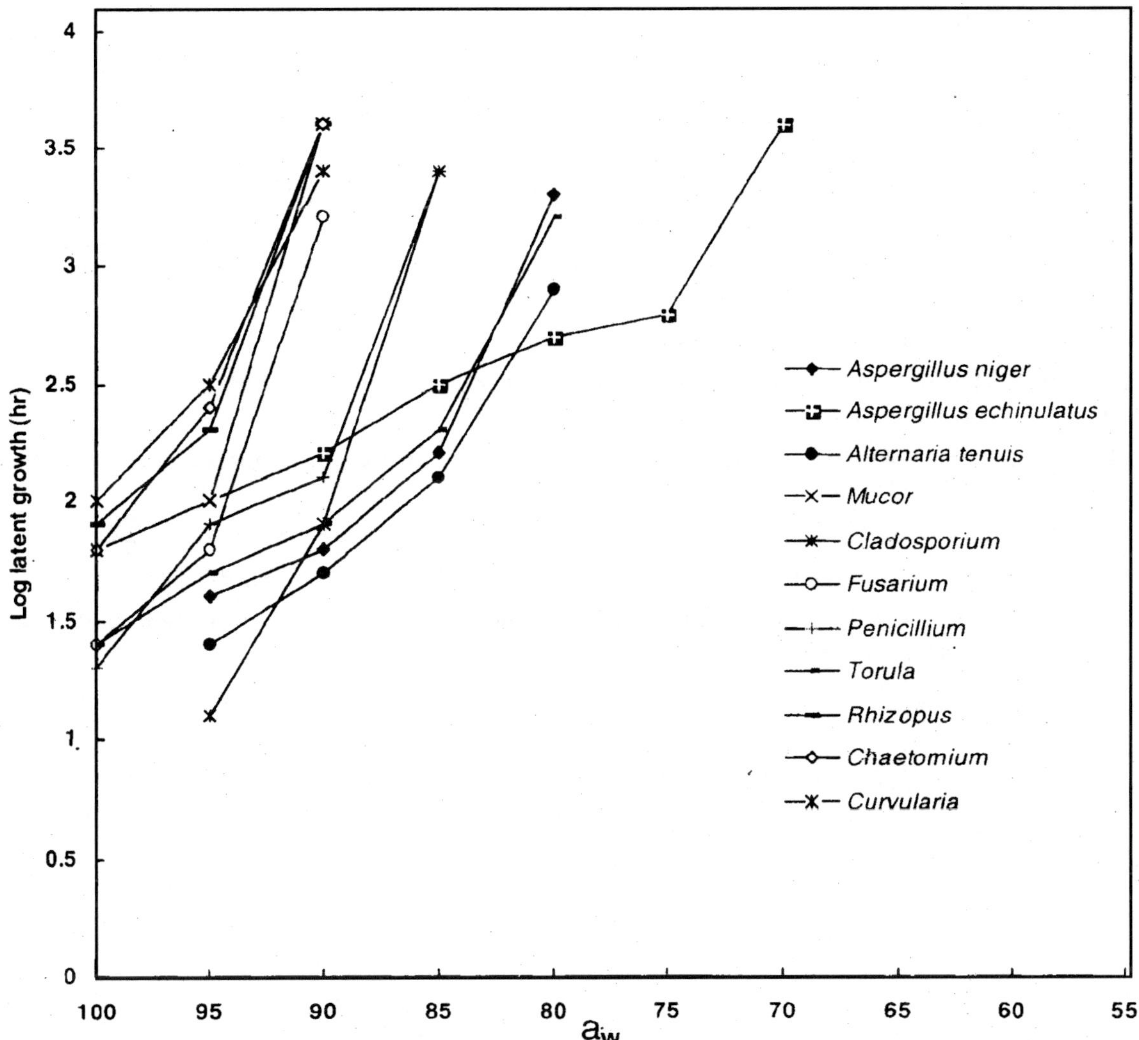

Figure 18.2: Correlation Between Water Activity (a_w) and Latent Growth Period for Germination of Certain Indoor Airborne Fungal Spores in Water

germinates and grow well with higher lag phase which ranges from 1.3 to 3.4 corresponding to 20 hours to four months in Czapek's medium and 1.7 to 3.6 in distilled water only corresponding to 50 hours to 5.5 months or 165.87 days. Again, six fungal spore types out of a total of eleven spore types (*i.e.* 54.5 per cent) were able to germinate and grow with higher lag phase of growth which ranges from 1.8 to 3.3 in Czapek's medium and 2.1 to 3.4 in water. Similarly, three spore types could germinate with higher lag phase of 2.2 to 3.0 in Czapek's medium and 2.7 to 5.3 in water.

Thus, the present findings clearly established the close relationship between the habitat of the fungi and the environment for spore germination and production.

Acknowledgement

A. Premila Devi is thankful to Head, Department of Life Sciences, Manipur University for providing laboratory facilities.

References

Chayton, C.N., 1942. The germination of fungus spores in relation to controlled humidity. *Phytopath.*, 32: 921–943.

Christian, J.H.B., 1965. Water activity in the growth of microorganisms. *Recent Advances in Food Science,* 3: 247–255.

Edmonds, R.L., 1979. The ecological systems approach. US/IBP Synthesis Series No.10. Dowden, Hutchinson and Ross Inc. Stroudshurg, Pennsylvania, pp. 386.

Garner, R.G., 1966. Water activity in relation to foods. *Agri. Sc. Review,* 4(3): 17–23.

Michael, P., 1984. *Ecological Methods for Field and Laboratory Investigations*. Tata McGraw Hill Publishing Co. Ltd., New Delhi.

Rockland, L.B., 1960. Saturated salt solutions for static control of relative humidity between 50 and 40. *Ann. Chem.,* 32: 1375–1376.

Scott, W.J., 1957. Water relations of food spoilage microorganisms. *Adv. Food Res.,* 7: 83–127.

Tomkins, R.G., 1929. Studies of the growth of moulds–I. *Proc. Roy. Soc. Lond., B.,* 105: 375–401.

Winston, P.W. and Bates, D.H., 1960. Saturated solutions for the control of humidity in biological research. *Ecology,* 41: 232–237.

Chapter 19

Acute Toxicity of Some Indigenous Plants against Pulse Beetle, *Callosobruchus chinensis* Linn. (Coleoptera : Bruchidae)

S.S. Patole[1] and R.T. Mahajan[2]*

[1]Department of Zoology, V.V.M's. S.G Patil Arts, Science and Commerce College, Sakri – 424 304, Dist. Dhule, M.S.

[2]Post Graduate Department of Zoology, Moolji Jaitha College, Jalgaon – 425 001, M.S.

ABSTRACT

Preliminary evaluation on acute toxicity of total aqueous extract (TAE) of thirty indigenous plants against stored grain pest *Callosobruchus chinensis* was carried out at laboratory conditions at doses ranging from 25 to 100 mg/ml during 48 h duration as per method described by De-Petro and De-Petro. The order of toxicity of TAE of various plants at high dose concentration was *Acorus calamus* (83.33 per cent) > *Azadirachta indica* (80.00 per cent) > *Vitex nigundo* (76.67 per cent) > *Sphaeranthus indicus* (73.33 per cent) > *Nicotiana tabacum* (70.00 per cent) > *Annona squamosa* (70.00 per cent) > *Pongamia glabra* (70.00 per cent) > *Melia azadiracta* (73.33 per cent) > *Randia dumentorum* (70.00 per cent). Other thirteen plants showed moderate toxicity (50–70 per cent), remaining plants recorded poor activity (50 per cent). The toxicity was dose dependent. Utility of these plants as excellent post-harvest stored grain protectants may be accounted for ideal formulation.

Keywords: *Acute toxicity, Indigenous plants, Callosobruchus chinensis, Total aqueous extracts.*

---

* E-mail: rt_mahajan@rediffmail.com

Introduction

In recent years, there has been an increased interest in natural, plant derived materials as alternative pesticides of plant origin to conventional broad-spectrum synthetic pesticides. The important aspect of these natural materials is because of their wide range of effects against insect pests such as repellence, antifeedant, oviposition deterrence, insecticidal, ovicidal and growth inhibitors (Khaire *et al.*, 1992; Ascher, 1993; Tripathy *et al.*, 2001; Ignacimuthu, 2004). The use of plant powders, ashes and minerals as traditional protectants of stored products is an old practice used all over the world. Farmers have largely neglected these traditions after the World War II, with the advent of synthetic or petroleum-based insecticides. However, the potential hazards for mammals from synthetic insecticides, the ecological consequences and the increase of insect resistance to pesticides has led to a search for new classes of insecticides with reduced human toxicity and less persistence in the environment and higher efficacy against pest (Isman *et al.*, 2002; Maheshwari *et al.*, 2003; Tandon, 2005; Dwivedi and Bhati, 2006).

Pulse beetles (*Callosobruchus* spp.) are internal feeder. Therefore, insecticides do not successively control them. Secondly, mixing of synthetic insecticides with food grains is not advisable. Fumigation being the most effective method cannot be practiced in rural areas because the storage structures are not airtight and are mostly built inside the residential areas. Plant materials that are being traditionally used by some farmers are quite safe and appear to be the most promising as grain protectants. To overcome this problem, we planned to evaluate locally available, indigenous plants for its insecticidal efficacy against pulse beetle *Callosobruchus chinensis* Linn. (*C. chinensis)* under laboratory conditions. Secondly, the plant materials are economical, perennial, non-hazardous, target specific, innate biodegradable, eco-friendly, non-persistent and easily available as compared to chemical insecticides (Saxena *et al.*, 1994; Ignacimuthu, 2004; Chaman and Verma, 2006; Patole *et al.*, 2008).

Materials and Methods

Plant Material

The plant material was collected locally and authenticated. It was thoroughly washed in water, air dried in shade and pulverized. TAB were prepared by treating the 10g powder in 300 ml water at 50–60°C on water bath for 5–6 h then filtered by Whatmann paper. The filtrate was concentrated on water bath, the final volume adjusted to 100 ml. The desired concentrations such as 100, 75, 50 and 25 mg/ml was prepared by diluting the TAB with distilled water and used for acute toxicity studies.

Pulse Seeds (Green gram, *Vigna radiata* L.)

Non-infested Green gram pulse seeds were procured from local farmers immediately after harvesting. Possible insect contamination was eliminated by thoroughly washed the seeds and dried in bright sunlight for 3 days having 6 to 7 per cent moisture content and stored in airtight containers until required for experiment.

Insect Rearing (Pulse beetle, *Callosobruchus chinensis* L.)

The pulse beetle, *C. chinensis* was identified by an expert entomologist and subsequently reared in plastic jars (Sun pet, India) of 1 kg capacity. Each jar containing 100 g of pulse seeds as a nutritional source and 20 pairs of pulse beetle of either sex were added to each jar, the mouth of jar was tied with muslin cloth, banded and labeled properly. All these rearing jars placed in growth chamber at 30±2°C temperatures and 70±5 per cent relative humidity.

Method

Acute toxicity of indigenous plants tested by using dipping method described by De-Petro and De-Petro (1994). Whatmann filter paper No.1 was dipped in to the TAE of test plant at various concentrations. Dipping time was maximum 5 seconds and maintained in the subsequent experiments. Excess solution was removed and the paper was dried at 40°C in oven for 30 min. For acute toxicity test, each filter paper was placed at the bottom of glass Petri plate; 10 adult pulse beetles, *C. chinensis* of 1–2 days old (either sex) released with a normal food and replicated thrice. Insect mortality was observed at 24 and 48 h after treatment control was run simultaneously in which paper was soaked with distilled water. The data for per cent mortality is corrected by Abbott's (1925) formula.

$$\text{Abbott's corrected mortality} = \frac{\%\text{ kill in treated} - \%\text{ kill in control}}{100 - \%\text{ kills in control}} \times 100$$

Results and Discussion

A result obtained on acute toxicity test of TAE of selected indigenous plants against *C. chinensis* is given in Table 19.1. Among four doses, the mortality was found to be positively correlated with the extract concentrations being highest at 100 mg/ml. As per this table amongst 30 plants, 10 of them showed highest mortality percentage *i.e.* above 70 per cent at the dose of 100 mg/ml for 48 h of exposure. The moderate activity (40–69 per cent) was observed in 13 plants. Remaining plants showed poor activity. The synthetic insecticides used for the protection of pulses as postharvest protectants; these pulses are not for consumption but uses only for agricultural seeds. These insecticides affect environment and mankind adversely. Therefore, alternative strategies are urgently required to combat attack of insect pest at various stages of protection during storage. In present study the TAE of some indigenous plants have proved effectively to arrest the pulse beetle damage in stored pulses. The above findings are corroborated with reports of earlier workers (Punam-Kumari *et al.*, 1999; Misra, 2000; Kokate and Chintalwar, 2003; Ignacimuthu, 2004; Jaykumar *et al.*, 2005; Patole *et al.*, 2008).

Table 19.1: Acute Toxicity Studies of Plant TAE Against Pulse Beetle, *C. chinensis*

Botanical Name	*Part Used*	*Per cent Mortality within 48 h Dose Concentrations*				
		25 mg/ml	*50 mg/ml*	*75 mg/ml*	*100 mg/ml*	*Activity*
Acorus calamus Linn	Rh	33.33±04.24	56.67±08.18	70.00±12.67	83.33±08.12	A
Adhatoda vasica Nees	L	10.00±00.00	40.00±04.94	60.00±05.06	70.00±14.14	A
Anagallis arvensis Linn	Wp	26.67±08.17	50.00±14.14	50.00±24.49	60.00±08.17	B
Andropogon schoenthus Linn	Wp	30.00±14.14	30.00±09.55	23.33±08.17	36.67±08.17	C
Annona squamosa Linn	L	23.33±08.17	40.00±14.14	60.00±05.06	70.00±24.49	A
Annona muricata Linn	L	13.33±08.17	26.67±08.17	50.00±24.49	66.67±08.17	B
Azadirachta indica A Juss	Sd	30.00±14.14	56.68±08.17	70.00±14.14	80.00±08.17	A
Barringtonia racemosa Linn	Fr	20.00±00.00	30.00±09.55	46.67±08.17	50.00±14.14	B
Brassica nigra Linn	Sd	00.00±00.00	20.00±09.43	33.33±08.17	40.00±20.00	C
Cocculus hirsutus Linn	Wp	13.33±04.71	20.00±00.00	40.00±04.71	50.00±14.14	B
Croton tiglium Linn	Sd	16.67±08.17	30.00±14.14	40.00±04.71	63.33±08.17	B

Contd...

Table 19.1–Contd...

Botanical Name	Part Used	Per cent Mortality within 48 h Dose Concentrations				
		25 mg/ml	50 mg/ml	75 mg/ml	100 mg/ml	Activity
Duranta repens Linn	L	20.00±00.00	36.67±08.17	50.00±14.14	63.33±08.17	B
Eucalyptus globule Labill	L	10.00±14.14	20.00±09.43	26.67±08.17	36.67±08.17	C
Lantana camera Linn	L	13.33±04.71	36.67±08.17	53.33±08.17	66.67±08.17	B
Lycopersicon esculentum Milll	L	16.67±08.17	30.00±04.71	46.67±08.17	60.00±24.49	B
Melia azadirachta Linn	L	20.00±08.17	43.33±08.17	60.00±14.14	73.33±08.17	A
Mentha arvensis Linn	Wp	10.00±00.00	23.33±08.17	33.33±16.33	40.00±24.49	C
Murraya koenigii Spreng	L	03.33±00.00	20.00±00.00	20.00±08.17	30.00±24.49	C
Nerium oleander Linn	L	30.00±8.17	46.67±08.17	50.00±24.49	63.33±12.90	B
Nicotiana tabacum Linn	L	20.00±24.49	30.00±28.28	53.33±08.17	70.00±37.42	A
Pongamia glabra Linn	L	30.00±09.55	43.33±08.17	56.67±08.17	70.00±24.49	A
Polygonum hydrapiper Linn	WP	23.33±08.17	30.00±14.14	43.33±08.17	60.00±14.14	B
Randia dumetorum Linn	Fr	26.67±08.17	40.00±14.14	60.00±24.49	70.00±14.14	A
Ricinus communis Linn	L	10.00±00.00	33.33±12.90	50.00±24.49	63.33±08.17	B
Spilanthes acmella Murt	St	10.00±14.14	23.33±08.17	30.00±14.14	30.00±14.14	C
Sphaeranthus indicus Linn	L	20.00±14.14	50.00±14.14	60.00±24.49	73.33±12.90	A
Tephrosia purpurea Pers	Rb	23.33±08.17	40.00±14.14	50.00±14.14	63.33±08.17	B
Thevetia nerifolia Juss	L	23.33±12.90	40.00±24.49	40.00±14.14	50.00±14.14	B
Trigonella foenum Graecum L.	L	10.00±00.00	20.00±00.00	26.67±08.17	30.00±14.14	C
Vitex nigundo Linn	L	30.00±14.14	30.00±14.14	60.00±14.14	76.67±21.60	A
Untreated Control	–	00.00±00.00	00.00±00.00	10.00±00.00	0.00±0.00	–

Values are mean±SD (n = 3). Rh: Rhizome; L: Leaves; Wp: Whole plant; Sd: Seeds; St: Stern; Fr: Fruit; Rb: Root bark.

A: Mortality above 70 per cent; B: Mortality between 40 to 69 per cent C: Mortality below 40 per cent.

The orders of toxicity of TAE of various plants against pulse beetle, *Callosobruchus chinensis* was *Acorus* > *Azadirachta* > *Vitex* > *Sphaeranthus* > *Nicotiana* > *Annona* > *Adhatoda* > *Pongamia* > *Melia*> *Randia*. From these findings we conclude that, the toxic effect of TAE of indigenous plants was time as well as dose dependent and pesticidal principle(s) may be hydrophilic in nature. Advantage of such hydrophilic nature of phytotoxicant may be useful in formulation of eco-friendly natural insecticide, which may prove its utility in controlling the pest generally in the field and particularly in the stored grains. This will help in reducing the load of insecticides on ecosystem.

In nutshell, our preliminary results are encouraging. Thus combination of these plants in appropriate proportion may be used to protect grains effectively, which warrant further experimentation either in the field to protect damage of vegetable crops or in the stored grains.

Acknowledgements

Authors are thankful to principal, S.G. Patil College, Sakri for providing necessary laboratory facilities. Grateful acknowledgement honored to the UGC for financial assistance.

References

Abbott, W.S., 1925. A method of computing the effectiveness of an insecticide. *J. Econ. Ent.,* 8: 265–267.

Ascher, K.R.S., 1993. Non-conventional insecticidal effects of pesticides available from the Neem tree, *A. indica. Insect Biochem. Physiol.,* 22: 433–449.

Chaman, L. and Verma, L.R., 2006. Use of certain bio-products for insect-pest control. *Indian J. Tradi. Knowledge,* 5(1): 79–82.

De-Petro, L.B. and R.C. Jr. De-Petro, 1994. Alternative control strategies against stored product insect pests. *Pest Management Council of Philippines,* 35.

Dwivedi, S.C. and Bhati, P.C., 2006. Antifeedant properties of four plant extracts against pulse beetle, *C. chinensis* L. *Nat. J. Life Sci.,* 3(2): 159–162.

Ignacimuthu, S., 2004. Green pesticides for insect pest management. *Curr. Sci.,* 86(8): 1059–1060.

Isman, M.B., Arnason, J.T. and Towers, G.H.N., 2002. Chemistry and biological activity of ingredients of other species of Meliaceae. In: *The Neem Tree,* (Ed.) H. Schmutterer. Neem Foundation, Mumbai, p. 652–666.

Jaykumar, M., John William, S. and Ignacimuthu, S., 2005. Evaluation of some plant extracts on the oviposition deterrent and adult emergence activity of *C. maculatus* F (Bruchidae : Coleoptera). *Pestology,* 29(1): 37–41.

Khaire, V.M., Kachare, B.V. and Mote, U.N., 1992. Efficacy of different vegetable oils as grain protectants against the pulse beetle, *Callosbruchus chinensis* L. in increasing storability of pigeon pea. *J. Stored Prod. Res.,* 28: 153–156.

Kokate, S.D. and Chintalwar, G.J., 2003. Insect repellent activity of certain plant extracts against pulse beetle, *Callosobruchus chinensis* L. *National Academy Science Letters,* 26(1/2): 44–46.

Maheshwari, V.L., Salunke, B.K., Upasani, S.M., Kotkar, H.M. and Mendki, P.S., 2003. Efficacy of flavonoids in controlling *Callosobruchus chinensis* (L.) (Coleoptera : Bruchidae), a post-harvest pest of grain legumes. *Crop Protection JCRP,* 1406: 1–6.

Misra, H.P., 2000. Effectiveness of indigenous plant products against the pulse beetle *Callosobruchus chinensis* on stored black gram. *Indian J. Ent.,* 62(2): 218 –220.

Patole, S.S., Chopda, M.Z. and Mahajan, R.T., 2008. Biocidal activities of a common weed, *Sphaeranthus indicus* Linn. *Uttar Pradesh J. Zool.,* 28(1): 67–72.

Punam, Kumari, Prem, Kumari, Verma, M.K., Kumari, P. and Kumari, P., 1999. Effect of *Acorus calamus* as pulse grain protectant against *Callosobruchus chinensis. J. Appl. Biol.,* 9(2): 188–189.

Saxena, R.C., Gupta, S., Gupta, S. and Tiwari, A., 1994. Insecticidal action of *Ageratum conyzoides* against *C. chinensis* L (Coleoptera: Bruchidae). *Bio. Sci. Res. Bull.,* 10(1): 21–24.

Tandon, V.R., 2005. Medicinal uses and biological activities of *Vitex nigundo. Nat. Prod. Radiance,* 4(3): 162–165.

Tripathy, M.K., Sahoo, P., Das, B.C. and Mohanty, S., 2001. Efficacy of botanical oils, plant powders and extracts against *Callosobruchus chinensis* L. attacking black gram. *Legume-Research,* 24(2): 82–86.

Chapter 20

Electron Spin Resonance Spectroscopic Technique for Dating of Fossil Bones

G. Velraj and K. Janaki
Department of Physics, Periyar University, Salem – 636 011

ABSTRACT

Electron spin resonance (ESR) spectroscopic technique is applied to estimate the age of bones collected from the archaeological sites Melchittamur and Sankari Durg. The radicals or defects are produced artificially by irradiating with gamma rays from a ^{60}CO source. With an external dose rate of 1.12mGy/a and DPPH standard 'g' value the rough estimates of the site have been done. The results of the investigation are in good agreement with the historical period given by Archaeologists.

Keywords: *ESR, Dating, Fossil bones.*

Introduction

The study of the human past is to understand the human activity in a sequential order. The archaeological data has to be placed in chronological order to realize this sequential cultural process. For this, dating methods are being used in the archaeological investigations. Dating a site or artifact has always been one of the most important elements in archaeological research. Dating is the backbone of archaeology. There are many methods developed for dating the archaeological finds (Taylor and Martin, 1997; Aitken Martin, 1990).

Electron Spin Resonance (ESR) is one of the spectroscopic technique which deals with the interaction between electromagnetic radiation and magnetic moments (Paramagnetic centers) which are found in many materials either as intrinsic constituents or as impurities. In the field of cultural heritage the most important and better known ESR contribution is certainly dating, which has played a key role in developing our present knowledge of the origin and evolution of modem humans (Gron, 1990). The method measures the concentration of paramagnetic centers induced by ionizing radiations (Grun, 1990; Ikeya, 1975) and operates in a time range where other dating techniques can not be applied or unreliable. Hence an attempt ha been made to date the fossil bones collected from Melchittamur in Villupuram district, Sankari Durg in Salem district in Tamil Nadu, India. These two sites were excavated by the Department of Ancient History and Archaeology, University of Madras, Chennai and Archaeological Survey of India (ASI), branch office, Salem respectively.

Sample Preparation and Dose Determination

The bones were collected from the field area by digging through the sand. The samples were dipped initially in 0.1 N acetic acid and then grain with agate mortar and sieved to a grain size of 125–250 microns with repeated soaking with acetone, washed with acetone, dried overnight and exposed to γ-rays from Cobalt source at a rate of 810Gy (Randhawa, 2003). The resulting artificial dose was labeled as 405Gy, 810Gy, 1620Gy, 2430Gy and 3240Gy respectively. The EPR defects, *i.e.,* unpaired electrons are created in the samples by α, β and γ-rays. The concentration of the unpaired electrons can be determined directly by microwave absorption giving the EPR signal intensity. These unpaired electrons are trapped by an impurity or left at a hole, either of which is simply called 'defects' (Ikeya, 1975). ESR signal intensity doesn't show the radiation dose or the age unless the intensity (defect concentration) is calibrated to the dose. The artificial irradiation is usually by γ-rays from a ^{60}CO or a ^{137}Cs source are used as a "time machine" to produce the defect concentrations which will be found in future if natural irradiation continued. The EPR spectrum was recorded on a JEOL TES 100 ESR spectrometer, having 100 kHz field modulation and phase sensitive detection. α,α'-diphenyl-β-picrylhydrazyl (DPPH) with a g value of 2.0036 was used as an internal standard for g factor calculations because it contains 1.53×10^{21} unpaired spins per gram (Aitken, 1985). All the EPR spectra were recorded at room temperature with an optimum microwave power (1.0 m W), sweep time (2 min) and time constant (0.03s) to get good resolution. The g factor was determined using the equation

$$g = hv/\mu_B B \tag{1}$$

here h is the Plank's constant and v is the microwave frequency in GHz and B is the magnetic field in mT and finally μ_B is the Bohr magneton. If the dose rate D is obtained, the age T_{EPR} may be expressed (in ka) using the relation:

$$T_{EPR} = [ED(Gy)/D(Gy)]ka \tag{2}$$

The values ED and D of the archaeological samples constitute the main part of the EPR dating. The intensities have been measured from EPR spectrum and are plotted against the artificial dose for the respective spectrum by using the linear least squares method and Non-linear fit methods, the past equivalent dose is estimated (Ikeya and Miki, 1980).

Results and Discussion

Bone is a special form of connective tissue which is rigid and supports the soft tissues of the body. It constitutes the principal component of almost all adult vertebrate skeletal structures, which protects vital organs, permits locomotion, and plays a vital role in calcium homeostasis (balance) in the body. A fossil bone is composed of an organic part called "collagen" and an inorganic part called "hydroxyapatite" $Ca_{10}(PO_4)_6(OH)_2$ (Brennan *et al.*, 1997). The signal of CO_2 produced from CO_3^- impurity ions in hydroxyapatite has been used for the ESR dating of fossil bones. Dating using ESR method is a analytical tool to explicate recent methods and analyses.

The basic principle is that when released, trapped electrons produce heat and light at quantifiable rates. This is the method based on the effect of ionizing radiation on the fossil material. The radiation creates paramagnetic centers and free radicals with long time, whose concentration is proportional to the total radiation dose absorbed by the sample (Taylor and Martin, 1997). The technique is based on the accumulation in the fossil material during the period of burial of a measurable absorbed radiation dose from cosmic rays, plus uranium, thorium and potassium and their decay products. The ESR technique can estimate the accumulated dose to which the sample was exposed up to the time of measurement by administering to the sample in the laboratory additive known radiation exposures (Grun, 1990). By measuring the signal at each additive dose, a growth curve can be constructed. Then, after assessing the ambient dose rate, an estimate of the burial period of the sample can be obtained. The fossil bones samples were collected from Melchittamur and Sankagiri sites and are named as MCB and SJB respectively.

These samples have been divided into six sets and five of which are given Additive Dose (AD) of 405 Gy, 810 Gy, 1620 Gy, 2340 Gy and 3240 Gy respectively (Randhawa, 2003). The EPR spectrum was recorded for all the five irradiated samples, and the graphs are shown in Figures 20.1 and 20.2.

From the figures Lande's splitting factors were calculated for each irradiation using the formula (1). The estimated 'g' values for each irradiation are given in Table 20.1. The average 'g' values are found to be 2.0185 and 1.975. The intensity of the peaks for all the signals is calculated from the EPR spectra and plotted against artificial dose and the graph for the samples MCB and SNB are shown in Figure 20.3 and Figure 20.4 respectively. The equivalent dose (ED) for a particular intensity versus the artificial dose plot is towards the negative direction and allowing it to intersect the negative x-axis. The negative value of the dose is ED of the sample and this value is substituted in equation (2) which gives the age of the samples in kiloannum and the ages in years are found to be 4424.33 and 397.41 years old (Dash *et al.*, 2002) and the results are tabulated in Table 20.2.

Table 20.1: The Estimated 'g' Values for each Irradiation

Artificial Dose (Gy)	*Sample Code*	
	MCB	*SNB*
0	2.0190	1.9756
405	20.191	1.9758
810	2.0184	1.9756
1620	2.0183	1.9756
2430	2.0183	1.9756
3240	2.0183	1.9751

Table 20.2: The Estimated Ages of the Samples

Sample Code	*Equivalent Dose (Gy)*	*Age in Years by ESR*	*Historical Age*
MCB	1.20176	4424.33	(4th and 5th ACE)
SNB	0.39984	397.41	(17th ACE)

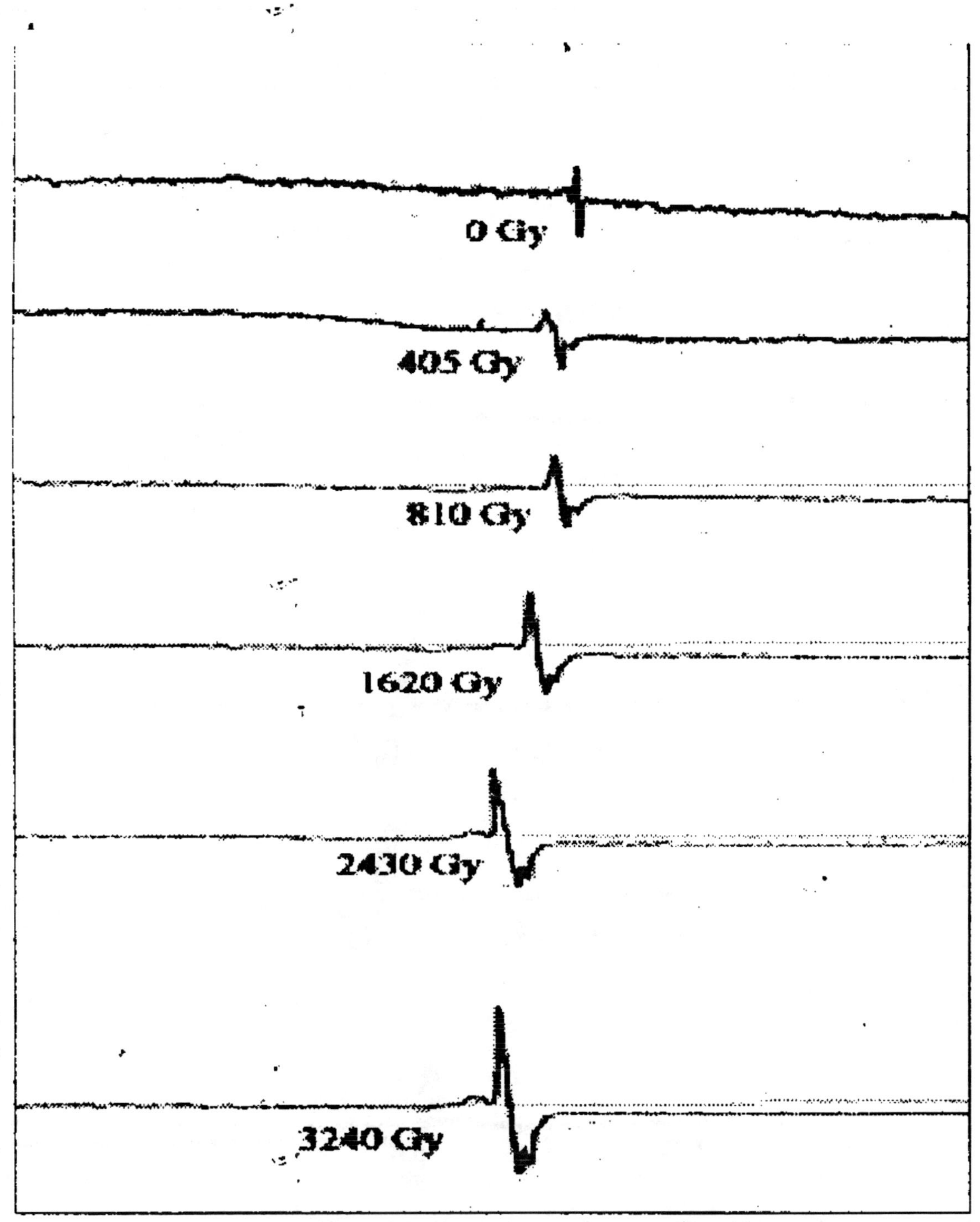

Figure 20.1: ESR Spectrum of the Sample MCB

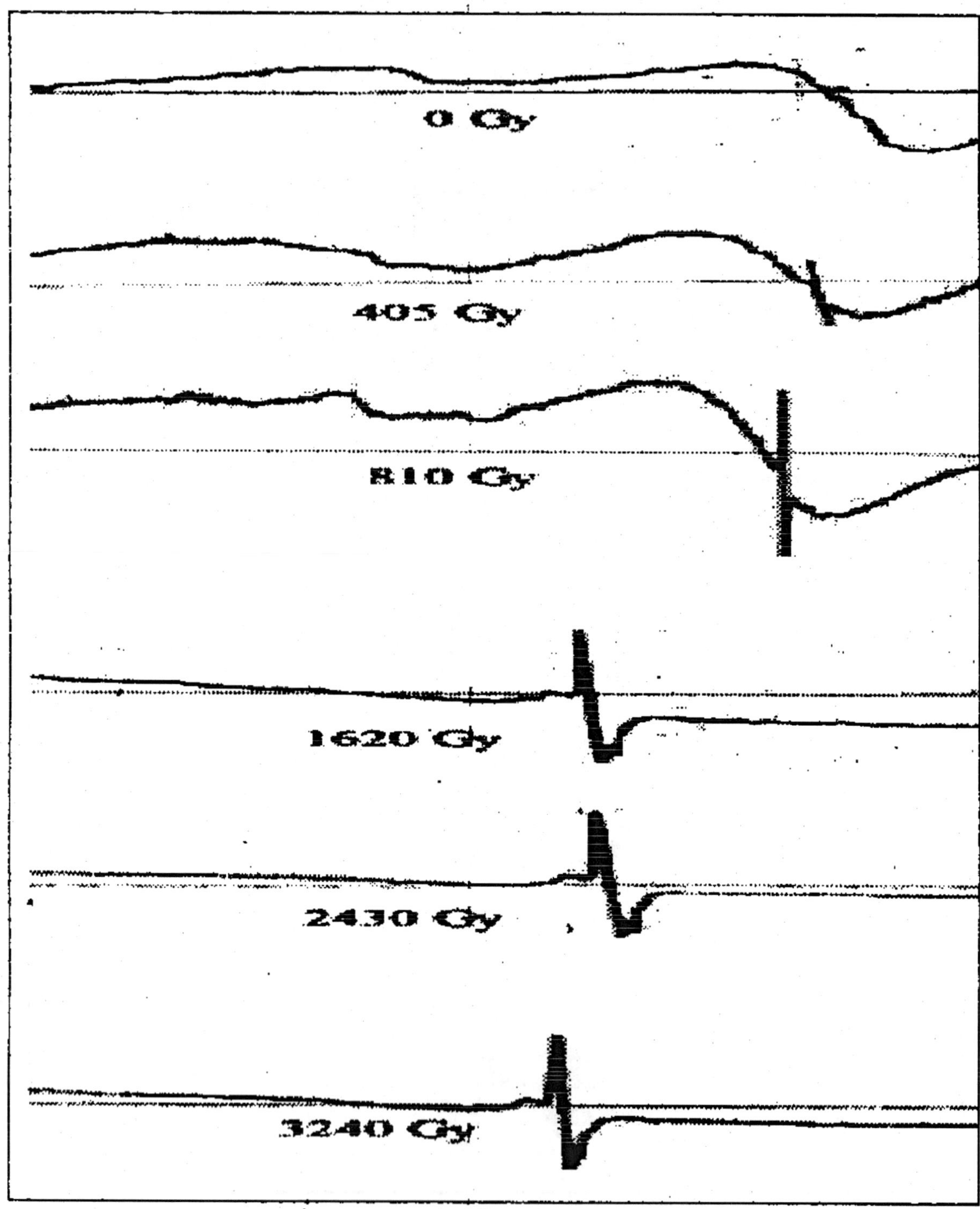

Figure 20.2: ESR Spectrum of the Sample SNB

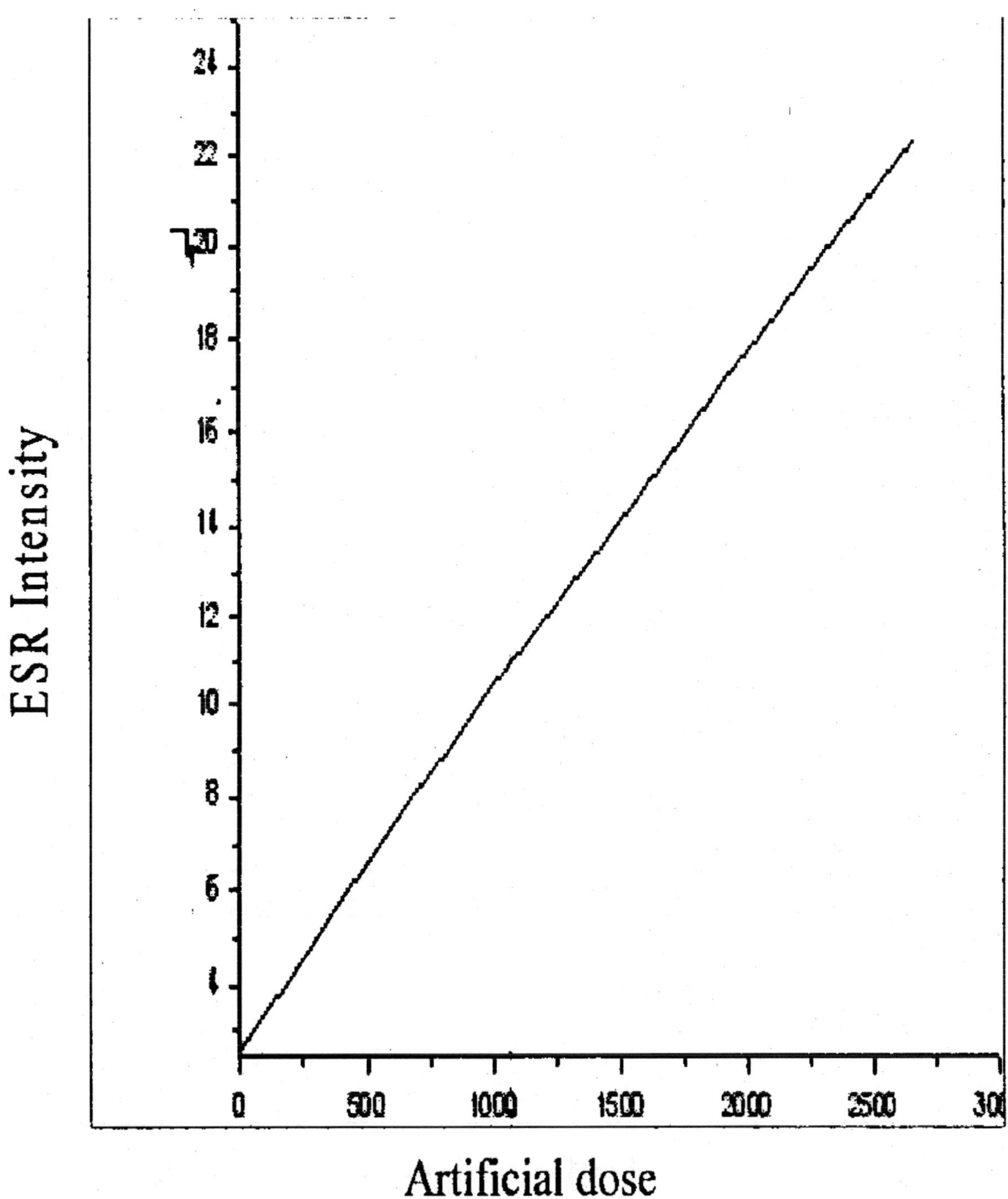

Figure 20.3: Artificial Dose Vs ESR Intensity Graph of the Sample MCB

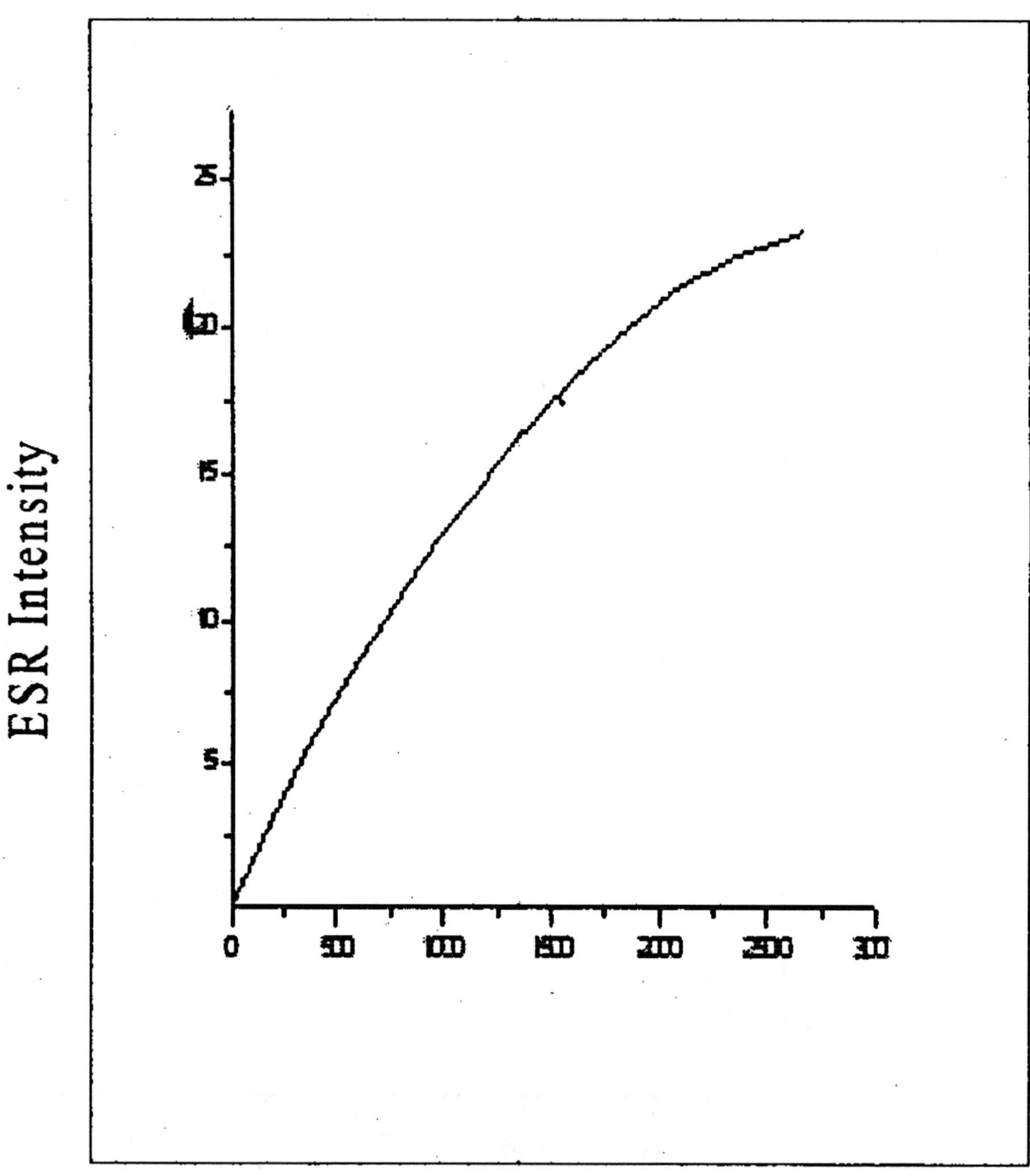

Artificial dose

Figure 20.4: Artificial Dose Vs ESR Intensity Graph of the Sample SNB

Conclusion

The samples collected from Melchittamur and Sankari Durg sites were given equivalent dose and ages were calculated in years old using ESR Spectroscopy. The calculated ages were in good agreement with the historical period given by Archaeologists.

Acknowledgements

The authors acknowledge with thanks to Professor and Head, Department of Ancient History and Archaeology, University of Madras and Mr. S. Mathivanan, Conservation Assistant, Archaeological Survey of India (ASI) for their timely help to collect the samples for the present work. And also the authors thankful to Dr. D. Ponraj, Scientific Officer-F, Indira Gandhi Centre for Atomic Research (IGCAR), Kalpakkam for his help to irradiate the samples. Authors very much thankful to Dr. P. Sambasiva Rao, Professor and Head, Department of Chemistry, Pondicherry University for his timely help to record ESR spectra of the samples.

References

Aitken, M.J., 1985. *Thermoluminescence Dating.* Academic Press, Oxford, London.

Aitken Martin, J., 1990. *Science-Based Dating in Archaeology*. Longman, New York and London, Longman Archaeology Series.

Brennan, B.J., Rink, W.J., McGuirl, E.L., Schwarcz, H.P. and Prestwich, W.V., 1997. Beta doses in tooth enamel by "one-group" theory and the ROSY ESR dating software. *Radiation Meas,* 27: 307–314.

Dash, J.K., Balakrishnan, S. and Rao, P.S., 2002. A preliminary study on application of ESR in dating marine shells along the Pondicherry coast, India. *Advances in ESR Applications*, 18: 41–46.

Grun, R., 1990. Dose response of the paramagnetic centre at g = 2.001 in corals. *Ancient TL,* 8: 20–22.

Ikeya, M., 1975. Dating a Stalacite by electron paramagnetic resonance. *Nature,* 255: 48–50.

Ikeya, M. and Miki, T,. 1980. Electron spin resonance dating of animal and human bones. *Science,* 207: 4434.: 977, 979.

Randhawa, H., 2003. *Modern Molecular Spectroscopy*. MacMillan India Ltd., pp. 520–523.

Taylor, R.E. and Martin J. Aitken (Eds.), 1997. Chronometric dating in archaeology. In: *Advance in Archaeological and Museum Science,* Vol. 2. Plenum Press, New York.

Chapter 21

Electricity Generation and Effect of Catholyte in a Dual Chambered Microbial Fuel Cell Using *Klebsiella pneumoniae* as Biocatalyst

M. Vijay* and Rm. Meyyappan
Department of Chemical Engineering,
FEAT Annamalai University, Annamalainagar – 608 002

ABSTRACT

Biofuel cells (BFC) are capable of converting chemical energy to electrical energy using microorganisms/enzymes (biocatalyst) using electrocatalytic technology. Microbial fuel cells (MFCs) are special types of biofuel cells, producing electric power by utilizing microorganisms, instead of isolated enzymes, to assist redox reactions. The performance of catholyte on the bioelectricity generation was studied in a dual chambered batch type biofuel cell utilizing *Klebsiella pneumoniae*, as anodic inoculum. In this study, microbial based biofuel cell (ie) MFC was operated separately with three configurations, permanganate catholyte (MFCPC), hydrogen peroxide catholyte (MFCHPC) and ferricyanide catholyte (MFCFC). The efficiency of catholyte in power generation under similar operating conditions was studied. From the results, it was clear that a maximum electrical power was obtained, when potassium permanganate was used as the catholyte when compared with other two catholytes (ie) MFCPC generated a maximum power density of 3.435mW/m^2, which was higher than MFCHPC and MFCFC respectively. This study

* Corresponding Author; E-mail: vjtech98@yahoo.co.in.

shows that in addition to ferricyanide, permanganate and hydrogen peroxide can also be used as catholyte in a biofuel cell, out of which, permanganate performed well in the generation of bioelectricity.

Keywords: *Biofuel cell, Biocatalyst, Klebsiella pneumoniae, Catholyte, Electron acceptor.*

Introduction

The forthcoming depletion of fossil fuel resources and increased global warming requires new methods for energy production by using renewable energy sources. The development of biological fuel cells (enzymatic and microbial) are rapidly gaining acceptance as an alternative, sustainable green energy technology in addition to the more conventional environmental energy sources such as wind, solar, and geothermal power. Among electrochemical cells, microbial fuel cells (MFCs) are special types of biofuel cells, producing electric power by utilizing microorganisms, instead of isolated enzymes, to assist redox reactions. They directly generate electricity from low-value biomass, such as agricultural wastes and sewage (Zhen He *et al.,* 2006). This potential combination of waste treatment and power generation by BFC has led to an increased attention and advancement in research progress.

Microorganisms are not designed to dispense energy to power a fuel cell. The majority of relevant redox processes take place buried within the microbial cells and it is a great challenge and a major research issue to find means to efficiently transport electrons from the metabolism to a fuel cell anode (Schroder *et al.,* 2007). To summarize the electricity production, MFC predominantly consists of five steps: (*i*) bio-catalyzed electrochemical oxidation of organic matter in the anode chamber; (*ii*) electron transfer to the anode via endogenous or exogenous mediators or nanowires; (*iii*) electron transport across the external circuit; (*iv*) diffusion of protons through proton exchange membrane and (*v*) reduction of oxidants in the cathode chamber. In MFC, the biochemical energy contained in the organic matter is directly converted into electricity, which can be called as microbiologically mediated "combustion" reaction. This signifies that the overall conversion efficiencies that can be reached are potentially higher for MFCs compared to other step processes (Aelterman *et al.,* 2006). The conventional MFC is a two-chamber system, in which proton exchange membrane (PEM) separates the anode and cathode chambers. This system is termed as half-biological, because the anode side contains electrochemically active microorganisms, whereas the cathode is abiotic. MFC converts chemical energy stored in the organic matter to electricity using microorganisms as the biocatalyst that produces energy, electrons and protons, by oxidizing the organic matter. The energy stored by microorganisms is used for the growth. The electrons are brought to the anode from the inside of microorganisms by the mediator, and flow to the cathode through copper wire. The protons diffuse through the proton exchange membrane (PEM) and enter the cathode chamber. They react with the oxygen and the electrons on the cathode, producing water. The potential between the respiratory system and electron acceptor generates the current and voltage thus converting the organic matter into electricity (Liu *et al.,* 2004). Our study was aimed to investigate the influence of catholyte on bioelectricity generation employing MFC using *Klebsiella pneumoniae* (biocatalyst). In this chapter, we had employed potassium permanganate(MFCPC) and hydrogen peroxide(MFCHPC) as catholytes in the constructed MFC and their efficiency in generation of electricity was investigated and compared with potassium ferricyanide (MFCFC).

The cathode reaction is an important factor in the performance of a MFC. At the cathode, an electron acceptor is reduced using the electrons delivered by the anode. Oxygen is the most sustainable electron acceptor in microbial fuel cells because of its high redox potential and availability in the environment. On the other hand, the following limitations reduce the efficacy of O_2 as electron acceptor. They are (*i*) low oxygen solubility in water limits the availability of the electron acceptor to the electrode and oxygen mass transfer limitation in the liquid phase, (*ii*) high activation energy of O_2 reduction to water at a graphite electrode leads to high cathodic activation over potential. By reducing the MFC voltage, over potentials strongly hinder electron transfer (Freguia *et al.*, 2007). Using more effective catholyte other than oxygen can substantially increase power output.

Materials and Methods

Growth Medium, Microorganism and Inoculum

In the anodic compartment of the fuel cell, we grew *K. pneumoniae* (MTCC 109) using the growth medium that was composed of tryptone, 10 g/L; yeast extract, 5 g/L; NaCl, 5 g/L; Na_2HPO_4, 1.825 g/L; $KH2PO_4$, 0.35 g/L; under aseptic microenvironment. As inoculum, we used 100 mL of microbial suspension in a 250 mL flask. The flask, which was placed in a "rotary shaker at room temperature, operated at 150 rpm and the microorganisms were grown for 24 h prior to inoculating the anodic compartment. The stock solutions necessary for the experiment, as follows; 1 per cent glucose solution; phosphate buffer with pH = 7 (using mono and dibasic sodium) 150 ml; potassium ferricyanide (0.05M), 100ml; potassium permanganate (0.05M), 100ml; and 3 per cent hydrogen peroxide, 100ml; were prepared.

MFC Construction and Operation

A two compartment MFC with nafion membrane as the separator was constructed and was shown in Figure 21.1. Rectangular graphite electrodes (31.7 cm^2) served as the anode and cathode and were suspended from the top cover of the stopper. The volumes of anolyte and catholyte were 200 ml. Anode compartment was filled with 100 ml bacterial suspension (*K. pneumoniae*), 50 ml of phosphate buffer [pH=7], 1 per cent glucose solution, 50 ml distilled water and was agitated using magnetic stirrer. In all the three experiments, the cathode compartment was filled with 100ml of respective catholyte and 100 ml phosphate buffer. Three experimental runs were conducted in the MFC. The first experiment represents 0.05 M of permanganate catholyte (MFCPC), second represents 3 per cent hydrogen peroxide catholyte (MFCHPC) and third represents 0.05 M of ferricyanide catholyte (MFCFC). During this experiment, in order to maintain anaerobic conditions, N_2 gas was sparged into the anodic compartment and get sealed. The distance between the electrodes was 16 cm approximately.

Chemical Analysis

From the anode compartment, 5 ml of anolyte was sampled out daily, and the glucose concentration was determined using DNS method. Based on the result, Glucose was aseptically added to the anolyte after autoclaving in order to maintain 1 per cent glucose concentration throughout the experiment and simultaneously cell mass was calculated using centrifuge method.

Electrical Measurements

An open circuit was used to measure potential, while current was measured in the closed circuit configuration with 100Ω resistance. Voltage and current were measured using a digital multimeter (Unity DT-830D). The maximum power generated was calculated using the following formula: $P_{max} = I^2 R$, where I is the corresponding maximum current obtained and R is the resistance. Current density was calculated using the formula I_{max}/A, where I_{max} is the maximum current generated. Power density

Figure 21.1: MFC Setup with Anode and Cathode Chamber Containing Graphite Electrodes

was calculated using the formula P_{max}/A, where A is the surface area of the anode and for calculations, we used surface area of 31.7 cm^2 for the graphite electrode.

Results and Discussion

A comparative study of three electron acceptors such as permanganate (0.05 M), hydrogen peroxide (3 per cent) and ferricyanide (0.05 M) were performed in the dual-chamber batch type MFC. Generally, the OCP indicates the potential difference between the anode and the cathode, in which the anode potential is generally determined by factors such as substrate conversion rate, electrons from microorganisms to anode transfer rate and surface characteristic of anode; while cathode potential depends on the types of cathodic electron acceptor used as well as the mass and electrochemical reaction rates. Assuming the redox potential of $NAD^+/NADH$ in the anode is a constant, OCP would be only dependent on the performance of cathode. Thus the OCP value actually depends on the type of electron acceptor used (Rabaey *et al.,* 2005). This concept was exactly reproduced in our results because the OCP values of MFCPC, MFCHPC and MFCFC are 512mV, 540mVand 535mV respectively shown in Figure 21.2. Growth curve obtained for different electron acceptors was shown in Figure 21.3. The change in catholytes does not affect the microbial growth factor because similar growth curves were obtained. It is evident from the Figures 21.4 and 21.5, a high power density (3.435m W/ m^2) was achieved in MFCPC at a current density of 104.10 mA/m^2, whereas MFCHPC and MFCFC produced maximum power densities of 2.508 mW/m^2 (at 88.95 mA/m^2) and 2.437 mW/m^2 (at 87.69 mNm^2) respectively, which are all less then MFCPC. Various parameters showing the performance of different catholytes used in the cathode compartment of a MFC was summarized in Table 21.1

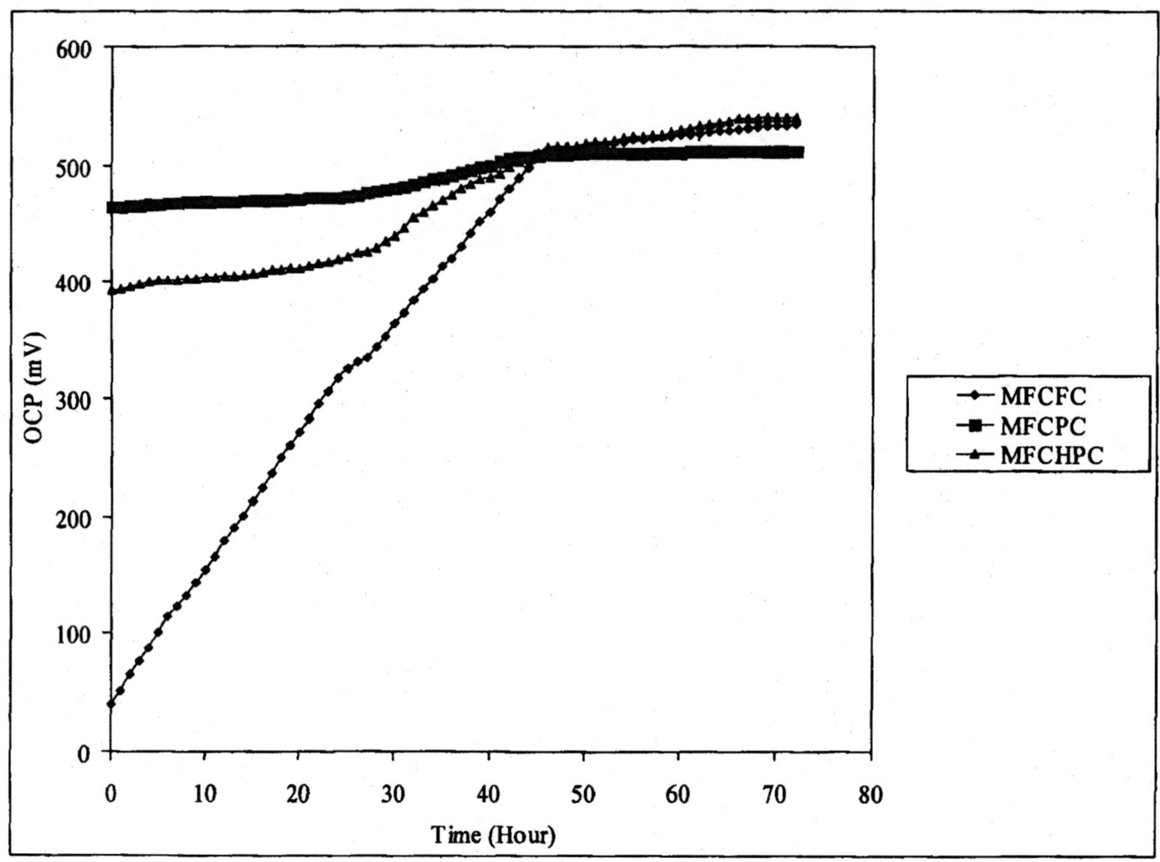

Figure 21.2: Open Circuit Potential of Different Catholytes in MFC

Table 21.1: Various Parameters Showing the Performance of Different Catholytes Used in the Cathode Compartment of a MFC

Parameters	*MFCFC*	*MFCPC*	*MFCHPC*
OCP (mV)	535	512	540
Current max (μA)	278	330	282
P_{max} (μW)	7.72	10.89	7.95
Current density (mA/m^2)	87.69	104.10	88.95
Power density (mW/m^2)	2.437	3.435	2.508

Conclusion

In this study, microbial based biofuel cell has been fabricated and the performance of the cell is evaluated with the effect of catholyte in electricity generation. This work demonstrates a simplest biological fuel cell, employing graphite electrodes that were cost-effective instead of platinum catalyst,

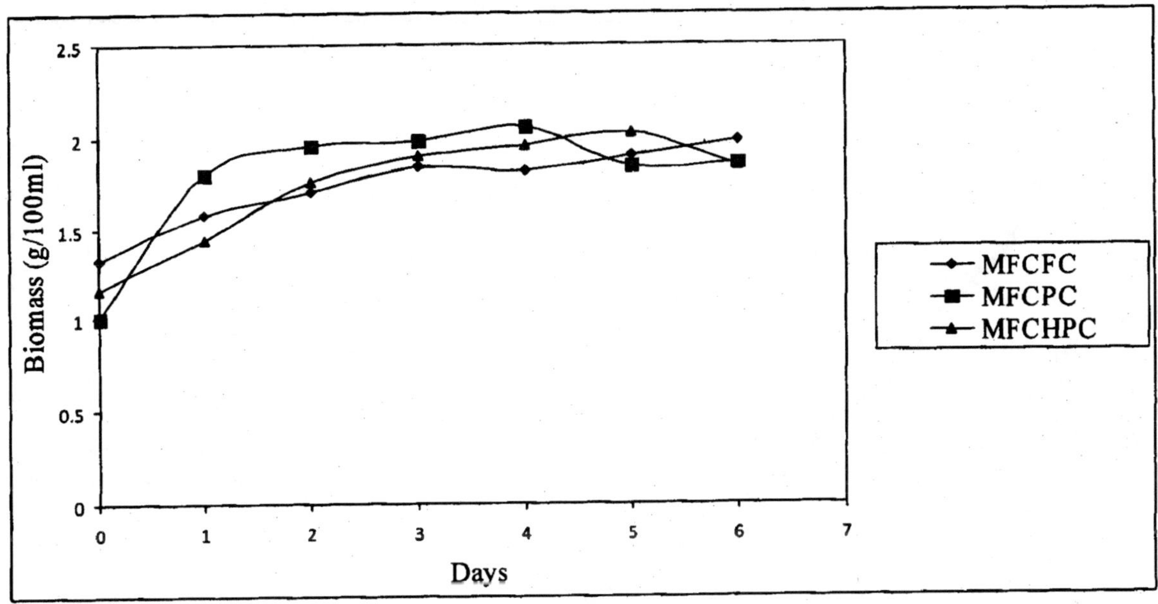

Figure 21.3: Growth Curve of *K. pneumoniae* Obtained in of Different Catholytes in MFC

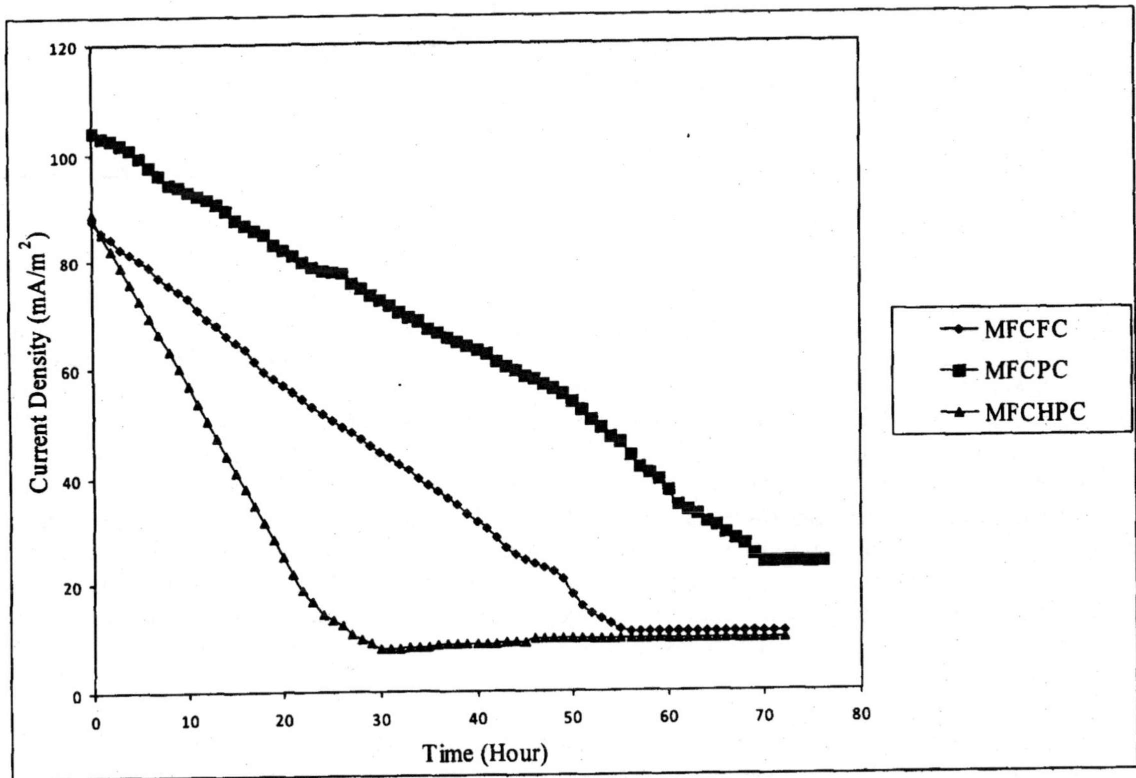

Figure 21.4: Current Density Variations of Different Catholytes in MFC Under a Fixed Load of 100 ohm

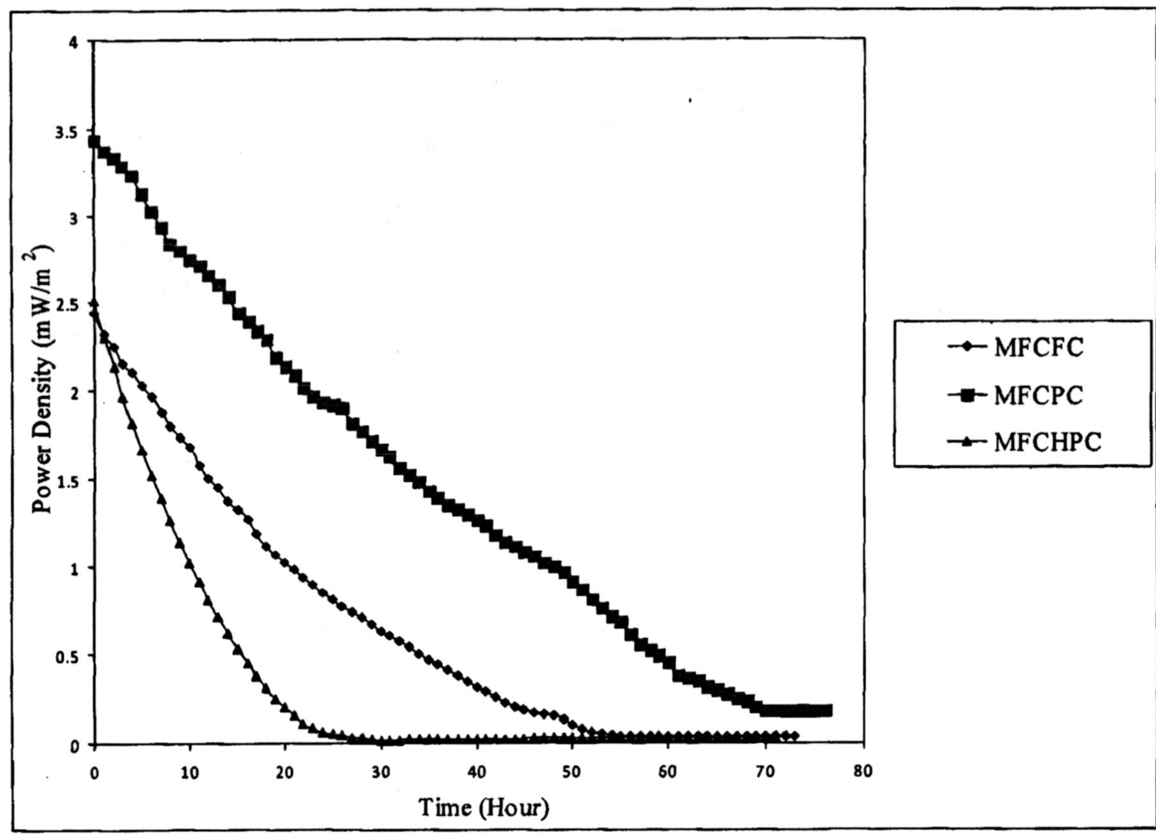

Figure 21.5: Power Density Variations of Different Catholytes in MFC Under a Fixed Load of 100 ohm

carbon cloth and non-coated electrodes. By concept, the efficiency of the cell is increased by various parameters such as (*i*) surface area of membrane, (*ii*) type of electrode and its size, (*iii*) distance between the electrodes, (*iv*) effect of mediator, (*v*) potential of the organism used and these factors need to be studied in future in order to attain high values. Based on the results, it is reasonable to conclude that permanganate performed well and can be used as an electron acceptor in MFC. It enhances a maximum power of 10.89 μW in contrast to potassium ferricyanide (7.72 μW) and hydrogen peroxide (7.95 μW). Like ferricyanide, permanganate MFC also requires liquid replacements to compensate its depletion and this factor limits its application to large scale power supplies. Microbial fuel cells represent an green technology for renewable energy production with simultaneous wastewater treatment. Researchers are paying attention, since it resolves the need of wastewater treatment, biohydrogen production (BEAMR process) and energy generation as an alternative to fossil fuels.

References

Aelterman, P., Rabaey, K., Clauwaert, P. and Verstraete, W., 2006. Microbial fuel cells for wastewater treatment, *Water Science and Technology*, 54: 9–15.

Freguia, S., Rabaey, K., Yuan, Z. and Keller, J., 2007. Non-catalyzed cathodic oxygen reduction at graphite granules in microbial fuel cells. *Electrochimica Acta*, 53: 598–603.

Liu, H., Ramnarayanan, R. and Logan, B.E., 2004. Production of electricity during wastewater treatment using a single chamber microbial fuel cell. *Environ. Sci. Technol.*, 38: 2281–2285.

Rabaey, K. and Verstraete, W., 2005. Microbial fuel cells: Novel biotechnology for energy generation. *Trends in Biotechnology*, 23: 291–298.

Uwe Schroder, 2007. Anodic electron transfer mechanisms in microbial fuel cells and their energy efficiency. *Phys. Chem. Chem. Phys.*, 9: 2619–2629.

Zhen He and Angenent, L.T., 2006. Application of bacterial biocathodes in microbial fuel cells. *Electroanalysis*, 18: 2009–2015.

Chapter 22

Selection of Potential Site for Solid Waste Disposal through GIS Technique: A Case Study of Tambaram Municipality, Kancepuram District, Tamil Nadu, India

R. Annaduri*

Professor and Head I/C, Remote Sensing and GIS, School of Civil Engineering, SRM University, Kattankulathur – 603 203, Tamil Nadu

ABSTRACT

Selection of suitable sites for solid waste disposal has been normally carried by throwing it at all types of Vacant Land in or/and around the city. Remote Sensing data and GIS Techniques can be provided information about the wasteland and other associated features, which help in selection of potentials sites for waste disposal. This paper emphasizes the assessment and systematic analysis the both spatial and non-spatial data through GIS Techniques for site selection for solid waste disposal of Tambaram Municipality Corporation (TML), Kancheepuram district, Tamil Nadu, India, Which is situated 40 kms away from Chennai. At present there is only one declared waste disposal site in this area. Based on the Site Sensitive Index (SSI) value and integration of various parameter using satellite data and GIS as a tool have been found to be very useful for identification of solid waste disposal site for the study areas. During analysis it was observed that the best site for disposal of waste is Vengedamangalam.

Keywords: *Remote Sensing and GIS, Potential site, Solid waste disposal, Site Sensitivity Index (SSI).*

---

* E-mail: annadurisrm@yahoo.com

Introduction

Earlier waste disposal did not pose problems due to less population and lack of public awareness. But due to rapid Urbanization proportionate increase domestic and industrial solid waste generation was observed. The collection, handling and disposal of these large wastes are disposed by some conventional methods have become increasingly difficult. So the efficiency of solid waste depends upon selection of proper site and there are several issues that have impact for site selection. Broadly they are divided into three categories, *i.e.* Economic, Social and Environmental. The Geological, Geotechnical and Hydrological parameters fall within the environmental category. The ultimate greatest protection of the environment is provided. Due to increasing Industrial stations and population in Tambaram Municipality Corporation (TMC), Kancepuram district, Tamil Nadu, India. Large quantities of waste are being generates its different forms such as solid, Liquid, sludge and Gases. Some of these are Bio-Degradable and same or not and hazardous wastes. Wastes are directly thrown away as the streets/roads, Garbage collecting places like bins, etc. Out of which most of it still remain there, which later pile up and Chocks TMC drainage lines. TMC has brought forth several maladies and suffering to human kind, besides bringing economic and cultural development in its fold.

Due to pressure of urbanization it is growing fast and sometime it's developed beyond the planned limits.

Materials and Methods

The site selection for the solid waste disposal facility involves comparison of different options based on social evaluation of each of them for detailed environmental, Social and Community impact and certain rules and regulation have to be followed as directed by the Government of India. The area under study is Tambaram Municipality Corporation (TMC), Kancepuram district, Tamil Nadu, India, lies between Latitude 12°15'38" N and Longitude 80°06'38"E covered by the SOI toposheet on the 1 : 25,000 scale. In this study, on FCC scene of IRS 1 D PAN + LISS III merged satellite imagery at 1 : 25,000 scale, covering a study area was used for identification of the site for disposal of solid waste. The non-spatial data such as number of houses (26500), [Censes data *i.e.* population (1,40347), number of wards (39), number of industries (365), number of hotels (26), Information about Geological Features, Groundwater Table Positions (pre and post-monsoon] data have been collected and analyzed using ARC GIS Software 9.1 Platform. In order to achieve the objective sets for selection of solid waste disposal sites, the Cross Parameter Relationships Method (CPRM) has been adopted (Figure 22.1). In which weightage of different parameters were assigned on the basis of their importance.

Results and Discussion

From this study, the quantity of MSW is 75 MT per day and number of Garbage bins 407 are generated in all 39 wards in the Tambaram Municipality Corporation (TMC).Parameter such as depth to bed rock, slope of the ground, type of rocks, Structural Features, water table, Fluctuation of water table, distance of surface drainage, Groundwater yield, Soil type, Permeability of soil, Population around the site in Kms radius, Land use of the area, Distance of the sites from the waste, Production of Units, present Transport routes, type of waste, etc., were used for the selection of solid waste disposal sites for TMC (Table 22.1).

As there were no wastelands in and around TMC and to go for other options like three sites namely Vengadamangalam, Nallur and Ottivakkam which are present outside the TMC. But these sites are selected on the aspects like Economical, Social and Environmental. On the basis of actual field data, Parameters of site sensitivity Index (SSI) are given weightage (Table 22.2).

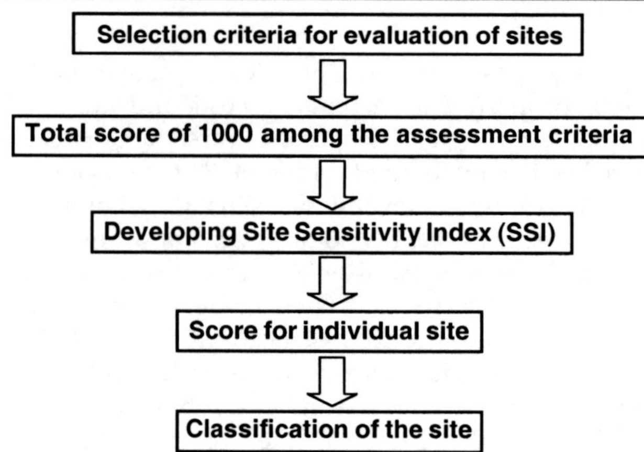

Figure 22.1: Flow Chart for SSI Development and Classification of the Site

Table 22.1: Criteria with Weightages for Municipal Waste Disposal Sites

Sl.No.	*Criteria*	*Weightages for Municipal Waste Disposal Sites*
1.	Geological	Total = 280
	✰ Bedrock depth	80
	✰ Slope of the ground	60
	✰ Type of rock	70
	✰ Structural features	70
2.	Hydrological (Surface and Groundwater)	Total = 260
	✰ Water table depth	90
	✰ Fluctuation of water table	80
	✰ Drainage distance (m)	50
	✰ Groundwater yield	40
3.	Geotechnical	Total = 80
	✰ Soil type	30
	✰ Permeability	50
4.	Socio-economical	Total = 350
	✰ Population within 1 Km radius	90
	✰ Land use	100
	✰ Distance from waste producing units	40
	✰ Transport route Airport	30
	✰ Type of waste	60

The value of Sensitivity Index is multiplied by the corresponding weight age, results in score of the parameters and different classes were ranked on the basis of overall score obtained for the three sites (Tables 22.3, 22.4 and 22.5). The total score for proposed disposal sites are varying from 550 to 660. On the basis of results obtained, the site Nallur scored lowest 557.5, and Ottvikkam site scored

585. Score of these two site areas are near site Vengadamanagalam scored highest 660. So from these results it is observed that even though the sites Nallur and Ottivakkam are capable for Municipal solid waste disposals, Vengadamangalam site which scored high is considered as the best site because it can create employment opportunities from near by villages. Further composting recycling and vermin-culture process can be carried on successfully in this site.

Table 22.2: Rating Parameters with Sensitivity Levels of Various sites

Sl.No.	*Criteria*	*Sensitivity*			
		0.25	*0.5*	*0.75*	*1.0*
1.	Geological				
	☆ Bedrock depth (m)	>100	100-60	60-20	<20
	☆ Slope of the ground	<1 per cent	1 to 3 per cent	3 to 5 per cent	>5 per cent
	☆ Type of the rock	Highly Compact (Massive and stable)	Compact	Weathered	Poor and highly weathered
	☆ Structural features	Without Fault and Joints	Small fault and joints	More fault and joints	Highly faulted and joined
2.	Hydrological (Surface and Groundwater)				
	☆ Water table depth (m)	>30	30–20	20–10	<10
	☆ Fluctuation of water table	<1	1–3	3–5	>5 to waterlog
	☆ Drainage distance (m)	>500	500–300	300–100	<100
	☆ Groundwater yield	Limited	Moderate	Good	Very good
3.	Geotechnical				
	☆ Soil type	Saline/Alkaline	Rocky stony	With Silt	Alluvium
	☆ Permeability	<10–5	10–5–10–4	10–4–103	> 1 0–3
4.	Socio–economical				
	☆ Population within 1 km radius	Negligible	Low	Moderate	High
	☆ Land use	Wasteland	Agricultural	Residential	Forest
	☆ Distance from waste producing units(Km)	<10	10–20	20–30	>30
	☆ Transport route present	Rail	Metalled road	Un metalled road	Nothing
	☆ Distance from Airport (Km)	>20	20–10	15–10	5
	☆ Type of waste	Organic	Municipal	Municipal with small amount of Hazardous	Hazardous

During analysis through ARC GIS 9.1 Software. It was observed that the best site for disposal of solid waste is vengadamangalam and the main advantage of this site is the availability of vast areas/ spaces of waste land and fully developed transportation network.

Table 22.3: Score Sheet for Site Vengadamangalam

Sl.No	*Criteria*	*Sensitivity Index*	*Weightage Score*
1.	Geological		
	☆ Bedrock depth(m)	60	40
	☆ Slope of the ground	1 to 3 per cent	30
	☆ Type of the rock	Poor and highly Weathered	70
	☆ Structural features	Small Fault and joints	.35
2.	Hydrological (Surface and Groundwater)		
	☆ Water table depth (m)	<10	90
	☆ Fluctuation of water table	3-5	60
	☆ Drainage distance (m)	600	12.5
	☆ Groundwater yield	Moderate	20
3.	Geotechnical		
	☆ Soil type	Hand Gravel Soil	15
	☆ Permeability	$> 10^{-3}$	50
4.	Socio-economical		
	☆ Population within 1 km radius	Moderate	67.5
	☆ Land use	Residential	75
	☆ Distance from waste producing units (Km)	14.5	20
	☆ Transport route present	WBM road	15
	☆ Distance from Airport (Km)	35	15
	☆ Type of waste	Municipal	30
	TOTAL		**660**

Table 22.4: Score Sheet for Site Nallur

Sl.No	*Criteria*	*Sensitivity Index*	*Weightage Score*
1.	Geological		
	☆ Bedrock depth (m)	60	40
	☆ Slope of the ground	1 to 3 per cent	30
	☆ Type of the rock	Poor and highly Weathered	70
	☆ Structural features	Small Fault and joins	35
2.	Hydrological (Surface and Groundwater)		
	☆ Water table depth (m)	20–10	67.5
	☆ Fluctuation of water table	>5	80
	☆ Drainage distance (m)	>500	12.5
	☆ Groundwater yield	Good	30

Contd...

Table 22.4–Contd...

Sl.No	*Criteria*	*Sensitivity Index*	*Weightage Score*
3.	Geotechnical		
	✰ Soil type	Gravel with Clay Soil	22.5
	✰ Permeability	< 10^{-3}	12.5
4.	Socio-economical		
	✰ Population within 1 km radius	Low	45
	✰ Land use	Waste Land	25
	✰ Distance from waste producing units (Km)	14	20
	✰ Transport route present	Unmetalled Road	22.5
	✰ Distance from Airport (Km)	45	15
	✰ Type of waste	Municipal	30
	TOTAL		**557.5**

Table 22.5: Score Sheet Site Othivakkam

Sl.No	*Criteria*	*Sensitivity Index*	*Weightage Score*
1.	Geological		
	✰ Bedrock depth (m)	55	40
	✰ Slope of the ground	1 to 3 per cent	30
	✰ Type of the rock	Poor and highly Weathered	70
	✰ Structural features	Small Fault and joints	35
2.	Hydrological (Surface and Groundwater)		
	✰ Water table depth (m)	20–10	67.5
	✰ Fluctuation of water table	3-5	60
	✰ Drainage distance (m)	>500	12.5
	✰ Groundwater yield	Very Good	40
3.	Geotechnical		
	✰ Soil type	Hard Gravel Soil	15
	✰ Permeability	> 10^{-3}	50
4.	Socio-economical		
	✰ Population within 1 km radius	Low	45
	✰ Land use	Agricultural	50
	✰ Distance from waste producing units (Km)	25	10
	✰ Transport route present	WBM road	15
	✰ Distance from Airport (Km)	32	15
	✰ Type of waste	Municipal	30
	TOTAL		**585**

On the basis of integration of various parameters using Remote Sensing data and GIS as a powerful tool have been found to be very useful for identification of solid waste disposal site. Thus use of these technologies management of municipal waste will no longer be a problem for city Administrations.

Acknowledgements

The author express sincere gratitude to Prof. P. Sathyanarayan, Vice-Chancellor, Prof. Dr. T.P. Ganaesan, Pro-Vice-Chancellor (P&D), Prof. Dr. S. Krishnan, Director (E&T), Prof. Dr. C. Muthamizhchelvan, Associate Director (E&T), and Prof. Dr. M. Laxmipathy, Dean, School of Civil Engineering, SRM University, Kattankulathur – 603 020, Chennai, Tamil Nadu, India for providing necessary facilities for carry out this work and also thank Mr. C. Venkatesh, Programmer (Rs&GIS lab) for his helping the processing and preparation a report.

References

Agarwal, C.S. and Garg, P.K., 2000. *Remote Sensing in Natural Resources Monitoring and Management.* Wheeler Publishing and Co. Ltd., New Delhi.

Rao, K.S., 1997. Site selection for Land fills, Narosa Publishing House, New Delhi, 56–64 p.

Sharma, B.K., 1995. *Design of Environmental Protection System for Containment of Municipal and Hazardous Waste: Geohydrological Aspects.*

Chapter 23

Studies on Growth and Biomass Production of *Alnus nepalensis* D. Don, as Affected by Crop and Mulches

P.C. Vanlalhluna and U.K. Sahoo*

Department of Forestry, School of Earth Sciences and Natural Resource Management, Mizoram University, Post Box 190, Aizawl – 796 009

ABSTRACT

Studies on growth and biomass production of *Alnus nepalensis* D.Don, was carried out during 2003 to 2005. Three mulch types (rice straw, weeds and subabul leaves) and doses (6,8 and 10 t/ha) under three crop fields (ginger, turmeric and maize) were tried to assess the growth and biomass production. Among the mulches, subabul leaves influenced better growth increments of tree (height, collar thickness, number of branches and leaf number) compared to other mulches and 10 t/ha mulch applied yielded better growth compared to other treatments (8 and 6 t/ha). The biomass production (shoot biomass, above ground and below ground biomass) increased with the seedling age. However, the differences were not statistically significant. The shoot biomass and total biomass of tree in plots was in the order of maize > turmeric > ginger > control. The vertical root spreading was found greater compared to the horizontal spreading. The taproot biomass and shoot biomass contributed maximally to the below ground and above ground biomass respectively.

Keywords: *Growth, Biomass production, Species mixture, Mulch material, Alnus nepalensis.*

---

* E-mail: sahoo_formzu@yahoo.co.in; uksahoo_2003@rediffmail.com.

Introduction

Alnus nepalensis D.Don., belong to the family Betulaceae. It is widely distributed in South Asia, predominantly a tree of Himalayas, South-western China, Northern Myanmar in between 600–3000 m asl. The species grow well in full light and is moderately shade tolerant, commonly early successional and nitrogen fixing (Boojh and Ramakrishnan, 1981). The drought and frost tolerant characteristic of the species make it a better choice for introduction in a wide range of soil. It is therefore, with those characteristics considered to be a typical pioneer species. Pre-monsoon defolition takes place sometimes due to attack by construction of works. The leaves and the twigs are mulched into excellent green manure, composed manure, and it is also burnt and the ash is spread as fertilizer in the fields. Thick bark are peeled and crush powder soaked in water is used for dyeing clothes and the bark is also burnt to charcoal. The juice extract, of the young barks are drunk to cure Dysentery. The leaves are crushed into paste and are applied to injury, and to wound as homeostatic (Changkija, 1992). Alder is a common early succession and nitrogen fixing tree species, which is innovated by the Nagas for its soil fertility management and biomass in this farming system since time immemorial. Naga farmers have learned to exploits its nitrogen fixing capacity by including it in the agroforestry farming system (Neil, 1989; Dhyani and Chauhan, 1990; Carter, 1992; Amatya *et al.*, 1993).

Materials and Methods

Study Area

The experiment was carried out in the research farm at Tanhril Campus of Mizoram University, Aizawl which is about 15 km on the south-western part of Aizawl city, the capital of Mizoram during 2003 to 2006. It is located between 23°42′ to 23°46′ N latitude and 92°38′ to 92°42′ E longitude with 845 m asl altitudinal range. The agro climatic condition is of humid tropical characterized by short and dry winter, long summer with heavy rainfall. The average annual rainfall of the area is 2500 mm. The temperature does not fluctuate very much throughout the year. In summer, the temperature is usually range between 20° to 30°C and in winter the temperature normally records between 8° to 18°C. Soil of the experimental plots was sandy loam, characterized by P-deficient acidic alfisol having 51.24 per cent sand, 20.71 per cent silt and 28.04 per cent. Soil pH ranged from 5.03–5.40 along the slope.

Treatments

The experiment had 10 treatments laid down in RBD (Randomized Block Design) with three replications totaling 30 subplots. The treatments consisted of three crops field (ginger, turmeric and maize), three mulch types (rice straw, weeds, subabul leaves) and doses (6 t/ha, 8 t/ha, 10 t/ha). One year old *Alnus nepalensi* seedling was transplanted to the experimental field by digging method in the third week of May, 2003 during rainy season at a distance of 2.5 m × 2.5 m a part. At the time of transplanting, the tree seedling height and collar thickness was recorded. The stem diameter at 5 cm from the base stem was measured. Observations were recorded in respect to the various growth attributes like plant height, collar thickness, number of branches, number of leaves, leaf biomass, tap root biomass, fibrous and lateral root biomass, shoot biomass, above ground and below ground biomass, horizontal root spreading and vertical root spreading of the seedling at 6-monthly intervals using standard methods. At a 6-monthly interval, 2 seedlings of each tree species grown under a particular treatment from each replicated plot larva of *Caleaptera* sp., *Oreina aurata foriopiencata* of Chrysomelidae (Phaloura and Singh, 1991). *Alnus,* a multipurpose tree is used as firewood and also used for were excavated from the field preferably near the border, washed gently with water to remove the adhering soil particles. Careful attempt was made to fill back the area with the soil. Since excavation was done during crop harvest time, there was no damage to the field crop. Various root characteristics like lateral root spreading and vertical root length were recorded. Seedlings were divided into various parts like shoots, leaves, tap root, fibrous roots and oven dried at 70°C until the constant weight was obtained. The above ground biomass and below ground biomass and root/shoot ratio on biomass was calculated. The seedling quality parameters such as sturdiness (the ratio height to collar thickness, root/shoot ratio and the ratio of fibrous root biomass to the total biomass were computed. Biomass of individual components was worked out by using formula, Dry matter = Dry weight of sample / Fresh weight of sample × Total fresh weight. The data were subjected to ANOVA (1-way) to seethe effect of growth performance due to different treatments.

Results and Discussion

Effect of Species Mixture on Growth Increments

The growth differences were varied between treatments. The various growth increments (height, collar thickness, number of branches and leaf number) of tree was higher in maize field, followed by turmeric, ginger and minimum in control (Table 23.1). Similar was the case for all the years. Growth increments were always higher in species mixture compared to control (pure tree). A relatively higher growth increments under species mixture could be due to the possibility of their beneficial compatibility, interaction and greater biological efficiency of crops grown in association (Moitra *et al.*, 2000).

Effect of Mulch on Growth Increments

Among mulches, subabul leaves influenced better growth increments of tree (height, collar thickness, number of branches and leaf number) compared to other mulches. This could have been due to quick decomposition of subabul leaves releasing/adding some nutrients to the soil with the mulch application causing better growth performance of the tree in the system. 10 t/ha mulch applied yielded better growth compared to other treatments (8 and 6 t/ha) (Table 23.2). These results indicated that mulches have beneficial effects on growth performance of tree. Similar findings are also reported by Moitra *et al.* (1994); Singh *et al.* (2002) and Sonia Aggarwal *et al.* (2003). Various workers like (Marumota *et al.*, 1991; Ducan *et al.*, 1992; Abdul-Saki and Teasdale, 1993) also mentioned that mulches influence plant growth and yield of crops.

Effect of Species Mixture on Biomass Distribution

The biomass production (shoot biomass, above ground and below ground biomass) increased with the seedling age (Table 23.2). However, the differences were not statistically significant (Table 23.4). The shoot biomass and total biomass of tree in plots was in the order of maize > turmeric > ginger > control. The vertical and horizontal root spreading follow the same pattern (Table 23.3). The vertical root spreading was found greater compared to the horizontal spreading. The taproot biomass also contributed maximally to the below ground biomass. Besides, the shoot biomass also contributed substantially to the above ground biomass in all treatments.

Effect of Mulch on Biomass Distribution

The biomass production over a 3-year period under various treatments was given in (Tables 23.2 and 23.3). Tree having subabul leaves mulch gave relatively higher biomass production (shoot biomass, aboveground, belowground biomass, leaf biomass, tap root biomass and total biomass) over rice straw and weeds mulch. Similarly, tree growing with higher mulch doses (10 t/ha) also gave higher biomass production than those of medium (8 t/ha) and low (6 t/ha) mulch dose. The vertical and

Table 23.1: Growth Performance of *Alnus nepalensis* Under Different Treatments

Treatments	*Growth Parameters*	*Months*					
		6	*12*	*18*	*24*	*30*	*36*
			Crop				
Control	Height (cm)	25.06±0.76	29.33±0.07	33.16±0.49	63.33±2.00	72.53±2.42	86.13±1.10
	Collar thickness (cm)	0.26±0.01	0.53±0.02	0.79±0.04	1.01±0.05	2.56±0.09	3.86±0.23
	Number of branches	1.0±0.02	1.2±0.01	2.9±0.03	3.5±0.03	4.9±0.07	5.6±0.01
	Leaf number	13.2±0.01	17.8±0.04	29.2±0.05	36.3±0.04	49.2±0.01	59.2±0.04
Ginger	Height (cm)	27.46±0.48	32.53±0.70	41.86±0.29	74.73±1.29	87.86±0.63	106.46±0.17
	Collar thickness (cm)	0.29±0.0 1	0.55±0.0 I	0.85±0.05	1.07±0.07	2.64±0.06	3.91±0.21
	Number of branches	1.0±0.03	1.2±0.02	3.1±0.05	3.7±0.05	5.1±0.04	5.7±0.03
	Leaf number	13.7±0.02	18.2±0.03	30.1±0.02	36.9±0.02	50.2±0.03	60.5±0.03
Turmeric	Height (cm)	29.46±1.56	34.26±0.59	46.46±0.63	77.73±3.84	92.73±0.58	110.93±0.67
	Collar thickness (cm)	0.31±0.01	0.58±0.02	0.92±0.08	1.11±0.06	2.67±0.07	3.95±0.20
	Number of branches	1.0±0.03	1.2±0.04	3.2±0.04	3.9±0.06	5.5±0.02	5.8±0.01
	Leaf number	13.9±0.05	18.9±0.02	30.8±0.02	37.3±0.03	50.8±0.02	61.2±0.05
Maize	Height (cm)	35.66±0.96	43.06±0.46	58.53±2.21	87.26±1.55	108.53±1.04	128.73±1.35
	Collar thickness (cm)	0.33±0.02	0.60±0.03	0.98±0.07	1.16±0.06	2.71±0.09	4.00±0.08
	Number of branches	1.0±0.02	1.5±0.01	3.4±0.02	4.2±0.04	6.8±0.02	7.1±0.03
	Leaf number	14.2±0.04	19.2±0.03	31.3±0.05	38.0±0.05	51.2±0.05	62.4±0.02
			Mulch types				
Rice straw	Height (cm)	37.62±0.92	43.21±0.48	58.58±2.36	89.71±1.23	112.65±1.02	130.39±1.29
	Collar thickness (cm)	0.34±0.04	0.62±0.05	0.99±0.06	1.18±0.08	2.73±0.01	4.75±0.28
	Number of branches	1.0±0.03	1.5±0.01	3.0±0.04	4.2±0.06	5.2±0.05	5.9±0.02
	Leaf number	12.7±0.01	15.9±0.02	26.5±0.08	33.1±0.07	46.8±0.02	55.3±0.01
Weeds	Height (cm)	29.48±0.75	35.28±0.09	46.46±0.65	77.79±3.87	92.74±0.48	110.19±1.18
	Collar thickness (cm)	0.32±0.02	0.59±0.04	0.94±0.07	1.13±0.02	2.68±0.07	3.97±0.22
	Number of branches	1.0±0.01	1.3±0.02	2.9±0.05	3.9±0.02	4.9±0.08	5.7±0.04
	Leaf number	12.5±0.04	15.7±0.03	25.8±0.01	32.9±0.05	46.0±0.05	54.9±0.05

Contd...

Table 23.1–Contd...

Treatments	Growth Parameters	Months					
		6	12	18	24	30	36
Subabul leaves	Height (cm)	39.74±0.81	45.28±0.52	61.27±0.74	91.65±2.14	115.32±1.13	134.28±1.31
	Collar thickness (cm)	0.37±0.03	0.65±0.11	1.72±0.03	2.15±0.02	3.99±0.07	5.02±0.14
	Number of branches	1.0±0.04	I. 9±0. 04	3.8±0.01	4.7±0.02	5.9±0.02	6.1±0.04
	Leaf number	12.9±0.02	16.1±0.03	26.8±0.03	33.9±0.01	47.3±0.05	56.8±0.09
			Mulch quantity				
6 t/ha	Height (cm)	29.52±0.71	35.56±0.03	46.66±0.43	77.91±3.82	92.99±0.52	110.89±1.12
	Collar thickness (cm)	0.44±0.03	0.72±0.07	1.02±0.03	1.49±0.05	2.82±0.12	4.01±0.02
	Number of branches	1.0±0.02	1.5±0.03	3.0±0.02	4.1±0.08	5.0±0.03	5.9±0.03
	Leaf number	12.7±0.06	15.8±0.0 1	25.9±0.05	33.0±0.07	46.5±0.02	55.1±0.02
8 t/ha	Height (cm)	36.99±0.83	42.23±0.43	56.58±2.16	87.58±1.34	11 0.60±1.09	127.32±1.52
	Collar thickness (cm)	0.33±0.02	0.60±0.02	0.96±0.01	1.17±0.01	2.71±0.06	4.73±0.20
	Number of branches	1.0±0.01	1.3±0.08	3.0±0.02	4.1±0.02	5.1±0.07	5.9±0.01
	Leaf number	12.7±0.03	15.8±0.06	26.0±0.02	33.0±0.04	46.6±0.06	55.3±0.06
10 t/ha	Height (cm)	38.71±0.69	43.75±0.71	59.62±0.53	89.37±2.33	113.39±1.70	132.21±1.50
	Collar thickness (cm)	0.35±0.07	0.64±0.19	1.71±0.11	2.13±0.11	3.82±0.02	4.99±0.12
	Number of branches	1.0±0.03	1.7±0.08	3.5±0.08	4.6±0.05	5.6±0.07	6.0±0.05
	Leafnumber	12.8±0.12	16.0±0.04	26.5±0.07	33.4±0.06	47.0±0.01	56.1±0.01

±S.E. m. n = 10.

Table 23.2: Biomass Production (g/plant) of *Alnus nepalensis* Under Different Treatments

Treatments	Growth Parameters	Months					
		6	12	18	24	30	36
			Crop				
Control	Shoot biomass	5.10±2.01	5.25±1.08	6.47±1.13	7.4 7±2.12	8.53±2.03	9.81±2.10
	AGB	7.15±1.02	7.33±3.02	9.31±0.23	10.70±2.10	11.52±1.13	12.76±2.09
	BGB	4.93±2.21	6.09±1.04	7.40±2.10	9.04±2.14	10.21±2.03	11.09±1.21
	ABG/BGB	1.45	1.20	1.25	1.18	1.12	1.15
	Total Biomass	12.08	13.42	16.71	19.74	21.73	23.85
	Plant sturdiness	96.38	55.34	41.97	62.70	28.33	22.31
Ginger	Shoot biomass	5.23±1.12.	5.59±2.04	6.65±2.11	7.82±2.01	8.62±2.10	9.93±2.03
	AGB	7.32±0.21	7.72±1.13	9.88±2.10	12.00±1.12	13.27±2.23	13.85±2.02
	BGB	5.98±1.21	6.57±2.15	8.04±2.12	9.71±1.13	11.07±2.10	11.21±1.20
	ABG/BGB	1.22	1.18	1.23	1.24	1.20	1.24
	Total Biomass	13.30	14.29	17.92	21.71	24.34	25.06
	Plant sturdiness	94.46	59.15	49.25	69.84	33.28	27.23
Turmeric	Shoot biomass	5.29±2.10	5.69±2.02	6.74±2.11	7.90±1.02	9.05±2.03	10.04±1.02
	AGB	7.40±0.12	7.83±2.02	10.00±2.22	12.11±2.1 0	13.84±2.20	14.25±3.05
	BGB	6.03±2.07	6.60±2.05	8.13±2.07	9.79±3.30	11.20±3.00	11.85±2.03
	ABG/BGB	1.23	1.19	1.23	1.24	1.25	1.20
	Total Biomass	13.43	14.43	18.13	2 I.90	25.04	26.10
	Plant sturdiness	95.03	59.07	50.50	70.03	34.73	28.08
Maize	Shoot biomass	5.31±0.12	5.74±1.10	6.88±1.23	7.95±1.33	9.12±2.0 I	10.24±3.02
	AGB	7.43±2.13	7.90±1.04	10.15±3.12	12.18±2.20	14.55±3.01	15.06±3.11
	BGB	6.08±3.12	6.80±3.02	8.29±2.17	9.90±3.01	11.52±3.12	12.03±3.02
	ABG/BGB	1.22	1.16	1.22	1.23	1.26	1.25
	Total Biomass	13.51	14.70	18.44	22.08	26.07	27.09
	Plant sturdiness	108.06	71.77	59.72	75.22	40.05	32.18

Contd...

Table 23.2–Contd...

Treatments	*Growth Parameters*	*Months*					
		6	*12*	*18*	*24*	*30*	*36*
			Mulch types				
Rice straw	Shoot biomass	6.09±3.01	6.61±2.17	6.98±1.03	8.22±2.10	9.67±2.17	10.99±2.04
	AGB	7.95±2.12	8.03±2.14	10.78±2.11	11.52±2.02	12.02±1.09	13.74±2.11
	BGB	6.00±2.18	6.89±1.21	7.97±2.12	9.69±2.04	10.97±2.12	11.98±1.22
	ABG/BGB	1.32	1.16	1.35	1.18	1.09	1.14
	Total Biomass	13.95	14.92	18.75	21.21	22.99	25.72
	Plat sturdiness	110.65	69.69	59.17	76.03	41.26	27.45
Weeds	Shoot biomass	5.29±2.09	5.69±3.07	6.74±[illegible].23	7.90±2.32	8.99±1.10	10.11±2.14
	AGB	7.40±3.01	7.83±2.03	10.32±1.13	10.99±2.12	11.87±1.09	13.06±2.12
	8GB	5.22±2.23	6.53±3.21	7.75±2.15	9.69±2.23	10.94±2.21	11.83±2.30
	ABG/BGB	1.41	1.19	1.29	1.33	1.08	1.10
	Total Biomass	12.62	14.36	18.07	20.68	22.81	24.89
	Plant sturdiness	92.13	59.80	49.43	68.84	34.60	27.76
Subabul leaves	Shoot biomass	6.72±3.08	6.99±2.22	7.52±2.14	8.85±2.17	10.02±2.25	11.68±3.41
	AGB	8.13±2.52	8.81±3.05	10.96±3.13	11.97±2.21	12.77±1.23	13.93±2.09
	BGP	6.39±2.11	7.02±1.01	8.22±3.12	9.97±1.07	11.15±2.02	12.91±1.34
	ABG/BGB	1.27	1.25	1.33	1.20	1.14	1.07
	Total Biomass	14.52	15.83	19.[illegible]8	21.94	23.92	26.84
	Plant sturdiness	107.41	69.66	35.62	42.63	28.90	26.75

Contd...

Table 23.2–Contd...

Treatments	*Growth Parameters*	*Months*					
		6	*12*	*18*	*24*	*30*	*36*
		Mulch quantity					
6t/ha	Shoot biomass	5.35±2.07	5.71±2.01	6.82±2.13	7.99±2.10	9.05±2.17	10.23±2.11
	AGB	7.51±3.02	7.92±2.07	10.46±2.05	11.02±2.14	11.97±2.01	13.19±3.10
	BGB	5.32±2.13	6.61±3.21	7.83±2.18	9.75±2.03	10.69±2.23	11.90±2.31
	ABG/BGB	1.41	1.19	1.33	1.13	1.11	1.10
	Total Biomass	12.83	14.53	18.29	20.77	22.66	25.09
	Plant sturdiness	67.09	49.39	45.75	52.29	32.98	27.65
8t/ha	Shoot biomass	6.05±3.07	6.52±2.07	6.81±2.04	8.18±2.15	9.60±2.05	10.81±2.11
	AGB	7.90±1.52	8.00±0.12	10.72±1.10	11.48±2.12	12.00±3.01	13.63±2.07
	BGB	5.86±2.18	6.80±1.21	7.91±2.12	9.82±2.04	10.72±2.12	11.93±1.22
	ABG/BGB	1.34	1.17	1.35	1.16	1.11	1.14
	Total Biomass	13.76	14.80	18.63	21.30	22.72	25.56
	Plant sturdiness	112.09	70.38	58.94	74.85	40.81	26.92
10t/ha	Shoot biomass	6.52±3.05	6.76±2.22	7.32±2.08	8.45±3.12	9.85±2.52	11.26±2.14
	AGB	8.02±2.10	8.56±1.56	10.81±1.18	11.73±2.15	12.59±2.04	13.88±2.10
	BGB	6.42±1.08	6.92±2.22	8.03±0.17	9.86±2.01	10.91±1.17	12.05±1.20
	ABG/BGB	1.24	1.23	1.34	1.18	1.15	1.15
	Total Biomass	14.44	15.48	18.84	21.59	23.50	25.93
	Plant sturdiness	110.60	68.36	34.87	41.96	29.68	26.49

±S.E. m, n = 10, AGB: Above ground biomass; BGB: Below ground biomass.

Table 23.3: Characteristics Rooting Ratio and Biomass (Dry Weight) of *Alnus nepalensis* Under Different Treatments

Parameters	*Crop*				*Mulch Types*			*Mulch Quantity*		
	Control	*Ginger*	*Turmeric*	*Maize*	*Rice Straw*	*Weeds*	*Subabul*	*6t/ha*	*8t/ha*	*10t/ha*
Vertical root spreading (cm)	18.44 (1.22)	21.20 (2.01)	22.19 (1.33)	23.85 (1.02)	23.82 (3.02)	19.22 (2.13)	32.05 (2.21)	19.24 (3.02)	24.52 (1.28)	32.01 (1.02)
Horizontal root spreading (cm)	10.10 (I. 19)	11.14 (0.20)	11.74 (1.29)	13.44 (0.10)	14.09 (2.05)	10.15 (2.11)	18.36 (2.10)	10.18 (1.07)	15.65 (1.20)	18.32 (1.29)
Shoot biomass (g/plant)	9.81 (2.10)	9.93 (2.03)	10.04 (1.02)	10.24 (3.02)	10.99 (2.04)	10.11 (2.14)	11.68 (3.41)	10.23 (2.11)	10.81 (2.11)	11.26 (2.14)
Leaf biomass (g/plant)	3.23 (0.52)	4.18 (1.25)	4.21 (2.03)	4.23 (2.11)	4.26 (0.03)	3.25 (0.22)	5.23 (1.00)	3.27 (1.02)	4.76 (0.10)	5.21 (0.11)
Above ground biomass(g/plant)	12.76 (2.09)	13.85 (2.02)	14.25 (3.05)	15.06 (3.11)	13.74 (2.11)	13.06 (2.12)	13.93 (2.09)	13.19 (3.10)	13.63 (2.07)	13.88 (2.10)
Tap root biomass (g/plant)	6.03 (1.01)	6.54 (1.04)	6.59 (2.01)	6.63 (1.20)	6.85 (1.10)	6.07 (1.04)	7.08 (1.03)	6.09 (2.02)	6.93 (0.10)	7.05 (0.11)
Fibrous and lateral root biomass(g/plat)	3.01 (0.02)	3.17 (1.10)	3.20 (1.02)	3.27 (1.02)	3.62 (2.10)	3.05 (1.02)	4.83 (0.03)	3.07 (1.00)	3.75 (0.12)	4.80 (1.10)
Shoot : root length ratio	1.95	3.43	3.48	3.55	3.62	2.06	5.23	2.11	3.74	5.21
Below ground biomass (g/plant)	11.09 (1.21)	11.21 (1.20)	11.85 (2.03)	12.03 (3.02)	11.98 (1.22)	11.83 (2.30)	12.91 (1.34)	11.90 (2.31)	11.93 (1.22)	12.05 (1.20)
AGB : BGB ratio	1.15	1.24	1.20	1.25	1.14	1.10	1.07	1.10	1.14	1.15
Total biomass (g/plant)	23.85	25.06	26.10	27.09	25.72	24.89	26.84	25.09	25.56	25.93
Plant sturdiness	22.31	27.23	28.08	32.18	27.45	27.76	26.75	27.65	26.92	26.49

± S.E. m, n = 10, AGB: Above ground biomass; BGB: Below ground biomass.

horizontal root spreading followed the same sequence of 10 t/ha > 8 t/ha > 6 t/ha. The higher in biomass production and the root spreading due to spread of mulch could be attributable to the beneficial influence of mulches on soil properties particularly the organic matter content favouring plant growth as also argued by Escober *et al.* (2002).

Table 23.4: Analysis of Variance (ANOVA, 1-way fixed effect model) Due to Different Treatments on Growth Attributes of *Melia azadirachta*

Parameters	Crop		Mulch Types		Mulch Quantity	
	F-ratio	P	F-ratio	P	F-ratio	P
Height	0.620907	0.609660*	0.951567	0.434630	0.838196	0.488804
Collar thickness	0.008942	0.998799*	0.308566	0.818909*	0.252683	0.858485*
Shoot biomass	0.061689	0.979389*	0.759850	0.529760*	0.504270	0.683697*
Leave biomass	0.619626	0.604697*	1.102356	1.101245*	1.101185	1.100256*
Above ground biomass	0.274255	0.843245*	0.364091	0.779629*	0.270834	0.845667*
Tap root biomass	0.437381	0.726967*	0.526897	0.213567	0.201356	0.214548
Fibrous and lateral root biomass	0.136591	0.937851*	0.120356	0.102124	0.111895	0.100321
Below ground biomass	0.192908	0.899983*	0.230445	0.874085*	0.159320	0.922418*
Total biomass	0.236631	0.869760*	0.292995	0.829960*	0.211345	0.887347*

* Significant at P = 0.05

Acknowledgements

We are grateful to the Indian Council of Agricultural Research, New Delhi for providing financial assistance to carry out this research work.

References

Abdul-Baki, A. and Teasdale, J.R., 1993. Ano tillage tomato production system using hairy vetch and subterranean clover mulch. *Hort. Sci.*, 28: 106–08.

Aggarwal, Sonia, Korla, B.N. and Raina, J.N., 2003. Effect of mulches on soil hydrothermal regimes, weed incidence, yield and quality of ginger. *J. Indian Soc. Soil Sci.*, 51: 65–67.

Amatya, S.M. and Newman, S.M., 1993. Agroforestry in Nepal: Research and practices. *Agr. Syst.*, 21: 217–222.

Boojh, R. and Ramakrishnan, P.S., 1991. Germination of seeds of *Alnus nepalensis* D.Don. *National Acad. Of Sci. Letters*, 4: 53–56.

Carter, E.J., 1992. Tree cultivation on private land in the middle hills of Nepal: Lessons from some village of Dolakha district. *Mountain Research and Development*, 12: 241–255.

Changkija, S., 1992. Ethnobotanical study of Mokokchung and Tuensang Districts of Nagaland (Unpublished) *Ph.D. Thesis*, North Eastern Hill University, Shillong.

Dhyani, S.K. and Chauhan, D.S., 1990. Nitrogen fixing trees suitable for agro-forestry in Meghalaya. *Indian J. Hill Farming*, 3: 65–68.

Duncan, R.A., Stapleton, J.J. and McKenry, M.Y., 1992. Establishment of orchards with black polythene film mulching. *J. Nematology*, 24: 681–687.

Marurnota, T., Aoki, M., Suzuki, Y., Kusaka, T., Kheng, J.C.W. and Higashi, T., 1991. Effects of rhizosphere conditions on the growth of strawberry. *Bulletin* 39. Faculty of Agriculture, Yamaguchi University, Yamaguchi, pp. 23–35.

Moitra, R., Ghose, D.C., Majumdar, S.K. and Sarkar, S., 1994. Influence of tillage and mulching on growth, yield and economics of rainfed yellow (*Brassica rapa* var. *glausa*) in late rite acid belt of West Bengal. *Indian J. Agric. Sci.*, 64: 156–159

Moitra, R., Ghose, D.C., Sounda, G., Jana, P.K. and Roy, D.K., 2000. Production, competition and economics of inter cropping legumes in finger millet (*Eleusine coracana*) at different fertility levels. *Indian J. Agric. Sci.*, 70: 824–828.

Neil, P.E., 1989. *Alnus nepalensis*: A multipurpose tree for the tropical highlands. *Nitrogen Fixing Tree-Highlights*, 90–106: 2.

Phaloura, S.P.S. and Singh, T., 1991. Coccinelid (*Coleoptera*) fauna associated with Indian Alder, *Alnus nepalensis*, D.Don. *Uttar Pradesh Journal of Zoology*, 11: 79–80.

Singh, K., Singh, U.N., Singh, Y., Singh, S.R., Chandel, R.S. and Singh, K.K., 2002. Effect of tillage, inter terrace and mulching on yield and nutrient uptake by maize (*Zea mays*), Chick pea (*Cicer arietinum*) cropping system. *Indian J. Agric. Sci.*, 72: 727–730

Chapter 24

Cryopreservation of Spermatozoa of an Endangered Fish *Osphronemus goramy* and a Common Table Fish *Puntius sarana*

L. Bindu*

Zoological Survey of India, M-Block, New Alipore, Kolkata – 700 053

ABSTRACT

Developed an extender-cryoprotectant system for the long-term storage of the spermatozoa of an endangered fish, *Osphronemus goramy* and a common table fish, *Puntius sarana*. Five extenders and three cryoprotectants were used in the present study. The cryoprotectants were used in three different percentages *viz.*, 5 per cent, 7.5 per cent and 10 per cent. The experiment was conducted for 90 days. The extenders and cryoprotectants were found to be species specific.

Keywords: *Cryopreservation, Extender, Cryoprotectant, Spermatozoa, Osphronemus goramy, Puntius sarana.*

Introduction

Cryopreservation of spermatozoa is an important component of genetic conservation programmes for fish. It is an alternative method of preserving endangered stock at low temperature and this is the

---

* E-mail: bindulajapathi@yahoo.com

method adopted successfully from animal husbandry by aquaculture industry. Storage of fish milt improves the induced spawning technology; permits the exchange, easy transport and distribution of sex products from one area to another and to spawn the fishes with differential maturity, thus making it possible to produce the fish juveniles year round (Stoss and Refstie, 1983). The long-term storage of fish gametes would enable the gap between the spawning season to be bridged and this opens the door to production, management and marketing potentials (Thorogood and Blackshaw, 1992). The present study was undertaken to develop a suitable extender cryoprotectant system for the long-term cryopreservation of the spermatozoa of *Osphronemous goramy* and *Puntius sarana*. Cryopreservation of spermatozoa of these fishes has got great importance as these are annual breeders and also *O. goramy* is an endangered species.

Materials and Methods

The milt used for the investigation was collected from 10 mature male specimens of each species having the following size range: *O. goramy*, 7 to 9 kg (50 to 55 cm); *P. sarana*, 110 to 115g (25 to 30 cm). A series of experiments were conducted to develop the best extender-cryoprotectant system for the long-term cryopreservation of spermatozoa of these fishes. Ten replications were maintained for each experiment.

Five different types of extenders were prepared for the preservation of sperm cell viability. Cryoprotective agents were incorporated into the extender media. In each extender, experiments were conducted using three different cryoprotectants *viz;* glycerol, dimethyl sulfoxide (DMSO) and methanol. Each of these cryoprotectant was used in three different concentrations *viz.*, 5 per cent, 7.5 per cent and 10 per cent. The extenders used and their composition are as follows:

Extender-I (Fishringer solution)	NaCl-6.5g, KCl-0.25g, $NaHCO_3$-0.2g, $CaCl_2.6H_2O$-3g, DW-1000ml
Extender-II	NaCl-6.5mg, KCl-0.25mg, $NaHCO_3$-0.20mg, $CaCl_2.6H_2O$-0.15mg, Milk powder-150mg, DW-1000ml
Extender-III	NaCl-750mg, KCl-20 mg, $CaCl_2$-20mg, $NaHCO_3$-20mg, DW-100ml
Extender-IV	NaCl-11.682g, KCl-5.115g, $CaCl_2.2H_2O$-0.205g, $MgCl_2.6H_2O$-0.440g, Na_2HPO_4-0.530g, Citric acid-0.200g, Bicine solution-40.00ml (5.3g bicine in 100ml distilled water), Glucose-20g, DW-2 litres
Extender-V	NaCl-11.682g, KCl-5.115g, CaCh.$2H_2O$-0.205g, $MgCl_2.6H_2O$-0.440g, Na_2HPO_4-0.530g, Citric acid-0.200g, Bicine solution-90ml, Glucose-20g, KOH(1.27g/100ml)-20ml, DW-2 litres

Collection and Extension of Semen

The method used by Erdahl (1986) was followed in the present study with slight modification. Samples of milt exhibiting more than 90 per cent motile spermatozoa were used for the study. The ratio of milt and diluent used in all experiments was 1 : 4.

Samples were immediately placed in dry ice for 15 minutes after 10 minutes equilibration period by which sample temperature had reached at –79°C and then the eppendorf tubes containing samples were transferred to cannisters in a cryocan. Cannisters holding the tubes were frozen by exposing them to liquid nitrogen vapours by holding them at about 2cm. above the surface of liquid nitrogen and then stored in liquid nitrogen at –196°C.

Thawing and Observation

After 90 days the preserved samples were thawed by immersion in a water bath at 30°C (warming rate of approximately 10°C/minute). The percentage of motile spermatozoa was observed under a prefocussed microscope. The viability was estimated by standard eosine-nigrosine dye exclusion method (Chao *et al.*, 1975).

Results and Discussion

In the present study percentage viability and motility of the spermatozoa are used to verify the effect of extender-cryoprotectant combination on pre-frozen and frozen thawed spermatozoa of 0. *goramy* and *P. sarana*. Spermatozoan motility was used to evaluate the quality of raw milt and to verify the effect of extendercryoprotectant combination on freezing and frozen thawed samples (Chao *et al.*, 1987). Percentage of motile spermatozoa has been used as criterion for assessing potential fertility of semen (Temer, 1986). Motility and fertility are reported to be well correlated (Stoss, 1983). Viability is the most reliable index of the physiological condition of the spermatozoa and correlates well with fertility (Chao *et al.*, 1987). A slight decrease in the motility and viability of spermatozoa from the fresh state was observed after preservation for 90 days. Similar decrease in the spermatozoan viability and motility after long-term preservation has also been reported by several workers in many species of fishes (Gwo *et al.*, 1991; Holtz *et al.*, 1991; Leung, 1991). Various extenders and cryoprotectants have been reported to be species specific for freezing fish spermatozoa (Tiantian *et al.*, 1993; Weismann *et al.*, 1995; Conget *et al.*, 1996). The present observation shows that extender II added with 10 per cent methanol appears to be a superior extender cryoprotectant combination for preserving the spermatozoa of *O. goramy* (Table 24.1). The present results are in agreement with the findings of Calvi *et al.* (1994) in *C. carpio*. In experiments with channel catfish Tiersch *et al.* (1994) found that motility was highest in samples cryopreserved with 5 per cent and 10 per cent solution of methanol. The post thaw motility and viability of spermatozoa of *P. sarana* was higher when the milt was diluted with extender-I with 7.5 per cent DMSO (Table 24.2). DMSO has been successfully used by many investigators (Kurokura *et al.*, 1986; Leung, 1987). It has also been suggested that DMSO is more suitable as a cryoprotectant for freshwater fish spermatozoa than glycerol (Scott and Baynes, 1980). Compared to glycerol DMSO has

Table 24.1: Percentage Motility and Viability of Spermatozoa of *O. goramy* Using Different Extenders and Cryoprotectants at 5 per cent, 7.5 per cent and 10 per cent After Preservation for 90 Days

Cryoprotectants	*Motility*					*Viability*				
	Extenders					*Extenders*				
	I	*II*	*III*	*IV*	*V*	*I*	*II*	*III*	*IV*	*V*
5 per cent DMSO	91.2	90.0	80.6	85.2	82.1	96.2	92.2	83.06	89.6	85.08
5 per cent Methanol	92.0	82.1	83.0	70.3	72.2	95.86	89.53	88.46	74.0'	75.4
5 per cent Glycerol	90.0	72.6	73.5	83.2	82.0	92.46	77.0	78.6	86.53	86.26
7.5 per cent DMSO	90.1	82.0	72.5	75.3	91.3	94.33	88.26	77.26	78.66	96.53
7.5 per cent Methanol	88.5	82.4	78.7	69.0	75.1	89.33	88.66	84.13	72.66	79.0
7.5 per cent Glycerol	73.2	80.1	71.0	68.2	85.5	77.01	84.13	73.86	69.73	87.66
10 per cent DMSO	80.0	71.1	68.0	70.2	90.1	84.8	76.33	71.06	74.8	93.6
10 per cent Methanol	90.0	98.0	82.0	71.0	83.0	92.46	98.46	89.0	76.13	86.0
10 per cent Glycerol	83.2	72.0	62.1	62.5	78.0	87.06	76.6	64.06	65	78.6

been proved as a better cryoprotectant for freezing spermatozoa of a variety of species (Doi *et al.*, 1982; Leung, 1987; Miyaki and Dotsu, 1987). Koldras and Bieniarz (1987) successfully preserved carp spermatozoa in 10 per cent DMSO.

Table 24.2: Percentage Motility and Viability of Spermatozoa of *P. sarana* Using Different Extenders and Cryoprotectants at 5 per cent, 7.5 per cent and 10 per cent After Preservation for 90 Days

Cryoprotectants	*Motility*					*Viability*				
	Extenders					*Extenders*				
	I	*II*	*III*	*IV*	*V*	*I*	*II*	*III*	*IV*	*V*
5 per cent DMSO	89.0	81.0	82.2	88.0	69.0	92.93	83.53	85.26	90.66	73.2
5 per cent Methanol	82.0	84.2	85.3	68.6	60.1	85.53	85.6	88.2	73.06	62.33
5 per cent Glycerol	79.8	78.4	71.7	73.0	70.1	82.86	80.4	74.66	76.93	72.26
7.5 per cent DMSO	93.86	85.0	88.2	88.4	89.0	94.93	85.93	92.06	91.26	86.06
7.5 per cent Methanol	82.9	78.0	82.0	78.2	81.5	85.26	83.46	85.6	82.06	84.4
7.5 per cent Glycerol	86.7	85.0	87.0	84.0	84.0	88.0	88.66	88.8	85.13	86.2
10 per cent DMSO	91.0	82.2	80.5	89.3	85.2	91.0	85.33	82.66	92.73	90.0
10 per cent Methanol	88.0	88.0	79.1	77.8	89.0	86.66	86.13	65.53	72.2	90.86
10 per cent Glycerol	89.2	88.3	87.8	90.0	85.5	91.13	91.8	90.86	92.0	89.46

In conclusion there is a pronounced species specificity in the requirements of extender, the freezing and thawing method and the freeze-thaw tolerance of fish spermatozoa. Once spermatozoa have been successfully frozen in liquid nitrogen, their viability upon thawing should not change appreciably with increasing storage time (Ashwood-Smith, 1980; Wittingham, 1980). The ability to freeze gametes and store them for long periods of time without drastic loss in fertilizing capacity, should allow the gap between spawning season to be bridged, and thus ensure the maintenance of adequate numbers of viable embryos from which to hatch new fish populations the world over.

Acknowledgement

The author is grateful to the Head of the Department, Aquatic Biology and Fisheries, University of Kerala for the facilities provided.

References

Ashwood-Smith, M.J., 1980. Low temperature preservation of cells, tissues and organs. In: *Low Temperature Preservation in Medicine and Biology*, (Eds.) M.J. Ashwood-Smith, and J. Farrant. Pitman Medical Ltd., Turnbridge Wells, Kent, pp.19–44.

Calvi, L.S., Zoccarato, I., Gasco, I. and Andrione, A., 1994. Motility of cryopreserved carp (*Cyprinus carpio* L.) milt: Effects of diluent, cryopreservation and thawing procedures. *Rivista Italiana Acquacoltura*, 28: 187–195.

Chao, N.H., Chao, W.C., Liu, K.C and Liao, I.C., 1987. The properties of Tilapia sperm and its cryopreservation. *Journal of Fish Biology*, 30: 107–118.

Chao, N.H., Chen, H.P. and Liao, I.C., 1975. Study on the cryogenic preservation of grey mullet sperm. *Aquaculture*, 5: 386–406.

Conget, P., Fernandez, M., Herrera, G. and Minguell, J.J., 1996. Cryopreservation of rainbow trout (*Oncorhynchus mykiss*) spermatozoa using programmable freezing. *Aquaculture,* 143: 319–329.

Doi, M., Hoshmo, I., Taki, Y. and Vgasawara, Y., 1982. Activity of the sperm of the bluefin tuna *Thunnus thunnus* under fresh and preserved conditions. *Bull. Jpn. Soc. Sci. Fish.*, 48: 495–498.

Erdahl, A.W., 1986. Factors affecting storage and fertilization of fish gametes. *Ph.D. Thesis*, University of Minnesota, MN, pp. 128.

Gwo, J.C., Strawn, K., Longnecker, M.T. and Arnold, C.R., 1991. Cryopreservation of Atlantic croaker spermatozoa. *Aquaculture,* 94: 355–375.

Holtz, W., Schmidt-Baulain, R. and Mimers-Getken, 1991. A simple saccharide extender for cryopreservation of rainbow trout (*Oncorhyncus mykiss*) sperm. In: *Proc. Fourth Int. Symp. Reprod. Physiol. Fish,* 7–12 July, pp. 250–252.

Koldras, M. and Bieniarz, K., 1987. Cryopreservation of carp sperm. *Pol. Arch. Hydrobiol.*, 34: 125–134.

Kurokura, H., Kumai, H. and Nakamura, M., 1986. Hybridization between female red seabream and male crimson seabream by means of sperm cryopreservation. In: *Proc. First Asian Fish Forum,* (Eds.) J.L. Maclean, L.B. Dizon and L.V. Hisillos. Manila, Phillippines, 26–31 May, pp. 113–116.

Leung, L.K.P., 1987. Cryopreservation of the spermatozoa of the barramundi, *Lates calcarifer* (Teleostei : Centropomidae). *Aquaculture,* 64: 243–247.

Leung, L.K.P., 1991. Principles of biological cryopreservation. In: *Fish Evolution and Systematics*, (Ed.) B.G.M. Jameson. Cambridge, Great Britain, pp. 213–214.

Scott, A.P and Baynes, S.M., 1980. A review of the biology, handling and storage of salmonid spermatozoa. *J. Fish Biol.,* 17: 707–739.

Stoss, J., 1983. Fish gamete preservation and spermatozoan physiology. In: *Fish Physiology,* Vol. 9, Part B, (Eds.) W.J. Hoar, D.J. Randall and E.M. Donaldson. Academic Press, Orlando, FL, pp. 305–330.

Stoss, J. and Refstie, T., 1983. Short-term storage and cryopreservation of milt from Atlantic salmon and sea trout. *Aquaculture,* 30: 229–236.

Terner, C., 1986. Evaluation of salmonid sperm motility for cryopreservation. *The Progressive Fish Culturist,* 18: 230–232.

Throgood, J. and Blackshaw, A., 1992. Factors affecting the activation motility and cryopreservation of the spermatozoa of Yellowfin bream, *Acanthopagrus australis* Gunther. *Aquaculture and Fisheries Management,* 23: 337–344.

Tiantian, Z., Rawson, D.M. and John Morris, G., 1993. Cryopreservation of prehatch embryos of zebrafish (*Brachydanio rerio*). *Aquat. Living Resour.,* 6: 145–153.

Weismann, T., Lahnsteiner, F. and Patzner, R.A., 1995. A uniform cryopreservation method for semen of salmonid fish and its adaptation for practical application. In: *Proc. of Fifth International Symposium on Reproductive Physiology of Fish,* Austin, pp. 90–92.

Whittingham, D.G., 1980. Principles in embryo preservation. In: *Low Temperature Preservation in Medicine and Biology,* (Eds.) M.J. Ashwood Smith and J. Farrant. Pitman Medical, Turnbridge, Wells, Kent, pp. 65–83.

Chapter 25

Bacteriological Assessment of the Marine Environment of Kollam (Quilon), Southern Kerala Coast of India

Wilma Cyril[1] and B.S. Sharma[2]

[1]**Head, Post Graduate Department of Chemistry, Fatima Mata National College, Kollam – 691 001, Kerala**

[2]**Head, Department of Environmental Studies, School of Life Sciences, Dr. Bhim Rao Ambedkar University, (Formerly Agra University), Khandari Campus, Agra – 282 002, U.P.**

ABSTRACT

Indicator bacteria total coliform and *E. coli* were enumerated in the marine environment of Kollam Coast, Kerala during the period June 2003 to May 2005 from the four sites Chavara, Neendakara, Tangasseri and Paravur on a monthly basis from the surface water. In addition, the population of total coliforms, *E. coli*, Fecal coliform, *Shigella* like organisms, *Salmonella* like organisms and *Proteus and Klebsiella* like organisms were examined twice during the study period from the surface water, bottom water and from the, sediment of the four sites of study during the peak of monsoon in July 2003 and July 2004. The highest proliferation of microbes were at Neendakara nearshore where the most probable number MPN/100ml of total coliform was 34800 and *E. coli.* 4260 during the monsoon. It was followed by Tangasseri near shore region where the values were 18600 MPN/100ml of total coliform and 13 02.50 MPN/100ml of *E. coli.* Paravur was found to be a microbially less contaminated zone and Chavara a clean zone. The offshore areas are not microbially contaminated which lead us to conclude that human activities which include activities at the fishing harbours, solid waste disposal and sewage accumulation lead to microbial contamination in the Kollam Coast.

Keywords: *Kollam coast, Microbial contamination, Bacteriological assessment.*

Introduction

Marine microbial communities play a fundamental role in the biogeochemical processes occurring in the oceans. An understanding of their abundance, type and distribution in the coastal zone are of importance in understanding the state of environmental health. Marine environmental contamination is an invariable consequence of anthropogenic activities and bacterial communities are the quickest to respond to such alterations.

Pollution in the Kollam Coast is mainly of microbial as the industrial development has declined in the region. The sea-food industry contributes to microbial pollution of Kollam as the wastewater from many factories which is rich in organic matter end up in the Arabian sea. The sanitation facilities at the fishing harbours are minimal and the loading, handling and distribution of fish lead to spill or waste. The waste dumpsite of Kollam Corporation at Mamoottilkadavu, very close to the banks of the Ashtamudi estuary (a Ramsar site) which enters the sea at Neendakara is another major contributor to the proliferation of microbes at the coastal area. Bacterial pollution is caused by discharge of mostly raw and partly treated sewage in to the sea and leaching from refuse dumps and other activities.

Kollam Coast has become the recipients of sewage and domestic waste. Borrego and Figueras (1997) had pointed out that sewage effluents contain a wide range of pathogenic microorganisms that may pose a health hazards to human population when they are discharged into recreational waters. The increased pollution of the natural waters in conjunction with the improved screening techniques has resulted in the detection of emerging pathogens in areas affected by sewage discharges (Morinigo *et al.*, 1992) with subsequent hazards to public health (Moe, 1997). Bacteriological assessment of Indian coastal waters have been reported earlier (Chandorkar and Raman, 1986; Hatha *et al.*, 2004; Lokabharathi and Chandramohan, 1986; Mohandass *et al.*, 2000; Naif and Lokabharathi, 1982; Nair *et al.*, 1994). Sarthre *et al.* (2000) had reported that the quantum of pollution caused by sewage into the marine system off Kerala Coast is planning. It receives large amounts of wastes both from domestic and industrial discharges causing deleterious effects on living marine sources. There is certain marginal hazard to human health as well as from sea-water contaminated by human excreta. The most direct effect of sewage bacteria on marine produce is the contamination of shell fish by pathogenic bacteria capable of causing various enteric diseases. Shell fishes concentrate microbes by their natural filtering of large quantities of sea-water. In addition, they provide a favourable environment for the continued growth of such microbes (Johnston, 1976). Background concentrations of indicator organisms in coastal waters are useful in the future guidelines for beach water, and to keep them in aesthetically pleasing conditions, hence the current study assumes importance.

Materials and Methods

Kollam, located at 8°45′ and 9°28° N latitude and 76°28′ and 77°17′E has a coastline length of 37 km. During the period June 2003 to May 2005 (the study area having three seasons June to September (monsoon), October to January (post-monsoon) and February to May (pre-monsoon), marine water sampling was done from a mechanized canoe on a monthly basis during the first week of every month from four sites of Kollam Coast namely Chavara near shore S_1, Chavara offshore S_2, Neendakara nearshore S_3, Neendakara offshore S_4, Tangasseri near shore S_5, Tangasseri offshore S_6, Paravur near shore S_7 and Paravur offshore S_8. The offshore sampling station represent 5 km zone away from the coast. Pre-sterilized at 121°C for fifteen minutes and paper-wrapped glass sampling bottles were used for collection. The bottles were lowered into water, opened under water, three fourths filled and closed under water. The staining was done within an hour after collection. The total coliforms and *E. coli* were determined as their most probable number MPN/100ml, APHA (1998). A lactose broth (Mac Conkey broth) culture medium (Hi Media) was used for presumptive test.

In addition marine water sampling was performed from bottom waters and sediments collected in sterilized petri dishes during the peak of monsoon season in July 2003 and July 2004 and investigated for total coliforms (TC), *E. coli* (ECLO), fecal colifonn (FC), *Shigella* like organism (SHLO), *Salmonella* like organism (SLO) and *Proteus* and *Klebsiella* like organisms (PKLO). Samples were placed for *Salmonella, Shigella* and *Proteus* and *Klebsiella* like organisms on xylose lysine deoxycholate agar. The samples were incubated on the respective selective media at 35±2°C. Fecal coliform was distinguished by further incubation at 44.5±2°C.

The location map of the study area along with sampling stations are shown in Figure 25.1. The atmospheric temperature, water temperature, relative humidity and pH were recorded. The mean seasonal atmospheric temperature, water temperature, relative humidity, and pH is given in Table 25.1.

Table 25.1: Average Value of Temperature (°C), Relative Humidity (Per cent) and pH in the Kollam Coast During June 2003 to May 2005

	AT	*WT*	*RH*	*pH*
Pre-monsoon	27.40	25.50	85.50	8.40
Monsoon	24.10	22.40	90.20	7.90
Post-monsoon	25.50	23.20	87.50	8.30

AT: Atmospheric temperature; WT: Water temperature; RH: Relative humidity.

Results and Discussion

The value of most probable number (MPN) of total coliforms is given in Table 25.2 and that of *E. coli* for the four sites of Kollam Coast are provided in Table 25.3.

Table 25.2: Average Value of Total Coliform of Kollam Coast (MPN/100 ml) During June 2003 to May 2005

	S_1	S_2	S_3	S_4	S_5	S_6	S_7	S_8
Pre-monsoon	412.50	15.20	16400	547.20	14200	936.20	55.60	75.80
Monsoon	492.50	21.40	34800	866	18600	1490.20	413.20	51.20
Post-Monsoon	266.20	41.60	24600	205.5	16200	843.60	471.20	56.20

Table 25.3: Average Value of *E. coli* of Kollam Coast (MPN/100 ml) During June 2003 to May 2005

	S_1	S_2	S_3	S_4	S_5	S_6	S_7	S_8
Pre-monsoon	175.5	10.5	2200	125	1250	128	155.50	208
Monsoon	155.5	12.50	4260	115	1302.50	120.40	164.20	38.20
Post-Monsoon	187.5	14.50	3600	302	1203	320.30	132.02	15.50

During the pre-monsoon the MPN of coliform bacteria ranged from 15.20 MPN/100ml at S_2 Chavara offshore to 16400 MPN/100ml at S_3 Neendakara nearshore. The values of *E. coli* ranged from 10.50 at S_2 Chavara offshore to 2200 MPN/100ml at S_3 Neendakara nearshore. During monsoon the

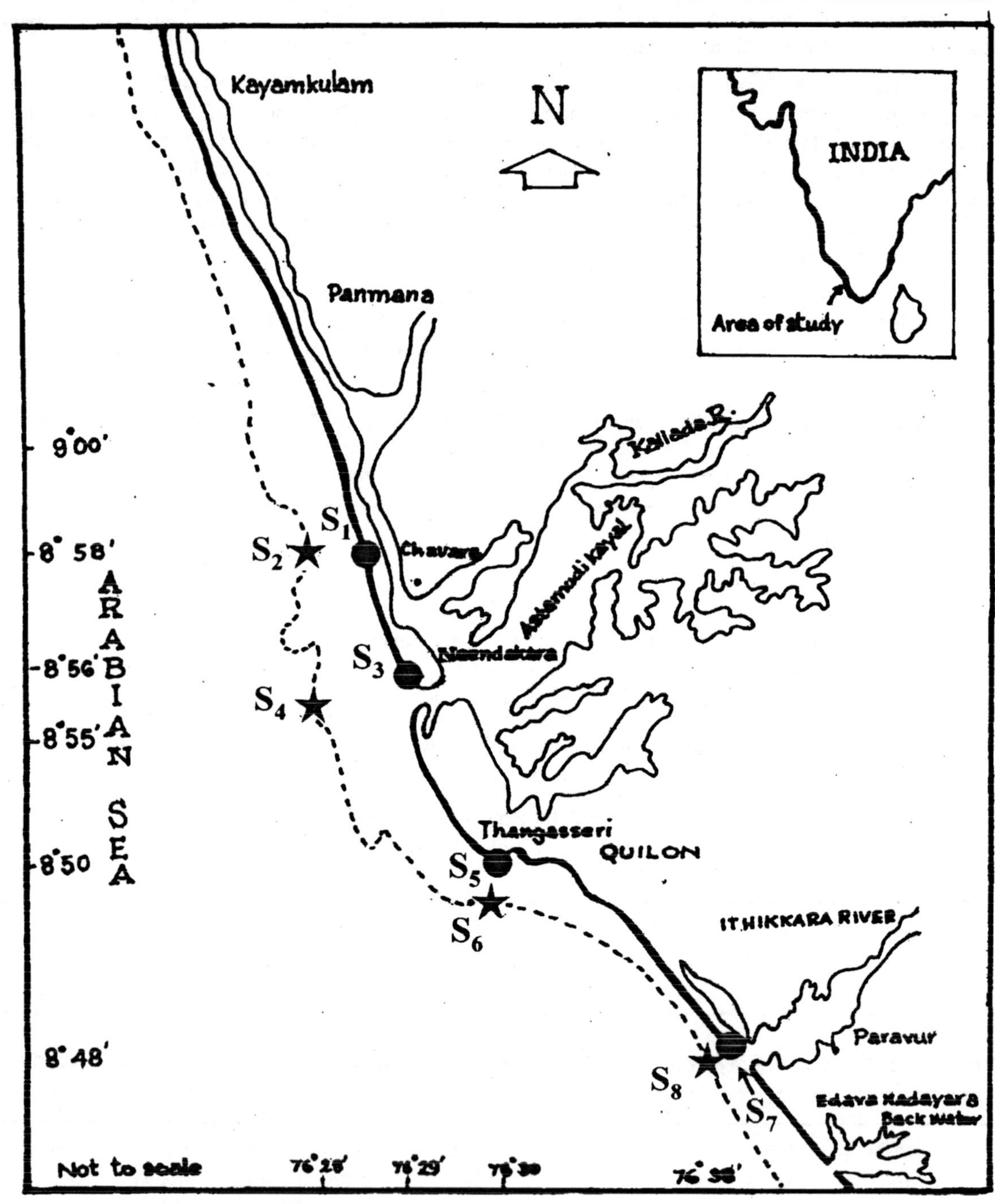

0 km zone of Kollam Coast ●

5 km Westward from the Kollam Coast ★

Figure 25.1: Map of Kollam Coast Showing the Sampling Stations

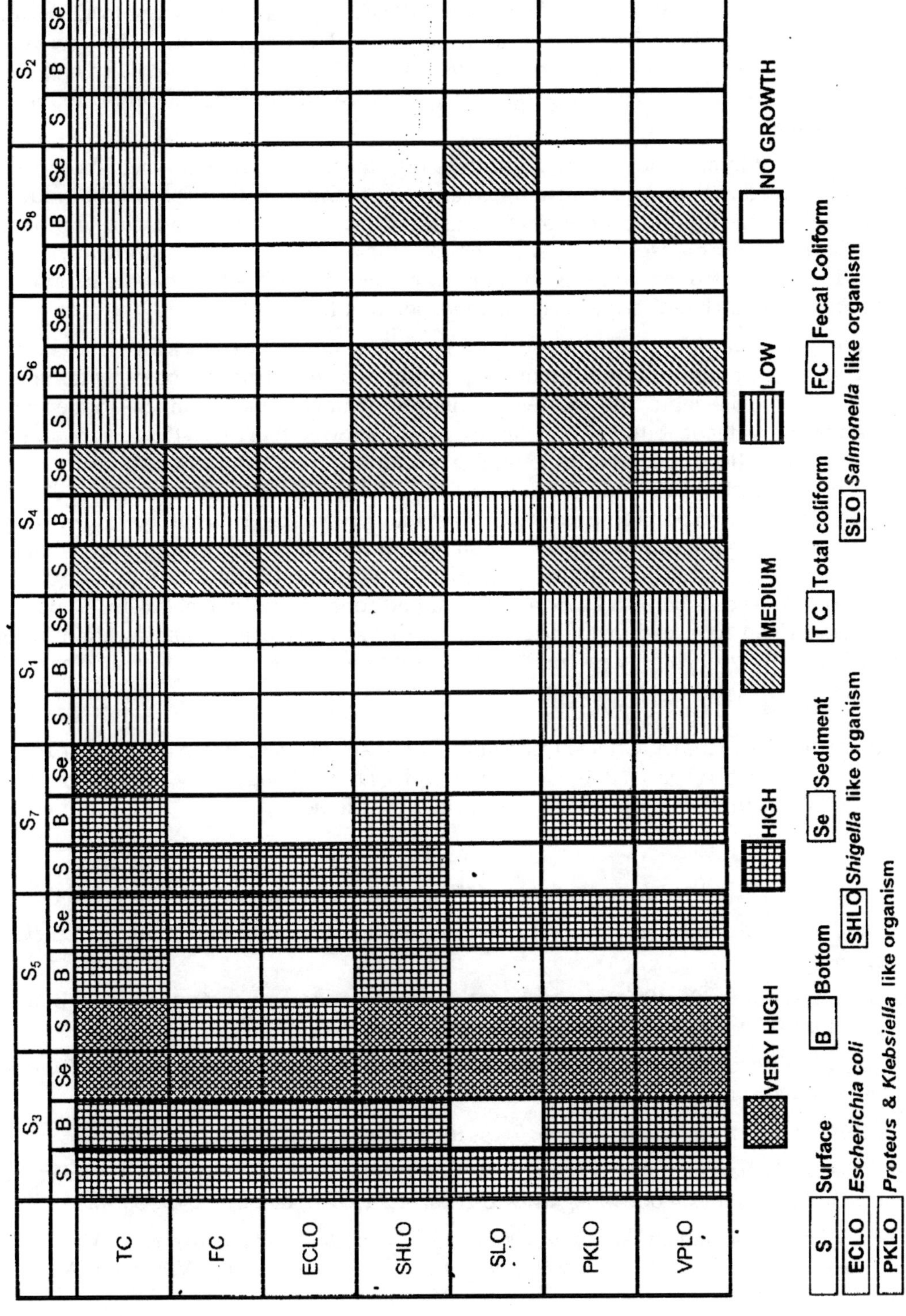

Figure 25.2: Relative Proliferation of Microbes at Kollam Coast During June 2003 to May 2005

total coliform ranged from 21.4 MPN/100ml at S_2 Chavara offshore to 34800 MPN/100ml at S_3 Neendakara nearshore. The count of *E. coli* ranged from 12.50 at S_2 Chavara offshore to 4260 MPN/100ml at S_3 Neendakara nearshore. During post-monsoon the MPN/100ml of total coliforms ranged from 41.60 MPN/100ml at S_2 Chavara offshore to 24600 MPN/100ml at S_3 Neendakara nearshore. The value of *E. coli* ranged from 14.50 MPN/100ml at S_2 Chavara offshore to 3600 MPN/100ml at S_3 Neendakara nearshore.

The proliferation of total colifoims and *E. coli* was highest at S_3 Neendakara nearshore. Direct sewage impact combined with low tidal flushing is causative to this. A comparison of the values for the three seasons show the highest proliferation during monsoon. The high load during monsoon months is the result of heavy monsoon showers bringing in heavy land run off resulting in increased levels of waste discharge into the marine environment and also due to the fact that the inactivating effect of sunlight is very much reduced during the monsoon months.

The offshore waters show low values there by showing that human activities including harbour activities, solid waste disposal and sewage accumulation are causative to the microbial contamination at Kollam Coast. Figure 25.2 depicts the relative proliferation of microbial organisms in the Kollam Coast in the decreasing order of abundance. Pillai *et al.* (1998) showed that the pathogenic counts had shown a proportional relation to coliform bacteria. The occurrence of bacterial indicators in the coastal water of Nagore on the east coast of India was reported by Mohandass and Lokabharathi (2003) and they have pointed out that non point sources of pollution may become a serious problem. The present study also conform to the above observations. The standards for total coliforms of 1000/100 ml and fecal coli forms 500/100 ml have been quite often used for beach waters and hence we grade the four sites at Kollam Coast as Neendakara and Tangasseri, polluted zones, Paravur a relatively less polluted zone and Chayara a clean zone leading to the conclusion that sanitation at the fishing harbours and proper disposal of city waste is highly essential in the fragile environment of the Kollam Coast.

Acknowledgement

We express gratitude to the UGC, New Delhi, India for financial assistance in the form of Teacher Fellowship (FIP), to Dr. J.S. Sharma, ONGC, New Delhi, and to Dr. V. Thangamany, Head, Department of Biotechnology Trivandrum, for help and encouragement rendered.

References

APHA, 1998. *Standard Methods for the Examination of Water and Wastewater*, 20th edn., Washington DC.

Borrego, J.T. and Figueras, M.T., 1997. Microbiological quality of natural waters. *Microbiologia*, 13: 413–426.

Chandorkar, A.A. and Raman, Y., 1982. Bacteriological assessment of coastal waters of greater Bombay. *Indian J. Environ. Hlth.*, 2: 107–123.

Hatha, M.A.A., Abirosh, C. and Mujeeb, R.K.M., 2004. Prevalence of diarrhegenic serotypes of *Escherichia coli* in the Cochin estuary, along west cost of India. *Indian J. Mar. Sci.*, 33(3): 238–242.

Johnston, R., 1976. *Marine Pollution.* Academic Press Inc., London, pp. 48–53.

Lokabharathi, P.A. and Chandramohan, D., 1986. Characterisation of purple and green photosynthetic bacteria isolated from the lagoon of Agatti Atoll (Lakshadweep sea). *Curr. Sci.*, 55: 426–427.

Moe, C.L., 1997. Waterborne transmission of infectious agents In: *Manual of Environmental Microbiology*, (Eds.) C.J. Hurst, G.R. Knudsen, M.J. Mcinerney, L.D. Stetzenbach and M.V. Walter. American Society for Microbiology, Washington, D.C., pp. 135–162.

Mohandass, C. and Lokhabharathi, P.A., 2003. Representation, dispersion and variation of bacterial indicators in the coastal water of Nagore (East coast of India). *Water Environ. Res.*, 75(1): 66–72.

Mohandass, C., Nair, S., Achuthankutty, C.T. and Lokabharathi, P.A., 2000. Pollution monitoring of coastal and estuarine areas. I. Bacterial indicators along the south Gujarat Coast. *Indian J. Mar. Sci.*, 29(1): 43–47.

Morinigo, M.A., Munor, M.A., Cornax, R., Martinez, M.E. and Borrego J.J., 1992. Presence of indicators and *Salmonella* in natural waters affected by outfall wastewater discharges. *Water Sci. Technol.*, 25: 1–8.

Nair, S. and Lokabharathi, P.A., 1982. Bacteriological studies of Mangalore coast. *Mahasagar*, 15(4): 215–222.

Nair, S., Lokabharathi, P.A. and Chandramohan, D., 1994. Culturable heterotrophic bacteria from the euphotic zone of the Indian Ocean during the summer monsoon. *Oceanol. Acta*, 17(1): 63–68.

Pillai, M.G., Bijumon, K.B., Satheeshkumar, C.S. and Ouseph, P.P., 1998. Assessment of indicator microbial population in relation to few possible human pathogens in surface waters along the South West Coast of India. In: *Proceedings of the Tenth Kerala Science Congress*, Kozhikode, pp. 205–207.

Sarthre, A., Ramalingam, P.A., Satheeshkumar, C.S. and Ajithkumar, A., 2000. Assessment of pollution status of Cochin and Neendakara estuaries in Kerala: A case study. In: *Proceedings of Twelfth Kerala Science Congress*, Kumily, pp. 595–597.

Chapter 26

Effect of Seed Stratification and Nursery Raising of *Quercus robur* (English Oak)

Gh. Mohi-u-din Bhat, A.H. Mughal, P.A. Sofi, J.A. Mugloo, A.A. Mir and T.A. Rather
Faculty of Forestry, SKUAST-K Shalimar, J&K

ABSTRACT

Studies conducted on stratification requirement to overcome dormancy in seeds of *Quercus robur* revealed that 45 days of cold stratification increased germination per cent to a maximum of 87.50 per cent. Nursery raising in open beds and polybags revealed that open nursery bed-sowing is far superior than raising in polybags. Germination of 83.50 per cent and survival of 95.80 per cent was recorded in open beds. Other growth parameters *viz.* height (24.87cm), collar diameter (3.78mm) and biomass (30.58gm) were also recorded higher in nursery raised seedlings.

Introduction

The genus *Quercus* known as red oak, includes world wide 500 species, of which 58 species are in United States, making it country's largest genus of native trees (Little, 1979). *Quercus robur* commonly known as English oak belongs to family Fagaceae. The tree is deciduous and reaches to a height and spread of 50–60 feet. The crown often appears open with large branches dominating the round crown. The tree is recommended for buffer strips around parking lots, median strip plantings on the highway, shade tree, and residential street tree. Different species of Quercus exhibit variable dormancy. Not only do species differ in dormancy but within a species difference can be due to geographical or altitudinal variations. Farmer (1974), for instance, found that chilling requirement for germination of northern red oak increased with latitude and altitude of seed source. Although seed dormancy is a natural protection for species regeneration, it must be understood and controlled for practical silvicultural management. Dormancy in *Quercus robur* has been attributed to the pericarp by Jones and Brown (1966). Hopper and associates (1985) also concluded that the pericarp was significantly involved in dormancy of *Quercus robur* and Bonner (1968) found pericarps to inhibit germination of four red oak species. Since dormancy in oaks may have been attributed due to mechanical strength of the pericarp and chemical inhibition by the pericarp and both these factors can be taken care off by cold stratification. Hence the present study on stratification requirement has been undertaken, besides the study was also taken to compare growth of seedlings in open bed nursery and polybags.

Materials and Methods

The study was conducted at Faculty of Forestry, SKUAST-K Shalimar, India during 2006–07. Acorns were harvested from phenotypically superior trees in the second fortnight of November 2006. Five seed lots were prepared and kept for stratification in moist sand at 4±1°C in refrigerator for 15, 30, 45, 60 and 75 days and germination of each seed lot was tried after completion of stratification, in the germinator at 25±1°C in Petri plates. The germination count was recorded daily up to 28th day and germination per cent was calculated as the number of seeds germinated to the number of seeds sown and expressed in percentage.

The seedling nursery of *Quercus robur* (English oak) was raised from collected seeds in open nursery beds and in the perforated polybags of size 5″ × 7″. Open beds of size 1 × 1 m were prepared, mixed with sand and dal weed. 400 seeds were sown in the first week of December 2006 and the beds were irrigated as and when required. The sowing in polybags was carried out at the same time after the filling of polybags with soil, dal weed and sand in the ratio of 2 : 1 : 1 respectively. The seeds after germination were watered as and when necessary. The performance of seedling nursery in both the cases in terms of germination percent, survival per cent, plant height and collar diameter during one growing season was recorded and monitored. Plant height was measured with the help of graduated scale in centimeters and the collar diameter was recorded in millimeter with the help of digital vernier caliper. Root shoot ratio was also calculated on dry weight basis.

Results and Discussion

The effect of stratification on the germination of *Quercus robur* is depicted in Table 26.1. The study revealed that with the increase in stratification duration there was significant increase in germination percent upto 45 days when maximum germination of 87.50 per cent was recorded. Stratification beyond 45 days did not bring about any further significant improvement in germination but of course was more than what was recorded in control. Effect of cold stratification on different temperate species have been studied by a number of authors and improvement in germination and other germination parameters have been recorded. Allen (1960) stated that longer the stratification period, the better was the germination of coniferous seed. West *et al.* (1970) reported 100 per cent germination of *Ginko biloba* after 12 weeks stratification period where as non stratified seeds remained dormant. Radwan and Crouch (1977) used stratification at 2°C to 5°C for 4 months to breaks the embryo dormancy in *Ceanothus sanguinus.* The stratified seeds of *Carpinus caroliana* for 18 weeks at 45°C resulted in 58 per cent germination but the unstratified seeds did not germinate (Bretzloff and Pellet, 1979), Singh (1989) reported 60 days stratification period resulted in an increase of 19.75 per cent more germination than control.

Table 26.1: Effect of Stratification on the Germination of *Quercus robur*

Sl.No.	*Stratification Duration*	*Germination (Per cent)*
1.	Control	62.50
2.	15 Days	68.00
3.	30 Days	73.00
4.	45 Days	87.50
5.	60 Days	78.00
6.	75 Days	75.00
	CD (0.05)	4.86

The improvement in seeds germination is probably associated with the increase in gibberellic acid level in the seed during stratification besides leaching out of chemical inhibitors present in pericarp and cotyledons also takes place during stratification. Regarding performance of seed and seedling in open nursery beds and polybags the results are depicted in Table 26.2. As far as germination is concerned, maximum germination of 83.50 per cent was recorded in open beds followed by 63.50 per cent in poly bags. Under natural conditions dormancy is overcome simply by cold, wet environment in which acorns over winter on forest floor or in nursery beds/containers.

Table 26.2: Effect of Nursery Type on the Germination and Seedling/Growth of *Quercus robur*

Type of Nursery	*Ger. (Per cent)*	*Plant Height (cm)*	*Collar Dia. (mm)*	*No. of Branches*	*Fresh Root Wt. (g)*	*Fresh Shoot Wt. (g)*	*Dry Root Wt. (g)*	*Dry Shoot Wt. (g)*	*Root Shoot Ratio*	*Total Fresh Biomass (g)*	*Survival (Per cent)*
Openbeds	83.50	24.87	3.78	2.00	19.11	11.41	13.09	7.28	1.80	30.58	95.80
Poly bags	63.50	17.30	3.12	1.00	14.28	8.28	8.17	9.99	1.61	22.66	92.04
CD (0.05)	7.39	2.45	0.24	0.74	2.15	1.64	2.12	2.15	0.14	3.06	2.56

The seedling growth parameters *viz.*, average plant height (24.87 cm) collar diameter (3.78mm) number of branches (2.07) Root shoot ratio (1.80) total biomass (30.58gm) survival per cent (95.80) after one growing season was recorded maximum in open beds and were statistically different from the values obtained for polybags.

The vigorous growth of seedlings raised in nursery beds may be probably due to abundance of available space for root system for drawing moisture and nutrients and transporting the same to the seedlings as poly bags provide limited space and nutrients for root development. Similar observations have been observed by Misra and Jaiswal (1993) obtaining maximum survival, height, collar diameter in Silver oak seedlings raised in bigger size polybags.

Mughal, (1996) reported that *Cupressus torulosa* and *Cerdrus deodara* seedlings attain optimum shoot and root development when grown in nursery beds, while as seedlings raised in polybags did not exhibit better growth in terms of height and root development.

References

Allen, G.S., 1962. Factors affecting the viability and germination behaviour of coniferous seeds. IV. Stratification period and incubation temperature *Pseadotruga menilesil* (Mirb.). *Franco For. Chron.*, 36: 18–29.

Bonner, F.T., 1968. Water uptake and germination of red oak acorns. *Botanical Gazette*, 129: 83–85.

Bretzloff, L.V. and Pellett, N.E., 1979. Effect of stratification and gibberellic acid on the germination of *Carpinus Caroliana* Walt. *Hort. Sci.*, 14(5): 621–622.

Farmer, Robert E., 1975. Effect of chilling and gibberellic acid on germination of northern red oak. In: *Proceedings of 8th Control States Forest Tree Improvement Conference.*

Hopper, G.M., Smith, D.W.M. and Parfish, D.G., 1985. Effect of stratification and pericarp removal on germination and seedlings growth of northern red oak. *Forest Science*, 31: 31–39.

Jones, L. and Brown, C.L., 1966. Cause of slow germination in cherry bark and northern red oak. *Proceedings, Association of Official Seed Analysts*, 56: 82–88.

Misra, K.K. and Jaiswal, H.R., 1993. Effect of size of polybags and potting mixture on survival and growth of silver oak. *Indian Forester*, 119: 941–943.

Mughal, A.H., 1996. Studies on container reared and nursery reared seedlings of deodar and cupressus. *Indian Forester*, 122: 760–762.

Paul, K.B., Patel, C.S. and Biswas, P.K., 1973. Changes in endogenous growth regulator in loblolly pine seed during the process of stratification and germination. *Plant Physiology*, 28: 530–534.

Radwan, M.A. and Crouch, C.L., 1977. Seed germination and seedlings establishment of red stem Caenothus. *J. Wildl. Manage.*, 41: 760–766.

Roos, J.D. and Bradfeer, J.W., 1971. Studies in seed dormancy. V. The content of endogenous gibberellins in seeds of *Corylus avelliana* L. *Planta*, 100: 288–302.

Singh, V., 1989. Role of stratification and gibberelic acid in spruce seed germination. *Ind. J. Fty.*, 12(4): 269–275.

West, W.C., Frattavelli, F.J. and Russin, K.J., 1970. Effect of stratification and gibberellin on seed germination in *Ginlko biloba. Bull. Tovey Bot. Club*, 97: 380–384.

Willemsen, R.W. and Rice, E.L., 1971. Mechanism of seed dormancy in *Ambrosia arteemisitolia. Am. J. Bot.*, 59: 248–257.

Chapter 27

Environmental Impact of Mine Spoils

M. Baskar[1], A. Solaimalai[2], P.T. Ramesh[3] and T.S. Muthukumar[1*]

[1]A.D. Agricultural College and Research Institute,
(*EID Parry Project), TNAU, Trichy -9
[2]Krishi Vigyan Kendra, RRS (TNAU), Vridhachalam
[3]Department of Environmental Sciences, TNAU, Coimbatore - 3

Introduction

The earth resources are extensively exploited to improve the quality of life. Every year one finds an increasing trend inspite of the fact that in recent decades much awareness campaign has been made to inform people that the energy and mineral resources are of finite nature. In India the mining of minerals bas been carried out on an extensive scale, but without giving any consideration to the adverse impacts, these activities are causing to the mined tract. This has led to the deterioration of the environmental conditions, especially around mining sites and adjacent areas. It is estimated that about 27,346 ha of spoils are available in India due to mines numbering 4082 (Dadhwal *et al.*, 1988). Mining in many areas has resulted in permanent non-restorable land surfaces. Majority of the mined areas are left with no productive value or with negligible vegetative cover of ephemetal/annuals and few perennials in isolated spots which fail to check wind and water erosion hazards. In rocky areas and sloping land, the drainage pattern is generally altered. All the non-scientitic activities lead to the formation of mined wastelands and virtually become the foci of desertification during and after the mining operation (Mann and Chatterji, 1978).

Mine spoils are defined here as the waste materials generated by the grinding and processing of ores and other materials containing economically retrievable minerals. Many processing method involve grinding of rock and ores, chemical and or physical removal of the desired commodity and transportation of the wastes, often as slurry, to a tailings or retention pond. These impoundments may range from a few to thousands of hectares in size. More than 99 per cent of the original material may finally become tailings when utilizing low quality ores (Gemmell, 1977). Disturbed land resulting from both surface and underground mining can result in major water quality problems as well as being unsightly and unproductive. Besides coal, sand, gravel, stone, clay, copper, iron ore, phosphate rock, and other minerals account for most of the mining. Abandoned mine tailings have extremely diverse physical, chemical and ecological conditions (Berg *et al.*, 1975; Gemmell, 1973; Hunter and and Whiteman, 1974; Ludeke, 1977; Shamshuddin *et al.*, 1986). The mine spoils am normally variable in physical composition particularly nitrogen, phosphorus and potassium. Acid drainage due to sulfide oxidation may be a consideration. Some mine spoils may have elevated levels of heavy metals or other toxic materials (Whiteby and Hutchinson, 1974; Hutchinson and Whitby, 1974). Plant uptake of potentially toxic chemicals or heavy metals and their incorporation into the food chain are real concerns.

Considering the ill-effects of mine spoils Government of India has framed some environment–related rules to be followed by the mine owners. All mining projects are now to be cleared by the Ministry of Environment and Forests before these are considered further. The projects are required to spell out how the mine-spoil areas are to be rehabilitated and what steps are contemplated for maintaining proper environment of the mining areas.

Mine Spoil Site Development Problems

Buildings, septic systems and other such features located on mine spoil may settle, move or have leachate problems.

1. Buildings can be damaged as a result of mine spoil settling under the foundation.
2. A building's footer drains can stop functioning as a result of mineral leachate clogging the drainage system. Coal refuse and certain types of mine spoils have high amounts of minerals, which are susceptible to leaching, when introduced to air and water.

The indiscriminate placement of steeply sloped unconsolidated mine spoil, prevalent on abandoned surface mines, can result in landslides that impact existing roads, structures and streams. Drainage from deep mines and strip mine impoundments can also saturate native soil units on non-mined slopes and result in the instability of these slopes.

Mine-Related Landslides Site Development Problems

Buildings and roadways if not constructed properly on or near mine spoil can cause hillside slippage also known as a landslide. Grading and/or removing the spoil material on hillsides can cause instability. Constructing a bench or a level area on a hillside composed of mine spoil for a road, house, trailer or other outbuilding can result in the following problems:

1. Undercutting the spoil material on a hillside may reduce toe or base support for the upper portion of the slope.
2. Excavating spoil material and moving it down slope can overload the hillside. This activity can cause a landslide because the excavated spoil material is adding additional weight, which can make the entire hillside unstable.
3. Subsurface waters or springs can cause mine spoil to become saturated and eventually unstable. Grading mine spoil on a hillside without installing tile proper surface and or under drain system can cause hillside instability.

Environmental Consequences of Mine Spoils

Acidity

Mining often exposes sulfide-bearing minerals (pyrite, pyrhotite, chalcopyrite, arsenopyrite, cobalitite) in the atmosphere. Overburden adjacent to mineral concentrations often contain low concentrations of pyretic minerals that are not economically retrievable. Metal processing also not remove all pyritic minerals and tailings, therefore, they often have significant sulfide concentrations (Berg *et al.*, 1975; Fuller and Lanspa, 1975; Sorensen *et al.*, 1980). Oxidation of pyrites and production of acid can be illustrated using iron (Fe) sulfide as an example in the following general equations:

$$2FeS_2 + 2H_2O + 7O_2 = 2FeSO_4 + 2H_2SO_4 \quad (1)$$

$$2FeSO_4 + H_2SO_4 + ½F_2(SO_4)_3 + H_2O \quad (2)$$

$$Fe_2(SO_4)_3 + 6H_2O = 2\,Fe(OH)_3 + 3H_2SO_4 \quad (3)$$

$$2FeS_2 + {}^{15}/_2O_2 + 7H_2O - 2Fe(OH)_3 + 4H_2SO_4 \text{ (Summary)}$$

Under conditions found in acid mine spoils, oxidation of ferrous iron by oxygen (Reaction 1) is much slower than the oxidation of iron disulfide by ferric iron (Reaction 4)

$$FeS_2 + 14\,Fe^{3+} + 5H_2O - 15Fe^{2+} + 2SO_2^{2-} + 16\,H^+ \quad (4)$$

This reaction is catalysed at pH values or approximately 3 by iron oxidizing *Thiobactillus ferroxidans* bacteria, which greatly speed the oxidation of ferrous [Fe^{2+} to ferric (Fe^{3+})] iron thereby regenerating the supply of ferric ions in solution (Nordstrom, 1982). The amount of acidity produced by a given sample of FeS_2 over time is a function of crystal structure, surface area, temperature, oxygen concentration, water partial pressure, pH, ferrous/ferric adsorption ratio and total Fe concentration, bacteria, absorbed impurities and flushing frequency (Caruccio *et al.*, 1988; Pugh *et al.*, 1981; 1984; Anna *et al.*, 1978). Formation of sulfuric acid decreases the pH of the mine spoil environment and results in increased solubility of metals and minerals that may be present. An additional process that may account for acidic tailings is the direct addition of acid in the processing of ore for product extraction. A baseline study in southern Arizona on abandoned mines which were worked from the 1600's to 1949 for silver (Ag), Pb and Zn revealed pH measurements of mine spoil drainage ranging from 2.9 to 4.2 directly below three sampled mines (Dean and Fogel, 1982). Drainage from the mine spoils contained maximum concentrations of 4200 mg sulfate L^{-1}, 1860 mg Fe L^{-1} and 286 mg Mn L^{-1}.

Salinity and/or Sodicity

Excess concentrations of soluble slats are present in many mine spoil materials. In some instances, the salt may consist largest of sodium (Na) which may introduce additional problems in reclamation of mine spoils. These salts commonly accumulate as the ore body is processed and are concentrated by recycling of water. Use of saturation percentage and saturation extracts for evaluating soil, overburden and mine spoils in mined-land studies has been recommended by a number of authors (Dollhopf *et al.*, 1980; Merrill *et al.*, 1980; Merrill *et al.*, 1983; U.S. Department of Agriculture, 1979; Schafer, 1979). Saturation percentage is a useful parameter for detecting low water holding capacity mine spoils and estimating the degree of sodic hazard as indicated by the sodium adsorption ratio (SAR). A saturation percentage of 25 has been used as an indicator point for low water-holding capacity and 80 per cent to 90 per cent saturation has been used as an indicator for swelling tenedency associated with sodic hazard (Merrill *et al.*, 1987).

Nutrient Deficiencies

Tailings are almost universally deficient in N. Many tailings are deficient in P and are commonly deficient in K. In addition, secondary and micro-nutrients are sometimes deficient.

Toxic Iones

Toxic ions are frequently present in mine spoils in sufficient concentrations to present revegetation unless considerable amelioration is undertaken. Heavy metals decrease root respiration, water and nutrient uptake, and inhibit cell mitosis in root meristematic regions (Gemmell, 1977). Heavy metals, reduce enzymatic activity and the microbial and microfaunal populations in soils so that the effects of heavy metals; (*i.e.* copper, iron, lead, manganese, nickel and zinc) on higher plants could be due partly to their inhibition of soil enzymes (Clark and Clark, 1981). Toxic ions may also contaminate soils and waters adjacent to smelters (Rutherford and Bray, 1979) or tailings through seepage, runoff waters, and eroded sediments (Sheppard *et al.*, 1984). Soils contaminated with heavy metals differ from each other and from normal soils in many ways other than metal content. The nutrient status, organic matter content and texture may all affect the number and types of plants growing on the soil. Metal toxicities or imbalances are common in mine spoils. Copper (Cu,) Pb, and Zn are the most extensively mined metals in the world and often occur in waste materials at concentrations toxic to plants (Jennett and Linneman, 1977; Wong, 1986). These metals are naturally present in the ore and extraction techniques commonly increase the acidity or alkalinity of the mine spoils and increases the conentration of metal ions. Some of the modern waste materials are less toxic that in the past. The total toxic metal concentration rarely exceeds 1 per cent by weight because of improvements in mining procedures and processing methods (Wong, 1986).

Table 27.1: Total Concentrations of Heavy Metals (mg kg^{-1}) in Mine Spoils Compared with the Normal Concentration Range of these Elements in Soil

Mine Spoils	*Cu*	*Pb*	*Zn*	*Ni*	*Mn*	*Reference*
Asbestos	4	–	70	462	–	Ellery and Walker (1986)
Lead	113	20530	1600	–	–	Simon (1978)
Lead	42000	1200	1250	–	–	Nunn and Riches (1978)
Lead–Zinc	–	12300	37940	–	–	Morrey *et al.* (1984)
Lead–Zinc	625	14000	34000	219	–	Johnson *et al.* (1977)
Copper	2060	327	124	–	–	Mcneilly and Johnson
Copper	1740	–	125	95	–	Lopex and Lee (1977)
Copper	890	7	73	33	490	Ruppin (1987)
Copper	15400	20900	20200	–	–	Simon (1978)
Copper–Nickel	440	–	43	330	250	Crowder *et al.* (1982)
Copper–Lead–Zinc	99	1125	255	–	1325	Dean and Foget (182)
Zinc	57	12930	22700	–	–	Simon (1978)
Metal	1329	–	101	1323	334	Hogan *et al.* (1977)
Metal	–	21320	1273	–	–	Williams *et al.* (1977)
Metal	–	38833	19376	–	–	Simon (1978)
Normal range in soil	2–100	2–200	10–300	5–500	20–300	Lindsay (1979)

Availability of metals to plants will depend on such properties as pH, redox potential, organic content, and crystalline properties of metal compounds.

Physical Limitation

Mine spoils generally possess one or more chemical and/or physical properties that limit successful revegetation. The texture of tailings ranges from sand to clay depending on the composition of the original material, stratification and method of slurry entry into the tailings pond. Coarser fractions generally are poorly buffered, devoid of organic matter deficient in nutrients, structure-less prone to crusting have a low water-holdings capacity (Verma *et al.*, 1977). Mine spoil exposed to direct solar radiation can have temperatures of 35 to 60°C at a 1 to 2 cm depth when air temperatures are 35 to 38°C (Shetron, 1983). High potential evapotranspiration and low water-holding capacity suggest that water deficits limit revegetation of coarser tailings, especially in arid regions. Finer textured tailings usually exhibit greater water-holding capacity, but water infiltration is often limited by poor structural characteristics.

Mine spoils are commonly handled as a water slurry. They are transported and deposited into dammed artificial ponds where the particle settle out on the basis of size (Murray, 1977). Stratification based on particle size frequently occurs within tailings pond. Larger, sand-sized particles settle near the slurry inlet, while clay-sized particles are located at the outlet end where water is ponded. Crusting, cracking and a general lack of structure are common characteristics of mine tailings brought about by differences in texture, lack of organic matter and variable mineralogy. Extremely high clay or sand contents are commonly encountered and add to the physical limitations of the tailings. The fines (slimes) have been found to be relatively impervious to water difficult to leach, and have poor physical properties. In contrast, the coarser fractions of tailings have a low available water capacity and are commonly deficient in plant nutrients. Bulk densities of mine tailings are sometimes elevated due to compaction.

Dust

An almost universal problem associated with unreclaimed mine spoil is blowing dust. These wind-blown silicious particles pollute air, streams and lakes, abrade vegetation, decrease esthetics and create health problems for communities near these wastes. Approximately 40 per cent of mineral waste in fine sized material that requires some kind of stabilization if air and water pollution are to be controlled (Day and Ludeke, 1973). Blowing dust has been reported as a problem for asbestos dumps (Ellery and Walker, 1986), red muds and sands from bauxite processing (Want, 1987; Coffey *et al.*) 1986) Cu and Ni mine spoils (Peters, 1984; McNeilly and Johnson, 1978; Peterson and Nielson, 1973; Day and Ludeke, 1913) Fe mine spoil (Jordan and Dewar, 1985) and for Au, Ag and Zn mine spoil (Harwood, 1979; Creswell, 1973). These wastes normally are susceptible to both wind and water erosion.

Hydrogeologic Characteristics

An understanding of the hydrology of surface mine spoil is important in predicting mine drainage quality. However, it is poorly understood and one of the least analyzed aspects of mine drainage prediction. Groundwater is a integral chemical component in acid mine drainage (AMD) formation and it serves as the contaminant transport medium. Therefore, prediction of post mining drainage quality requires the inclusion of a surface mine spoil groundwater hydrology component in the process. The nature, degree, and duration of groundwater and spoil interactions need to be factored into any comprehensive mine drainage predictive method or model. Recharging waters, moving through the unsaturated portion of backfill, will have intermittent episodes where discrete areas of the spoil are briefly contacted, whereas groundwater within the saturated zone will have a considerably more consistent and longer contact time with that portion of the spoil. Groundwater within the zone of water table fluctuation will contact nearly all of that spoil zone periodically during brief episodes when water levels rise. Groundwater will, in both the unsaturated and saturated zones, chemically and physically react with the spoil material that it contacts. However, mine drainage quality prediction is often based on the assumption of uniform contact with 100 per cent of the spoil material and does not take into consideration that groundwater only contacts a limited fraction of the spoil. Also, under differing hydrologic conditions, the sections of the spoil contacted by groundwater can change. It is important to determine what portions of the spoil are contacted by the groundwater and what is the nature of this contact. Spoil excavations and aquifer testing indicate that there are areas within backfills that, because of very low permeability, allow very little groundwater flow through them. These relatively "dead" areas contribute little to the groundwater system and to the associated mine drainage quality.

Groundwater Flow Systems

Most individuals have made the assumption that groundwater flow in mine spoil is a porous media system, similar to flow through unconsolidated alluvium. Recent field work and testing of surface mine spoil indicate that thus assumption is not completely valid. Caruccio *et al.* (1984) noted that groundwater flow in that backfill of a surface mine in Central West Virginia was highly channelized and that it was not observed until one of these randomly located channels was intercepted. Based on their physical observations during excavations in mine spoil, they referred to the groundwater flow regime as pseudokarst, where groundwater flows mainly through large voids and conduits. Pseudokarst hydraulic characteristics are similar to the characteristics observed in some karst (carbonate underlain) terrains, however, the mechanism of channel and void formation differs. These types of systems are more descriptively defined by the term double-porosity. Hawkins and Aljoe (1990) noted that mine spoil exhibits characteristics of both porous medium and double-porosity aquifers. Under steady-state conditions spoil behaves mainly as a porous medium aquifer. For example, the presence of a relatively continuous water table in the backfill and perennial consistent-flowing mine discharges are indicative of an overall porous media system. Hydraulic conductivity values measured in mine spoil are substantially below values expected for open conduit flow.

Conversely, when a spoil aquifer is stressed (for example, during an aquifer test) or subjected to transient conditions, double-porosity characteristics become more pronounced and can briefly dominate the hydrologic regime. For example, multiple water tables and temporary flowing artesian conditions have been observed during substantial recharge events (Hawkins and Aljoe, 1990). Monitoring and testing indicate that groundwater is stored in and flows through large voids or conduits in spoil; however, these voids are not always well interconnected across a mine site. Therefore, diffuse groundwater flow through the interstices of the fine-grained material between the voids exerts significant control on the overall site hydrology. Slug tests performed in wells penetrating surface mine spoil empirically illustrate the bimodal nature of the groundwater flow regime. These tests yield two distinctly different types of responses in mine spoil. One type of response is indicative of a porous media system. The actual displacement and the projected displacement are similar, indicating gradual and diffuse flow from the well into the aquifer as expected in a porous medium. The second type indicates the presence of large voids and conduits of a double-porosity system. The actual displacement is substantially less than the projected displacement, indicating large voids adjacent to the well rapidly fill with water flowing from the well. Once the voids are filled, there is a gradual and diffuse flow from the well into the spoil (Hawkins, 1993).

These dual aquifer flow characteristics exist because mine spoil is an extremely heterogenous and anisotropic material. The heterogeneities are created by the processes of mining and reclamation. During mining and subsequent reclamation, spoil becomes sorted to some extent. When dumped by a rock truck or dragline and regarded by bulldozers, the larger spoil particles tend to roll toward the base of the spoil ridges into the valley between the ridges, while the midsized and smaller fragments tend to stay on the sides and top of the spoil piles (Rehm *et al.*, 1980). Groenewold and Bailey (1979) observed that in Western North Dakota, monitoring wells completed in the spoil valleys exhibit more variable hydraulic conductivity than wells completed in the spoil ridges. Aquifer testing (constant-discharge tests) of spoil in Northern West Virginia and Western Pennsylvania indicates that linear zones of high hydraulic conductivity tend to parallel spoil ridge orientation. Hydraulic conductivity perpendicular to the spoil ridges appears to be significantly lower (commonly by several orders of magnitude) than that parallel to the ridges (Hawkins and Aljoe, 1991).

Factors Influencing Hydraulic Characteristics

Lithologic Controls

Lithology of the spoil can influence the hydraulic conductivity in reclaimed mines. Parent rock (overburden) of surface mine spoil in Northern Appalachia is comprised primarily of sandstone, siltstone, and shale. In some areas, limestone may occur in significant quantities, as may glacial sediments. The increasing percentages of sandstone (and decreasing percentages of shales) in mine spoil appear to yield higher median hydraulic conductivity values and a narrower range of values (Aljoe and Hawkins, 1994). However, the median hydraulic conductivity increases observed were not statistically significant. The apparent trend is explained by the hydraulic properties of the different lithologies and by mechanisms of mining and reclamation. Sandstone-rich spoil zones tend to have larger fragments than shale-rich zones. This is because sandstones of this region tend to be well cemented and are better able to resist breakage and weathering. Shales tend to break into smaller fragments during mining and more readily weather and break down to silt and clay-sized particles, which decreases the hydraulic conductivity (Aljoe and Hawkins, 1994). An accumulation of clay and silt toward the base of the spoil is often observed in monitoring wells that are purged (pumped or bailed) infrequently, conforming the breakdown.

The processes of mining and reclamation may further facilitate spoil heterogeneity by creating zones comprised predominantly of one lithology. During mining, a dragline or frontend loader often will remove the overburden in layers, spoiling strata composed mainly of one lithology at a time. Monolithic zones are also created by the tendency of large spoil fragments (mainly sandstone) to roll to the base of spoil ridges, while the medium and smaller fragments (shale and some sandstone) tend to remain on the sides and top. This appears to account for the observation of Groenewold and Bailey (1979) that spoil valleys had a higher mean and greater range of hydraulic conductivity than the ridge areas in the Northern Great Plains. A test well drilled randomly into spoil with a lithology of 50 per cent sandstone and 50 per cent shale should have an equal chance of intersecting a shale-rich or a sandstone-rich saturated zone. Therefore, the hydraulic conductivity is expected to range more widely when the sandstone content is 50 per cent than when it approaches 100 per cent. As the sandstone content of spoil increases, the number of wells that will intersect sandstone-rich zones likewise; increases, thus causing the median hydraulic conductivity to increase. Phelps (1983) observed that spoil bulk density generally decreases with depth. This appears to be caused by the creation of a significant volume of interstitial voids when the large spoil fragments, commonly sandstone, roll to the bases of spoil piles.

Mining Methods and Topography

Topography influences groundwater flow in surface mine spoil, because groundwater flows down the hydraulic gradient and topography directly influences the hydraulic gradient. Although, spoil aquifer can exhibit multiple water tables for brief periods under transient conditions (Hawkins and Aljoe, 1990), surface mine spoil generally exhibits a single continuous water table with a moderate hydrologic gradient. Thus, the water table tends to reflect the overlying topography. However, the water table is also influenced by other geologic and hydrologic conditions, such as permeability variations, local structure, and the adjacent unmined areas. The structural dip of the pit floor is a major influence on the direction of groundwater flow in spoil aquifers. Groundwater tends to flow down dip and perpendicular to the strike of the pit floor. Toe-of-spoil discharges will commonly form at the structural low point of the pit floor outcrop. Influences from the groundwater system in adjacent unmined areas can cause groundwater in spoil to disregard structural dip and other hydrologic factors. Localized rolls or swales in the pit floor can affect the direction of groundwater flow. When the strata dip toward the final highwall the groundwater tends to saturate spoil behind the high wall. This is because the hydraulic conductivity of the undisturbed aquifer in the high wall is commonly at least 2 orders of magnitude less than in the spoil and the pit floor material usually has significantly lower permeability than the spoil (Hawkins, 1995). Groundwater impounding at the highwall may flow in any of several directions. Depending on the hydraulic gradient, groundwater may enter the unmined strata and flow down the structural dip, or it may flow laterally, parallel to die highwall, and discharge where the pit floor is exposed at die surface along the structural strike. The spoil may become sufficiently saturated to permit discharges to emanate at the topographic low point of the mine along the original coal cropline, opposite to the dip direction. Depending on the permeability, significant amounts of groundwater may leave the backfill by downward flow through the pit floor. Depending on the configuration and size of adjacent unmined areas, the groundwater gradient may cause flow from the highwall into the spoil, regardless of the structural dip.

Direction of mining and the configuration of the backfill can dramatically impact the direction of groundwater flow. The highly permeable zones that form in the valleys between spoil ridges permit substantial groundwater flow parallel to the ridges. Because groundwater tends to follow the path of least resistance, groundwater flow perpendicular to the spoil ridges is considerably less than flow parallel to them. Constant-discharge testing in a reclaimed surface mine in Central West Virginia indicates that that hydraulic conductivity difference between buried spoil valleys and ridges can exceed 2 orders of magnitude. Groenewold and Winczewski (1977) observed that the surface over these highly transmissive spoil valleys is more susceptible to subsidence from piping of fine grained spoil materials because of the substantial amount of groundwater movement.

The location of haul roads across the backfill also can influence groundwater flow. Spoil underlying haul roads can become highly compacted (less transmissive) from the traffic of vehicles and heavy equipment. The spoil on each side of the haul road will be substantially more transmissive than the spoil under the haul road. The level of spoil compaction is related to the lithology of the spoil material and the amount of equipment traffic. Groundwater may flow along the road edge until a pathway through exists or impound behind these haul roads which may be buried and hidden in the reclaimed backfill.

Impacts of Spoil Age

Shortly after regrading of the spoil, differential settling and piping of the finer material begin in the backfill (Groenewold and Bailey, 1979). These subsequent processes contribute greatly to the heterogeneity of spoil and are facilitated by infiltrating surface waters and the water table re-establishment. The "uplift" pressure provided by the rebounding groundwater table may aid the shifting and repositioning of spoil fragments (Sweigard, 1987). Sweigard observed significant settling within a year after reclamation on Illinois surface mines and noted that considerable settling may continue at least 2 to 3 years after reclamation. Settling within spoil continues, apparently at a lesser rate, even 12 years after reclamation. Hydraulic conductivity has been observed to change as the age of the spoil increases. Aljoe and Hawkins (1994) observed that reclaimed surface mine spoil that was 30 months or less old had a significantly lower median hydraulic conductivity than reclaimed mine spoil that was over 30 months old. The manner by which hydraulic conductivity increases with time may be caused by the improved interconnectedness of the voids that were created during backfilling. Piping and differential compaction of fine-grained spoil material in response to vertical movement of recharging waters through the unsaturated portion and horizontal movement of groundwater in the saturated portion of the spoil may be the cause of increased void communication. Bulk density also may change as the spoil settles and fine grained materials migrate toward the base of the spoil.

Changes in hydraulic conductivity in spoil are directly related to the mechanisms and timing of the post-mining water table re-establishment. In eastern Ohio, water table re-establishment at three reclaimed surface mines was observed to be nearly complete approximately 22 months after reclamation was completed (Helgesen and Razem, 1980). The rate of water table recovery is related to several factors including the precipitation rate, recharge and discharge rates, porosity, topography, and geologic structure. Rehm *et al.* (1980) and Moran *et al.* (1979) stated that spoil permeability decreases with age. This may be caused by differences in physical and chemical properties of the rock units that they encountered in the coalfields of the Northern Great Plains compared to the units of the eastern coal fields. Overburden for the western coal fields are mainly comprised of weakly cemented units, which tend to form few large voids. Permeability may be further decreased by swelling clays common to overburden in the Northern Great Plains. In the Northern Great Plains, post-mining hydraulic conductivity values are very similar to pre-mining values (Rehm *et al.*, 1980).

Hydraulic Conductivity and Transmissivity

Performing aquifer tests to determine hydraulic properties (hydraulic conductivity and transmissivity) of spoil can be a difficult procedure. Assumptions must be made as to the homogenity and isotropic nature of the aquifer. However, testing indicates that spoil is highly heterogeneous and anisotopic. Hawkins (1993) detailed some of the problems incurred while conducting slug tests and during the subsequent data analysis to determine the hydraulic properties of surface mine spoil. Large voids within mine spoil may permit rapid, possibly turbulent groundwater flow during aquifer testing. If turbulence actually occurs, Darcian methods of data analysis cannot be used. Additional problems are created because unknown surface area of the adjacent voids is added to the known surface area of test well. This situation can cause the hydraulic conductivity to be overestimated. Conversely, aquifer modeling (Mineflo and Modflow) of mine spoil indicates that the effective site hydraulic conductivity may be 1 to 2 orders of magnitude higher than the field determined values at individual test wells. This was determined by error reduction during model calibration. The modifications to the hydraulic conductivity may vary substantially under differing hydrologic conditions because of extreme heterogeneity and anisotropy common to reclaimed mine spoil (Hawkins and Aljoe, 1990; Hawkins, 1994). Despite the problems involved in conducting aquifer tests, aquifer testing to determine hydraulic properties is still an integral part of the hydrologic characterization of mine spoil. The hydraulic conductivity and other site-specific hydrologic data can be used to predict the post-mining water table elevation within the spoil, which in turn is important for alkaline addition and special handling techniques. Transmissivity, in the strictest sense, refers only to confined aquifers. However, many researchers have reported transmissivity for mine spoil which is mainly unconfined. These hydraulic parameters are important for mine drainage prediction because they are used to determine groundwater velocity, water table fluctuation groundwater storage turnover rates as well as other factors influencing groundwater contact within different zones in the spoil. Knowledge of these and other hydrologic parameters is especially important to special handling and alkaline addition techniques and other aspects of mine drainage prediction.

Table 27.2: Ranges of Hydraulic Conductivity and Transmissivity in Surface Mine Spoil

Geographic Area Conductivity in m/s	*Hydraulic*	*Transmissivity in m²/s*
Western Pennsylvania	4.2×10^{-9} to 7.6×10^{-2}	1.2×10^{-9} to 2.0×10^{-1}
Northern West Virginia (Hawkins, 1995)	9.5×10^{-8} to 2.7×10^{-4}	1.4×10^{-7} to 2.9×10^{-1}
Eastern Ohio (Weiss and Razem, 1984; Bonta *et al.*, 1992)	5.4×10^{-8} to 1.9×10^{-5}	–
Western Kentucky and Southern Illinois (Hogan *et al.*, 1977; Lindoff, 1980)	2.4×10^{-5} to 4.1×10^{-5}	2.2×10^{-4}
Western North Dakota (Groenewold and Bailey, 1979)	2.9×10^{-3} to 4.6×10^{-3}	–
Northern Great Plains (Rehm *et al.*, 1980)	8.0×10^{-7}	–
Wyoming (Moran *et al.*, 1979)	1.9×10^{-6} to 2.1×10^{-1}	–
Edmonton, Albem, Canada (Moran *et al.*, 1979)	1.5×10^{-6}	–

Hydraulic conductivity and transmissivity of surface mine spoil exhibit a very broad range of values from location to location and within a mine site. Table 27.2 illustrates the range of hydraulic conductivity and transmissivity values in different regions of North America. Hydraulic conductivity values generated from testing of 103 monitoring wells from 15 surface mines in northern West Virginia and western Pennsylvania ranged from 4.2×10^{-9} to 7.6×10^{-2} m s^{-1} (Table 27.2). The data exhibited a geometric mean of 2.5×10^{-5}, a median of 2.8×10^{-5}, and a standard deviation of 9.6×10^{-3} m/s. Hydraulic conductivity within a single mine site ranged over 5 orders of magnitude (6.6×10^{-9} to 9.3×10^{-4} m/s). Hydraulic conductivity range exceeding 3 orders of magnitude within a mine site were common when more than 4 wells were tested. Aquifer testing of the 15 mine sites exhibited a range of transmissivity values exceeding 8 orders of magnitude (1.2×10^{-9} to 2.0×10^{-1} m^2/s). Rehm and others (1980) reported a hydraulic conductivity range of 6 orders of magnitude with a geometric mean of 8×10^{-7} m/s for the northern Great Plains region (North Dakota, Montana, Wyoming, and Alberta, Canada). Similar hydraulic conductivity values were observed by Moran *et al.* (1979) in the northern Great Plains. They recorded hydraulic conductivities of 1.9×10^{-6} and 2.1×10^{-1} m/s from aquifer testing on a Wyoming surface mine. Spoil aquifer testing near Edmonton, Alberta, Canada yielded a mean hydraulic conductivity of 1.5×10^{-6} m/s.

The hydraulic conductivity of mine spoil in the Appalachian coal fields is considerably greater than that of the undisturbed rock. Hawkins (1995) analyzed data from five Northern Appalachian surface mines and observed that the hydraulic conductivity of mine spoil (1.2×10^{-5} to 1.4×10^{-1} m/s) ranged from nearly 1 to over 2.5 orders of magnitude greater than adjacent bedrock (3.8×10^{-8} to 4.1×10^{-6} m/s) with a geometric mean 2 orders of magnitude greater. In other words, mine spoil tends to be approximately 100 times more conductive than undisturbed bedrock. The differences in hydraulic conductivity cause differences in water levels between mined and unmined areas. In some cases, perched aquifers on the unmined rock call also account for these water level differences. Water levels in spoil welts were 50 per cent lower than those measured in wells of similar elevation in adjacent unmined aquifers. This illustrated that premining water level measurements may not be indicative of post-mining levels.

Porosity

Spoil porosity is important to mine drainage quality prediction in terms of determining groundwater storage volumes and predicting water level changes stemming from recharge or discharge. These characteristic impact the nature and scope of groundwater contact with different spoil zones. Wells *et al.* (1982) repented laboratory porosities of 25 to 36 per cent for surface mine spoils from Eastern and Western Kentucky. The spoils tested were composed mainly of shale and sandstone. Laboratory-measured values on eastern Ohio spoil samples ranged from 41 to 48 per cent with a mean of 44 per cent (Mezga, 1973). Field tests indicate that the laboratory-generated values are significantly greater than actual field conditions. Cederstrom (1971) estimated that the porosity values of east spoil ranges between 15 and 25 per cent. He stated that this range was 7 to 25 times greater than the porosity of undisturbed strata. The magnitude of increases in porosity depends greatly on the pre-mining aquifer porosity, which is determined mainly by lithology and fracture density. Based on pumping tests, storage coefficients of 17 and 23 per cent were determined for spoil in Wyoming (Rahn, 1976, reported in Moran *et al.*, 1979). For unconfined conditions, storage coefficient is roughly equivalent to the effective porosity. The field-determined porosity values approximate that percentage of backfill volume increase (swell) created when overburden is spoiled during mining. Herring (197) stated that the swell for the Illinois Basin is about 20 per cent. Van Voast (1974) estimated that the swell is roughly 25 per cent. He stated that the backfill volume increase was accompanied by an increase in porosity and vertical permeability. Given the relatively high porosity values of spoil, reclaimed surface mines are capable of storing large volumes of groundwater. For example, a 10 ac (4.05 ha) reclaimed surface mine with a 10 ft (3.05 m) saturated zone and 18 per cent porosity will have nearly 6 million gallons (0.023 million m^3) of groundwater in storage. When making this type of calculation, a range of effective porosity values is better than a simple value. Effective porosity values for reclaimed surface mine spoil should be based on field testing or measured swell as opposed to laboratory determinations.

Groundwater Velocity

Aquifer testing indicates that the groundwater velocity in surface mine spoil is substantially greater than that of the undisturbed overburden. Average groundwater velocities are affected by recharge rate, effective porosity, hydraulic conductivity, and head differential, and can vary widely depending on site-specific conditions. Determination of groundwater velocity is important in mine drainage prediction because it relates directly to groundwater contact with the spoil and groundwater storage turnover rates. Hawkins and Aljoe (1991) measured a groundwater velocity range in the backfill of a reclaimed surface mine in Central West Virginia of 1.2×10^{-5} to 4.9×10^{-5} m/s. Caruccio *et al.* (1984) observed similar velocities, ranging from 1.4×10^{-5} to $1.8 \times$ to 10^{-5} m/s, for another reclaimed mine in Central West Virginia. A groundwater velocity of 2.0×10^{-5} m/s was determined for a surface mine in eastern Ohio (Mezga, 1973). Ladwig and Campion (1985) observed a groundwater velocity of 6.1×10^{-4} m/s at a surface mine in Pennsylvania. A groundwater velocity range of 2.7×10^{-3} to 4.3×10^{-3} m/s was measured in surface mine spoil in eastern Kentucky (Wunsch and others, 1992). Most of these velocities are below the groundwater velocities commonly measured for the true karst aquifers, underground mines, and accentuated fractured rock aquifers. However, the measured groundwater velocities were similar to velocities in unconsolidated glacial sands and gravels (Hawkins and Aljoe, 1992).

Groundwater Recharge

Initially, after reclamation, diffuse recharge from the surface is generally well below premining levels because of the destruction of soil structure, soil compaction by mining equipment, and low vegetative growth, which tend to promote surface water runoff rather than infiltration (Razem, 1983; Rogowski and Pionke, 1984). Wunsch *et al.* (1992) noted that during re-excavation, spoil within a few inches of the surface was dry indicating little infiltration was occurring. Decreases in recharge may also be facilitated by increases in porosity in the unsaturated zone (Razem, 1984). Flow-duration curves show that receiving streams after mining have reduce base flows, which indicate that recharge is decreased (29 per cent less than pre-mining levels) and surface runoff is increased (Weiss and Razem, 1984). After this initial period, as soil structure and vegetation re-establishes, diffuse recharge from the surface begins to increase. This may coincide with the observed increases in hydraulic conductivity after 30 months, as previously mentioned. The slow recovery of the water table during this period may be linked to the decreased recharge shortly after reclamation and the increased effective porosity and permeability of the spoil. Some of the recharge from the surface during this early period occurs through discrete openings of voids exposed at the surface (Hawkins and Aljoe, 1991; Wunsch *et al.*, 1992). Groenewold and Bailey (1979) observed that surface water naming into these swallets may enlarge them and cause considerable subsidence from piping of the finer grained materials. However, this has not been observed to be a significant problem in the Appalachian coal fields. Others contend that mining may improve the recharge potential from undisturbed areas (Cederstrom, 1971). Herring (1977) observed that the overall recharge and surface water runoff to reclaimed surface mines in the Illinois Basin were greatly increased. He attributes the increased recharge to the dramatic increase in permeability of the cast overburden. He observed a four fold increase in recharge from mining one half of a watershed in Indiana. Those two studies did not factor in the impact of mining on the soil horizon as discussed by Razem (1983, 1984). Once the infiltrating water has passed through the soil horizon, it appears that the recharge potential is dramatically increased. In the Appalachian basin, surface mine spoil aquifers receive a substantial amount of lateral in-flow from adjacent areas (Wunsch and Dinger, 1994). Adjacent unmined areas (low walls and high walls) as well as previously reclaimed areas will contribute groundwater to the newly reclaimed site. Groundwater modeling by Hawkins and Aljoe (1990) and Hawkins (1994) reported that groundwater flowing from adjacent areas may be the main source of recharge to the spoil aquifer and that this type of recharge occurs on a more continuous basis at a more consistent but lower rate than the recharge through the exposed surface voids. The surface voids will only recharge the spoil when runoff is occurring. In contrast, lateral recharge is controlled primarily by the hydraulic properties of the adjacent aquifer and the hydraulic gradient. Under certain conditions, spoil can be recharged from groundwater flow from the underlying strata. If artesian conditions exist beneath the stratum underlying the coat (seat rock), groundwater can flow under pressure via fractures in the intervening strata or through brocholes drilled through the pit floor and recharge the spoil. Artesian-induced recharge has been observed in a reclaimed surface mine in Southern Tennessee.

Thus the hydrologic information used in conjuction with the overburden geochemical data can be used to improve mine drainage predictive models and methods. Hydrologic data will give individuals involved with mine drainage prediction a better understanding of the spoil material that is contacted by the groundwater and the physical, spatial and temporal nature of this contact. This information is directly applicable to the use of special handling of acid forming materials and placement of alkaline materials to prevent acid mine drainage.

Environmental Impact of Specific Mine Spoils

Lignite/Coal

Lignite Mine Spoils

Lignite mining is a rapidly expanding industry, particularly in the context of our emphasis on the utilization of alternate sources of fuel. In the actual process of mining the area is completely stripped off without vegetation and subsoil is removed to reach lignite. This overburden spoil is dumped on barren land. Coal/lignite refuse is waste coal, with some crushed rock impurities, left as a result of coal processing. In its post-mining state, mine spoil and coal refuse, if unvegetated, can be highly erosive, and can be a source of significant sediment and acid mine drainage to dreams. If the flow capacity of a stream is significantly reduced due to sediment accumulation, the stream will flood more frequently, possibly damaging structures and overflowing roadways. Mine spoil and coal refuse, even if reclaimed, are prone to settlement and are subject to movement by freeze-thaw cycles. Refilling operations differs from mine to mine and depend largely on the extent, manpower and machinery.

Table 27.3: Characteristics of Lignite Mine Spoil and Normal Garden Soil

Parameters	*Normal Garden Soil*	*Lignite Mine Spoil*
Structure	Blocky to crumb	Massive
Texture	Sl to C	C
pH	6.95	6.52 (6.48–6.56)
EC (dSm^{-1})	0.58	0.37 (0.35–0.39)
Bulk density ($Mg\ m^{-3}$)	1.33	1.64 (1.58–1.70)
Water retentivity ($kg\ kg^{-1}$)	0.20	0.23
CEC [cmol (p+) kg^{-1}]	20.7	9.9
Organic C ($g\ kg^{-1}$)	0.55	0.17 (0.07–0.27)
Available N ($kg\ ha^{-1}$)	250.0	81.2 (77.2–85.2)
Available P ($kg\ ha^{-1}$)	15.0	3.05 (296–3.14)
Available K ($kg\ ha^{-1}$)	300.0	178.7 (171.7–183.6)
Total Ca ($g\ kg^{-1}$)	0.040	0.02 (0.02–0.02)
Total Mg ($g\ kg^{-1}$)	0.019	0.017 (0.016–0.017)
Total Mn ($mg\ kg^{-1}$)	2.37	5.64 (5.50–5.78)
Total Zn ($mg\ kg^{-1}$)	0.96	1.08(0.99–1.17)
Total Cu ($mg\ kg^{-1}$)	1.61	2.22(2.15–2.29)
Total Fe ($mg\ kg^{-1}$)	27.00	11.25 (11.17–11.33)

Senthil Kumar and Sekar, 1997.

The massive structure of mine spoil reveals (Table 27.3) the poor state of aggregation. The higher value of bulk density (1.64 Mg m^{-3}) in mine spoil implies the compaction which ultimately leads to poor aeration. The low organic matter (0.17 g kg^{-1}) in mine spoil might be responsible for the poor aggregation and compaction (Howarth *et al.*, 1996). Available N (81.23 kg ha^{-1}) and P (3.0 kg ha^{-1}) were comparatively low. But available potassium was medium (178.7 kg ha^{-1}). The mixture of different genesis levels of suit (Fehrunbacher, 1982) and absence of adequate microbial population (Balasubramanian, 1995) might have resulted in reduced fertility. Among the micronutrients, mine spoils are higher content of Cu, Zn and Mn whereas Ca, Mg, and Fe were found to be low. Lignite mine spoils are low in organic matter content, have poor physical properties, especially compaction and low nutrient status. Similar results have also been reported by Dadhwal *et al.* (1992) for limestone mine spoils of outer Himalayas. Thus, the success of revegetation of these areas depends largely on various measures of amendments to improve the organic matter content, physical and chemical properties.

Coat Mine Spoils

Every million of coal extracted by surface mining methods damages a surface area of 4 hectares in India (Ghosh, 1990). The coal industry rendered about 1400 hectares of area biologically unproductive and 500 M m^3 of overburden a year by 2000 AD (Chari *et al.*, 1989). Topsoil is an essential component for land reclamation in coal mining areas (Kundu and Ghose, 1994). The physical characteristics and water retention properties of unmined soil and mined soil are given in Table 27.4. The sand sized particles increased from 61.2 (in unmined soil) to 66.3 per cent (in mined spoils), whereas silt decreased from 27.7 to 25.1 per cent and clay from 11.1 to 8.8 per cent. This trend may be possible due to erosion and leaching in mined-spoil dumps. Dominance of sand particles in mined spoil indicates low stability of aggregates and consequently high rate of infiltration. The bulk density of mined soil was found to increase from 1.39 to 1.66 Mg m^{-3}. High value of bulk density of mined soil may be due to the influence of the use of heavy machinery. This has serious implications for subsequent changes in soil properties, because gaseous diffusion becomes more difficult. Thus, such high bulk density would pose restrictions for growth of deep-rooted plants and may be one of the reasons for stagnation of plant succession at shrub stage. Pore space of mine soil decreased in mined soil from 47.2 to 37.4 per cent. The field moisture, field capacity and water holding capacity decreased. It may be due to the decrease of organic carbon in mined soil. The pH of the unmined soil was found to be acidic in reaction and decreased in mined soil. Under acidic conditions, H^+ toxicity, high availability of Al and Mn and unavailability of Mo are the principal deterrents for plant growth. Electrical conductivity and CEC showed increase. The decrease of exchangeable Ca and Mg may be due to leaching loss. Organic Carbon in mined soil was low. The deficiency of nitrogen in mined soil is often due to lack of adequate amount of mineralizable organic nitrogen and lower mineralization rate. Optimum range of C : N ratio in soil is 10 : 1 to 12 : 1. Once C : N ratio exceeds 25 : 1 there will be a demand for available nitrogen by microbial population. Nitrogen accumulation and building up of a nitrogen cycle is the most crucial factor in soil development in the mined areas. Unless the soil maintains the available pool of each nutrient, there will be no agriculture. The available macro-nutrients (NPK) exhibited decrease in mined soil. It may be due to the reduction of soil microbes by stock pilling and excessive leaching. Available N was not deficient but available P and K were deficient in mined mil. The available (DTPA extractable) Fe and Mn increase but were not toxic to plants. Copper and Zn were also found to be in acceptable range. The number of microorganisms in mined soil decreased markedly in comparison to unmined soil. The decrease may be due to the presence of lower concentration of organic matter, moisture is an other edaphic factors (Ghose and Kundu, 1998). The major qualitative deterioration in the excavated soil is due to the complete loss of soil profile due to mixing of soil horizons. Dominance of sand particles in mine soil indicates tow stability of aggregates and consequently high rate of infiltration. The change in soil quality was found to be drastic.

Table 27.4: Characteristics of the Mined Soil

Parameters	*Unmined Soil*		*Mined Soil*	
	Range	*Mean*	*Range*	*Mean*
Particle Size distribution				
Sand (per cent)	55.7–65.9	61.2	65.1–67.5	66.3
Silt (per cent)	24.5–31.3	27.7	24.3–26.1	25.1
Clay (per cent)	8.6–15.1	11.1	8.1–9.2	8.6
Bulk density (Mg m^{-3})	1.38–1.42	1.39	1.65–1.67	1.66
Pore space (per cent)	46.4–47.9	47.2	37.0–37.7	37.4
Field moisture (per cent)	11.2–16.6	14.8	8.6–10.5	9.5
Field capacity (per cent)	18.2–22.1	20.6	13.5–15.6	14.7
Maximum WHC (per cent)	38.2–45.8	42.7	35.9–37.5	36.6
Wilting capacity (per cent)	5.11–5.82	5.45	5.62–6.05	5.81
pH (1 : 2 : 5)	0.14–6.65	6.38	6.16–6.35	6.24
EC (dSm^{-1})	0.17–0.37	0.24	0.32–0.41	0.36
CEC [cmol (p^+) kg^{-1}]	9.5–13.9	11.8	9.27–10.14	9.61
Ca^{2+} [cwol (p^+) kg^{-1}]	5.31–7.23	6.30	4.74–5.41	5.05
Mg^{+2} [cmol (p^+) kg^{-1}]	1.71–2.63	2.1	1.52–2.05	1.73
Mg^{+} [cmol (p^+) kg^{-1}]	0.28–0.33	0.30	0.22–0.26	0.24
K^{+} [cmol (p^+) kg^{-1}]	0.17–0.23	0.19	0.11–0.14	0.12
SAR	0.13–0.17	0.14	0.11–0.14	0.12
Base saturation (per cent)	72.2–8.12	76.0	72.8–76.1	74.3
Organic carbon (per cent)	0.57–0.88	0.72	0.35–0.41	0.38
Total nitrogen (per cent)	0.07	0.03	–	–
C/N	11.1	13.7	–	–
N (kg ha^{-1})	173.5–260.1	221.0	143–159	151.7
P (kg ha^{-1})	6.7–10.6	8.4	6.1–7.0	6.5
K (kg ha^{-1})	198.5–254.1	223.6	143.6–181.3	162.0
Fe (mg kg^{-1})	26.5–31.3	28.9	19.4–21.8	20.5
Mn (mg kg^{-1})	16.7–20.8	18.6	15.7–16.5	16.1
Cu (mg kg^{-1})	0.63–0.87	0.71	0.48–0.57	0.52
Zn (mg kg^{-1})	0.41–0.66	0.53	0.27–0.34	0.31
Bacteria (X $l0^5$ g^{-1})	640–920	775	81–106	92.2
Actinomycetes (X 10^4 g^{-1})	470–610	536	52–87	73.4
Fungi (X 10^3 g^{-1})	200–330	272	42–65	53.8

Asbestos

Chrysotile of fibrous morphology, is the mineral normally mined for asbestos production. The combination of this mineral's fiberous, morphology, low heat conductivity, high electrical resistance and chemical inertness have contributed to its wide industrial application. The Thetford mines Black Lake and asbestos districts of southeastern Quebec are the largest producers of chrysotile in the world. The main problems associated with vegetation establishment in asbestos mine spoil are extremely alkline conditions, low nutrient concentrations and surface crusting. Blowing dust has been reported as a problem for asbestos dumps (Ellery and Walker, 1986), as it has for many types of mine spoils. Because of health problems associated with prolonged asbestos exposure, dust from chrysotile mine spoil can have considerable impact on nearby communities and should be abated. Although asbestos mine spoil in South Africa had pH values of 9.6, available heavy metal concentration in me mine spoils were still greater than the limits normally tolerated by plants (Ellery and Walker, 1986). Decreasing the pH to 8.0 further exacerbated the problem. Nutrient imbalanccs may have been present because magnesium (Mg) was in extremely high concentration compared with Ca and K (Proctor and Woodwell, 1975).

Bauxite

Bauxite is a heterogeneous mixture of aluminum oxides and various impurities of iron oxides, silica, aluminum silicates and titanium oxides. Because of its versatility Al consumption world wide is exceeded only by iron (Fe) (McCawley and Baumgardner, 1985). Overburden spoils from bauxite mining are often acidic. White residue sands and red muds from bauxite processing are highly alkaline. The low pH was caused by iron sulfide oxidation. Soluble salts were also high. The spoil was low in fertility and microbial population, coarse textured and droughty. Red muds and residue sands are common waste products of bauxite processing. The Bayer process, the most common method of bauxite processing utilises NaOH to dissolve alumina and form a sodium aluminate solution. Insoluble impurities are filtered and transported to retention ponds as slurries. Depending on texture, these impurities are referred to as residue sands and red muds. Both wastes are highly alkaline (pH often greater than 10) and are also often saline (electrical conductivity commonly 30 dSm^{-1} or greater) and sodic (Meecham and Bell, 1977; Fuller *et al.*, 1982; Fuller and Richardson, 1986; Ward, 1987) (Table 27.5). Both Meecham and Bell (1977) and Ward (1987) reported residue sands with pH values of almost 10 and red muds with pH > 11. Red mud from the former study had an electrical conductivity (EC) of 30.2 dSm^{-1} and an exchangeable sodium percentage (ESP) of 81 while red mud from the latter study exhibited an EC of 36 dSm^{-1} and an ESP of 96. Residues sands generally have low water holding capacities while red muds have low infiltration and hydraulic conductivity because of fine texture and high Na content. Both residue sands and red muds; are low reported to be low in plant available N, P, and several micro-nutrients. Potassium can also be moderately limiting (Ward 1983).

Clay

The primary clays mined on a significant scale are bentomite and kaolinite (Kaolin). Bentonite is a 2 : 1 expanding clay formed from the alteration of volcanic ash deposited in shallow marine seas. The United States is the world's largest producer of bentonitic clay, with Wyoming accounting for almost 70 per cent of its production. Kaolinite is a 1 : 1 silicate clay mineral formed in acid environments. The soluble Al and Si released in such environments can recrystalize to form kaolinite. Kaolinite minerals are mined worldwide. The inner coastal plain of the South Eastern United States contains one of the world's largest Kaoline reserves as does South West England (Bradshaw and Chadwick, 1980). Alkalinity, salinity and sodicity can be problems in certain clay mine spoils. Smite *et al.* (1985)

worked with bentonite mine spoils in Wyoming with electrical conductivity of 13.4 dSm^{-1} and sufficient Na to be classified as sodic. Bentonite also may contain high Pb, titanium, Mn and other trace elements. China clay is composed chiefly of the silicate clay, kaolinite. Compared with metal mine spoils, China clay wastes normally do not contain heavy metals or large quantities of oxidizable sulfides (Gemmell, 1977). Deficiencies of plant nutrients especially N, P and K are often the principal limiting constraints to revegetation of such wastes.

Table 27.5: Physico-chemical Properties of Residue Sand and Red Mud

Parameters	*Residue Sand*	*Red Mud*
pH	9.75	11.2
EC (dSm^{-1})	30.2	37.7
LM NH_4OAC extractable cations		
Na^+ [cmol(p^+)kg^{-1}]	40.3	93.5
Ca^{2+} [cmol(p^+)kg^{-1}]	4.6	0.4
Mg^{2+} [cmol(p^+)kg^{-1}]	1.0	2.4
K^+[cmol(p^+)kg^{-1}]	5.3	10.4
Cation exchange capacity [cmol(p^+)kg^{-1}]	11.1	17.3
Sodium adsorption ratio	52.8	28.3
ESP	88	81

Meecham and Bell, 1977.

Copper, Gold, Nickel and Silver

Mining of Silver, (Ag), Gold (Au) and Copper (Cu) resulting in numerous waste dumps and mill mine spoil ponds that were abandoned after the ore was exhausted. Drainage from the orphaned mine sites contributes toxic mineral and sediment pollution into Lynx Creek and eventually into Lynx Lake (Verma and Felix, 1977). Acid mine drainage and elevated levels of Cu, Fe, Mn, Zn and SO_4 in the creek were the primary problem areas identified for treatment.

Table 27.6: Chemical Characteristics of Water Saturation Extracts of Copper Tailings

Sample Location	*EC (dSm^{-1})*	*pH*	*B*	*Cu*	*Soluble Ions*		*Zn*	*Pb*
					Fe	*Mn*		
New Mexico	9.5	2.2	10	>600	3000	24	21	>5
Montana	3.2	5.2	0.4	195	0.6	154	2.4	>5
Montana	2.1	7.6	0.2	<1	<0.5	<0.2	0.1	<5
Arizona	3.4	7.4	0.1	<1	<0.5	<0.2	0.1	<5
Arizona	3.1	7.3	0.1	<1	<0.5	<0.2	0.2	<5
Utah	9.5	7.0	0.4	<1	0.6	<0.2	<0.6	<5

Peterson and Nielson, 1973.

Heavy metal mine spoils frequently are become highly acidic over time because of the presence of sulfide-containing minerals. Overburden adjacent to mineral concentrations often contain low concentrations of iron sulfide minerals that are not economically retrievable or metal processing does not remove all iron sulfide minerals from the mine spoils. Therefore they often exhibit significant sulfide concentrations (Berg *et al.*, 1975; Sorensen *et al.*, 1980). Additional acidity may be contributed by acid extraction of these ores during processing. Metal ions are frequently present in sufficient concentrations in acidic Cu and Au tailings to inhibit or terminate plant growth (Wong, 1986) (Table 27.6). Heavy metals in tailings decrease plant root respiration, water and nutrients uptake and inhibit cell mitosis in root meristematic regions (Gemmell, 1977). Toxic ions may also contaminate soils and waters adjacent to mine spoils through seepage, runoff waters, and eroded sediment. Peterson and Nielson (1973) reported that Cu tailings in the western United States contained up to 5700, 875, 420 and 140 mg kg^{-1} of water soluble Cu, Fe, Mn, and Zn respectively. A portion of these tailings also exhibited electrical conductivities as high as 150 dSm^{-1}. Soil reactions ranged from 2.25 to 7.1 but most were pH 4.0 or lower. Shetron (1983) reported available Cu, Fe, Mn and Zn concentrations to be as high as 2444, 445 668 and 32 mg kg^{-1} in Cu tailings from Michigan. This author also suggested the difficulty in containing sufficiency and toxicity levels in tailings with those derived for soil. Lopez and Lee (1977) found that Cu could be approaching concentrations toxic to fish and aquatic invertebrates in a lake receiving Cu tailings and sediments. McNeilly and Johnson (1978) reported available Cu and Ph concentrations of 2060 and 327 mg kg^{-1} for metalliferous tailings in Great Britain. Crowder *et al.* (1982) reported higher concentration of Fe, Cu, and nickel (Ni) in tailings and vegetation growing on tailings than for soils. Metal concentrations were generally greatest after summer storms because of the flushing effect of rainfall on tailings that became well oxidized during the summer. Lampking and Sommerfeld (1981) reported on metal contaminated drainage from abandoned Au tailings in the Bradshaw Mountains of Arizona (Table 27.7). Drainage from the Sheldon Mine tailings from pond contained average Cu, Fe, Mn and Zn concentration of 21, 109, 16 and 116 mgl^{-1} respectively, and was implicated in the failure of a trout fishery located in the same watershed. Iron sulfide bearing tailings in Canada still contained DTPA-extractable concentrations of Cu, Fe, Mn and Zn as high as 310, 364, 8, and 40, 987 mg kg^{-1} respectively, after cropping in the greenhouse (Mac-lean and Dekker, 1976). The pH of these tailings was 2.5

Table 27.7: Total (Dissolved + Suspended) Mean Element Concentrations of Waters Entering Lynx Creek

Element	*Sheldon Pond Seep*	*Spruce Creek (mg l^{-1})*	*Rich Gulch Greek*
P	0.4	0.1	0.2
Ca	176.4	35.4	90.1
Mg	53.1	12.2	26.6
Na	13.4	9.6	14.6
K	1.5	1.6	3.2
Cu	21.0	1.2	6.5
Fe	109.1	3.8	6.9
Mn	16.1	0.3	2.3
Zn	115.6	2.0	24.3

Lampkin and Sommerfeld, 1981.

Iron

India has large resources of good quality iron ore located in a number of states such as Andhra Pradesh, Bihar, Goa, Karnataka, Madhya Pradesh, Maharashtra, Orissa, Rajasthan etc. Recoverable reserves measure upto 13 billion tonnes which make India the 6th richest country in iron ore resources during 1999–2000. The iron ore mines are causing various environmental and social problems to the natural resources and living things. There are some problems associated with active mining operations such as air emissions release and ore dispersal during transportation. The major problems associated with iron ore mining start from the disposal of mining related wastes. The waste piles associated with mining operations consists of waste rock, substrate ore and heavy metals. The metals of concern in mine waste piles are primarily iron and manganese. Iron ore mines operate on small leases of approximately 100 hectares. This results in a lack of space for the disposal of mined wastes. Very few of the pits are fully exhausted which means that old pits are not available for the disposal of mine wastes. In some cases, the wastes disposal are located away from the mine lease areas. This necessitates the transport of the mines waste in trucks from the mining areas to the waste disposal areas, leading to the disposal of some waste during transportation. The waste piles created are often unstable with a steep slope. In high rainfall areas, the combination of washable piles and heavy rains leads to the increased possibility of run off. Many mines and their waste piles are located close to residences and agricultural fields. The silt transported from the waste piles are being accumulated in the adjacent agricultural fields and the land becomes unsuitable for cultivation. In the monsoon season (June to September) most mining activities ate stopped and the mine pits are filled with water. Later, this water should be pumped out when mining operations recommence. The surface level of the mine pits in well below the water table and the pits are continuously fitted with water. The pH of the pit water is highly acidic and it pollutes the groundwater in the adjacent areas.

Most mining operations in North Goa are located such that the drains leading, runoff from the mining areas are approximately 5 to 10 kilometers from the Mandovi river. In the lean season prior to the onset of the monsoon, the flow of water in the river drops sharp, because, the residents of the downstream of the river are exploiting groundwater for domestic and other agricultural uses. Sediment load in the river varies from 10 to 400 mg/litre. Deforestation and cultivation of plantation crops without proper soil and water conservation measures in the upper watershed regions of the river also play a role in increasing the sediment loads. The silts accumulated in the river also affecting the surface and groundwater quality in the region.

The waste materials that emerges from the excavation of ore contains a type of material that does not allow vegetation to grow on the waste piles. A portion of the material that is excavated is a magniferrous clay substance that does not allow any plant growth. This material is impervious in nature and water does not penetrate easily. Some Fe tailings however, are neutral to alkaline with infertility and low organic matter, Cation Exchange Capacity (CEC) and Water Holding Capacity (WHC) being major constraints to revegetation. Sheleron and Spindler (1983) studied alfalfa establishment on alkaline ($pH \geq 8$) Fe tailings in the Iron Range of Michigan. Greatest concentrations of alfalfa roots and nodules were observed in fine textured strata within the tailings, presumably because of greater nutrient and water availability. Clay strata also exhibited lower bulk densities than sand or interstratified sand/silt tailings.

Lead and Zinc

Mine wastes from Pb-Zn processing often have low pH, excessive concentrations of metals and are devoid of vegetation. Rutherford *et al.* (1982) studied Pb-Zn mine spoils in Ontario Canada, that had remained devoid of vegetation for >40 years. Tailings profiles were strongly acidic, low in organic matter and exhibited profile characteristics of acid sulfate weathering (Van breeman, 1982; Brinkman and Pons, 1973). Total S was about 20 times higher in the tailings compared with normal soils. Abandoned Pb-Zn mine sites in Oklahoma characterized by pH values ranging from 7.0 to 7.4 and toxic levels of Pb (568 to 2575 mg kg^{-1}) Zn (2667 to 13090 mg kg^{-1}) and Cd (11 to 27 mg kg^{-1}) were naturally colonized by a wide variety of species from Oklahoma forests and prairies (Gibson, 1982). These species were characterised by efficient seed dispersal mechanisms and generally broad edaphic tolerance. Lead concentrations of sycamore (*Platanus occidentalis* L.) foliage ranged from 1.3 to 1120 mg ka^{-1} while that of twigs ranged from 1.8 to 320 ma kg^{-1} near Pb smelters and mines in eastern Missori (Palmer and Kucera, 1980). Nitric acid extractable soil Pb ranged from 7 to 62 000 mg kg^{-1}. Particulate size, chemical form, level of atmospheric Pb and the chemically active fraction of soil Pb were factors identified as playing important roles in the absorption of Pb into vegetation. Thompson and Proctor (1982) found *Agrostis capillaries* and *Festuca rubra* to be the most common species on a mine spoil having potentially toxic concentrations of Cu, Pb and Zn.

Phosphate

Mining operations for phosphate may produce a number of waste products (1) normal overburden, (2) sand tailings, (3) phosphate clay slimes, and (4) waste gypsum. Each waste has its own revegetation and reclamation problems. The chemical composition of phosphorites has been extensively studied and well documented. For reclamation purposes, two elements associated with phosphorites continue to be of environmental concern-uranium (U) and fluorine (F). These elements tend to be retained in phosphate products, by-products, and wastes. Radioactivity associated with phosphate tailings is a primary environmental concern. Radium activity as ^{226}Ra. of phosphate slimes is reportedly 24 to 50 picocuries (pCi) g^{-1} of dry material (Roessler *et al.*, 1979a). Phosphate deposits throughout the world contain U and thorium (Th) series radionuclides in concentrations greater than natural background levels. The ^{238}U series including ^{22}Ra and its decay products is the primary radiological concern. Generally, the upper 20 ml of undisturbed overburden has relatively low radioactivity (Environmental Protection Agency, 1977). The radionuclide content below about 2m increases, and Ra than the phosphate matrix (Clark, 1975). Mixing of the deeper material with the surface material during overburden stripping results in the increased radionuclide concentrations in the disturbed overburden. These nuclides are redistributed during processing among the products and wastes and have potentially adverse effects on plants, animals and man. Radium –226 in forage materials grown on dried phosphate clay ponds was nearly six times higher (0.23 p Cl g^{-1}) than in plants from an umined surface spodosol (0.04 $pCig^{-1}$) (Mislevey *et al.*, 1989).

Uranium

One of the largest uranium (U) ore reserves in the world is located in the Elliot Lake area of Ontario, Canada. It is estimated that uranium tailings from previous mining operations amount to about 80×10^6 t (Moffett and Telier, 1977), covering more than 400 ha. Uranium tailings contain about 85 per cent of the initial radioactivity of the source ore. It is estimated that 1 t of tailings has a radioactivity of at least 5 m Ci (Moffett and Telier, 1977). About 90 per cent of this activity is due to the ^{238}U series (Bryant *et al.*, 1977). The most troublesome daughter isotopes of ^{238}U are ^{230}Th, ^{226}Ra and ^{210}Pb. The solid tailings at Elliot Lake are made up of quartz, pyrite, calcium sulfate (from the neutralized reaction caused by lime addition at discharge), metal hydroxides, precipitated Th and rare earths (Vivyurka, 1975). The liquid portion of the mine spoil is a neutral to alkaline saturated calcium sulfate solution containing Ra, NH_4 and NO_3. Fresh mine spoil areas have a pH of 6.5 to 9.0 whereas seepage and runoff from abandoned mine spoils have a pH of 2 to 3 due to oxidation of pyrite. Liming percolation flow through mine spoil piles to reduce the bacterial oxidation responsible for the pH decrease and solubilization of radioactive constituents has been proposed.

The Biological Costs of not Reclaiming Mine Spoils

Until the late 1960's, mine spoil materials were left in steep piles and no effort was made to restore biological productivity to these disturbed sites. As a result, unreclaimed spoils are barren and eroded. The biological costs of not reclaiming these spoils are examined by Carolyn Hull *et al.* (1982) in terms of plant canopy cover, plant standing crop, small mammal productivity and invertebrate numbers. Vegetation, small mammals and ground-dwelling macro arthropods were samples for two years on unreclaimed spoils mined in the 1950'a and in adjacent unmined Sagebrush grasslands in southeastern Montana. The most significant biological cost of not reclaiming these old spoils was the loss of plant cover and production, followed by a decrease in macro-arthropod numbers, and, finally, the loss of small mammal habitat and a decrease in small mammal numbers. The annual costs appear insignificant in some cases, but when summed over 30 years, become very significant. Natural plant growth on unreclaimed spoils is limited by low soil pH, excessive levels of salinity and sodium, plus severe soil compaction. Total plant canopy cover on the unreclaimed mine spoils averaged 29 per cent less than on the unmined sagebrush-grasslands, 25 fewer plant species were present on the spoils, and the plant standing crop deficit on the unreclaimed spoil was 372 kg/ha. A spoil left unreclaimed for 30 years would result in an estimated accumulated loss of 11,160 kg/ha in plant standing crop. Plant cover and standing crop deficits reduce the wildlife and livestock carrying capacity of the range. The actual loss would be ever higher when the standing crop of big sagebrush and forage consumed by livestock are accounted for. Loss of total numbers of ground-dwelling macroarthropods comprised another biological cost of not reclaiming mine spoils. Macroarthropod numbers averaged 163 less per hectare than on the unmined sagebrush-grasslands. The accumulated loss on a 30-year-old unreclaimed spoil would be 4,890 macroarthropods per hectare. Not reclaiming mine spoil also resulted in the loss of small mammal numbers and habitat. An average of 19 fewer animals per hectare were captured on the unreclaimed mine spoils, compared to the rangelands. At an annual loss of 19 animals per hectare, a spoil left unreclaimed for 30 years would result in a deficit of 570 animals per hectare. The loss of small mammal burrowing habitat was an added cost of not reclaiming spoils. Small mammal burrows were limited to sagebrush-grass range lands. An average of 50 burrows per ha were found on the sagebrush-grass sites, whereas small mammal burrows were not detected on the unreclaimed mine spoil habitats. Burrowing habitat would be eliminated on unrcelaimed spoils, at a loss of 50 burrows per hectare. Because spoils are small, the low biological productivity on unreclaimed areas is compensated to some degree by surrounding unmined Sagebrush-grasslands. However, in areas where unreclaimed spoils are numerous, rangeland productivity is severely reduced and unmined sagebrush-grasslands do not compensate for the resulting biological deficits. Reclamation efforts may reduce the biological costs of mining but several authors have indicated that reclamation techniques used in the past have not been successful (Bjugstad, 1979). Research is needed to identify methods that compensate for adverse physical and chemical properties of the spoils materials and sparse top soil. Use of salt-tolerant plant species, such as fourwing saltbrush (*Atriplex canescens*), along with soil fertilization, mulching, and physical amendments have shown some promise in increasing the effectiveness of reclamation efforts (Bjugstad *et al.*, 1979). However, further information is needed to correctly assess and compensate for the biological costs of mining.

Faunal Succession in Mine Spoil

While faunal succession will vary many aspects from site to site many features of the process are well illustrated by data from rehabilitated coal mining dumps in the former German Democratic Republic (Curry and Good, 1992). Forty dumps were studies over a period of 25 years; most had been afforested with *Populus, Alnus* and *Robinia* and seeded with clover and lupin. Some of the main conclusions are as follows:

1. The most important requirements for annual rehabilitation of coal dumps were: (*a*) stabilisation of the water regime; (*b*) nutrient build up facilitated by rapid production of organic matter by N fixing plants; (*c*) detoxification *e.g.* Neutralization of mineral acids by timing.
2. Successions proceeds at different rates above ground and in the wit; the process is generally slower below ground.
3. Succession does not proceed uniformly across all groups with a steady increase in density and diversity. For some groups a pioneer peak may be followed by decreasing density and diversity.
4. Succession process are modified by individual site factors and rehabilitation procedures. Moisture was the most important factor, the amount of dead organic input was of secondary importance.
5. Five stages of succession could be recognized:
   (*a*) The pioneer stage characterized by hot, my conditions and the absence of organic matter and colonization by strategies. This stage could last up in 30 years in harsh sites but as little as 2 years under favorable conditions. Collembolans tend to be the most abundant micro-arthropods.
   (*b*) The second stage is influenced by the method of rehabilitation and the degree of natural colonization by various organisms. Increasing litter accumulation from the herb layer occurs and high densities of dipterous larvae, Collembola Predatory Carabidae and spiders are common. Minimal earthworm populations may be present. This stage may end at that 5th year or may last until the 9th year or longer on acidic dumps.
   (*c*) The third stage is characterized by a shrub layer without a closer canopy and a well developed litter layer which was colonized and reduced by the epigcic earthworms *Dendrobaena* spp. and later *Lumbricus rubellus* Hoff. This stage may last from about 5 years on better Pleistocene sites to 10 or more years on Tertiary acid dumps
   (*d*) This stage begins with the closure of the tree canopy. Earthworms dominate faunal activity and surface organic matter is incorporated into the soil by the activity of anecique species such as *Lumbricus terrestris.*
   (*e*) During the transition to the fully developed woodland stage there is gradual immigration of further species, especially K strategists, and a marked increase in the activity of saprophagous fauna. Under optimum conditions the woodland stage may commence 20 to 25 years after rehabilitation. This stage coincides with the end of soil zoological development.
6. Many animal groups have adequate powers of dispersal to allow rapid colonization at the appropriate stage of reclamation but this was not the case for certain earthworms, snails, isopods and centipedes.

Natural Recovery of Mine Spoils

Mining activities drastically alter the physical, chemical and biological status of an ecosystem. Natural recovery of mine spoils is a slow process. However, with time the natural plant invasion starts, accumulation of nutrients takes place, activity of microorganisms increases and ultimately a self-sustaining ecosystem develops. The time requirement for natural recovery of the iron mine spoils of Dalli-Rajhara has been assessed by Banergee (2001) considering the time-dependent changes of physical, chemical, nutritional and biological characteristics of overburden materials. The study revealed that the number of species increased as a function of age of sites. Twenty five year old site had maximum number of species (58) having 27 herbs, 11 shrubs and 20 trees (Table 27.8). In one and 2 year old sites there was no tree species but in other sites tree species were found. The presence of tree species in a community is an indication of its stability and maturity. The herbs including grasses were maximum in all the sites. The data indicate that herbs and grasses were the major contributors to natural re-vegetation. Total number of species recorded in adjacent natural forest was 43 out of which 21 were trees, 8 shrubs and 12 herbs. Number of shrubs and herbs in natural forest was comparatively less in comparison to 8, 10, 15, 20 and 25 year old sites because of closed canopy obstructing sunlight and thus many species could not thrive, with an increase in age of overburden dumps, not only the number of trees, shrubs and herbs increase but their density also increased. The proportion of legumes at different sites was 7.3, 9.1, 18.2, 17.6, 13.5, 17.8, 16.16, 16.9 and 16.2 per cent of flora in 1, 2, 5, 8, 10, 15, 20 and 25 year old sites and natural forest area, respectively. A good number of leguminous species in different sites is an indication of gradual improvement of the fertility status of the spoils because majority of them are potential N-fixer. The phyto-sociological parameters (Table 27.8) provide some idea about the dynamics of developing communities. As the age of the soil increase, the diversify index also increased with an exception of 15-year-old site. It is also interesting to not that eight sites studied showed a very regular relationship between diversity index and concentration, of dominance ($r- 0.865$, $p < 0.01$). Diversity index bore significant positive correlation with age of spoils ($r = 0.798$, $p < 0.01$).

The proportion of coarse fractions (> 2.0mm) in mine spoils of different sues was much higher in comparison to native form soil and decreased with the increase of age of mine spoils and the proportion of < 0.1 mm particles increased with age. The increases of finer fractions with age indicates the development of soil processes. Water stable aggregates (0.25 mm) also increased with age of spoils which indicates that fragmentation, redistribution and aggregation of particles gradually took place with the passage of time. Total and available N increased with age of mine spoils but the values were much less than those present in native forest soil even after 25 year of succession. Total soil N ranged from 120 to 600 kg ha^{-1} in 1 to 25 years old mine spoils.

There was a regular increase of the organic carbon with the increase in the age of the overburden. Down (1975) found 0.79, 1.52 and 1.81 per cent organic matter on sites of age 0,5 and 12 years, respectively in mine spoils at Somerset coalfield; whereas Rimmer (1982) reported organic matter accumulation at the rate of 0.8 per cent per annum in northern England in 1–8 years old spoils. Banerjee *et al.* (2000) noted 2.5 times increase between 4 to 10 years and 1.5 times increase between 10 to 14 years in coal mine overburdens at Gevra colliery of Bilaspur. The accumulation of organic matter in mine spoils begin once a grass stand is established and available carbon source is a critical factor to stimulate soil microflora. Available P, available K, exchangeable Ca and Mg also increased significantly with the age of the mine spoil. The population of all the organisms increased sharply with the increase in the age of overburdens.

Table 27.8: Physico-chemical Properties of Iron Mine Spoils and Native Forest Soil

Parameters	*Mine Spoils (Age in Years)*								*Natural Forest*
	1	*2*	*5*	*8*	*10*	*15*	*20*	*25*	
pH	6.3	6.5	6.6	6.4	6.4	6.7	6.5	6.6	6.3
Particle size (per cent)									
> 2.0 mm	29.2	27.0	25.2	24.6	18.8	20.2	16.9	17.8	6.8
20–0.2 mm	6O.3	61.2	60.4	60.6	64.5	62.8	62.8	58.8	63.0
0.2–0.1 mm	4.3	5.5	5.2	5.2	7.4	6.1	6.1	7.2	13.3
< 0.1 mm	6.2	6.3	9.6	9.6	9.3	10.9	16.6	16.2	16.9
Aggregates (per cent > 25 mm)	8.8	11.6	15.8	15.8	16.0	21.2	22.9	27.3	42.3
Org. C (g kg^{-1})	0.86	1.02	2.52	2.52	2.50	3.87	4.21	5.82	9.82
Total N (g kg^{-1})	0.06	0.10	0.16	0.16	0.14	0.21	0.25	0.30	0.94
Available N (mg kg^{-1})	7.2	18.6	51.6	51.6	63.3	77.6	88.3	98.8	167.8
Available P (mg kg^{-1})	0.4	1.4	2.8	2.8	2.6	3.5	3.6	4.2	8.6
Available K (mg kg^{-1})	16.6	28.4	37.5	37.5	57.6	72.8	74.3	70.2	206.6
Ex. Ca^{2+} [cmol (p^+) kg^{-1}]	0.3	0.6	1.9	1.9	2.2	2.2	2.5	2.8	6.4
Ex. Mg^{2+} [cmol (p^+) kg^{-1}]	0.1	0.2	0.6	0.6	0.4	0.5	0.5	0.9	2.2
Bacteria (x$10^5 g^{-1}$ soil)	2.12	5.87	32.5	32.47	42.2	71.3	88.7	106	481.2
Actinomycetes (x 10^5 g^{-1} soil)	0.96	3.30	21.1	21.10	32.6	52.3	60.2	74.4	381.1
Fungi (x10^4 g^{-1} soil)	0.72	2.10	19.9	19.90	30.7	48.9	51.3	65.5	210.3
No. of species	8	11	32	34	'37	45	51	58	43
Species diversity	1.51	2.23	2.50	2.97	2.98	2.81	2.99	3.22	2.72
Conc. of dominance (Cd)	0.27	0.12	0.10	0.098	0.09	0.16	0.09	0.08	0.173

Benergee, 2001.

Improvement of spoil condition promotes plant succession and microbial activity and vice-versa. There is a gradual development of the ecology and during ecosystem development accumulation of nutrients take place. However, to predict the time required for the total reclamation of mine spoils, multiple regression was computed considering the physical (particle size < 0.1 mm and aggregates > 0.25 mm) and nutritional (organic carbon, available N, P and K) characteristics of the mine spoils and the equation was

$$Y - 6.498 + 0.502X_1 - 0.280X_1 + 25.60X_3 - 0.078X_4 - 1.971X_5 + 0.016X_6 \ R2 - 0.998$$

where,

Y: Age in years (1, 2, 5, 8, 10, 15, 20 and 25)

X_1: Particle size <0.1 mm

X_2: Aggregata >0.25

X_3: Organic carbon (per cent)

X_4: Available N (ppm)

X_5: Available P (ppm)

X_6: Available K (ppm)

The time required on the basis of the above equation for the total reclamation was computed by taking values of X_1, X_2, X_3, X_4, X_5 and X_6 of the natural forest soil, and found to be 38.4 years. However, the allogenic inputs like Physical reclamation, organic and inorganic amendments, moisture conservation measures etc. will catalyse the process.

Mining Guidelines and Norms

Considering the ill-effects of mine spoils Govt. of India has framed some environment-related rules to be followed by the mine owners. All mining projects are now to be cleared by the Ministry of Environment and Forests before these are considered further. The projects are required to spell out how the mine-spoil areas are to be rehabilitated and what steps are contemplated for maintaining proper environment of the mining areas. The guidelines provided by the Ministry of Environment in 1982 are significant and worth mentioning here.

Our Burden

1. Proper area for the disposal should be identified. The site should be located on impervious base and away from water resources.
2. Piles and dumps should not exceed a mature tree top (10 m).
3. Mining muck should be used for filling up the mining pits.
4. Mining waste may also be utilised to fill the low tying areas.
5. Over-burden in case of strip mining may be preserved at suitable site for reuse.

Tailing Material

1. The tailing material from the benefication plant should be dumped on a hard and impervious base to avoid any leaching and thus keeping the aquifer safe.
2. Tailing dames will be subject to wind and water erosion hazards and therefore its location should be surrounded by hills. The wind breaks of tall trees are needed to be raised to check the flow of tailing dust.
3. Tailing dam needs quick vegetation by non-palatable woody or semi woody material as it contains several toxic matter like zinc, copper, lead, cyanide, etc. Once it enters food chain it is bound to have its impact on human beings.

Control of Gases

1. Minimize gaseous pollution and check suspended particulate matter.
2. Prevent all kinds of airborne dust.
3. Have large scale tree plantation all around the site to check the dust.

Action Plan

Land use plants, to minimize the adverse environmental impacts, should be drawn before undertaking the mining.

1. General survey of the area to be carried out to record geomorphic, hydrological, pedological and biologal features, together with climatic aspects.
2. The available information shall become the bases to formulate action plans.
3. Effective and comprehensive rehabilitation plan should be drawn. All these should be carried out in a phased manner.
4. In the action plant, the management and maintenance aspects should also be spelled out clearly.
5. A management schedule should be drawn and made an integral part of the revegetation programme.
6. Known methodology of mined land reclamation is very limited and hence more emphasis on research should be laid.
7. The reclamation plan should be produced as per the mining code and laws of the state/ centre.

Post Mining Care

The vegetation plan should be in accordance with stringent environmental and ecological requirements. Plant species selected should be capable to check soil erosion tolerant to drought, salinity and alkalinity and should be fast growing. Department allotting the land on lease to mining agency should be made responsible to examine that the proposed rehabilitation plan, cleared by department (now Ministry) of Environment, is property executed after mining. Many such sites may be reclaimed as National Heritage/wildlife/energy plantation/agroforestry/silvipastoral system.

Surface Mining Control and Reclamation Act of 1977 (U.S)

Revegetation of disturbed land is currently an area of environmental concern and active research, as well as practical application, particularly since the 1971 federal Surface Mining Control and Reclamation Act established strict regulations for the revegetation of currently mined land. The act (PL 95-87 Sec 515) requires that a diverse, effective and permanent vegetative cover of the same seasonal variety native to the area of land to be affected must be established and must be capable of self-regeneration and plant succession at least equal in extent of cover to the natural vegetation of the area. In March 1982 amended regulations were published. These rules are currently set forth in 30 CFR 816 and 817 (Federal Register, 1982) Specifically, they state that:

1. The permanent vegetative cover of the area must be at least equal in extent of cover to the natural vegetation of the area and must achieve productivity levels compatible with the approved post mining land use. Both native and introduced vegetation species may be used.
2. The period of responsibility initiates after the last year of augmented seeding, fertilizing, irrigation or other work which ensures revegetation success.
3. In areas of more than 66 cm of average annual precipitation, the period of extended responsibility will continue for not less than 5 years. In areas with 66 cm of precipitation or less, the period of responsibility will continue for not less than 10 years.
4. Normal husbandry practices essential for plant establishment would be permitted during the period of responsibility so long as they can reasonably be expected to continue after bond release.

5. In areas of more than 66 cm of precipitation, the vegetative cover shall be equal to the success standard only during the growing season of the last year of the responsibility period unless 2 years would be required by the regulatory authority. In areas with less than 66 cm, the vegetative cover must be equal to the success standard for the last 2 years of the responsibility period.
6. The ground cover, productivity or tree stocking of the revegetated area shall be considered equal to the success standard approved by the regulatory authority when the parameters are fully equivalent with 90 per cent statistical confidence.

It will be difficult to meet these requirements using current reclamation techniques. New methods will have to be developed and larger amounts of lime, fertiliser and seed will undoubtedly be needed. Soil amendments, mulching and even irrigation may be required on some sites.

Summary and Conclusion

Open cut and underground mining are the two major types of mining. Surface mining operation enable man to obtain many of the minerals such as coal, lignite, copper, iron etc. Sustainability science for mine-spoil restoration generation of mine waste threatens the economic, ecological and social sustainability of our country. This is one of the visible consequences of imperiled nature-society interactions. Various forms of minig regardless of whether acknowledged or omitted from studies on land use change, ecological degradation and biodiversity loss have a role in ecological problems in India. Unsustainable mining, in conjunction with other threats to ecosystems exacerbates the local environmental degradation, livelihood insecurity and global climate change. Abandoned mine tailings have extremely diverse physical, chemical and ecological Conditions. The mining operation leave their detrimental effects on the ecosystem by way of (1) removal of the entire vegetation from mining spot, (2) the overburden is removed nod placed at a nearby point. (3) Low grade ore/mineral which do not find market is to be dumped on the surface near the mining site. (4) Over-burden and great volume of debris, if spread unsystematically will disrupt surface and groundwater circulation. (5) When overburden of mineral is stored on adjacent land then that land also become unproductive. (6) Many strip mining are carried out on agricultural land. These include marble, bentonite brick clay etc., causing reduction of arable land. (7) Mining activities cause several pollution hazards. Fine mineral particulates are released to the atmosphere which on settling cause loss of land productivity and spoil the surface water resources. In many cases toxic and harmful minerals and metals get dissolved in the rainwater and percolate down to the adjoining aquifer of mining site, thereby creating health hazard. Mining activities disturb the natural drainage and reduce the recharge of aquifer, resulting in low water yield and poor water quality. Many industries produce fine dust while in operation. It causes respiratory hazard to its workers and neighbouring inhabitants, *i.e.* lungs silicosis (8) Underground mining *i.e.* coal and mica cause land subsidence. (9) In many mining sites there is development of salinity. (10) The mining sites loose the entire vegetation, whereas there is unchecked degradation of vegetation in the adjoining area, especially when the mining activity is labour intensive. Much of the vegetation is for hut construction and fuel purpose. (11) Many allied mining activities like benefication of mineral, crushing, grinding, polishing and chemical processing also affect the working habitat at ground, underground and atmospheric levels. (12) Plant uptake of potentially toxic chemicals or heavy metals and their incorporation into the food chain are real concerns. Some mine spoils contain radioactive nuclides which can pose long-term health considerations. Erosion by wind and water with the associated environmental degradation is a universal concern associated with mine spoil materials.

References

Aljoe, W.W. and Hawkins, J.W., 1994. In: *Proceedings of the 5th International Mine Water Congress,* Nottingham, U.K., pp. 3–21.

Arora, H.S. *et al.*, 1978. *Soil Sci.*, 125: 151–159

Balasubramanian, A., 1995. In: *Nat. Symp. Org. Farming,* Abst., TNAU, Madurai.

Banerjee, S.K., 2001. *J. Indian Soil Sci.*, 49: 380–382

Banerjee, S.K. *et al.*, 2000. *J. Indian Soc. Soil Sci.*, 48: 63.

Berg, W.A. *et al.*, 1975. In: *The Ecology of Resource Degradation and Renewal,* (Eds.) M.J. Chadwick and G.T. Goodman. Blackwell Scientific Publications, Oxford, pp. 207–222.

Bjugstad, A.J., 1979. Bentonite mine spoil and pit reclamation.

Bonta, J.V. *et al.*, 1991. *Water Resource Bulletin,* 18(3): 577–593.

Bradshaw, A.D. and Chadwick, M.J., 1980. *The Restoration of Land.* Blackwell Scientific Publications, Osncy Mead Oxford England.

Bradsbaw, A.D. *et al.*, 1978. In: *The Breakdown and Restoration of Ecosystems,* (Eds.) M.W. Holdgate and M.J. Woodman. Plenum Press, New York, pp. 249–278.

Brinkman, R. and Pons, L.J., 1973. In: *Acid Sulphate Soils,* (Eds.) H. Dosl. ILRI Pub 18. Institute for land reclamation and improvement. Wageningen, The Netherlands, 1: 169–203.

Brown, R.L. and Parizck, R.R., 1971. In: *Special Report of Research.* The Pennsylvania State University, University Park, PA, SR–84, 207 p.

Bryant, D.N. *et al.*, 1977. Technology Development, Report No. EPS 4–WP–79–4. Environmental Protection Service, Ottawa, Canada.

Carolyn, Hulls, 1982. In: *Proceedings of the Annual Conference of the Western Association of Fish and Wildlife Agencies,* July 18–22, Las Vegas.

Caruccio, F.T. *et al.*, 1984. In: *Proceedings of the 1984 Symposium of Surface Mining, Hydrology, Sedimentology and Reclamation*, Lexington, KY, pp. 43–47.

Caruccio, F.T., 1988. In: *Reclamation of Surface Mined Lands, Vol. 1.* CRC Press, Boca Raton, Fl., pp. 159–190.

Cederslnnn, D.J., 1971. *Mining Congress Journal,* March, pp. 46–50.

Chai, K.S.R. *et al.*, 1989. Dept. of Env. Forest and Wildlife, New Delhi.

Clark, R.K. and Clark, S.C., 1981. *U.S. FDA 7th Annual Conference on Radiation Control,* pp. 166–169, Report No DHEW/FDA 76–8026 Conf. 7504107.

Clark, V., 1975. *U.S. FDA 7th Annual Conference on Radiation Control,* pp. 166–169.

Coffey, P.S. *et al.*, 1986. *J. Environ. Qual.*, 13: 168–172

Craze, B., 1977a. *J. Soil Cons.*, News South Wales, 33: 190–199.

Creswell, C.F., 1973. In: *Ecology and Reclamation of Devastated Land, Vol. 1*, (Eds.) R.J. Hutnik and G. Davis. Gordon and Breach, New York, 1: 335–359.

Crowder, A.A., *et al.*, 1982. *Reclamation Revegetation Res.*, 1: 77–193.

Curry, J.P. and Good, J.A., 1992. *Advances in Soil Sci.*, 17: 171–215.

Dadhwal, K.S. *et al.*, 1988. *Indian J. Soil Conserv.*, 2: 51.

Dadhwal, K.S., *et al.*, 1992. *Indian Forester*, 118: 650..

Day, A.D. and Ludeke, K.L., 1973. *J. Environ Qual.*, 2: 314–315

Dean, S.A. and Fogel, M.M., 1982. Symposium on surface. In: *Mining Hydrology Sedimentology and Reclamation*, pp: 269–276.

Dollhopf, D.J. *et al.*, 1980. Bulleting 736 Montana Agricultural Experiment Station Bozeman MT.

Down, C.G., 1975. *J. Ecol.*, 12: 613.

Ellery, K.S. and Walker, B.H., 1986. S. Afr. J. Bot., 52: 201–206.

Environmental Protection Agency (U.S), 1977. U.S. Office of Radiation Program, Report No. ORPILV–77–3.

Fehrumbacher, 1982. *Soil Sci. Soc. Am. J.*, 46: 353.

Fuller, W.H. and Lanspa, K., 1975. J. Environ. Qual., 4: 417–422.

Fuller, R.D. and Richardson, C.J., 1986. Environ. Top Chem., 5: 905–915

Fuller, R.P. *et al.*, 1982. J. Environ. Qual., 11: 533–539

Gemmell, R.P., 1973. *Environ. Pollution.*, 5: 181–197.

Gemmell, R.P., 1977. *Colonization of Industrial Wasteland.* Camelot Press, South Amptom, England.

Ghose, M.K. and Kundu, N.K., 1998. *J. Indian Soc. Soil Sci.*, 46: 15.

Ghose, A.K., 1990. *J. Inst. of Engg.*, 39: 1.

Gibsen, D.J., 1982. *Southwestern Naturalist*, 27: 425–436.

Groenewold, G.H. and Dailey, M.J., 1979. In: *Ecology and Coal Resource Development*, (Ed.) M.K. Wali, pp. 685–692.

Groenewold, G.H. and Winczcwsn, L.M., 1977. In: *Annual Proceedings of the North Dakota,* Academy of Science, pp. 160–167.

Harwood, G., 1979. USDA Forest Service, General Technical Report INT 57 Ogden UT.

Hawkins, J.W., 1993. In: Society for Mining, Metallurgy and Exploration Inc., Transaction, 294: 1810–1813.

Hawkins, J.W., 1994. In: *Proceedings of the International Land Reclamation and Mine Drainage Conference and Third International Conference on the Abatement of Acidic Drainage*, (Ed.) P.A. Pittsburgh, 2: 265–272.

Hawkins, J.W., 1995. In: *Proceedings of the 1995 Annual Meeting of the American Institute of Hydrology,* Denver, CO.

Hawkins, J.W. and Aljoe, W.W., 1991. In: *Proceedings of the Second International Conference on the Abatement of Acidic Drainage*, Montreal Quebec, Canada, Tome 1, pp. 47–68.

Hawkins, J.W. and Aljoe, W.W., 1990. In: *Proceedings of the National 1990 Symposium on Mining,* Lexington, KY, pp. 43–51.

Hawkins, J.W. and Aljoe, W.W., 1992. *Mine Water and the Environment*, 11(2): 37–52.

Helgeson, J.O. and Razem, A.C., 1980. In: *Proceedings of the Symposium of Surface Mining Hydrology, Sedimentology and Reclamation*, Lexington, KY, December 1–5, pp. 351–360.

Herring, W.C., 1977. In: *Proceedings of the Seventh Symposium on Coal Mine Drainage Research*, Louisville, KY, pp. 77–87.

Hogan Gary D. *et al.*, 1977. *Can. J. Bot.*, 55: 1038–1042

Howarth *et al.*, 1996. *J. Hort. Sci.*, 3: 225.

Hunter, G. and Whitetuan, P.C., 1974. *J. Aust. Inst Agric. Sci.*, 40: 270–278.

Hutchison, T.C. and Whitby, L.M., 1974. *Environ. Conserv.*, 1: 123–132.

Jennett, J.C. and Linnemann, S.M., 1977. *Water Pollut.* Control Fed., 49: 1842–1856.

Johnson, M.S. *et al.*, 1977. *Environ. Pollution*, 12: 261–277.

Jordan, J.E. and Dewar, S.W., 1985. In: *Symposium on Surface Mining Hydrology Sedimentology and Reclamation,* Dec. 9–13, University of Kentucky, Lexingtion, KY.

Kundu, N.K. and Ghose, M.K., 1994. *Environmental Conservation*, 21: 126.

Ladwig, K.J. and Campion, P., 1985. In: *Proceedings of the 1985 Symposium on Surface Mining Hydrology, Sedimentology and Reclamation*, Lexington, KY, pp. 121–130.

Lampkin, A.J. III and Sommerfeld, M.R., 1981. *J. Arizonanevada Acad. Sci.*, 16: 165–169.

Lindorff, D.E., 1980. In: *Proceedings of the Symposium on Surface Mining Hydrology, Sedimentology and Reclamation*, Lexington, KY, pp. 33–40.

Lindsay, W.L., 1979. *Chemical Equilibria in Soils.* Wiley, New York.

Lopez, J.M. and Lee, G.F., 1977. *Water Air and Soil Pollut.*, 8: 373–385

Ludeke, K.L., 1977. In: *Reclamation and Use of Disturbed Land in the Southwest,* (Ed.) J.L. Thames, University of Arizona Press, Tucson, pp. 262–271.

MacKay, D.M. and Cherry, J.A., 1989. *Environ Sci. Tech.*, 23(6): 630–636.

Maclean, A.J. and Dekker, A.J., 1976. *Can J. Soil Sci.*, 56: 27–36.

Mann, H.S. and Chatterji, 1978. *Ecology and Coal Resource Development,* 2: 615–625.

McCawley, F.X. and Baumgardner, L.H., 1985. *USDI Bureau of Mines Publ. Bull.*, 675.

McNeilly, T. and Johnson, M.S., 1978. *J. Environ. Qual.*, 7: 483–486.

Meecham, J.R. and Bell, L.C., 1977. *Australia J. Exp. Agric. Animal Husb.*, 17: 169–688.

Merrill, S.D. *et al.*, 1987. In: *Reclaiming Mine Soils and Overburden in the Western United States Analytic Parameters and Procedures,* (Eds.) R.D. Williams and G.E. Schmnan. Soil Conservation of American Academy, IA, pp. 35–58.

Merrill, S.D. *et al.*, 1980. *North Dakota Agric. Exp. Sta. Farm Res.*, 37: 13–16.

Merrill, S.D. *et al.*, 1983. *Soil Sci.*, 136: 308–316.

Mezga, L.J., 1973. *Master's Thesis.* Kent State University, 82 p.

Mislvey, P. *et al.*, 1989. *J. Environ. Qual.*, 18: 95–100.

Moffett, D. and Tellier, M., 1977. *Can. J. Soil Sci.*, 57: 417–424.

Moran, S.R. *et al.*, 1979. In: *Proceedings of the Symposium on Surface Mining Hydrology, Sedimentology and Reclamation*, Lexington, KY, pp. 57–65.

Morrey, D.R. *et al.*, 1981. *J. Environ. Mgt.*, 19: 99–105.

Murray, D.R., 1977. *CANMRT* (Can Centre Min. Energy Technol.), Toronto, Canada, Report No. 77–31.

Nordstrom, Darrell Kirk, 1982. In: *Acid Sulfate Weathering*, (Eds.) J.A. Kittrick *et al.* Soil Science Society of America Madison WI, pp. 37–56.

Nunn, R.M. and Riches, J.R., 1978. In: *Rehabilitation of Mined Lands in Western Australia*, (Ed.) J.R.D. Fox. Western Australian Institute of Technology Perth, Australia, pp. 113–116.

Palmer, K.T. and Kucera, C.L., 1980. *J. Environ. Qual.*, 9: 106–111.

Peters, T.H., 1984. In: *Effects of Pollutants at the Ecosystem Level*, (Eds.) P.J. Sheehan *et al.* John Wiley and Sons, New York.

Peterson, H.R. and Nielson, R.F., 1973. In: *Ecology and Reclamation of Devastated Land*, (Eds.) R. Huntnik and G. Davis. Gordon and Breach Sci. Publ., New York, 1: 15–25.

Pehlps, L.B., 1981. In: *Proceedings of the 1983 Symposium on Surface Mining Hydrology*, Sedimentology and Reclamation, Lexington, KY, pp. 47–51.

Proctor, J. and Woodell, S.R.J., 1975. *Adv. Ecol. Res.*, 9: 256–366.

Pugh, C.E. *et al.*, 1984. *Soil Scl.*, 137: 309–314.

Rahn, P.H., 1976. In: *Report on Potential of Coal Strip-mine Spoils as Aquifers in the Powder River Basin*. Old West Regional Commission, Billings, Montana, Project No. 10470025, 108 p.

Razem, A.C., 1984. In: Groundwater hydrology and quality before and after strip mining of a small watershed in Jefferson County, Ohio. U.S. Geological Survey, Water Resources Investigation Report 84–42.15. Columbus, OR, p. 39.

Rehm, B.W. *et al.*, 1980. *Groundwater*, 20: 217–236.

Rmmer, D.L., 1982. *J. Soil Sci.*, 33: 567.

Rogowski, A.S. and Pionke, H.B., 1984. Hydrology and water quality on stripmined lands. U.S. Environmental Protection Agency, EPA–IAG–D5–E763, Washington, D.C., p. 183.

Ruppin, P.A., 1987. In: *Mining Rehabilitation*, (Ed.) T. Farrell. Australian Mining Industry Council, Canberra, Australia, pp. 71–87.

Rutherford, G.K. and Bray, C.R., 1979. *J. Environ. Qual.*, 8: 219–222.

Rutherford, G.K. *et al.*, 1982. *J. Environ. Qual.*, 11: 511–518.

Schafer, W.M., 1979. *Reclamation Rev.*, 2: 67–74.

Senthil Kumar, R. and Sekar, K., 1997. *J. Indian Soc. Soil Sci.*, 45: 829–830.

Shamshuddin, J. *et al.*, 1986. *Pertabuja*, 9: 89–97.

Sheppard, M.L. *et al.*, 1984. *J. Environ. Qual.*, 13: 357–361.

Shettron, S.G. and Spindler, J.J., 1984. *Pl. Sci.*, 73: 239–246.

Shettron, S.G., 1983. *Pl. Soil*, 73: 227–237.

Simon, R., 1978. *New Phytologist*, 91: 175–188.

Smith, J.A. *et al.*, 1985. *J. Environ. Qual.*, 14: 575–580.

Sorensen, D.L. *et al.*, 1980. *J. Environ. Qual.*, 3: 162–166.

Sweigard, R.J., 1987. In: *Proceedings of the 1981 National Symposium on Mining, Hydrology, Sedimentology, and Reclamation*, Texington, KY, pp. 183–88.

Thomson, J. and Protctor, J., 1982. New Phytologist, 94: 297–308.

United State Department or Agriculture, 1979. General Technical Report INT–68 Intermountain Forest and Range Experiment Station, Forest Service Logna UT.

Van Breeman, N., 1982. In: *Acid Sulfate Weathering*, (Eds.) J.A. Kilrick *et al.* Soil Science Society of American Madison WI, pp. 95–108.

Van Voast, W.A., 1974. Bulletin 93, Montana College of Mineral Science and Technology, Butte, MI, 24 p.

Verma, T.R. *et al.*, 1977. *Hydrology and Water Resources in Arizona and the Southwest*, 7: 61–68.

Verma, T.R. and Felix, R.N., 1977. *Hydrology and Water Resources in Arizona and the Southwest*, 7: 49–59.

Vivyurka, A.J., 1975. 7th *Annual Meeting of the Canadian Mineral Processors*, January 21–23, Ottawa, Ontario, Canada.

Ward, S.C., 1983. *Reclamation Reveg. Res.*, 2: 177–190.

Ward, S.C., 1987. In: *Mining Rehabilitation*. (Ed.) F. Farrell. 87. Australian Mining Industry Council, Canberra, Australia, pp. 61–70.

Weiss, J.S. and Razem, A.C., 1984. In: *Simulation of Groundwater Flow in a Mined Watershed in Eastern Ohio.*

Wells, L.G. *et al.*, 1982. In: *Proceedings of the 1982 Symposium on Mining, Hydrology, Sedimentology, and Reclamation*, Lexington, KY, pp. 445–456.

Whitby, L.M and Hutchinson, T.C., 1974. *Environ. Conserv.*, 1: 191–200.

Williams, S.T. *et al.*, 1977. *Soil Biochem.*, 9: 271–275.

Wong, M.H., 1986. *Environ. Mgt.*, 10: 707–713.

Wunsch, D.R. and Dinger, J.S., 1994. In: *Proceedings of the International Land Reclamation and Mine Drainage Conference and Third International Conference on the Abatement of Acidic Drainage*, Pittsburgh PA, 2: 138–191.

Wunsch, D.R. *et al.*, 1992. In: Report of Investigations 6 Series XI, Kentucky Geological Survey, Lexington, NY, 16 p.

Chapter 28

Growth Studies in Groundnut (*Arachis hypogaea*) Under the Influence of Fluctuating Groundwater Table and Levels of Irrigation

J.M.L. Gulati[1], M.M. Mishra[2] and D. Lenka[3]
Regional Research and Technology Transfer Station, OUAT, Chiplima, Sambalpur, Orissa – 768 025
[1]Professor, [2]Associate Professor, Department of Agronomy, OUAT, Bhubhaneswar – 3
[3]Retired Professor in Agronomy, 156 Nuasahi, Suryanagar, Bhubaneswar – 751 003

ABSTRACT

Growth studies in groundnut under fluctuating groundwater table and levels of irrigation during summer revealed that Net Assimilation Rate (1.99 g/m²/day) and Crop Growth Rate (3.44 g/m²/day) were the least in plants grown under shallow water table condition and increased to 2.31 and 6.10 g/m²/day respectively due to lowering of groundwater table. The maximum value of NAR (3.71) and CGR (10.49 g/m² day) were recorded between 69 to 75 days and 76 to 82 days of growth respectively irrespective of water table depth. Leaf Areas Index of the groundnut was the minimum (0.13) at 27 to 33 days of growth with its peak (5.34) at 97 to 103 days and then declined to 2.5 at harvest. Mean LAI under shallow (50±10 cm bgl), medium (90±10 cm bgl) and deep (110±10cm bgl) water table situation was 1.79, 2.47 and 3.03 respectively with corresponding leaf area duration of 163, 224 and 275. All the growth parameters showed positive response to irrigation.

Keywords: *Net assimilation rate, Leaf area index, Crop growth rate, Leaf area duration, Relative growth rate, Groundwater table.*

Introduction

Groundnut, though a C_3 plant, has the photosynthetic characteristic of C_4 grasses and thus, is a good harvester of solar energy. The complex yield phenomenon is the out come of interplay of genotypes and environmental factors. A soil moisture potential above –0.04 Mpa at 20cm depth and no stress moisture regime at 1.21W : CPE has been reported to cause an excessive growth due to elongation of main stem and the number of branches there in groundnut (Garbat and Rhodes, 1975 and Gulati *et al.*, 1999). On the other hand moisture stress has. been found to reduce the formation of leaves and their expansion. Chapman *et al.* (1993) observed that reduction in leaf water potential was quicker under moisture stress than under the unstressed conditions. Duncan *et al.* (1978) found various growth indices soil and moisture dependant. Groundnut when grown under terraced landscape gets interfered from fluctuating groundwater table under varying supply of irrigation water and produced pod yield of different quantum (Gulati and Lenka 1999). The present investigation aimed at studying the various growth parameters in groundnut under the influence of fluctuating groundwater table and irrigation water supply in Hirakud irrigation command in Sambalpur, Orissa.

Materials and Methods

Field study was carried out on sloppy terraced clay loam soil at Regional Research and Technology Transfer Station, Chiplima, Sambalpur in the year 1992 and 1993 during summer with groundnut OG 85-1, a Spanish bunch cultivar. The soil of the experimental site has 21–23 per cent moisture at field capacity and 8–10 per cent at permanent wilting point. The bulk density of the soil was 1.50 to 1.56 g/cc. and the hydraulic conductivity was 1.33 to 1.36 cm/hr. The crop was sown in last week of January with a crop geometry of 25cm row to row and 10cm seed to seed under a nutrient schedule of 20 : 40 : 40 kg N, P_2, O_5 and K_2O/ha, respectively, all applied as basal through straight fertilizers. Six irrigation schedules such as 0.2, 0.4, 0.6, 0.8, 1.0 and 1.2 IW : CPE (IW = 6.0cm) were laid out in randomized block design with 4 replications in each of the 3 groundwater table conditions where the groundwater table fluctuated between 50±10 cm (W_1), 90±10 cm (W_2), and 110±10 cm (W_3) below the ground level (bgl). Crop matured in 118 days in both the years. After the establishment of the crop, 3 plants at random were selected and uprooted at 7 days interval and data on various growth parameters such as net assimilation rate, leaf area index, leaf area duration, crop growth rate and relation growth rate were obtained as per the methods given by (Watson, 1952), correlation study was made as per the method given by Cocharan and Cox, (1950).

Results and Discussion

Data on growth parameters under different treatment have been presented in Tables 28.1 and 28.2.

Net Assimilation Rate (NAR)

The mean NAR of groundnut crop was 2.13 g/m^2/day. It increased with age of the crop uptill 69 to 75 days of crop growth and declined thereafter towards maturity at 110 to 116 days after sowing. NAR was the maximum (4.25 g/m^2/day) under medium water table situation of 90±10 cm bgl and reduced by 17.4 per cent and 8.5 per cent at higher (50±10 cm bgl) and lower (110±10 cm bgl) water table situations respectively. Mean NAR was also the maximum (2.31g/m^2/day) under medium water table situation. However, at initial stage the NAR was more (0.48 g/m^2/day) under W_1 and reduced to 0.33 g/m^2/day under W_2 and 0.30g/m^2/day under W_3 situation. NAR responded differently under the combined influence of water table depths and irrigation levels. The higher NAR (2.28 g/m^2/day) under W_1 was observed at 0.8 IW : CPE whereas under comparatively lower water table situation of W_2, the increase in NAR was observed up to more frequent irrigation level of 1.0 IW : CPE, the values being 2.80 g/m²/day. The net assimilation rate is a measure of excess of dry matter gained over the loss by the process of respiration. Since respiratory losses depend upon crop canopy, leaf temperature and atmospheric vapour gradient, the values of NAR under varying water table situations and levels of irrigation also differed. The difference of magnitude was, however, less under shallow water table situation.

Table 28.1: Two Way Table on Growth Parameters in Groundnut as Influenced by Irrigation Levels and Water Table Depth (Mean of two seasons)

IW : CPE	a: NAR (g/m²/day)				b: LAI				c: CGR(g/m²/day)			
	Water Table Depth			Mean	Water Table Depth			Mean	Water Table Depth			Mean
	W_1	W_2	W_3		W_1	W_2	W_3		W_1	W_2	W_3	
0.2	1.65	1.93	1.62	1.73	1.49	1.67	2.18	1.78	2.50	3.03	3.35	2.96
0.4	1.89	2.02	1.86	1.92	1.66	2.07	2.56	2.10	2.86	4.13	4.93	3.97
0.6	1.97	2.12	2.28	2.12	1.81	2.57	2.81	2.40	3.50	4.92	5.93	4.78
0.8	2.28	2.56	2.32	2.39	1.94	2.67	3.34	2.65	4.40	5.90	7.38	5.89
1.0	2.18	2.80	2.26	2.41	1.9	2.87	3.55	2.80	4.15	7.79	7.68	6.54
1.2	1.96	2.45	2.15	2.19	1.87	2.95	3.37	2.85	3.23	7.25	7.32	5.93
Mean	1.99	2.31	2.08	2.13	1.79	2.47	3.03	2.43	3.44	5.50	6.10	5.01

IW : CPE	d: LAD				e: RGR (g/cm²/day)			
	Water Table Depth			Mean	Water Table Depth			Mean
	W_1	W_2	W_3		W_1	W_2	W_3	
0.2	136	152	198	162	0.0536	0.0445	0.0432	0.0471
0.4	151	188	233	191	0.0464	0.0459	0.0485	0.0469
0.6	165	234	256	218	0.0478	0.0456	0.0439	0.0458
0.8	177	243	304	241	0.0480	0.0448	0.0452	0.0460
1.0	181	261	323	255	0.0437	0.0445	0.0451	0.0444
1.2	170	269	339	259	0.0406	0.0446	0.0457	0.0436
Mean	163	224	275	221	0.0467	0.0450	0.0453	0.0457

IW : CPE	f: Pod yield (q/ha)			
	Water Table Depth			Mean
	W_1	W_2	W_3	
0.2	9.7	11.4	13.6	11.6
0.4	11.8	14.8	16.6	14.4
0.6	14.9	19.6	18.9	17.8
0.8	17.7	21.9	21.8	20.5
1.0	16.0	24.8	25.1	21.9
1.2	15.9	25.7	25.8	22.5
Mean	14.3	19.7	20.3	

CD (0.05)
Irrigation (I) = 0.62
Water Table (W)=0.44
Interaction (IXW)= 1.09

Table 28.2: Growth Parameters in Groundnut Under Different Water Table Conditions Over Irrigation Levels (Mean of two seasons)

Particulars	*Water Table Depth*	*Days After Sowing*												
		27	*34*	*41*	*48*	*55*	*62*	*69*	*76*	*83*	*90*	*97*	*104*	*111*
NAR	W_1	0.48	1.00	1.81	2.38	3:14	3.51	3.28	3.24	2.63	1.09	0.57	0.70	
(g/m^2/day)	W_2	0.33	1.65	2.40	3.72	3.45	3.51	4.25	3.44	1.88	1.47	1.04	0.60	
	W_3	0.30	1.17	2.15	2.88	2.88	3.89	3.61	2.91	2.35	1.18	0.94	0.74	
Mean		0.37	1.27	2.12	2.99	3.16	3.64	3.71	3.20	2.29	1.25	0.85	0.68	
	W_1	0.14	0.35	0.84	1.01	1.18	1.51	1.70	2.16	2.72	3.37	3.79	2.95	1.75
LAI	W_2	0.10	0.28	0.65	1.27	1.70	1.89	2.52	3.24	3.99	4.72	5.45	3.58	2.67
	W_3	0.16	0.39	0.62	1.16	1.58	2.38	3.52	4.39	4.65	5.34	6.78	5.20	3.20
Mean		0.13	0.34	0.70	1.15	1.49	1.93	2.58	3.26	3.79	4.48	5.34	3.91	2.54
	W_1	0.07	0.38	1.69	2.37	3.82	5.28	5.56	7.05	7.14	3.77	2.15	1.98	
CGR	W_2	0.03	0.48	1.63	5.01	5.78	6.54	10.63	11.90	8.10	6.83	5.51	3.57	
(g/m^2/day)	W_3	0.05	0.45	1.35	3.51	4.65	9.29	13.06	12.51	11.39	6.36	6.33	4.23	
Mean		0.05	0.44	1.56	3.63	4.75	7.03	9.75	10.49	8.88	5.65	4.66	3.26	
RGR	W_1	0.0296	0.0714	0.1087	0.077	0.0709	0.0602	0.0555	0.0258	0.0258	0.0162	0.0112	0.0066	
(g/cm^2/day)	W_2	0.0186	0.1045	0.1068	0.0987	0.0769	0.0440	0.0299	0.0164	0.0133	0.0139	0.0101	0.0057	
	W_3	0.0172	0.0770	0.1022	0.0953	0.0589	0.0589	0.0399	0.0302	0.0291	0.0106	0.0092	0.0073	
Mean		0.0218	0.0843	0.1059	0.0906	0.0689	0.0544	0.0418	0.0241	0.0227	0.0136	0.0102	0.0065	

Leaf Area Index (LAI)

The mean LAI of the crop during the period of growth from 27 to 116 days. was 2.43. It increased with lowering of groundwater table. LAI increased by 40 per cent from W_1 to W_2 and by 22.7 per cent from W_2 to W_3 situation. The increase from W_1 (1.79) to W_3 (3.03) was 69.3 per cent. The maximum LAI was observed between 97 to 103 days of crop growth in all the water table situations. But the values differed markedly due the water table depths. Under W_3 situation it was 6.78 which reduced to 5.45 (19.6 per cent reduction) at W_2 and 3.79 (44.1 per cent reduction) at W_1 situation. LAI reduced towards maturity and it was in the order of 53.8 per cent (1.75), 51.0 per cent (2.67) and 52.8 per cent (3.20) under W_1, W_2 and W_3 water table situation, respectively. McCloud (1974) also observed LAI of 3.0 during initial stage with a peak of 7.1 towards cessation of flowering and the lowest of 1.7 at harvest. LAI increased with increase in irrigation levels, the peak being at 1.0 IW : CPE under W_1 and at 1.2 IW : CPE under W_2 and W_3 situations. Leaf area index was found to be a function of moisture level. Sivakumar and Sarma (1985) found maximum LAI of 4.4 in the irrigated plots compared to 1.7 under continuous stress treatment.

Leaf Area Duration (LAD)

Mean leaf area duration was 221. It increased from 163 at W_1 to 224 at W_2 situation. Further lowering of water table to W_3 situation increased the LAD by 68.7 per cent over W_1 and 22.8 per cent over W_2 situation. LAD also found to increase with increased IW : CPE. It was the minimum (162) under 0.2 IW : CPE and the maximum (259) under 1.2 IW : CPE ratio. The lower LAD under stress regime of 0.2 IW : CPE is indicative of early senescence of the crop.

Crop Growth Rate (CGR)

Data revealed that the CGR (8.44 $g/m^2/day$) increased from shallow water table) situation to 5.50 $g/m^2/day$ at W_2 with a maximum of 6.01 $g/m^2/day$ under W_3 situation. The maximum crop growth rate of 7.38, 11.9 and 13.1 $g/m^2/day$ were observed during 83 to 89, 76 to 82 and 69 to 75 days after sowing under W_1, W_2 and W_3 water table situations, respectively. Mean CGR increased with increase in frequency of irrigation from 0.2 IW : CPE (2.96 $g/m^2/day$) to 1.0 IW : CPE (6.54 $g/m^2/day$) and then decreased by 9.3 per cent at 1.2 IW : CPE. Under W_1 situation, the maximum CGR of 4.40 $g/m^2/day$ was observed at 0.8 IW : CPE whereas under W_2 and W_3 situation the maximum CGR of 7.79 and 7.68 $g/m^2/day$, respectively was observed at 1.0 IW : CPE. Murthy *et al.* (1983) and Murthy and Rao, (1986) has also reported a CGR value of 10.0 and 15.3 $g/m^2/day$ for different varieties of virginia and spanish types. Chapman *et al.* (1993) observed a decrease in CGR in virginia as well as spanish type due to water stress. Padmalatha *et al.* (2001) also reported lesser rate of crop growth (5.50 $g/m^2/day$) under 0.6 IW : CPE which increased to 7.83 at higher moisture content under 1.0 IW : CPE ratio of irrigation.

Relative Growth Rate (RGR)

The mean relative growth rate of the crop was 0.0457 $g/cm^2/day$ and it was maximum at shallow water table situation (W_1) with a value of 0.0467 $g/cm^2/day$. It reduced by 3.6 and 3.0 per cent when water table was lowered to W_2 and W_3 situations, respectively. Maximum RGR of 0.1087 in W_1, 0.1068 in W_2 and 0.1022 in W_3 was recorded during 41–47 days after sowing which decreased with the age of the crop. Mean RGR was more under stress at 0.2 IW : CPE (0.0471 $g/cm^2/day$) and reduced towards higher IW : CPE with lowest of 0.0436 $g/cm^2/day$ at 1.2 IW : CPE.

Table 28.3: Correlation Studies Between Growth Parameters and Pod Yield at Different Growth Stages in Groundnut as Influenced by Water Table Situation

Water Table Situation	Particulars	Stage–I					Stage–II				
		NAR	CGR	RGR	LAI	LAD	NAR	CGR	RGR	LAI	LAD
W_1	Yield	–0.1 34	0.273	–0.053	0.927**	0.928**	0.618	0.588	0.639	0.765	0.714
	NAR		0.904*	0.760	–0.220	–0.328		0.985**	–0.177	0.950**	0.961**
	CGR			0.771	0.220	0.079			0.864 *	0.955**	0.976**
	RGR				0.025	–0.65				0.864*	0.885*
	LAI					0.926**					0.994**
W_2	Yield	–0.928**	–0.688	–0.966**	0.894*	0.719	0.444	0.873*	0.163	0.717	0.772
	NAR		0.850	0.993**	–0.772	–0.623		0.924 **	0.848*	0.281	0.502
	CGR			0.804	–0.369	–0.234			0.597	0.507	0.673
	RGR				–0.760	–0.674				0.025	0.261
	LAI					0.791					0.963**
W_3	Yield	0.776	0.862*	0.488	0.863*	0.801	0.802	0.950**	–0.007	0.958**	0.961**
	NAR		0.984**	0.904*	0.545	0.560		0.927**	0.556	0.722	0.776
	CGR			0.859*	0.687	0.683			0.210	0.928**	0.920**
	RGR				0.372	0.441				–0.166	0.020
	LAI					0.915**					0.920**

Contd...

Table 28.3–Contd...

Water Table Situation	*Particulars*	*Stage–III*					*Stage–IV*					*Stage–IV*				
		NAR	*CGR*	*RGR*	*LAI*	*LAD*	*NAR*	*CGR*	*RGR*	*LAI*	*LAD*	*NAR*	*CGR*	*RGR*	*LAI*	*LAD*
W_1	Yield	0.960**	0.921 **	–0.095	0.823*	0.841*	0.106	0.758	–0.806	0.954**	0.954*	–0.004	0.178	–0.413	0.449	0.420
	NAR		0.960**	0.071	0.851*	0.880*		0.711	0.284	0.260	0.257		0.976**	0.606	0.044	0.130
	CGR			0.008	0.939**	0.962**			–0.392	0.848	0.845*			0.538	0.172	0.259
	RGR				–0.223	–0.173				–0.804	–0.809				0.359	0.447
	LAI					0.996**					0.999**					0.988**
W_2	Yield	0.662	0.730	0.579	0.916*	0.929**	0.810	0.922**	–0.672	0.987**	0.987**	0.059	0.629	–0.221	0.990**	0.990**
	NAR		0.982**	0.554	0.792	0.740		0.961**	–0.334	0.761	0.763		0.508	0.944**	–0.037	0.373
	CGR			0.647	0.814	0.803			–0.421	–0.525	0.885			0.299	0.591	0.563
	RGR				0.777	0.786				–0.748	–0.745				–0.295	0.918*
	LAI					0.996**					0.999**					0.999**
W_3	Yield	0.751	0.904*	0.837*	0.992**	0.995**	0.314	0.901 *	–0.727	0.951 **	0:953**	0.361	0.557	0.949**	0.991**	0.990**
	NAR		0.955**	0.979**	0.679	0.689		0.663	–0.002	0.375	0.376		0.955**	0.285	0.375	0.373
	CGR			0.972**	0.860*	0.726			–0.513	0.912*	0.913*			0.515	0.563	0.563
	RGR				0.769	0.781				–0.644	–0.645				0.916*	0.918*
	LAI					0.991**					0.999**					0.999**

Stage–I: Germination to flower initiation (26 days); Stage–II: Flower initiation to 50 per cent flowering (14 days); Stage–III: 50 per cent flowering to peg formation (15 days); Stage IV–Peg formation to pod development (35 days); Stage–V: Pod development to maturity (28 days).

Correlation Studies

Correlation of pod yield of groundnut with different growth parameters differed in nature and magnitude and was found to be influenced by crop phenophases and depth of groundwater table (Table 28.3).

Under shallow water table condition NAR was found to be positively and significantly correlated with pod yield (r = 0.960)** during 50 per cent flowering to peg formation stage. Whereas, under medium and deep water table condition LAI and LAD were more closely related with pod yield with correlation coefficient 'r' of 0.990**, and 0.991**, respectively. Relative growth rate showed negative correlation with pod yield and other growth parameters at different time of crop growth.

References

Chapman, S.C., Ludlow, M.M., Blamey, F.P.C and Fisher, K.S.1993. Effect of drought during pod filling on utilization of water and on growth of cultivars of groundnut (*Arachis hypogaea* L.) *Field Crops Research,* 32: 243–255.

Cocharan, W.G and Cox, G.M., 1950. *Experimental Designs.* John Wiley and Sons, Inc., New York.

Duncon, W.G., McCloud, D.E., McGraw, R.L and Boote, K.G., 1978. Physiological aspects of peanut yield improvement. *Crop Science,* 18: 1015–1020.

Garbet, D.W. and Rhode, F.M., 1975. A non-nodulating peanut. *Crop Science,* 19: 727–728.

Gulati, J.M.L., Lenka, D. and Mishra, M.M., 1999. Growth and yield of summer groundnut (*Arachis hypogaea*) as influenced by irrigation schedule under different water table conditions. *Indian Journal of Agronomy,* 44(4): 805–808.

Gulati, J.M.L and Lenka, D., 1999. Response of groundnut (*Arachis hypogaea*) to irrigation in different water table conditions. *Indian Journal of Agronomy,* 44(1): 141–143.

McCloud, D.E., 1974. Growth analysis of high yielding peanuts. In: *Proceedings of Soil and Crop Science Society of Florida,* 33: 24–26.

Murthy, P.S.S and Rao, R.C.N., 1986. Physiological basis of variation in pod yield of rainfed groundnut (*Arachis hypogaea* L.) under different depths of sowing. *Indian Journal of Agronomy,* 31(1): 106–108.

Murthy, P.S.S., Reddy, P.J.R and Sankara Reddy, G.H., 1983. Variation in the physiological parameters of popular groundnut varieties. *Madras Agriculture Journal,* 70(9): 603–610.

Padmalatha, Y., Rami Reddy, S. and Yellamanda Reddy, T., 2001. Influence of weather and soil moisture regime on winter groundnut (*Arachis hypogaea*) productivity. *Indian Journal of Agronomy,* 46(4): 712–720.

Sivakumar, M.N.K. and Sarma, P.S., 1985. Studies on water relations of groundnut. In: *Proceedings of an International Symposium on Agro-meteorology of Groundnut*, ICRISAT, Sahelian Centre, Niames, Niger, 21–26 August, p. 83–98.

Watson, D.J., 1952. Physiological basis of variation in yield. *Advances in Agronomy,* 4: 101–145.

Chapter 29

Field Culture of Fairy Shrimp *Streptocephalus dichotomus* Baird

B. Jesi Selvarani*

Kongu Nadu Arts and Science College, G.N. Mills, Coimbatore – 29

ABSTRACT

Freshwater anostracans (fairy shrimps) are or growing interest for various types of applications. The present chapter deals with the field culture of fairy shrimp biomass in a cost effective way using low cost feeds such as rice bran with groundnut oil cake (10 : 1) and cow dung with urea (10 : 1). The physico-chemical parameters such as temperature (min. 24°C, max. 36°C), pH (min. 7.2, max. 7.2), dissolved oxygen (min. 1.0 mg/lit, maxi. 4.5 mg/lit) ideal water depth between 90 cm–120 cm which are found to be optimum parameters under mass production of fairy shrimp could be done.

Keywords: *Anostracans, Low cost feeds, Mass production.*

Introduction

In recent years, there has been an upsurge of interest in the use of freshwater anostracans (fairy shrimps) in various types of applications *e.g.*, as live food for fresh water fish (Mura *et al.*, 1994, Prasath *et al.*, 1994) as test organisms in ecotoxicological testing of wastewater (Brendonck and Persoone, 1993 and Centeno *et al.*, 1993) and in the recycling of organic wastes and tertiary treatment

* E-mail: drbjs@rediffmail.com

of wastewaters (Mitchell, 1990, 1991). Recent studies tend to suggest their ability to filter facal bacteria, for example, *E. coli* in high levels, upto 2.5×10^6 ml^{-1} h^{-1} (Vandenberghe, 1994).

It is likely that fairy shrimp biomass will in future years become increasingly important as a feed supplement in aquaculture systems. In crustacean and fish farming, adult fairy shrimp are considered to be better reference diet than formulated feeds. Since this animal is a non-selective filter-feeder a wide variety of very inexpensive natural feeds have been considered for the possible mass culture of fairy shrimps. For large scale production of fairy shrimp adults natural dried by-products are less expensive than live algae and unlike algae they can be stored for long periods of time.

The information on culture methods for freshwater anostracans is very limited. Some small scale culture experiments, mainly for fundamental research are described and without giving any technical details. Prophet (1963) raised a small number of animals in garden soil extract on a diet of yeast. Recently, *S. proboscidens* was mass cultured in the closed recirculatioin system (Ali and Brendonk, 1995). Through this system, maintaining good rearing conditions, low mortality rates (5 per cent per week) a high cyst production (200 cysts per brood cycle) and total production yielding an average of 18,000 cysts per day and per culture unit, have been obtained.

Fairy shrimp farming in India seems to be expanding only very slowly due to the lack of information on culture techniques as well as its uses.

In the present chapter the methods of rearing fairy shrimp was investigated in field level to work out the basis for cultivating fairy shrimp on a commercial scale. The main aim of the present study is to standardize field culturing techniques, mass production of biomass in a cost effective way using cheap dry feeds.

Materials and Methods

Fairy shrimp culture tank is a rectangular cement tank with a length 42′2″ breadth 28′10″ and height 2′5″ provided with aeration. The top layer of soil with resting eggs were collected from carp fish fingerling tanks, Government Fisheries Department, Bhavani Sagar, Erode District, and transported to the culture spot, Surya Agro-farm, Erode.

The tank was provided with inlet and outlet with aeration and water exchange facility. The starting material being the top layer of soil with resting eggs of fairy shrimp requires previous drying for hatching. On the second day after hydration of soil with borewell/pond water, positively phototactic nauplii could be identified and sampled. On the third day itself, young fairy shrimp population juvenile) along with nauplii, was led with rice bran powder with groundnut oilcake powder in the ratio of 10 : 1. They were fed twice a day adequately. The ideal feeding time was morning between 7–8 am, evening 5–6 pm. The feed amount was increased according to the length of the animal. Finally, the total biomass was calculated from the initial nauplii population. Hand nets were used to harvest the fairy shrimps from the culture tank for further use.

Tank water parameters such as water-temperature, dissolved oxygen, pH transparency and depth were monitored daily. pH was measured using a digital pH meter dissolved oxygen using modified winkler's method. Phyto- and zooplankton communities were also studied. Plankton samples were taken by filtering the pond water through different mesh sizes (10 and 50μ).

Feeding Method or Recommended Feeding Level

Rice Bran with Groundnut Oil Cake (10 : 1 Ratio)

Finely ground rice bran and groundnut oil cake were homogenized and sieved through 50μ mesh net were poured in different corners of the tank. Feed was provided in the early morning and late evening hours.

Fresh Cow Dung with Urea or Super Phosphate in the Ratio of 10 : 1

Cow dung was added in a slurry form, poured into four comers of the tank.

Results

Abiotic Factors Influencing Population Density

Temperature

Temperature is one of the most important factors for ecological anostracans (Sluzhevskaya, 1981). Rising temperature results in an increased metabolism. The air temperature varied from 26°C to 38°C and water temperature 24°C to 36° C. Optimal range of temperature fluctuated between 28°C to 32°C.

Dissolved Oxygen

Anostracan species are more resistant to low oxygen concentration. From the present study, the minimum and maximum range of dissolved oxygen fluctuated between 1.0 to 4.5 mg/lit.

pH/Hydrogen Ion Concentration

As per mass culture point of view, it is essential to keep the water level high. High water depths are essential not only to prevent lethal high temperature but also to promote the development of phytoplankton which through its shading effect inhibits the development of phytobenthos. Ideal water depth in culture tanks ranges between 90 cm to 120 cm.

Factors Affecting Population Growth

In fertilized tanks, operated under optimal conditions and presence of sufficient quantities of particulate feed, especially phyto- and zooplankton sexual maturity attains between 7–11 days, resulting in a fast growth of the population. The size to which the population will grow is determined by the carrying capacity of the pond. The major factors which affect this carrying capacity are pond depth and food availability. In cow dung fertilized tank, following species are available in abundant quantities. *Chlorococcum* sp., *Closterium* sp., *Navicula* sp., *Amphora* sp., *Aphanocapsa* sp. In rice bran fertilized tank, zooplankton population dominates ciliates, rotifers, crustacean nauplii, *Daphinia* sp., *Diaplonnrs* sp. These plankters might have played an essential role in fairy shrimp population growth.

Discussion

Generally, fairy shrimps mature rather quickly when the conditions are suitable. The growth and development are most rapid at high temperature and low water hardness 55 ppm. From this study, egg formation of fairy shrimps may occur in 7–11 days, under optimal ecological factors and with an adequate diet.

This species starts to release eggs at an age of 14–17 days at 30°C water temperature. It was found that *S. dichotomus* was successfully cultured using rice bran with groundnut oil cake/or fresh cow dung with urea composition as feed. But rice bran with groundnut oil cake yielded the best results in terms of growth and cyst production under field condition.

A basic prerequisite for correct pond management implies regular evaluation of the environmental conditions or the tank. Maintenance or healthy plankton population is considered to be one of the most important factors for a successful fairy shrimp production. In cow dung with urea fertilised pond, phytoplankton population promoted, zooplankton population dominated while the pond was fertilised with rice bran and groundnut oil cake (Santiago *et al.*, 1989).

In a previous study, Venkatraman (1989) also reported a faster growth and higher fecundity rate in *S. dichotomus* fed rice bran than in animals fed algae. The success of inert feed application depends on nutritional value, particle size and nutritional requirement or the organisms being cultured (Doulliet, 1987).

Reproductive output generally raised with an increase in length or the organism (Brendnock, 1990). The present study 33–36 mm size animals yielded average 5123 eggs brood $^{-1}$ animal^{-1} which agrees with the earlier results (Brendnock, 1990, Ali and Brendonck, 1995).

Recent studies (Vanderberge, 1994) suggest an ability of fairy shrimps to filter feacal bacteria in high levels (upto 2.5×10^6 ml^{-1} indi^{-1}). Bacterial analysis of water in culture tanks, containing rice bran revealed the presence of a few bacterial species, such as *Bacillus, Aeromonos Acinetobacter* and *Pseudomonas.* At this stage, it is not clear if the successful culturing *S. dichotomous* fed rice bran may be related to their nutritional quality or to the colonization of the medium by adventitious micro organisms, enhanced by the chemical composition of these diets. These microflora may either increase the digestability of the feeds or act as a supplement (Ali, 1995).

Cowdung with urea fertilized tank favours the production of phytoplankton and also enhance the growth of undesirable filamentous algae. Light penetration was prevented by the blooming of these filamentous algae, and also it facilitates the growth of phytobenthos, adversely effect the somatic growth of *S. dichotomus.*

Acknowledgement

My earnest gratitudes are due to Prof. (Dr.) Abdul A. Rahaman, (Retd.), Research guide, AVVM Sri Pushpam College, Poondi for his assistance and Department of Science and Technology, Govt. of India for the fellowship.

References

Ali, A.J., 1995. Aspects of the biology of the fresh water fairy shrimp *Streptocephalus prohoscideus. Ph.D. Thesis,* Universiteit, Ghent, Ghent.

Ali, A.J. and Brendonck, L., 1995. Evaluation of agro-industrial wastes as diets for the fairy shrimp *Streptocephalus proboscideus* (Crustacea : Branchiopoda : Anostraca). *Hydrobiologia,* 298: 167–174.

Brendonck, L., 1990. Redescription of the fairy shrimp *Streptocephalus proboscideus* (Frauenfeld, 1893) (Crustacea : Anostraca). *Bull. K. Belg. Inst. Nat. Wet. Biol.,* 59: 49–57.

Brendonck, L., Centeno, M.D. and Persoone, G., 1993. Fecundity and resting egg characteristics of some subtropical fairy shrimp and calm shrimp species (Crustacea : Branchiopoda), reared under laboratory conditions. *Arch. Hydrobiol.,* 12b: 445–459.

Centeno, M.D.F., Brendonck, L. and Persoone, G., 1993c, Influence of production, processing and storage conditions of resting eggs of *Streptocephalus proboscideus* (Crustacea: Branchiopoda : Anostracea) on the sensitivity of larvae to selected reference toxicants. *Bull. Environ. Contam. Toxicol.,* p. 927–934.

Douillet, P., 1987. Effect of bacteria on the nutrition of the brine shrimp Artemia fed on dried diets. In: *Artemia: Research and its Applications,* Vol. 3, (Eds.) P. Sorgeloos, D.A. Bengtson, W. Decleir and L. Jaspers. Universa Press, Wetteren, Belgium, pp. 295–308.

Mitchell, S.A., 1990. Factors affecting the hatching of *Streptocephalus macrourus* Daddy (Crustacea : Eubranchiopoda) eggs. *Hydrobiologia,* 194: 13–22.

Mitchell, S.A., 1991. The growth rate and growth efficiency of *Streptocephalus macrourus* (Crustacea : Anostraca) cultured on micro-algae. In: *Studies on Large Branchiopod Biology and Aquaculture* (Eds.) D. Belk, H.J. Dumont and N. Munusamy. *Hydrobiologia*, 212: 1–10.

Mura, G., Venanazi, P., Avalle, V. and Quaglia, G.B., 1994. Fatty acid and aminoacid composition of two fairy shrimp species (Crustacea : Anostraca) from Italy: *Chiorcephalus diophonous* and *Chiorcephalus erkyrensis. Hydrobiologia*, 286: 149–154.

Prasath, E.R., Munusamy, N. and Abdul Nazar, A.K., 1994. Preliminary studies on the suitability of a fairy shrimp *Streptocephalus dichotomus* as live food in aquaculture. *J. World Aquacult. Soc.*, 25: 204–207.

Prophet, C.W., 1963c. Some factors influencing hatching of anostacans eggs. *Trans. Kans. Mad. S.*, 66: 150–159.

Santiago, C.R., Pantastico, I., Baldia, S.F. and Reyes, O.S., 1989. Milk fish (*Chanos chanos*) fingerlings production in freshwater ponds with the use of natural and artificial feeds. *Aquaculture*, 77: 307–318.

Sluzhevskaya, E.R., 1974. Some characteristics of the biology of the *Streptocephalus torvicornis* (Wega) and its role in intensifying the growth of young sturgeon in ponds. In: *Tezisyotchetnoy sessi TsNIORKh. (Proceed. of a TSNIORKH Conf),* Astrakham, pp. 146–147.

Vandenberghe, J., 1994. Filter-feeding Nj roeipoot kreeftijes: Functionele respons op experimentelange boden micro-organisms. *Diploma Thesis,* University of Ohent, 80 pp.

Venkataraman, K., 1989. Laboratory studies on growth and fecundity of *Streptocephalus dichotomus* using rice bran and algae as food. In: *Book of Abstracts, Euphyllopod Symposium,* 9–12 August, Ohent, Belgium.

Chapter 30

RAPD Analysis of F_1 Diallel Cross Population of Chickpea for Drought Tolerant Trait

P.K. Lokhande, U.S. Dalvi, S.V. Damame, R.M. Naik and L.B. Mhase
Department of Biochemistry, Mahatma Phule Krishi Vidyapeeth, Rahuri – 413 722, M.S.

ABSTRACT

The diallel F_1 cross populations *viz.*, PG. 96006 X ICCC-42 and PG-96006 X Vijay were analyzed by randomly amplified polymorphic DNA (RAPD) with two informative primers (OPA-14 and OPA-08). The RAPD patterns with OPA-14 showed a distinct band of 0.68 kb of parent PG96006 in both the crosses. However, with OP A-08, the specific locus sequence and 1.03 kb and parent PG-96006 appeared in PG-96000 X Vijay cross. With these two primers further segregation population of chickpea will be detected for drought tolerant trait.

Keywords: *Chickpea, F_1 diallel cross, Drought tolerance, RAPD.*

Introduction

Chickpea (*Cicer arietinum* L.) is an ancient crop that is considered to have been domesticated in West Asia in the Neolithic period, about 10,000 gears ago (Lev-Yadum *et al.*, 2000; Abbo *et al.*, 2003). It is an important pulse (grain legume) crop still widely cultivated in several regions of the world. Dryland (rainfed) chickpea are grown during the cool-season months either on stored soil moisture in post-rainy season in South Asia and north-east Australia. In both sub-tropical and Mediterranean-type climates, whether sown in autumn or spring. Chickpea is subjected to terminal drought.

The modem molecular techniques like Random Amplified Polymorphic DNA (RAPD) are necessary for evaluating the heritability of drought tolerant trait in chickpea. Good and unambiguous morphological markers are rare, and their expression depends on developmental and growth processes in the plant. In contrast to markers based on morphological and biochemical characteristics, DNA markers are independent of environmental conditions and are present in all plant cells. The introduction of restriction fragment length polymorphism (RFLP) techniques in to plant research has significantly improved the ability to discover genetic molecular markers (Beckmann and Soller, 1980). In comparison to RFLP, the RAPD technique is relatively fast and simple to operate, requires a small amount of DNA, and does not require the use of radioactive material (Williams *et al.*, 1990). RAPDs have been demonstrated as useful genetic DNA markers in various plants and crop species (Martin *et al.*, 1991; Paran and Michelmore, 1993). The present investigation is aimed to employ the RAPD approach in evaluating the drought tolerant heritability in diallel breeding population of chickpea.

Materials and Methods

The F_1 diallel cross population and drought tolerant parent of chickpea were subjected to RAPD analysis (Table 30.1). The genomic DNA of three drought tolerant cultivars as parent was initially amplified by using different random primers. The primers OPA-14 and OPA-08 showed distinct bands and hence these two informative primers were used for amplification.

Table 30.1: Drought Tolerant Parents and F_1 Diallel Cross Population of Chickpea

Parents		*F_1 Diallel Cross Population*
Female	*Male*	
PG-96006	ICCC-42	PG-96006 x ICCC-42
	Vijay	PG-96006 x Vijay

DNA Extraction and RAPD Analysis

gDNA was isolated from 5 g of germinated seedling of chickpea following the protocol of Saghai-Maroof *et al.* (1984). PCR was performed in 25 µl reaction volumes containing 1.5 U *Taq* DNA polymerase dNTPs mixture (0.2 mM each of dCTP, dGTP, dATP and dTTP), 2.5 µl of PCR-buffer, 50 mM KCl, 1.5 mM $MgCl_2$, 0.5 µM primer (Operon Technologies, Alameda, USA) and 50 ng of genomic DNA. PCR reactions were carried out on a Perkin Elmer 9600 DNA thermal cycler with the following temperature profile. The first cycle of the amplification consisted of an initial denaturation step of 4 min at 94°C. This was followed by 39 cycles programmed at 94°C for 45 sec., 34°C annealing for 1 min and primer extension at 72°C for 1 mm. An additional cycle of 8 min at 72°C was used for primer extension. The amplification products were resolved by using 1.5 per cent agarose gel in 1 x TBE buffer at 80 V. Gels were stained with ethidium bromide (1 µg of ethidium bromide per ml of 1 x TBE), visualized with UV transilluminator and photographed using ALPHAIMAGER gel documentation system.

Results and Discussion

The informative primers *viz.*, OPA-14 and OPA-08 were used for RAPD analysis of F_1 diallel cross population and their parents. RAPD detected consist polymorphism segregation in cross (Figures 30.1 and 30.2).

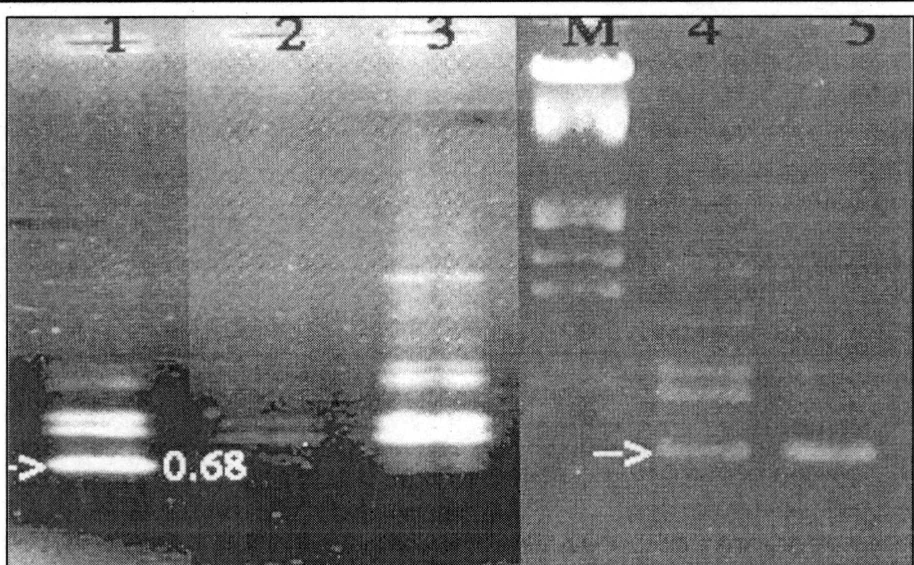

Figure 30.1: RAPD Banding Pattern of Parents and Diallel F_1 Cross Population for Drought Tolerance in Chickpea Using Primer OPA-14

Lane 1 to 5 represents: 1: PG-96006; 2: ICCC-42; 3: Vijay; 4: PG-96006 X ICCC-42; 5: PG-96006 X Vijay respectively.

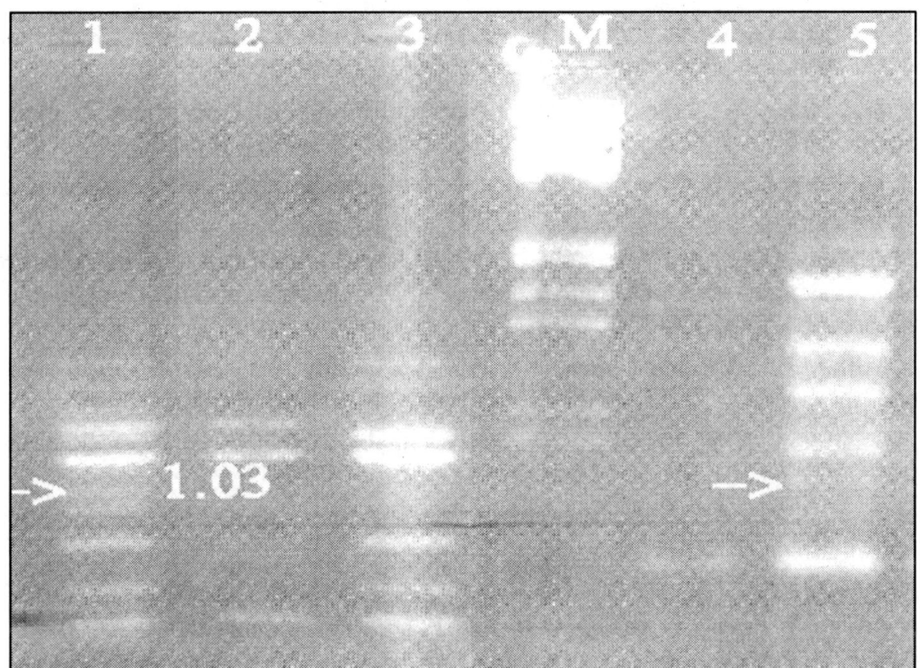

Figure 30.2: RAPD Banding Pattern of Parents and Diallel F_1 Cross Population for Drought Tolerance in Chickpea Using Primer OPA-08

Lane 1 to 5 represents: 1: PG-96006; 2: ICCC-42; 3: Vijay; 4: PG-96006 X ICCC-42; 5: PG-96006 X Vijay respectively

The heritability in F_1 diallel cross is shown on the presence (+) or absence (–) of RAPD based product of parent in F_1 cross. The primer OPA-14 showed a distinct band of 0.68 Kb of parent PG-96006 in PG-96006 x ICCC-42 and PG 96006 x Vijay cross population (Table 30.2). However, with OPA-08, the specific locus sequence of 1.03 Kb of parent PG-96006 appeared only in PG 96006 x Vijay.

Table 30.2: RAPD Patterns Depicting the Presence of Specific Locus Sequence in F_1 Diallel Cross

Primer	*Sequence*	*RAPD (Kb)*	*PG-96006 x ICCC-42*	*PG-96006 x Vijay*
OPA-14	TCTGTGCTGG	0.68	+	+
OPA-8	GAGCGAGGCT	1.03	–	+

+: Indicates the presence of amplified fragment; –: Indicates absence of fragment.

Similar sort of results has been reported by Moussa *et al.* (1997) for confirmation of heritability in F6 population of 80 recombinant inbred lines derived from (A2139 x JG62). The primers that produce polymorphism among the cross population is also documented (Tulla *et al.*, 1999). Further, Teran and Singh (2002) reported in particular that existence of useful genetic variation for specific traits related to drought resistance in parental germplasm is crucial for successful improvement of crop cultivars.

With these two primers by RAPD analysis, further segregation population of chickpea will be detected for drought tolerant trait.

References

Abbo, S., Berger, J. and Turner, N.C., 2003. Evolution of cultivated chickpea: our bottlenecks limit diversity and constrain adaptation. *Funct. Plant Biol.*, 30: 1081–1087.

Beckman, J.S. and Soller, M., 1986. Restriction fragment length polymorphism in plant genetic improvement. In: *Oxford Surveys of Plant Molecular and Cell Biology*, (Ed.) B.J. Miflin, 3: 197–250.

Lev-Yadun, S., Gopher, A. and Abbo, S., 2000. Respectively cradle of agriculture. *Science*, 288: 1602–1603.

Martin, G.B., Williams, J.G.K. and Tanksley, S.D., 1991. Rapid identification of markers linked to a *Pseudomonas* resistance gene in tomato by using random primers and near isogenic lines. *Proc. Nati. Acad. Sci., USA*, 88: 2336–2340.

Moussa, E.H., Millan, T., Gil, J. and Cubero, J.I., 1997. Variability and genome length estimation in chickpea (*Cicer arietinum* L.) revealed by RAPD analysis. *J. Genet. Breed.*, 51: 83–85.

Paran, I. and Michelmore, R.W., 1993. Development of reliable PCR-based markers linked to downy mildew resistance gene in lettuice. *Theor. Appl. Genet.*, 80: 437–448.

Teran, H. and Singh, S.P., 2002. Comparison of sources and lines selected for drought resistance in common beak. *Crop Sci.*, 42: 64–70.

Tulla, A., Kaiser, W.J., Kraft, J.M. and Mueblbauer, F.J., 1999. A second gene in resistance to race 4 of *Fusarium* wilt in chickpea and linkage with a RAPD marker. *Euphytica*, 109: 43–50.

Williams, J.G.K., Kubelik, A.R., Livak, K.J., Rafalski, J.L. and Tingey, S., 1990. DNA polymorphisms amplified by arbitrary primers are useful as genetic markers. *Nuci. Acids Respectively*, 18: 6531–6535.

Chapter 31

Identification of Molecular Marker Associated with Chickpea for Drought Tolerance Using RAPD Analysis

P.K. Lokhande, U.S. Dalvi, S.V. Damame, R.M. Naik and L.B. Mhase

Department of Biochemistry,
Mahatma Phule Krishi Vidyapeeth, Rahuri – 413 722, M.S.

ABSTRACT

Seven drought tolerant genotypes of chickpea (*Cicer arietinum* L.) were evaluated by randomly amplified polymorphic DNA (RAPD) analysis. These genotypes were evaluated using 12 different 10-mer primers of arbitrary sequences. All 12 primers were polymorphic in total, 124 amplified bands were observed in these patterns. Out of the 124 amplified bands, 100 were polymorphic accounting for 80.64 per cent. OPA 14 generated only one unique band of 0.68 Kb and OPA 8 generated three unique bands of 0.95, 1.03 and 1.56 Kb associated with PG 96006 as national check for drought tolerant genotype. With these two primers further selection of lines in breeding programme of chickpea will certainly help in detecting drought tolerant trait.

Keywords: *Cicer arietinum L., Drought tolerance, RAPD marker.*

Introduction

Chickpea (*Cicer arietinum* L.) is the world's second most important pulse crop that is cultivated in more than 41 countries. The mean protein content of chickpea is about 24 per cent that serve as an important source of dietary nutrition to poor of the semi-arid tropics. It also has tremendous beneficial effect on increasing the productivity of succeeding crops in rotation besides enhancing the sustainability and productivity of production systems (Kumar *et al.*, 1998). It is predominantly a rain fed crop grown after the end of the rain in south Asia and eastern Africa. This makes the crop to complete its life cycle on stored and receding soil moisture where it is exposed to increasing drought (Sxena, 2001). Global yield losses due to drought in chickpea are estimated to be around 3.7 million tones and around 2.1 million tones of these can be recovered through crop improvement efforts (Johansen *et al.*, 1993). An attempt in the present investigation was, therefore, made to understand the association of RAPD marker to the drought tolerance trait involved in drought tolerant chickpea genotypes.

Materials and Methods

Pure seeds of chickpea genotypes were procured from Pulses Improvement Project, Mahatma Phule Krishi Vidyapeeth, Rahuri and used for RAPD analysis. Genomic DNA was isolated from 5g of germinated seedling of chickpea genotypes, following the protocol of Saghai-Maroof *et al.* (1984). PCR was performed in 25µl reaction volume containing 1.5U *Taq* DNA polymerase, dNTPs mixture (0.2 mM each of dCTP, dGTP, dATP and dTTP), 2.5µl of PCR-buffer, 50 mM KCl, 1.5 mM $MgCl_2$ 0.5 µM primer (Operon Technology, Alameda, USA), and 50 ng of genomic DNA. Amplification was performed in thermal palm cycler with the following temperature profile. The first cycle of the amplification consisted of an initial denaturation step of 4 min at 94°C. This was followed by 39 cycles programmed at 94°C for 45 sec., 34°C annealing for 1 min and primer extension at 72°C for 1 min. An additional cycle of 8 min at 72°C was used for primer extension. The amplification products were resolved by using 1.5 per cent agarose gel in IX TBE buffer (10X TBE stock: 1M Tris-HCl, 10mM EDTA, pH 8.0) at 80V. Gels were stained with ethidium bromide (1µg of ethidium bromide pre ml of IX TBE), visualized with UV transilluminator and photographed using ALPHAIMAGER gel documentation system.

Data Analysis

The DNA fragment amplified by a given primer were scored as present (1) or absent (0) for all the genotypes studied. Both monomorphic and polymorphic primers were included in the analysis. Only reproducible fragments were considered.

Results and Discussion

A total of 12 random decamer primers (Table 31.1) of arbitrary sequences were used for analyzing the RAPD patterns of 7 genotypes. All 12 primers were polymorphic-in total, 124 amplified bands were observed. Out of these, 100 were polymorphic accounting for 80.64 per cent of the total number of amplified bands (Table 31.2). Williams *et al.* (1990) reported that the RAPD patterns generated from individual decamer primers varied for the same genotype indicating that the nucleotide changes in the primer, and in turn that of the template DNA, determines whether a given DNA segment will be amplified thereby contributing to the polymorphism. The total number of bands resolved per primer ranged from a minimum of four (OPA 09) to a maximum of 16 (OPE 06) with an average number of 10.3 bands per primer. OPE 06 was most informative primer with 15 polymorphic bands while OPA 09 was least informative with only 3 polymorphic bands. Average polymorphism was 80.64 per cent and monomorphism was 19.35 per cent employed PCR based RAPD markers to assess the genetic diversity in 23 chickpea genotypes (Singh *et al.*, 2001). It was revealed that 40 out of the 100 random primers detected polymorphism among the genotypes. Most of the primers showed single polymorphic band and only 14.12 per cent of the products were polymorphic, however, in the present investigation the polymorphism to the extent of 50 to 94 per cent was observed. OPA 14 generated only one unique band of 0.68 Kb and OPA 08 generated three unique bands of 0.95, 1.03 and 1.56 Kb associated with PG-96006 as national check for drought tolerant genotype (Figure 31.1).

Table 31.1: Operon Primers used in RAPD Analysis of Chickpea and their Base Sequences

Sl.No.	*Primer*	*5' to 3'*
1.	OPA-07	GAAACGGGTG
2.	OPA-08	CTGGCGAACT
3.	OPA-09	CTGGGCACGA
4.	OPA-13	CAGCACCCAC
5.	OPA-14	TCTGTGCTGG
6.	OPB-04	GGACTGGAGT
7.	OPB-07	GGTGACGCAG
8.	OPC-06	GAACGGACTC
9.	OPC-20	ACTTCGCCAC
10.	OPD-05	TGAGCGGACA
11.	OPD-08	GTGTGCCCCA
12.	OPE-06	GAGCGAGGCT

Table 31.2: Summary of RAPD Amplicons and Polymorphism Detected by 12 Random Decamer Primers among 7 Chickpea Genotypes

Sl.No.	*Primer Code*	*Maximum No. of Amplified Bands for Each Primers*	*Total No. of Polymorphic Band*	*Per cent Polymorphism*
1.	OPA-07	12	9	75
2.	OPA-08	13	12	92
3.	OP A-09	4	3	75
4.	OPA-13	11	10	91
5.	OPA-14	9	6	67
6.	OPB-04	9	8	89
7.	OPB-07	12	10	83
8.	OPC-06	8	7	88
9.	OPC-20	10	5	50
10.	OPD-05	10	7	70
11.	OPD-08	10	8	80
12.	OPE-06	16	15	94

The general conclusion from this study is that the two primers *viz.*, OP A 14 and OPA 08 can be utilized for further selection of lines in breeding programme of chickpea for drought tolerant trait

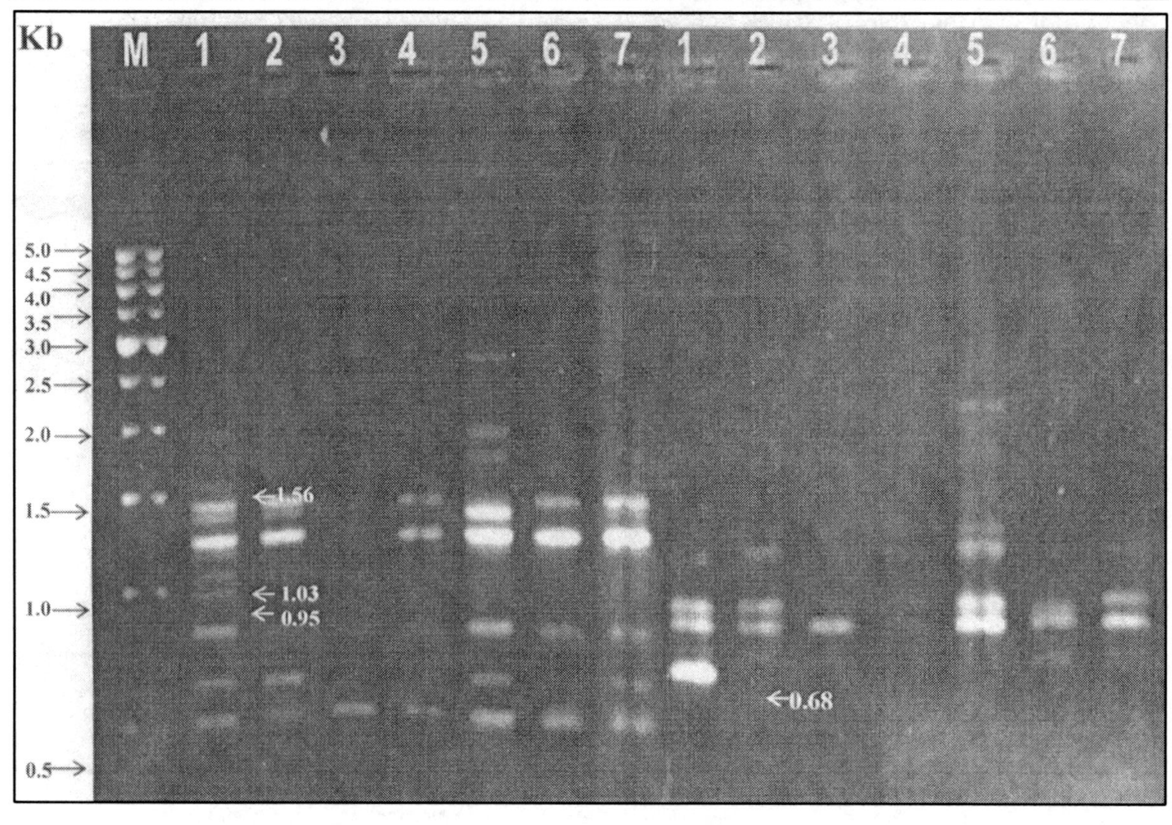

Figure 31.1: Amplification Profile of Seven Drought Tolerant Cultivars of Chickpea with OPA-08 and OPA-14 Primers

M: DNA Ladder (500 bp)

Drought Tolerant Cultivars of Chickpea

1: Phule G-96006; 2: PG-00109; 3: ICP-92-39; 4: ICCC-42; 5: Vijay; 6: G-26; 7: GNG-469.

References

Johansen, C., Baldev, B., Brouwer, J.B., Eriskine, W., Jermyn, W.A., Lang, L.J., Maik, B.A., Ahad, M.A. and Silim, S.N., 1993. Biotic and abiotic stresses constraining productivity of cool season food legumes in Asia, Africa, and Oceania. In: *Proc. Intl. Food Legumes Research Conference II*, F.J. Mehbauer and W.J. Kaiser, 12–16 April, Kluwer, Dordetcht.

Kumar Rao, J.V.D.K., Johansen, C. and Rego, T.J., 1998. Concluding remarks. In: *Residual Effects of Legumes in Rice and Wheat Cropping Systems of the Indo-Gangetic Plain*, (Eds.) J.V.D.K. Kumar Rao, C. Johansen, and T.J. Rego. Oxford and IBH Publ. Co., New Delhi, pp. 237–240.

Saghai-Maroof, M.A., Soliman, K.M., Jorgensen, R.A. and Allard, R.W., 1984. Ribosomal DNA spacer-length polymorphism in barley: Mendelian inheritance, chromosomal location, and population dynamics. In: *Proc. Natl. Acad. Sci., USA*, 81: 8014–8018.

Saxena, N.P., 2001. Management of drought in chickpea a holistic approach. In: *Management of Agricultural Drought: Agronomic and Genetic Options*, (Ed.) N.P. Saxena. Oxford and IBH Publ. Co. Pvt. Ltd., New Delhi, pp. 338–341.

Singh, R., Durga Prasad, C., Singhal, V. and Randhawa, G.J., 2002. Analysis of genetic diversity in *Cicer arietinum* L. using random amplified polymorphic DNA markers. *J. Plant Biochem. Biotech.*, 11: 109–112.

Williams, J.G.K., Kubelik, A.R., Livak, K.J., Rafalski, J.L. and Tingey, S.V., 1990 DNA polymorphism amplified by arbitrary primers are useful as genetic markers. *Nucl. Acids Res.*, 18: 6531–6535.

Chapter 32

Modelling on Risk of Vulnerability to STDs/HIV Infection Among Pre-marital Female Migrants in Urban Area

Himanshu Pandey[1], J.N. Mishra[2] and Umashankar Shukla[3]

[1]Department of Mathematic and Statistics,
DDU Gorakhpur University, Gorakhpur, U.P.
[2]Department of Mathematics, Ganpat Sahai P.G. College, Sultanpur, U.P.
[3]Department of Mathematics, Rajeev Gandhi Shiksha Degree College,
Paisia, Laxmipur, Maharajganj, U.P.

ABSTRACT

The present chapter deals with probability model based on some assumptions for the unmarried female migrant workers having number of closed boy friends, with more vulnerable to STDs and HIV transmission. Estimated parameters involved in the model. The suitability of proposed model have been tested by observed set of data.

Keywords: *HIV/STDs, Probability distribution, Method of moments.*

Introduction

In India, it is well known that females have been participating equally as males now days, women have started taking up professional roles and they are now entering new fields such as administration, science, technology, medicine, journalism and etc. We all known that India is a male dominated country, but we should remove this word from our dictionary. Historically females are totally dependent on the males. Now, we see that females have been migrating on their own irrespective of their marital status. The question arises, why do we need such study? Answer is simple that STDs and HIV/AIDS are increasing. It is not a particular answer of this question. It is major problem of world and not of India. HIV/AIDS is not clearly related to female migrants, it is related to make migrants also, but some how related to migrants (Reddy, 2004). A study of 120 samples of HIV/AIDS infected people was taken from Sir Sunderlal Hospital, Banaras Hindu University, Varanasi, India, the study reported that about 70 per cent of them were migrants. Now (Jain, *et al.*, 2007) National AIDS Control Organization (NACO) estimated around 0.6 million HIV infections in 2002. They also estimated that between 3.8 and 4.5 million Indians were living with HIV/AIDS during 2002, of who around 39 per cent were women. The epidemic continues to shift towards women and young people.

According to an estimate of UN AIDS (1998) although HIV prevalence rate is low (around 1 per cent), the overall number of people with HIV infection too high. The majority of the reported AIDS cases have occurred in the sexually active and economically productive age group. Earlier men were the main transmitters of the disease but now studies are showing that females are also transmitting the disease to males. A study conducted among 379 HIV/infected people in 1991, reported in the journal of the American Medical Association, observes an evidence of female-to-male transmission.

The real world phenomenon indicates three distinct type of female migration (Fawcett *et al.*, 1984).

Autonomous Female Migration

Many middle and upper middle class women migrate to cities for improving their educational credentials and also to get suitable employment apparently in a quest for social advancement and also to enhance their status in the marriage market. Among the semi-literate, young girls migrating to towns/cities to work in export processing units, garments industry, electronic assembling and food processing units is continuously on the increase in the recent years.

Relay Migration

To augment family income, families which have some land holdings in the rural area, send the daughters to work mostly as domestic servants where they are safe in the custody of a mistress. First the elder daughter is sent out and she is replaced by the second, third and so on, as one by one get married.

Family Migration

Here the wife instead of staying back in the village prefers to join her husband in the hope of getting some employment in the destination area. Family migration among agricultural wage labourers who have no land or other assets to fall back at times of crisis is becoming increasingly common.

In the Indian context women in the migrant households do play an important role in family survival but unfortunately they remain invisible in the official data because of the way the concepts are defined and data is collected. But the limited research studies that are available in this concern for the earlier periods indicate that these women are exposed more to the risk of sexual harassment and exploitation (Acharya, 1987 and Saradamani, 1995). Women migrant workers in sugarcane cutting, work almost twenty hours a day (Teerink, 1995). Female labour mostly from Kerala in the fish processing industries in Gujarat are subject to various forms of hardship and exploitation at the hands or their superiors (Saradamani, 1995).

Among females, the proportion of migrant and non-migrant workers in white-collar jobs was almost similar in 1971 but the same became smaller in 1991 than that of the non-migrant workers. There are more migrant women than the non-migrant women in the category of blue-collar jobs (Premi, 2001). A study conducted by Dholakia and Dholakia (1971) for 20 major Indian states showed that per capita income, average size of households and overall literary rates were the main factors explaining the variations in female participation rates across the states.

A model is a symbolic and simplified presentation of reality, which aids in the patterning of observed behaviour (Speare, 1974). Model can be clarified in to two categories as deterministic and stochastic. The deterministic models assumes a functional relation between their input and output variables, whereas in stochastic models the input variables are treated as probablistic distribution and their relation between the input and output variables are left to probability (Sarin, 1993). Formulation of models and their application is becoming very important in the recent times due to its usefulness and applicability in both natural and social science.

The objective of the present work to proposed probability model based on the above studies and applied to risk for sexually transmitted infections of HIV transmission or unwanted pregnancies due to change in sexual behaviour of single female migrants. They have wanted to make different close boy friends and most of them have been taking interest to go out with friends for movies or drama or to restaurants or hotels, while some of them go to night clubs, discos, bars, pubs or attend late night parties.

Model

The present model for the number of closed boy friends to describes. The distribution of single unmarried female migrants, which based on displaced geometric distribution under the following assumptions:

1. Let α be the proportion of female migrants having at least one close boy friend.
2. Number of close boy friends attached with the female migrants follows a displaced geometric distribution.
3. Let P be the probability of close boy friends attached with young unmarried female migrants, they are more vulnerable to STD/HIVs infections.

Under the above assumptions, the proposed probability model for the number of close boy friend X, is given by

$$P[X = 0] = 1 - \alpha, K = 0$$

$$P[X = K] = \alpha p q^{K-1}, \quad K = 1,2, \qquad (1)$$

Estimation of Parameters

The proposed model (1) involves two parameters α and p, to be estimated by method of moment, by taking zeroth cell and the theoretical mean to their observed values.

That is

$$1 - \alpha = f_o/f \qquad (2)$$

$$\alpha\left(\frac{1}{p}\right) = \text{Mean} = (\bar{X}) \qquad (3)$$

Data

The study is based on surveyed of 362 unmarried working women, less than 30 years of age in Delhi urban area, while they have wanted lavish life styles and having number rich boy friends, randomly selected from 12 working women's hostels in Delhi. The list of the hostels was obtained from Social Welfare Department, YWCA and Warden's of the hostels. Details about the data are given in Jain *et al.*, 2007.

Results and Discussion

Table 32.1 gives the distribution of observed and expected number of young female migrants and their number of closed boy friends. The value of $\chi^2 = 2.08$ with 3 degree of freedom at 5 per cent level of significance indicates that proposed model fitted well to the distribution of female migrants. The estimated value (0.8039) of the proportion of female migrants having at least one boy is very high because most of the young migrants believe that their friends watch pornographic materials with us. It also indicates that number of female (0.8039) migrants one having close boy friends. They are more vulnerable to STDs and HIV/AIDS. The probability of closed boy friends attached with young unmarried female migrant is 0.4788, which is also near to fifty per cent.

Table 32.1: Observed and Expected Number of Unmarried Single Female Migrants According to their Closed Boy Friends

No. of Closed Boy Friends	*Observed o_i*	*Expected e_i*
0	71	71.00
1	127	139.34
2	80	72.59
3	55	47.86
4	19	19.73
5	10	11.48
Total	362	362.00

α: 0.8039; p: 0.4788; χ^2: 2.048.

The first sexual event has clear health implications since it mark-unitiation into the sexual act which if unprotected, an carries a risk of adverse outcomes such as unplanned pregnancy, HIV and sexually transmitted infections (Wellings *et al.*, 1994).

Conclusions

The proposed model is tested set of data on the number of young unmarried female migrants. Work as less than 30 years of age having number of closed boy friends. They are clearly attached with their boy friends. Proposed model is fitted satisfactorily. Estimated values indicate an increasing the living and working conditions and they need adequate support should be provided to single young migrant women that might be make them vulnerable to STDs and HTV/AIDS infection.

References

Anand, N., 2003. Working women: Issues and problems. *Yojana*, 47(3): 11–14.

Binoy, Acharya, 1987. *Beyond Bricks and Stone*. PRIA, New Delhi.

Dholakia, R.H. and Dholakia, R.H., 1978. Interstate variations in female labour force participation rates in India. *Journal of Labour Economics.*

Fawcett, J.T., Khoo, S. and Smith, P.C., 1984. *Women in the Cities of Asia: Migration and Urban Adaptation.* Westview Press, Boulder Colorado.

Hugo, G., 2000. Migration and women's empowerment. In: *Women's Empowerment and Demographic Processes,* (Eds.) H.B. Pressler and G. Sen. Oxford University Press, Oxford.

Jain, Ruchi, Gupta, Kamla and Singh, Ajay, K., 2007. *Sexual Risk Behaviour and Vulnerability to HIV Infection Among Young Migrant Women Workers in Urban India*. Access on July 17, 2007. http:// paa2007. princeton.edu/download.aspx?submissionID=72114.

Reddy, C.R., 1986. *Changing Status of Educated Working Women: A Case Study*. B.R. Publishing Corporation, Delhi.

Reddy, D.C.S., 2004. *Prevention of HIV/AIDS in Uttar Pradesh*, Organized by the Status Innovatrions in Family Planning Services Agency (SIFPSA), U.P. State AIDS Control Society and the Policy Project, The Futures Group International, Agra, January 29–31.

Saradamani, K., 1995. Crisis in the fishing industry and women's migration: The case of Kerala. In: *Women and Seasonal Labour Migration,* (Ed.) Schenk Sanbergen. IDPAD Sage, New Delhi, pp. 145–152.

Sarin, D., 1993. Some mathematical models for the relationships among migration: Fertility and population growth. *Ph.D. Thesis in Statistics,* BHU, India.

Spear, Jr.A., 1974. Residential satisfaction as an intervening variable in residential mobility. *Demography,* 2: 173–188.

Terrink, Rensj, 1995. Migration and its impact on Khandeshi women in the sugarcane harvest. In: *Women and Seasonal Labour Migration,* (Ed.) Schenk Sandbergen. IDPAD Sage, New Delhi.

UNAIDS, 1998. *Migration and AIDS: International Migration*, 36(4): 445–466.

Willings, K., Field, J., Johnson, A. and Wardsworth, S.J., 1994. *Sexual Behavioural and Lifestyle in Britain: The National Survey of Sexual Attitudes and Lifestyle*. Penguin Books Ltd., London, England.

Chapter 33

Diversity and Impact of Macrophytes in Bhandardaha Beel, Murshidabad, West Bengal

Anita Chakraborty[1], B.C. Jha[1] and R.K. Bhakat[2]
[1]Central Inland Fisheries Research Institute, Barrackpore, Kolkata – 700 120
[1]Department of Botany and Forestry, Vidyasagar University, Mednipur, W.B

ABSTRACT

Extensive survey of macrophytes was conducted in a floodplain wetland (Bhandardaha Beel) of West Bengal to understand the status of species diversity, pattern of distribution and their impact on ecological processes, including the fishery. Increase in emergent and marginal macrophytes has been the hallmark in this wetland, indicating decrease in depth profile. Greater colonization of macrophytes has also affected the fish grazing chain like plankton, benthos and other communities, which in turn has affected the fishery thereby livelihood support to traditional fishers, adversely.

Keywords: *Wetland, Macrophyte, Biodiversity, Fishery, Livelihood.*

Introduction

India has highly varied natural and man-made aquatic ecosystems, including the wetlands, distributed widely, from the cold arid zone of Ladhak to the wet Manipur, farm warm and arid zone of Rajasthan to tropical monsoonic central India and humid southern peninsula. The Gangetic and Brahmaputra river basins, however, has the largest congregation of floodplain wetlands in the country, which have great potential for culture-based fisheries development. The area under wetland in India has been estimated at 4.1 million ha (MoEF, Govt. of India, 1992), comprising natural (1.5 million ha) and man made (3.1 million ha).

These wetlands sustain a very diverse assemblage of macrophytes, such as spermatophytes (seed bearing plants), pteridophytes (ferns), bryophytes (mosses and liverwort) and charophytes (algae) that spend whole or at least a part of their life cycle in aquatic environment. Macrophytes are at the base of the food chain, as such a major conduit for flow of energy in an aquatic system, providing critical habitat for a number of biotic groups, such as epiphytic bacteria, periphyton, macro-invertebrates and fish. They also strongly influence the water quality, acting either as nutrient sink through uptake or pumping of nutrients from the sediment to the water column.

Greater proliferation of certain macrophytes, assuming the status of weed, has, however, become a global problem in recent times, as affecting the ecological functions including the biodiversity adversely. Most of the natural floodplain lakes in India have dense colonization of macrophytes, indicating high level of eutrophication.

The information on the distributional pattern of macrophytes in various floodplain wetlands, located in different geographical of the country, is scanty (Jhingran, 1989; Jha, 1995, 1996; Jha and Chandra, 1997; Yadava, 1989, 1995; Mitra, 1989; Sugunan, 1995, 2000; Parmeswaran and Vass, 1995; Sinha and Jha, 1997, a and b.), Gupta *et al.* (2006) and Devi and Sharma (2007, a, b, c). The present chapter is an effort to unravel the floristic composition and distribution of macrophytes and its impact in Bandardaha beel, Murshidabad and West Bengal, which has been degraded of late due to closure of its connecting channel with river.

Study Area

Bandardaha beel (23°50′N Lat and 88°45′E Long) is a typical oxbow lake, a cut-off meander of river Jalangi, a tributary of river Bhagirathi. The lake has an effective water area of 440 ha with an average depth of 12.63 m and was maintaining its connectivity with river Jalangi, but since last five years the connecting channel has been blocked, making the system a closed type leading to decline in fish catch besides greater colonization of macrophytes.

Materials and Methods

Sampling of macrophytes was done at quarterly intervals from arbitrarily divided five different sites, of the beel. The macrophytes thus collected were identified up to species level following the standard keys, such as Biswas (1936), Subramanyam (1962), Prescott (1980) etc. The growth habits were assigned following Muller-Dombois and Ellen berg (1974).

Results and Discussion

Floristic Composition

During the present study (2005–2007) a total of 30 species of macrophytes were recorded (Table 33.1) from Bhandardaha beel. The distributional pattern of macrophytes indicated the presence of all the 30 species at sites–I and II, while 26 species were recorded from site–IV, 21 from site–III, and 20 from site–V. Out of the 30 species, 18 species (*Eichhornia crassipes, Pistia stratiotes, Azolla pinnata, Nymphoides indicum, Nelumbo nucifera, Nymphaea nauchali, Hydrilla vertic illata, Ceratophyllum demersum, Vallisneria spiralis, Marsilea quadrifolia, Marsilea minuta, Ipomoea aquatica, Enhydra jluctuans, Centella asiatica, Alternanthera philoxeroides, Jussiaea repens, Colocasia esculenta, Cyperus difformis*) were common at all the five sites (Table 33.1). The plants have been categorized into four different groups *viz.*,

Emergent (15 species), Rooted with floating leaves (4 species), Free-floating (6 species) and Submerged (5 species) (Table 33.2).

Table 33.1: Floristic Composition of Macrophytes in Bhandardaha Beel

Name of Species	*Family*	*Site I*	*Site II*	*Site III*	*Site IV*	*Site V*
Free floating						
Eichhornia crassipes	Pontederacea	+	+	+	+	+
Pistia stratiotes	Araceae	+	+	+	+	+
Azolla pinnata	Azollaceae	+	+	+	+	+
Lemna minor	Lemnaceae	+	+	–	–	–
Spirodella polyrhiza	Lemnaceae	+	+	–	–	–
Trapa natans	Trapaceae	+	+	–	+	+
Rooted with floating leaves						
Nymphoides indicum	Nymphaeaceae	+	+	+	+	+
Nelumbo nucifera	Nymphaeaceae	+	+	+	+	+
Nymphaea stellata	Nymphaeaceae	+	+	–	–	+
Nymphaea nauchali	Nymphaeaceae	+	+	+	+	+
Submerged						
Hydrilla verticillata	Hydrocharitaceae	+	+	+	+	+
Ceratophyllum demersum	Hydrocharitaceae	+	+	+	+	+
Vallisneria spiralis	Hydrocharitaceae	+	+	+	+	+
Potamogeton pectinatus	Naiadaceae	+	+	–	–	+
Najas minor	Naiadaceae	+	+	–	–	–
Emergent						
Marsilea quadrifolia	Marsilaceae	+	+	+	+	+
Marsilea minuta	Marsilaceae	+	+	+	+	+
Ipomoea aquatica	Convolvulaceae	+	+	+	+	+
Enhydra fluctuans	Asteraceae	+	+	+	+	+
Hyworhiza aristata	Gramineae	+	+	–	–	+
Centella asiatica	Apiaceae	+	+	+	+	+
Alternanthera philoxeriodes	Amaranthaceae	+	+	+	+	+
Jussiaea repens	Onagraceae	+	+	+	+	+
Polygonum barbatum	Polygonaceae	+	+	+	–	+
Polygonum gllabrum	Polygonaceae	+	+	–	+	+
Aeschynomene aspera	Leguminosae	+	+	+	–	+
Colocasia esculanta	Araceae	+	+	+	+	+
Cyperus difformis	Cyperaceae	+	+	+	+	+
Typha angustata	Typhaceae	+	+	–	–	–
Cyperus procerus	Cyperaceae	+	+	+	–	+

+: Present, –: Absent.

Table 33.2: Group-wise Distribution of Macrophytes in Bhandardaha Beel, WB

Sl.No.	Groups	Site I	Site II	Site III	Site IV	Site V	Lake as a Whole
1.	Emergent	15	15	12	10	14	15
2.	Rooted with floating leaves	4	4	3	3	4	4
3.	Free floating	6	6	3	4	4	6
4.	Submerged	5	5	3	3	4	5
	Total No. of species	30	30	21	20	26	30

Information on the macrophytes from the wetland of West Bengal in general and Bhandardaha beel in particular is rare, restricted to the study made by Sugunan and Mukhopadhaya (1995), who reported the existence of only five species *viz. Hydrilla verticillata, Vallisneria spiralis, Najas minor, Ceratophyllum demersum* (submerged) and *Eichhornia crassipes* (free floating). During the present investigation, however, as high as 30 species, have been recorded. The increasing incidences of macrophyte diversity in recent years may primarily be attributed to the closure of the connecting channel between the beel and the river Jalangi, converting the lake into a closed system and paving ways for greater colonization of macrophytes. Over these years, especially after the closure of connecting channel, the abundance and diversity of emergent plants have increased considerably, suggesting decline in water depth, from an average of nearly 2.0 m (1995) to 1.0 m (2007). Further, the colonization of marginal plants like *Marsilea quadrifolia, Centella asiatica, Ipomoea aquatica, Diplazium esculantum, Nymphaea nauchali etc.,* which are being used regularly, either as food or medicine, by the riparian community, has also increased, indicating changing ecological profile of the lake, from aquatic habitat to marshy land. In general, however, the enhanced colonization of macrophytes of all groups, especially the *E. crassipes,* has made the system vulnerable from biodiversity vie,w point, as the planktonic organisms and economically important fish fauna like Indian major carps have declined significantly while the smaller fish species of relatively less economic value have occupied the niches with vengeance (Table 33.3). This singular factor has affected the livelihood support of the traditional fishers, significantly. Evidently, the Bhandardaha beel needs appropriate restoration measures, not only to keep the infestation of macrophytes under manageable level, but also to restore its utility functions like fisheries and biodiversity.

Table 33.3: Fish Catch Structure of Bhandardha Beel, West Bengal

Fish Groups	1995 (per cent)	2007 (per cent)
Indian Major carp	15–20	5–11
Minor carp	25–30	18–22
Murrels	4–6	12–15
Feather–backs	1 0–12	15–18
Catfish	8–10	17–21
Minowos	27–30	45–64

Acknowledgement

The authors are grateful to Dr. K.K. Vass, the Director, Central Inland Fisheries Research Institute,

Barrackpore for encouragement and providing the necessary laboratory facilities to carryout the investigation. They also thankfully acknowledge the financial assistance received from the ICAR, New Delhi.

References

Devi, U.S. and Sharma, M.B., 2007. Investigation on pattern of distribution of macrophytes in Hidenkompat lake, Bishnupur (Manipur). *Indian Journal of Environment and Ecoplanning,* 14(1–2): 109–114.

Devi, U.S. and Sharma, M.B., 2007. Life-form analysis and biological spectrum of the macrophytes of the Laisoipat lake, Manipur. *Indian Journal of Environment and Ecoplanning,* 14(1–2): 153–159.

Devi, U.S. and Sharma, M.B., 2007. Studies on phytosociology of the macrophytes in the Oksoipat lake, Bishnupur (Manipur). *Indian Journal of Environment and Ecoplanning,* 14(1–2): 65–72.

Gupta N., Anthwal, A. and Bahuguna, A., 2006. Biodiversity of Mothronwala swamp, Doon Valley, Uttranchal. *The Journal of American Science,* 2(3).

Jha, B.C. and Chandra, K., 1997. Kusheshwarastha Chaur (North Bihar): Status and prospects for fisheries development. *Central Inland Fisheries Research Institute Bulletin* No. 71.

Jha, B.C., 1989. Beel fisheries resources in North Bihar. *Central Inland Fisheries Research Institute Bulletin* No. 63

Jha, B.C., 1995. The flood plain fishery recourses of Gandak basin, Bihar. In: *Conservation and Sustainable Use of Flood Plain Wetlands,* Asian Bureau, Kaula Lumpur, Malaysia

Jha, B.C., 1996. Role of macrophytes in fisheries development of flood plain lakes. In: *Procd. Weed Control,* Central Irrigation Directorate, New Delhi

Jhingran, A.G., 1989. Strategies for development in beel fisheries. *Central Inland Fisheries Research Institute Bulletin* No. 63.

Mitra, K., 1989: Limnological features of beels: Macrovegetation dynamics. *Central Inland Fisheries Research Institute Bulletin* No. 63.

Parmeswaran, S. and Vass, K.K., 1995. Floodplain Wetlands of West Bengal. In: *Conservation and Sustainable Use of Floodplain Wetlands,* Asian Bureau, Kaula Lumpur, Malaysia

Sinha, M. and Jha, B.C. 1997a. Ecology and fisheries of Ox-bow Lakes (Maun) of north Bihar: A Threatened ecosystem. *Central Inland Fisheries Research Institute Bulletin* No. 74.

Sinha, M. and Jha, B.C. 1997b. Uttar Bihar Ke Matsayaki Jal Sansadhan: Vartaman Awastha ava Bhavi Sambhayene. *Central Inland Fisheries Research Institute Bulletin* No. 69.

Sugunan, V.V. *et al.,* 2000a and b. Ecology and fisheries of beels in West Bengal and Assam. *Central Inland Fisheries Research Institute Bulletin* Nos. 101, 105.

Sugunan, V.V., 1995. Floodplain lakes: A fisheries prospective. In: *Conservation and Sustainable Use of Floodplain Wetlands,* Asian Bureau, Kaula Lumpur, Malaysia

Yadava, Y.S., 1989. Beel fisheries resources of Assam. *Central Inland Fisheries Research Institute Bulletin* No. 63.

Yadava, Y.S., 1995. Fisheries of floodplain lakes: A management perspective. In: *Conservation and Sustainable Use of Floodplain Wetland,* Asian Bureau, Kaula Lumpur, Malaysia

Chapter 34

Ecology and Biodiversity of Belloon Beel, Murshidabad, West Bengal with Special Emphasis on Macrophytes

Anita Chakraborty[1], B.C Jha[1] and R.K. Bhakat[2]
[1]Central Inland Fisheries Research Institute, Barrackpore, Kolkata – 700 120
[2]Department of Botany, Vidyasagar University, Mednipur, West Bengal

ABSTRACT

Ecology and Biodiversity study a degraded flood plain wetland (Belloon beel) was conducted over the period 2006–07. It indicated dense proliferation of macrophytes comprising of 34 species. The phytoplankton diversity showed the dominance of diatoms, followed by green-algae, blue-greens, euglenoids and dinophyceae. The lake found supporting 24 species of zooplankton and 17 species of macrobenthic invertebrates. 41 species of fish fauna, mostly smaller varieties, were also recorded from the beel. The abundance and diversity of macrophytes indicated a stressed ecosystem, which needs restoration measures to revive the utility functions.

Keywords: *Wetlands, Ecology, Biodiversity, Macrophytes.*

Introduction

Wetlands are highly productive aquatic ecosystems, which have been used extensively and exploited recklessly, throughout the globe, being associated intimately with mankind. The problem is more acute and intense in thickly populated countries like India. Being the repositories of aquatic biodiversity, including the avifauna, these ecosystems provide livelihood support to a sizeable section of the society through fishing, agriculture and a number of other ancillary activities. In recent times, however, the wetlands are in a critical stage of ecological transition, as evidenced from thick to very thick stands of macrophytes, indicating advanced stage of eutrophication.

The State of West Bengal is bestowed with very rich wetland resources (6049 nos.) with an estimated area of 42,000 ha, supporting thousands of traditional fishers through fisheries activities.

The biodiversity and production functions of floodplain wetlands of Ganga basin are reeling under serious threat and needs proper evaluation. It is unfortunate, however, that the detailed account of biodiversity from these water bodies hitherto remains under explored, barring Jha (1995), Jha and Chandra (1997) Sinha and Jha (1997), Sugunan (2000), Sugunan and Mukhopadhya (1995) Yadava (1995).

Study Area

Belloon beel (23°55′N lat and 87°55′E long) is located 52 km north of Baharampur Town, Murshidabad. It is by and large a closed beel, as cut-off from the parent river Dwarka (Brahmani), except retaining connectivity for a brief period during the high flood. In absence of effective water renewal, it is densely infested with floating, submerged, emergent and marginal vegetations. It was a relatively large wetland (929 ha), till 1970s, but presently been reduced to 200 ha, due to various man-induced activities, especially reclamation of the marginal areas for agriculture.

Materials and Methods

The beel was arbitrarily divided into four sampling zones *viz.* East, West, North and South for the collection of water and biological samples. Physico-chemical analysis of water was done as per CIFRI Bulletin 12. Plankton samples were collected with the aid of plankton net made up of bolting silk (No. 25, aperture size 0.64 mm). 50 litres of water was sieved and preserved in 5 per cent formalin. Macro-benthic invertebrates were collected with the aid of an Ekman dredge (15 × 15 cm and 22 × 22cm). The organisms were cleaned and segregated individually before being preserved in 5 per cent formalin. Quadrate samplers (1xl m) were used to collect the macrophyte samples randomly from different sectors of the beel for arriving at to their relative frequency. Digital photograph of each species have been taken, besides making herbariums for record in CIFRI Laboratory. Fish fauna were collected from the fishers while netting. The collection of biological samples was done as per the methodology of Jhingran *et al.* (1967). The identification of biological materials was done as per the standard keys following Ward and Wipple (1908), Fritsch (1935), Prescott (1980, 1982), Subramanyam (1962), Sculthrope, (1967), Denny (1995) and others.

Results and Discussion

The water quality of Belloon beel indicated productive aquatic regime, especially in relation to its alkaline pH (7.82–8.65) and reasonably high total alkalinity (145.6–150.5 mg/l). The dissolved oxygen, however, indicated wider fluctuation (3.0–12.0 mg/l), a typical example of macrophyte infested aquatic system. Relatively high concentration of specific conductance (> 300 11 mhos/cm) was indicative of eutrophic characteristics of the beel (Table 34.1).

Plankton

A total of 127 species of phytoplankton (Myxophyceae–29, Clorophyceae–38, Bacillariophyceae–55, Euglenophyceae–6, Dionphyceae–5) and 24 species of zooplankton (Rotifers–12, Copepods–5, Cladocera–3, Protozoa–4) were recorded during the period 2006–07. The diversity of dominant plankters is given in Table 34.2.

Table 34.: Core Physico-chemical Properties of Belloon Beel (2006–07)

Parameter	*Range*
Transparency (cm)	1.2–1.6
Water temperature °C	19.6–30.6
Dissolve oxygen (mg/l)	3.0–12.0
pH	7.82–8.65
Total alkalinity (mg/l)	145.6–50.5
Hardness mg/l	90.3–10 1.5
Sp. conductance (µmhos/cm)	310.5–397.5
PO_4 (mg/l)	0.032–0.051
NO_3 (mg/l)	0.030.033
Chloride (mg/l)	20.3–25.25

Table 34.2: Plankton Diversity in Belloon Beel (2006–07)

Plankton	*Diversity*
Phytoplankton	*Rhabdoderma lineare, Dactylococcopsis irregularis, Merismopedia glauca, M. elegans, Microcystis aruginosa, Aphanothece stagnia, Gloeocapsa mimima, Coleospherium lucustris, Nostoc* sp., *Anabena affinis, A. planctonica, Anabeannopsis arnoldi, Gleotrichia schinulata, Spirulina major, Phormidium tenue, Phormidium* sp., *Lyngbea limentica, Raphidiopsis curveta, Chlamydomonas globosa, Gonium sociale, G. pectorale, Volvox aureus, Golenkinia radiata, Schorederia setigera, Pediastrum calthratum, P. duplex, P. boryanum, P. simplex, P. tetras, Treubaria crassipina, Chlorella vulgaris, Tatraedron gracile, T. trigonum, T. regulare, Oocystis lacustris, O. elliptica, Scenedesmus dimorphus, S. quadricauda, S. obliquus, S. armatus, S. bijugatus, S. abundans, Crucugenia rectangularis, Closterim lunula, Euastrum oblongum, Cosmarium circulare, C. granatum, Xanthidium cristatum, Melosira granulata, M. varians, Cyclotella comta, Synedra ulna, S. acus, S. affinis, Tabellaria fenestrata, Eunotia lunaris, Achnanthes exigua, A. lanceolata, Cocconeis placentula, C. diminuta, Navicula gracilis, N. cryptocephala, N. pusila, N. dicephala, N. gastrum, Pinnularia major, P. divergense, P. virdis, Diploneis elliptica, Gyrosigma atteuatum, G. acuminatum, Cymbella turgida, C. tumida, C. lanceolata, C. cistula, Gomphonema acuminatum, G. intricatum, G. angustum, G. parvulum, Epithemia turgida, Nitzschia palea, N. closterium, N. linearis, N. dissipata, N. acicularis, Rhopalodia gibba, Peridinium pusillum, P. cinctum, Gymnodium excavatum, G. neglactum, Ceratium hirundiella, Euglena acus, E. sanguinea, Phacus, caudatus, P. longicauda, pleuronectus, Trachelomonas armata*
Zooplankton	*Rotaria rotaria, Philodina roseola, Clourella obtusa, Brachionus angularis, B. clayciflorus, B. falcatus, Keratella quadreta, Lecane luna, Monostyla cornuta, Asplanchana girodi, Trichocera elongata, Filinia longiseta, Sinoclanus dorii, Macrocyclops distincus, Cyclops strenuous, Chydorus* sp., *Alonella excisa, Alona costata, Ceriodaphnia setosa, Diphanosoma* sp., *Centropyxis aculeata, Arcella costata* and *Difflusia urceolata*

Benthos

During the period 2005–07 a total of 17 species of macro-benthic invertebrates were collected and identified, gastropoda (9 species), bivalvia (4 species) and others (5 species). The dominant forms were *Assiminea francesiae, A. woodmensonia, Gabbia orcula producta, pila globosa, Lymnaea acuminata, Balamya bengalensis, Gyrulus abiatus, segmentia calatha, Parreysia fevidens-plagiosoma* and *Parreysia* (*radiatula*) *caerulea gaudichaudi*. The dominance of molluscs in Bellon beel, as compared to other biota, indicates the characteristics of a macrophyte infested aquatic system.

Fish Fauna

A total of 41 species of fish were recorded, including one exotic species (*Cyprinus carpio*). The fish species diversity with local name is given in Table 34.3.

Table 34.3: Fish Diversity in Belloon Beel (2006–07)

Species	*Local Name*
Labeo rohita	Rohu
Catla catla	Catla
Cirrihinus mrigala	Mirig
Cyprinus carpio	Common carp
Channa striatus	Sol
Channa orientalis	Cheng
Channa punctatus	Lata
Channa marulis	Gajal, Sal
Channa gauchua	Chang
Colisha fasciatus	Cheli
Colisa lalia	Khalisa
Mystus tenf!:ara	Tenwa
Heteropeunius fossilis	Singhi
Pseudeutropius atherinoides	Potasi
Clarius batrachus	Magur
Anabus testudineus	Koi
Wallago attu	Wallago
Aorichthys aor	Aar
Nandus nandus	Nadosh
Chanda nama	Nama chanda
Chanda ranga	Ranfla chanda
Puntius saphore	Puti
Amblypharynodon mola	Maurala
Notopteris chitala	Chital
Notopteris notopteris	Phouli
Xenentodon cancila	Kankley
Notopteris cuchia	Cuchia
Mastercembalus armatus	Ban
Puntius ticto	Pothia
Chela bacaila	Ghora-chela
Glossoobius iuris	Bele

Contd...

Table 34.3–Contd...

Species	Local Name
Mastercembalus punculas	Pakal
Salmostoma bacaila	Chela
Gudusia chapra	Khoira
Oxygaster bacila	Cheli
Badis badis	Bhada
Lepidocephalus guntea	Guntea loach
Bagarius bagirius	Baghaar
Ailia colia	Kaiuli-vacha
Ompak bimaculatus	Puffta
Ompak bacaila	Pabda

Macrophytes

The proliferation of macrophytes remains to be the most dominant biotic communities of Belloon beel, indicating advanced level of eutrophication. A total of 34 species of macrophytes have so far been collected from the beel during the period 2006–07 (Table 34.4). Among the species thus recorded five

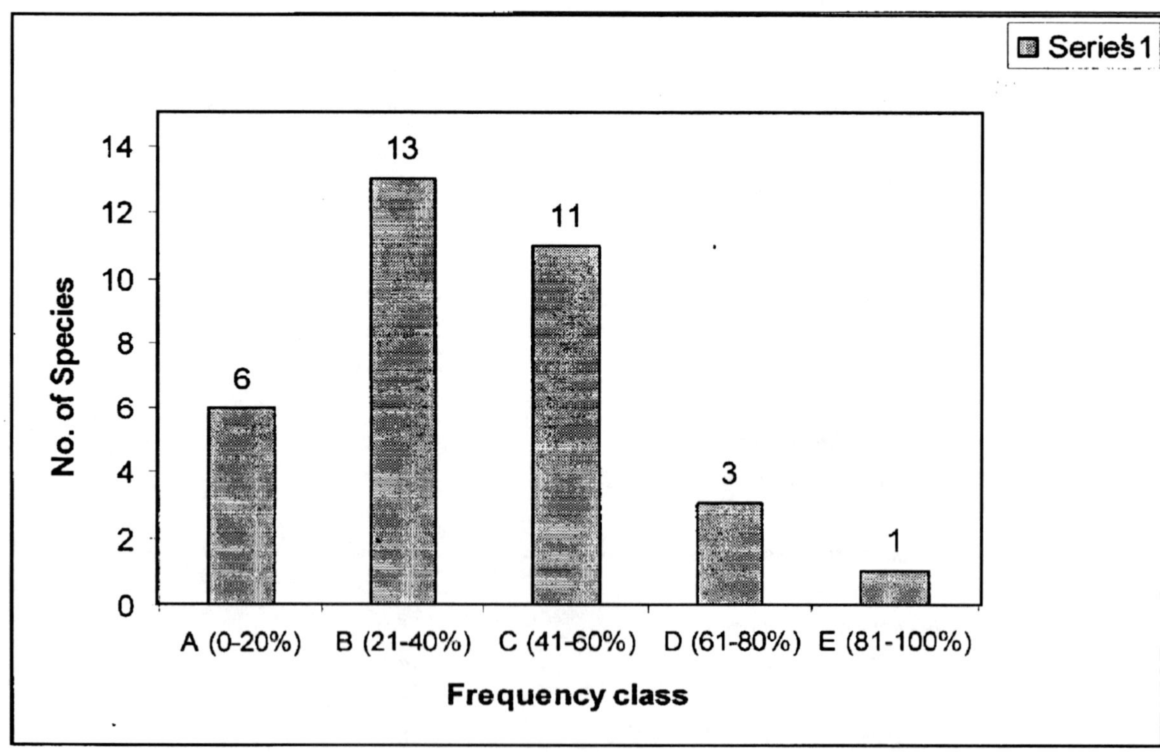

Figure 34.1: Frequency Class of Macrophytes in Belloon Beel (2006–07)

species *viz. Eichhornia crassipes, Hydrilla verticillata, Ceratophyllum demersum, Vallisneria spiralis* and *Najas minor* had much higher per unit biomass, assuming the status of weeds. Moreover, the colonization of *Typha angusta* and *Cyperus procerus,* restricted to certain pockets of the wetland were indicative of degraded aquatic regime, especially changing habitat, from aquatic regime to marshy land. The relative frequency of various macrophytes, calculated as per Curtis (1959) and Misra (1968), was the highest for *E. crassipes* (6.17 per cent), while it was the lowest for *Marsilea minuta, Lemna perpusila, Nymphaea nauchali, Mariophyllum tuberculatum,* and *Potamogeton pectinatus* (1.2 per cent).

Table 34.4: Diversity of Macrophytes in Belloon Beel, Murshidabad, West Bengal

Sl.No.	*Scientific Name*	*Family*	*Habitat*
1.	*Eichhornia crassipes*	Pontederaceae	Free floating
2.	*Pistia stratiotes*	Arecaceae	Free floating
3.	*Azolla pinnata*	Salvanaceae	Free floating
4.	*Lemna minor*	Salvanaceae	Free floating
5.	*Lemna perpusilla*	Azollaceae	Free floating
6.	*Spirodella polyrhiza*	Lemnaceae	Free floating
7.	*Trapa natans*	Lemnaceae	Free floating
8.	*Nymphoides indicum*	Lemnaceae	Free floating
9.	*Nelumbo nucifera*	Gentianaceae	Rooted floating
10.	*Nymphaea stellata*	Nymphaceae	Rooted floating
11.	*Nymphaea nauchali*	Nymphaceae	Rooted floating
12.	*Euryaleferox*	Nymphaceae	Rooted floating
13.	*Potamogeton nodosus*	Nymphaceae	Rooted floating
14.	*Mariophyllum tuberculatum*	Trapaceae	Rooted floating
15.	*Hydrilla verticillata*	Hydrocharitaceae	Submerged
16.	*Ceratophyllum demersum*	Hydrocharitaceae	Submerged
17.	*Vallisneria spiralis*	Ceratophyllaceae	Submerged
18.	*Potamogeton pectinatus*	Najadaceae	Submerged
19.	*Najas minor*	Najadaceae	Submerged
20.	*Marsilea quadrifolia*	Marsileaceae	Prostrate emergent (Marginal)
21.	*Marsilea minuta*	Marsileaceae	Prostrate emergent (Marginal)
22.	*Ipomoea aquatica*	Convolvulaceae	Prostrate emergent (Marginal)
23.	*Enhydra fluctuans*	Asteraceae	Postrate emergent (Marginal)
24.	*Hygrorhiza aristata*	Poaceae	Postrate emergent (Marginal)
25.	*Centella asiatica*	Apiaceae	Postrate emergent (Marginal)
26.	*Alternanthera philoxeroides*	Amaranthaceae	Postrate emergent (Marginal)
27.	*Jussiaea repens*	Onagraceae	Postrafe emergent (Marginal)
28.	*Polygonum barbatum*	Polygonaceae	Erect emergent (Marginal)
29.	*Polyonum labrum*	Polygonaceae	Erect emergent (Marginal)

Contd...

Table 34.4–Contd...

Sl.No.	Scientific Name	Family	Habitat
30.	*Aeschynomene aspera*	Leguminosae	Erect emergent (Marginal)
31.	*Colocasia esculanta*	Araceae	Erect emergent (Marginal)
32.	*Cyperus difformis*	Cyperaceae	Erect emergent (Marginal)
33.	*Typha angustata*	Pontederaceae	Erect emergent (Marginal)
34.	*Cyperus procerus*	Cyperaceae	Erect emergent (Marginal)

The diversity of macrophytes in Belloon beel followed a frequency sequence of A < B > C > D > E (Figure 34.1), which is far from the normal frequency of A > B > C > = < D < E (Raunkiaer, 1934; Ellenberg and Muller Dombois 1967; Muller Dombois and Ellenberg, 1974). The frequency class of macrophytes in Belloon Beel thus indicated stressed aquatic regime, due to greater abundance of *E. crassipes* (81–100 per cent frequency), affecting the normal biodiversity, adversely.

Acknowledgement

The authors are thankful to Dr. K.K. Vass, Director, Central Inland Fisheries Research Institute, Barrackpore, for encouragement and providing the necessary laboratory facilities to carryout the investigation. They also thankfully acknowledge the financial assistance received from the ICAR, New Delhi.

References

Curtis, J.T., 1959. *The Vegetation of Wisconsin.* Univ. of Wisconsin Press, Madison.

Deny, P., 1995. Wetlands and biodiversity. In: *Conservation and Sustainable Use of Floodplain Wetlands.* Asian Wetland Bureau, Kuala Lumpur, Malaysia

Ellenberg and Muller Dombois, 1967. A key to Raunkier's Plant life form with revised subdivisions. Ber. Geobot. Inst.. ETH Stiftg. Rubel., Zurich, 37: 56–73

Frttsch, F.E., 1935. *Structure and Reproduction of Algae.* Cambridge University Press, London.

Jha, B.C., 1995. The floodplain fishery of Gandak basin, Bihar. In: *Conservation and Sustainable Use of Floodplain Wetlands.* Asian Wetland Bureau, Kuala Lumpur, Malaysia.

Jha, B.C. and Chandra, K., 1997. Kusheshwarasthan Chaur (North Bihar): Status and prospects of fisheries development. *CIFRI Bulletin* No. 71.

Jhingran, V.G., Natarajan, A.V. and Banerjee, S.N., 1967. Methodology for reservoir fisheries. *CIFRI Bulletin* No. 12.

Misra, R.T., 1968. *Ecology Workbook.* Oxford and IBH, Publ. Co., New Delhi.

Muller Dombois, D. and Ellenberg, H., 1974. *Aims and Methods of Vegetation Ecology.* John Wiley and Sons, New York.

Prescott, G.W., 1980. *How to Know the Aquatic Plants?* The Picture Key, Nature Series, Dubuque, Iowa.

Prescott, G.W., 1982. *Algae of the Western Great Lake Area.* WmC Brown, Dubuque, Iowa.

Sinha, M. and Jha, B.C., 1997. Ecology and fisheries of Ox-bow lakes (Maun) of North Bihar: A threatened ecosystem. *CIFRI Bulletin* No. 74.

Sugunan, V.V., 1995. Floodplain Lakes: A fisheries perspective. In: *Conservation and Sustainable Use of Floodplain Wetlands.* Asian Wetland Bureau, Kuala Lumpur, Malaysia.

Sugunan, V.V and Mukhopadhyaya, M.K., 1995. Conservation and sustainable use of the floodplain wetland: A case study of Bandardaha and Belloon beel. In: *Conservation and Sustainable Use of Floodplain Wetlands.* Asian Wetland Bureau, Kuala Lumpur, Malaysia.

Subramanyam, K., 1962. *Aquatic Angiosperms: A Systematic Account of Common Indian Aquatic Angiosperms.* Council of Scientific and Industrial Research, New Delhi, India.

Sculthrope, C.D., 1967. *The Biology of Aquatic Vascular Plants.* Edward Arnold (Publ.) Ltd., London, 517–520 pp.

Raunkiaer, C., 1934. *The Life Forms of Plants and Statistical Plant Geography*. Clarrendon Press, Oxford.

Ward, H.B. and Whipple, G.C., 1908. *Freshwater Biology.* John Wiley and Sons, Inc., New York.

Chapter 35

Effect of Farm Yard Manure and Lime on Soil Adsorption and Plant Uptake of Phosphorus

H.K. Senapati, A.K. Dash, B.B. Dash, D.K. Behera and P.K. Das
Department of Soil Science and Agricultural Chemistry,
College of Agriculture, Orissa University of Agriculture and Technology,
Bhubaneswar – 751 003, Orissa

ABSTRACT

A field experiment was carried out to study the effect of farm yard manure (FYM) and lime on phosphate adsorption in soil and uptake of phosphorus by rice in an Inceptisol of Bhubaneswar. Twelve treatments comprising of optimal, suboptimal and super optimal doses of NPK alone, along with S, B, Zn, FYM and lime were undertaken in a rice–rice cropping system. The uptake of P was more in all the treatments than the control and it ranged from 5.58 to 9.59 kg ha $^{-1}$. The treatment receiving 100 per cent NPK with lime and FYM recorded the highest P uptake value of 9.59 kg ha^{-1} in rice grain and 2.52 kg ha^{-1} in rice straw as against 5.58 and 127 kg ha $^{-1}$ in the grain straw respectively with control. The soil samples from the selected treatments were incubated in the laboratory condition with the addition of known concentrations of phosphorus to study the adsorption behavior of P. The highest P adsorption (80 per cent) was observed in control with added P concentration 0.5 µg ml^{-1} and the value gradually decreased to 46 per cent with increased concentration of added P up to 12 µg ml^{-1}. On the other hand the lowest amount of Padsorbed (20 per cent) in 100 per cent NPK + FYM + lime treated plot where the concentration of P added was 12 µg ml^{-1}. Phosphorus adsorption at different concentration of added P was maximum in the control (No fertilizer)and was in the order of control > 100 per cent NPK > 100 per cent NPK + FYM > 100 per cent NPK + lime > 100 per cent NPK + FYM + lime.

Keywords: *Farm yard manure, Lime, Phosphate adsorption, Phosphorus uptake, Inceptisol.*

Introduction

Phosphorus is an essential plant nutrient but its availability is decreased in most of the acid soils. The grain yield and P uptake increased significantly with P alone or with FYM irrespective of sources over control and maximum values were recorded with SSP@60 kg P_2O_5 ha^{-1} + FYM in case of soyabean whereas P use efficiency was higher with 30 kg P_2O_5 ha^{-1} + FYM treatment (Majumdar *et al.*, 2007). Significant increase in P uptake and phosphorus use efficiency by maize crop with the application of 60 kg P_2O_s with 5 t FYM ha^{-1} in acid Alfisol was also reported by Venkatesh *et al.* (2002). The P use efficiency was found to be the highest with the integrated application of NPK and green manuring (40.4 kg kg^{-1}) with highest contribution of green manuring towards the P use efficiency (43.5 kg kg^{-1}) in comparison to other organic sources (Laxminarayana and Patiram, 2006).

Liming is the most important practice required in such soil in order to control acidity and increase phosphorus availability to plants (Sanchez and Vehare, 1980). Debnath *et al.* (2000) reported that liming increased the saloid bound P, Al-P and Ca-P but decreased the Fe-P fraction with a few exception. Westermann (1992) observed that plant P uptake was increased by P fertilization and decreased by increasing lime concentration.

Phosphate adsorption is an important property of soil that controls the phosphate concentration in the soil solution for subsequent utilization by plant Addition of 100 per cent N alone resulted in maximum phosphorus adsorption (98 per cent) and soils with 100 per cent NPK + lime resulted in lowest Phosphorus adsorption (Subehia *et al.*, 2005). The lime application increased and gypsum application decreased Phosphorus adsorption (Paliyal and Verma, 2002). The adsorption maxima were highest in unfertilized soil and lowest in NPK + FYM plot followed by NPK + lime (Prasad and Mathur, 1997). Earlier studies have shown that addition of organic matter decreased P adsorption in soil (Sibanda and Young, 1986).The decreased in P adsorption is due to (1) Organic acid block exposed hydroxyl on the surface of Fe and Al oxide (Appelt *et al.*, 1975) and decreased p fixation capacities of these soils (Dalton *et al.*, 1952). The competition of organic acids released during mineralization for same sites of fixation of P and complex formation of exchangeable Al by organic acids (Hue 1991, Sing and Jones, 1976).

Materials and Methods

The field experiment was conducted in the Central Research Station of Orissa University of Agriculture and Technology, Bhubaneswar, which lies between the longitude of 85°48′ to 85°49′ and latitude of 20°16′ to 20°17′N and comes under agroecological region of 12.2. The rice (Cv-Swarna) was grown during kharif 2006 with a spacing of 20 × 10 cm. The initial soil samples before start of the experiment were collected and analyzed for different physico-chemical parameters as per the standard procedure (Jackson, 1973).The soil was sandy loam in surface but clay loam to clay in lower depth. The soil of the experimental site was lateritic *Inceplisol* (*Udic ustochrept*) having pH 5.80, EC 0.12 $dSmol^{-1}$, Organic Carbon 4.3g kg^{-1}, CEC 3.75 C mol(P^+) kg^{-1} with available N, P, K, S were 187, 19.4, 43.4 and 22 kg ha^{-1} respectively. The following treatments were imposed to study the effect of FYM and lime on phosphate adsorption by the soil and uptake by rice. The experiment was designed with randomized block design (RBD) having 12 treatments and 4 replications. The treatments were, T_1: 50 per cent NPK; T_2: 100 per cent NPK; T_3: 150 per cent NPK; T_4: 100 per cent NPK + Zn; T_5: 100 per cent NPK + FYM; T_6: 100 per cent NPK + lime + FYM; T_7: 100 per cent NPK + B + Zn; T_8: 100 per cent NPK + S + Zn, T_9: 100 per cent N; T_{10}: 100 per cent NPK; T_{11}: 100 per cent NPK + lime; T_{12}: Control. The 100 per cent NPK = 80 : 40 : 60 kg ha^{-1} (N : P_2O_5 : K_2O). FYM = 10 t ha^{-1}, Lime = 1 t ha^{-1}, S = 30kg S ha^{-1} as gypsum, Zn =12.5 kg ha^{-1} from $ZnSO_4$ B = 0.25 per cent as borax.

For the phosphate adsorption study, soil samples were collected from selected treatments like control. 100 per cent NPK, 100 per cent NPK + FYM, 100 per cent NPK + lime, 100 per cent NPK + FYM + lime from the experimental site. Duplicate soil samples of 5 g were taken in a series of 100 ml conical flask to which 50 ml of 0.01 M $CaCl_2$ containing 0, 0.5, 1, 2, 4, 6, 8, 10, 12, µg P ml^{-1} as KH_2PO_4, were added. Two drops of toluene were added to check the microbial activities. The flasks were tightly covered with polythene to check evaporation of the added phosphate solutions. The contents were shaken in a reciprocating shaker continuously for 2 hours and equilibrated for 48 hours at 25°C in a 800 incubator with intermittent shaking. At the end of equilibration period the suspensions were filtered and the concentration of phosphorous in the filtrate was determined by developing blue colour. Intensity of colour was measured at 660 mµ wavelength in a UV-VIS spectrophotometer. The amount of phosphate adsorbed was calculated as P adsorbed µg g^{-1} = [concentration of P added (µg ml^{-1}) – equilibrium concentration of P (µg ml^{-1})] × 10. The grain and straw samples were collected and analysis for P uptake as per Piper (1950). The soil samples after harvest of rice crop was collected and analysed as per the standard procedure (Jackson, 1973).

Results and Discussion

The grain and straw yields of different treatments were recorded and presented in Table 35.1. The results revealed that the mean highest grain (29.7 q ha^{-1}) and straw (34.0 q ha^{-1}) yields of rice were observed where 100 per cent NPK + lime + FYM (T_6) was applied where as mean lowest grain (12.3 q ha^{-1}) and straw (18.8 q ha^{-1}) yields were observed with control (T_{12}). It was observed that there was significant response to the application of plant nutrients over the control. It was also revealed that due to application of FYM along with inorganic fertilizers; mean grain yield was increased slightly than application of only inorganic fertilizers. Similar trend was also observed with straw yield.

Table 35.1: Effect of Treatments on Grain and Straw Yield (q ha^{-1}) Uptake (kg ha^{-1}) and Uptake Rate (kg t^{-1}) of Phosphorus by Rice

Tr.No.	*Treatments*	*Grain Yield*	*Straw Yield*	*Uptake of P*			*Uptake Rate of P*	
				Grain	*Straw*	*Total*	*Grain*	*Straw*
T_1	50 per cent NPK	*22.5*	25.0	6.64	1.47	8.11	2.95	0.59
T_2	100 per cent NPK	243	28.8	7.77	2.12	9.89	3.19	0.74
T_3	150 per cent NPK	22.1	26.4	8.40	237	10.77	3.80	0.89
T_4	100 per cent NPK + Zn	19.4	25.4	8.64	2.06	10.70	4.45	0.81
T_5	100 per cent NPK + FYM	24.7	29.8	8.24	2.29	10.53	3.33	0.77
T_6	100 per cent NPK + lime + FYM	29.7	34.0	9.59	2.52	12.11	3.23	0.74
T_7	100 per cent NPK + B + Zn	212	262	8.36	2.47	10.84	3.94	0.94
T_8	100 per cent NPK + S + Zn	19.4	24.7	8.04	2.38	10.42	4.14	0.96
T_9	100 per cent N	19.8	25.4	6.41	1.93	8.34	3.24	0.76
T_{10}	100 per cent NP	20.9	273	7.52	228	9.79	3.60	0.83
T_{11}	100 per cent NPK + lime	24.1	30.7	8.22	2.44	10.67	3.41	0.79
T_{12}	Control	123	18.8	5.58	127	6.85	4.53	0.67
	C.D. (0.05)	3.4	4.1	1.05	0.33	1.08		

The effect of different treatments on the uptake and uptake rate of phosphorus was presented in Table 35.1. It was observed that the uptake of P was more in all the treatments than the control. The uptake of P by grain ranged from 5.58 to 9.59 kg ha^{-1}. The treatment receiving 100 per cent NPK with FYM and lime (T_6) recorded the highest value of 9.59 kg ha^{-1} of P by the rice grain. The treatments which received FYM and lime along with 100 per cent NPK (T_5 and T_{11}) resulted almost equal values of 8.24 and 8.22 kg ha^{-1}, respectively. The micronutrient treatments (T_4, T_7 and T_8) were at par ranging from 8.04 to 8.64 kg ha^{-1} of P by the rice grain. The individual and combined effect of P, FYM and Lime were observed which had a positive influence on the uptake of P by the rice grain.

The uptake of P by the rice straw ranged from 1.27 to 2.52 kg ha^{-1}. The highest value of 2.52 kg ha^{-1} was observed in the treatment of 100 per cent NPK with FYM and lime (T_6). The lime treatment was superior to 100 per cent NPK (T_2) and FYM treatment (T_5) and was at par with the micro nutrient treatments of T_7 and T_8. The application of nitrogen alone (T_9) suppressed the uptake of P both by grain and straw. The total highest uptake of P both by grain and straw was recorded to be 12.11 kg ha^{-1} in recommended dose of fertilizer along with FYM and lime (T_6). The treatments receiving FYM, lime, secondary (S) and micro-nutrients (Zn, B) resulted almost similar total P uptake values by the rice, grain and straw.

The uptake rate of P by rice grain varied from 2.95 to 4.53 kg t^{-1} and by rice straw it varied from 0.59 to 0.96 kg t^{-1}. The beneficial effect of soil submergence on the availability of P explains the lack of wide spread response of rice to P fertilizers. In acid soils the moderating effects of submergence on soil pH and E_h influence P availability in two ways. First there is reduction in levels of soluble forms of elements such as Fe, Mn, Al, which rendered P unavailable. Secondly there is a reduction in the sorption and occlusion of P on soil solids.

The equilibrium concentration of phosphorus at different concentration of P added in the incubation study was presented in Table 35.2. The concentration of P added were 0.5, 1, 2, 4, 6, 8, 10, 12 in 1 : 10 soil solution ratio which corresponds to 5, 10, 20, 40, 60, 80, 100, 120 µg P added per gram of soil. The equilibrium concentration of P increased with increase in concentration of P added (Table 35.2). With the addition of same quantity of P, the equilibrium concentration is increased with addition of amendments like fertilizer, FYM and lime. The highest value of 9.59 µg ml^{-1} was observed in 100 per cent NPK + FYM + lime treatment with the addition of 12 µg ml^{-1} of P followed by 8.99 µg ml^{-1} in 100 per cent NPK + lime treatment. The effect of FYM and lime on percentage of phosphate adsorption presented in Table 35.3 revealed that the amount of P adsorbed increased with increase in concentration of P added where as percent adsorption of P decreased with increase in concentration of P added. This might be due to increase in concentration of P with increasing concentration of P added to occupy the same number of adsorption sites. P adsorption at different concentration of P added was maximum (80 per cent) in the control and decreased in the order of control > 100 per cent NPK > 100 per cent NPK + FYM > 100 per cent NPK + lime > 100 per cent NPK + FYM + lime. Similar results were also reported by (Prasad and Mathur, 1997). In control P adsorption was highest because no phosphorus was added and there was exhaustion of native phosphorus from soil causing an increase in the number of unoccupied adsorption sites. Adsorption decreased after application of NPK, due to addition of phosphorus some of adsorption sites might have been occupied by P causing decrease in number of available adsorption sites. These results are in agreement with the findings of Subehia *et al.* (2005) and Dalton *et al.* (1952).

Table 35.2: Equilibrium Concentration of Phosphorus at Different Concentration of P Added

Sl.No.	*Treatments*	*Equilibrium Conc. of P (µg ml^{-1})*							
		Conc. of Added P (µg ml^{-1})							
		0.5	*1.0*	*2.0*	*4.0*	*6.0*	*8.0*	*10.0*	*12.0*
1.	Control	0.1	0.24	0.46	1.21	2.27	3.43	4.78	6.42
2.	100 per cent NPK	0.12	0.27	0.69	1.59	2.68	3.96	5.47	7.19
3.	100 per cent NPK + FYM	0.16	0.37	0.84	1.88	3.18	4.56	6.19	8.04
4.	100 per cent NPK + Lime	0.18	0.42	0.94	2.08	3.59	5.12	6.79	8.99
5.	100 per cent NPK + FYM + Lime	0.22	0.49	1.08	2.37	3.89	5.56	7.45	9.59

Table 35.3: Effect of FYM and Lime on Phosphate Adsorption

Sl.No.	*Treatments*	*P Adsorbed (µg g^{-1} Soil)*							
		Conc. of Added P (µg ml^{-1})							
		0.5	*1.0*	*2.0*	*4.0*	*6.0*	*8.0*	*10.0*	*12.0*
1.	Control	4.00 (80)	7.60 (76)	15.34 (78)	27.82 (70)	37.23 (62)	45.62 (57)	52.12 (52)	55.81 (46)
2.	100 per cent NPK	3.75 (75)	7.23 (72)	13.25 (66)	24.05 (60)	33.13 (55)	40.35 (50)	45.29 (45)	48.08 (40)
3.	100 per cent NPK + FYM	3.35 (67)	6.30 (63)	11.6 (58)	21.2 (53)	28.2 (47)	34.4 (43)	38.01 (38)	39.6 (33)
4.	100 per cent NPK + Lime	3.11 (62)	5.8 (58)	10.6 (53)	18.2 (46)	24.01 (40)	28.8 (36)	32.05 (32)	31.56 (26)
5.	100 per cent NPK + FYM + Lime	2.75 (55)	5.01 (50)	9.13 (46)	16.21 (40)	21.09 (35)	24.32 (30)	25.43 (25)	24.06 (20)

Figures in parentheses represent the percent P adsorbed.

Application of FYM along with NPK decreased the phosphate adsorption at different concentrations of P added, as computed to 100 per cent NPK. Application of FYM reduced P adsorption which might be due to coating of the soil particles by organic matter and blocking the adsorption sites (Table 35.2). The organic anions also compete with phosphate to occupy the adsorption sites causing the reduction in P adsorption. Similar results have also been reported by Reddy *et al.* (1999).

Addition of lime along with 100 per cent NPK decreased the P adsorption. This is due to the fact that liming increases the hydroxyl ion concentration in soil solution which might be competing with phosphate to occupy the adsorption site causing a reduction in P adsorption. P adsorption was also minimum in the treatment receiving FYM and lime for the same reason stated earlier which corroborated the finding of Prasad and Mathur (1997). The lowest adsorption of 20 per cent was observed in 100 per cent NPK with FYM and lime treatment at a concentration of added P @ 12.0 µg ml^{-1} as against 26 per cent in 100 per cent NPK with lime, 30 per cent in 100 per cent NPK + FYM, 40 per cent in 100 per cent/ NPK and 46 per cent in control (Figure 35.1).

Table 35.4: Effect of Treatments on Postharvest Properties of Soil

Sl.No.	Treatments	pH	EC (ds m^{-1})	OC (g kg^{-1})	Available Macro-nutrient (kg ha^{-1})				Available Micro-nutrients (ppm)			
					N	P	K	S	Fe	Cu	Mn	Zn
1.	Control	5.23	0.010	4.12	194.14	9.74	59.84	22.40	20.90	2.22	46.28	0.90
2.	100 per cent NPK	5.20	0.110	4.34	224.73	18.98	66.52	44.43	21.11	1.02	60.96	1.64
3.	100 per cent NPK + FYM	S.22	0.084	4.84	226.65	22.35	69.43	53.03	42.12	3.17	48.56	2.32
4.	100 per cent NPK + Lime	6.12	0.083	4.38	228.78	16.14	69.38	57.47	25.40	2.22	50.20	2.22
5.	100 per cent NPK + Lime + FYM	6.13	0.074	6.15	235.74	25.86	69.64	55.93	28.78	2.98	54.88	2.72

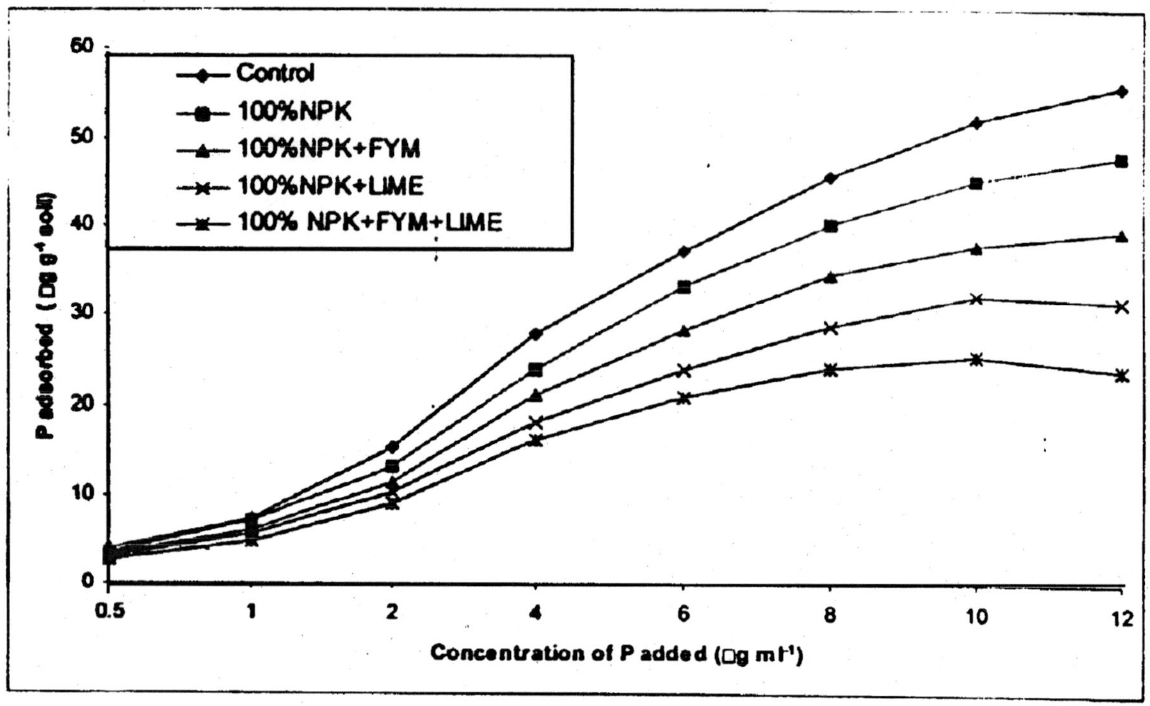

Figure 35.1

From the post harvest soil properties (Table 35.4) it was observed that when FYM was applied along with recommended dose of NPK the pH was reduced to 5.22 from the initial value of 5.80. FYM on decomposition and subsequent mineralization produced humus and a wide variety of organic acids by microorganism which helped in reducing the pH. The experimental soil is an acid laterite soil and the application of lime could neutralize the pH dependant acidity, thus the soil pH was increased which was observed from the treatments receiving 100 per cent NPK + lime (T_{11}) and 100 per cent NPK + FYM + lime (T_6). The inorganic chemical fertilizer treated plots resulting higher EC value where as the addition of FYM and further addition of lime reduced the value. The treatment receiving 100 per cent NPK + FYM + lime recorded the highest soil organic carbon of 6.15g kg^{-1} as against 4.12 g kg^{-1} in control. The application of recommended dose of fertilizer along with FYM and lime encouraged the availability of different nutrients in general and phosphorus in particular. This result corroborates the finding of Choudhury *et al.* (1981), Patnaik *et al.* (1989) and Debnath *et al.* (2000).

References

Appelt, H., Colemen, N.T. and Pratt, P.E., 1975. Interaction between organic compounds, minerals and ions in volcanic-ash-derived soils. Effects of organic components on adsorption of phosphate. *Proc. Soil Sc. Soc. of America*, 39: 628–631.

Dalton, J.D., Russel, G.C. and Seiling, D.H., 1952. Effect of organic matter on phosphate availability. *Soil Science*, 73: 173–177.

Debanath, A., Bhattacharya, T.K. and Debanath, N.C., 2000. Behaviour of inorganic phosphate fractions in limed acid soils. *J. of Indian Soc. Soil Sci.*, 48: 829–831.

Debanath, N.C. and Mandal, S.K., 1982. Evaluation of the relationship of inorganic phosphate traction of soil with available P determined by some chemical and biological methods. *J. of Indian Soc. Soil Sci.*, 30: 489–492.

Hue, N.V., 1991. Effects of organic acid/anions on P sorption and phyto availability in soils with different minerologies. *Soil Science*, 152: 463–470.

Jackson, M.L., 1973. *Soil Chemical Analysis*. Prentice Hall Inc. Englewood Cliffs, New Jersey, USA.

Laxminarayana and Patiram, 2006. Effect of integrated use of inorganic, biological and organic manures on rice productivity and soil fertility in Ultisols of Mizoram. *J. of Indian Soc. Soil Sci.*, 54: 213–220.

Majumdar, B., Venkatesh, M.S., Kailash Kumar and Patiram, 2007. Effect of rock phosphate, super phosphate and their mixtures with FYM on Soyabean and soil P pools in a Typic Hapludalf of Meghalaya. *J. of Indian Soc. Soil Sci.*, 55: 167–174.

Patnaik, S., Panda, D. and Dash, R.N., 1989. Long term fertilizer experiments with wetland rice. *Fertilizer News*, 34: 47–52.

Piper, C.S., 1950. *Soil Chemical Analysis*. Prentice Hall Inc., Englewood Cliffs, New Jersey, USA.

Prasad, J. and Mathur, B.S., 1997. Influence of long term use of fertilizers, manures and lime on phosphate adsorption parameters in acid alfisols of Ranchi. *J. of Indian Soc. Soil Sci.*, 45: 24–27.

Reddy, D., Subba Rao, A. and Takkar, P., 1999. Phosphate adsorption behaviour of a typic haplsutert in relation to long term use of fertilizer phosphorus and manure under soybean–wheat rotation. *J. of Indian Soc. Soil Sci.*, 47: 425–430.

Reddy, K.R., Overcash, M.R., Khaleel, R. and Westerman, P.W., 1980. Phosphorus adsorption desroption characterstics of two soils utilized for disposal of animal waste. *J. Environ. Qual.*, 9: 86–92.

Chapter 36

Mineral Characterization Studies of Sediment Samples of Tuna, Gujarat by Spectroscopic Studies

R. Ravisankar[1]* A. Rajalakshmi[1], D. Sahayathna Selvi[1], K. Vijay Anand[1] and Prince Prakash Jeba Kumar[2]

[1]Department of Physics, S.S.N. College of Engineering, Kalavakkam – 603 110, Tamil Nadu

[2]National Institute of Ocean Technology, Pallkaranai, Chennai – 601 302, Tamil Nadu

ABSTRACT

Qualitative analysis was carried out to determine the major and minor constituent minerals present in sediment samples collected at the coastal area of Tuna, Gujarat by FT-IR and XRD technique. From the I.R. absorption band or the locations of the different peaks, the minerals are identified with the available literature. In addition to the band positions, the sharpness or diffuseness of bands is helpful in the identification of mineral components. The I.R. study on sediment samples was found to be highly useful in identifying the various minerals in sediments. Further the representative sediment samples are analyzed by XRD technique to yield more information about the minerals. Both of these methods were non-destructive and can be used in the identification of mineralogical composition. These results confirmed that the applied techniques are relatively quicker and more reliable in mineral analysis.

Keywords: *Sediment samples, Mineral Analysis, FT-IR and XRD technique.*

* E-mail: ravissn@gmail.com; ravisankarr@ssn.edu.in

Introduction

Sediments are principal carriers of the trace elements in the hydrosphere. Sediment particles are made up of materials derived from rock, soil, biological and anthropogenic inputs. The basic structural unit of inorganic sediment is silicate and aluminosilicate. Major components of sediments include clay, quartz, feldspars, various silicate minerals, gibbsite and calcium carbonate. They have been classified in a variety of ways according to different criteria such as their source, particle size and their composition .The size of sediments ranges from large size (>256 mm) called boulders down to < 2μm called clay. The more important size groupings are sand–63 to 2000 μm; silt–2 to 63 μm, The properties of sediment depend on its mineral composition, percentage of organic matter, sorption capacity for pollutants, porosity and particle size distribution. Properties of the sediments decide the concentration of the pollutant in the sediment. In the present study an attempt has been made to and investigate the qualitative minerals analysis of coastal sediment samples collected at Tuna, Gujarat using FT-IR and XRD techniques.

Infrared spectroscopy as an analytical tool presents a lot of advantages as it is accurate, inexpensive and reliable. The necessary sample amount is in the sub milligram range and sampling procedure is simple. As a major advantage, infrared spectroscopy permits the identification of the changes in both crystalline and amorphous samples. Infrared spectroscopy has been utilized significantly in quantitative analysis in the minerals is less attempted.

The natural vibration frequencies of atoms in molecules and crystals fall in the infrared range. Certain frequencies corresponding to the energy levels of electrons in solids and in large molecules fall in the infrared. Infrared method is sensitive to short range ordering or nearest-neighbor relation; while X-ray analysis is responsive to periodic arrangement of atoms, *i.e.*, long-range order. Infrared spectroscopy has been widely used in the characterization of a large number of organic and inorganic compounds. Absorption of infrared energy by a mineral is associated with the vibrational and rotational motion of molecules within it.

Commonly the infrared analysis employs the group vibration concept to ascertain the presence or absence of various functional groups in the molecule. Certain chemical groups have characteristic absorption bands that are consistent among mineral containing groups. The spectra originate primarily from vibrational stretching and bending modes within molecules. The method offers a fingerprint for identification of molecular structures.

Empirical correlations of vibrating groups with specific observed absorption bands offer the possibility of chemical identification, and possibly quantitative analyses through intensity measurements. The band positions are noted in μm with corresponding wave number in cm^{-1}. In the IR study of inorganic compounds and minerals, the spectra usually run through the range of 2.5 to 50 μm (*i.e.*, 4000 to 200 cm^{-1}). On the other hand, UV-visible spectra generally cover 0.2 to 2.5 μm, termed as near IR (NIR), can be obtained in most of the modem UV-visible spectrometers.

This technique is a rapid, economical and non-destructive physical method universally employed in the study of structural analysis. It is versatile that it can be used both as a source of the physical parameters for crystal lattice determinations and as a means of eliciting purely empiliced qualitative relationships between specimens. It is a simple technique and is widely used in clay mineralogy, soil science and rock study.

Description of the Study Area

Tuna is located on the northern shore of Gulf of Kachchh at the innermost tip in the open sea just at the entrance to the Kandla Creek. The location is in a tidal mudflat adjacent to the Nakthi Creek typical to Gujarat coast. The tidal range in the location is about 6m. As of now, the location in Tuna is being used by KPT (Kandla Port Trust) as outer anchorage for ships waiting to be berthed in Kandla. Vast areas of salt pans occupy intertidal areas surrounding Tuna Island. The Tuna Island is approachable by kuchaa roads constructed by salt pan owners.

Mangroves are found in the intertidal areas of Tuna with a density of 1000 to 2000 plants/100 m^2. These are found in hypersaline conditions with heights of about 1.5–2.0 m only. Mangroves are also found behind the proposed berthing areas. There are no corals found in this part of the Gulf of Kachchh. Seaweeds/sea grasses are absent or of low occurrence due to unsuitable conditions such as muddy substratum, high turbidity and high velocity currents.

Geology of the Study Area

The physiography is characterised by sandy plains and bare rocky hills. This area (Kachchh district, Gujarat) has very dry climate with very mild monsoon of about 340 mm (average rainfall), for about 30 days in a year. The soil of this region is highly saline with salt encrusted surface, which are incapable of supporting any vegetation. The soils of the Banni grasslands are classed as aridisols represented by solorthids and Camborthids. They are very deep, very poorly drained, fine loamy, calcareous with very severe salinity and slight erosion.

The soils of the Kachchh dominantly belong to aridisol and entisol orders. The soils of the hilly terrain are characterised by varying depths and textures. They are moderately alkaline, and calcareous.

Materials and Methods

Sample Collection

Sediment samples were collected using Peterson grab at all the designated locations during low tide. The stations were oriented to get the flesh out from Kandla Creek, Nakthi Creek and coastal in puts. The distance between each station falls about 1.7 kms. At each sample site, covering a sampling area of 1 m^2, five wet samples were collected, each weighing about one kilogram. Among these five samples from a site, four were from the corners and one from the center. The other one of the sediment samples were also collected from a depth of 5 cm from the surface. The samples were placed in plastic pouches and transported to the laboratory. The map showing the study area and sample collection locations are indicated in Figure 36.1. The samples collected from different sites under study were labelled as T_1–T_{10}.

Sample Preparation

Sample of 2 mg is mixed with 40 mg of spectroscopic KBr in the ratio 1 : 20 using a mortar and pestle. Before mixing, necessary amount of KBr powder is dried at 120°C for six hours in an oven. Otherwise the broad spectral peak due to free OH will seriously affect the interpretation on the bound hydroxyls associated, with any of the minerals. The mineral sample was weighed in a microbalance and placed in a clean agate mortar along with the proper amount of dry KBr to prepare sample pellet A pellet of 1 mm in thickness and 13 mm in diameter is prepared. A small camel's hair brush is used to transfer the mixture to the die for pressing the pellet. The die is cleaned with water and acetone, and dried before another pellet is prepared. This procedure is followed for the preparation of every pellet. The prepared pellet is preseryed in a moisture tree glass container before it is placed in a suitable sample' holder and introduced in the infrared beam for analysis. For each site, five to six samples were collected and for each sample five to six pellet specimens were prepared.

Figure 36.1: Location Map

Sample Analysis

The ABB BOMEM 104.MB series available in Dr. Ceeal Analytical Laboratory, Thorappakkam, Chennai, Tamil Nadu, India is made use of in the present work for recording the I.R. spectra of the samples at room temperature. The KBr pellet technique (1 : 20) pellets were followed for the mineral analysis. The I.R. spectra were taken in the mid region of 4000–400 cm^{-1}. The instrument scans the spectra 16 times in minute and the resolution is 5 cm^{-1}. This instrument is calibrated for its accuracy with the spectrum of a standard polystyrene film. Every time, before the spectrum of sample is obtained; the spectrum of the polystyrene film is taken and checked for the accuracy and transmittance. The best spectrum for each site was considered as a representative spectrum of the site. The typical I.R. spectrum is given in Figure 36.2.

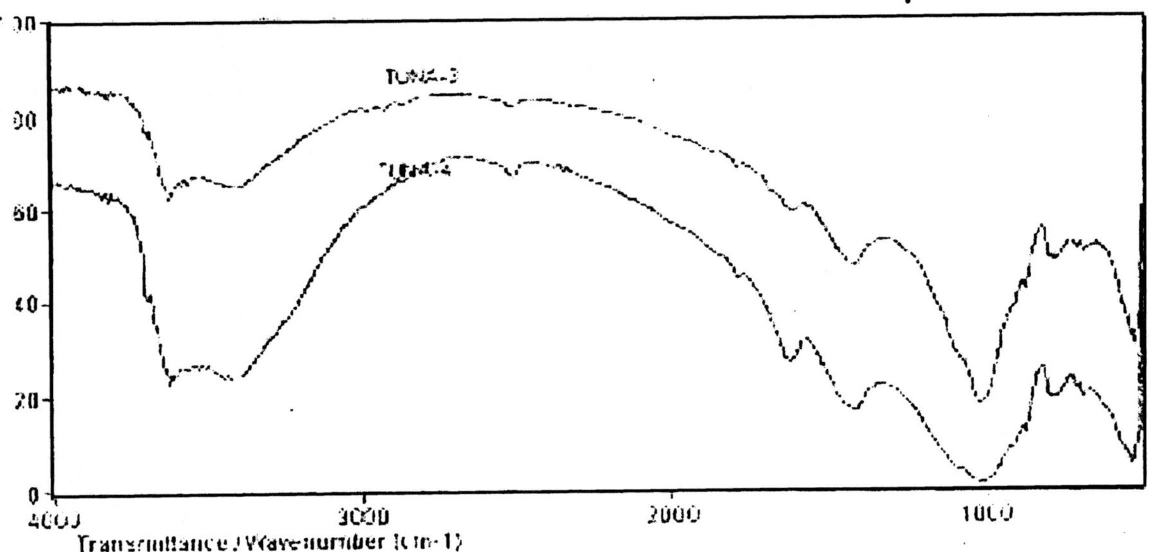

Figure 36.2: A Typical FT-IR Spectrum of Sediment Samples

Results and Discussions

The absorption frequencies of the peaks in the spectra of each site in wave number unit (cm^{-1}) are reported in Table 36.1. By comparing the observed frequencies with available literature, the minerals such as quartz, orthoclase, kaolinite, and montmorllinite have been identified. The mineral wise discussion is as follows.

Quartz

Quartz, one of world's most abundant crustal minerals and also important constituent of the Earth crust. Chemically, it is silicon dioxide, SiO_2. It is the second most abundant mineral in the Earth's crust. It occurs in crystals of the-hexagonal prism commonly having the form of a six-sided prism terminating in a six-sided pyramid; the crystals are often distorted and twins are common. It occurs in hydrothennal veins and pegmatites. Quartz is a common constituent of granite, sandstone, limestone and many other igneous, sedimentary and metamorphic rocks.

Table 36.1: Observed IR Absorption Frequencies

Sample Number	*Observed IR Absorption Frequencies*
T1	408, 471, 508, 525, 535, 630, 713, 776, 875, 925, 1025, 1437, 2516, 3619
T2	408, 455, 471, 506, 525, 536, 714, 796, 876, 1024, 1438, 3617
T3	475, 509, 523, 535, 630, 69713, 779, 798, 874, 922, 1023, 1437, 1632, 3441, 3623
T4	410, 475, 505, 522, 693, 712, 779, 797, 874, 923, 1025, 1436, 1635, 1796, 2516, 3448, 3620, 3695
T5	455, 509, 523, 540, 630, 694, 713.779, 797, 874, 1025, 1435, 1632, 2516, 3435
T6	407, 476, 506, 525, 694, 713, 778, 796, 874, 925, 1025, 1436, 1635, 1794, 2520, 3447, 3619
T7	408, 505, 525, 535, 628, 694, 713, 778, 796, 874, 920, 1025, 1635, 1795, 2518, 3447, 3619
T8	454, 477, 505, 524, 535, 626, 694, 713, 778, 798, 874, 1025, 1435, 1635, 1795, 2518, 3439, 3620, 3690
T9	410, 508, 525, 625, 694, 713, 779, 797, 874, 922, 1021, 1438, 1636, 1793, 2518, 3448, 3623
T10	455, 475, 505, 525, 540, 626, 694, 713, 779, 796, 874, 920, 1025, 1435, 1633, 1793, 2520, 2926, 3446, 3621

It can be observed from the Table 36.2, that i.r. absorption bands appearing at or around 455–450, 510–505, 695–690, 780–775 and 800–795 cm $^{-1}$ suggest the presence of quartz in the samples (Lyon, 1967; Chester and Green, 1968; Farmer, 1974; Clearance Karr, 1975; Coates, 1977; Hlavay *et al.*, 1978; Ghosh, 1978; Russell, 1987; Benedetto *et al.*, 2002; Reig *et al.*, 2002; Ramasamy *et al.*, 2003, 2004 and 2005; Ravisankar *et al.*, 2006).

Table 36.2: IR Absorption Frequencies cm^{-1} of Quartz Present in the Samples

T1	*T2*	*T3*	*T4*	*T5*	*T6*	*T7*	*T8*	*T9*	*T10*
–	455	–	–	455	–	–	454	–	455
	–	694	693	694	694	694	694	694	694
776	–	779	779	779	778	778	778	779	779
–	796	798	797	797	796	796	798	797	796

The characteristic feature of quartz is a doublet appearing at or around 800 and 780 cm $^{-1}$. Such a clear observation of a doublet was noticed in the samples (T_3–T_{10}). The doublet, at 800 and 780 cm^{-1} though not convincingly appeared at the positions mentioned above and seems to have their positions slightly shifted towards the lower frequency region.

Any slight shift in the peak may be due to impurities, crystal defects or isomorphous substitutions (Yariv and Mendelovici, 1979). This has also been confirmed by many workers (Hunt and Turner, 1953; Tuddenham and Lyon, 1960; Chester and Green, 1968; White, 1971; Farmer, 1974; Hlavay *et al.*, 1978; Ganguly and Banerjee, 1985; Boruah *et al.*, 1999).

The doublet at 800 and 780 cm^{-1} may be used for the identification and quantitative estimation of quartz. Usually, the intensities of the peaks of the doublet are not comparable. Hence in the samples (T_1 and T_2) the 800 cm^{-1} peak is either absent or very weak. For the same samples other than this doublet, two more peaks were additionally observed at or around 695–690 cm^{-1} and 455–450 cm^{-1}, which are also characteristic of quartz.

For every sample a minimum of two to maximum of four peaks ate obtained. The pattern of absorption in quartz can be explained by ascribing the 455 cm^{-1} region (Si-O asymmetrical bending vibrations), the bands in the region 690 cm^{-1} (Si-O symmetrical bending vibrations), the bands in the region 780 cm^{-1} (Si-O symmetrical stretching vibrations), and 800 cm^{-1} (Si-O symmetrical stretching vibrations).

The crystallinity index of quartz using the intensities of symmetrical stretching mode lying at 800–795 cm^{-1} and the symmetrical bending mode at 700–695 cm^{-1} has been calculated by many workers.

The crystallinity can be defined as the fraction of crystalline materials in a mixture of crystalline and non-crystalline materials. It is otherwise called as degree of disorder. If the crystallinity is minimum, then the minerals are said to be in disordered state. If it is maximum then the minerals are considered to be in ordered state. The crystallinity of quartz may give clear indication on the crystalline forms of other minerals also. If this crystallinity is maximum then the other minerals may also be expected to be equally in well-crystallized state.

It may be essential to find out the crystallinity of quartz in the samples. The 695 cm^{-1} peak, present in most of the samples gives useful information on the nature of mineral with regard to the structural stability. The intensity of this peak gives the type of crystallinity. The symmetrical bending vibrations of Si-O group obtained at 695 cm^{-1} is more useful to find whether it is crystalline or amorphous. In the amorphous state, 695 cm^{-1} peak will be missing. This is a well known fact (Hlavay *et al.*, 1978). As the 695 cm^{-1} peak is exhibited by the spectra of all the samples (expect T_1 and T_2) the quartz mineral in these samples are in crystalline form.

Feldspar

Feldspar is the name of an important group of rock forming minerals which make up perhaps as much as 60 per cent of the Earth's crust. The general formula for feldspars can be written as WZ_4O_8 in which W may be a Na, K, Ca, and/or Sa and Z is Si and/or Al, the Si : A; ratio ranging from 3 : 1 to 1 : 1. Since all feldspars contain a certain minimum amount of Al, the general formula may be somewhat more specifically stated as WAI (Al, Si) Si_2O_8, the variable (Al, Si) being balanced by variation in the proportions of univalent and bivalent cations. The potassic feldspar (KAl Si_3O_8) is classified into sanidine, orthoclase and microcline. Sanidine and orthoclase are monoclinic and microcline is triclinic. The plagioclase feldspar albite (Na Al Si_3O_8) and anorthite ($CaAl_2Si_2O_8$) have a triclinic crystal structure and form end members of solid solution.

From Table 36.3, the I.R. absorption peaks appearing at 410–405 cm^{-1}, 540–535 cm^{-1} and 630–625 cm^{-1} were assigned to feldspar (White, 1971; Farmer, 1974; Clarence Karr, 1975; Ghosh, 1978; Hlavay *et al.*, 1978; Russell, 1987; Neog *et al.*, 1999; Xu *et al.*, 2001; Ramasamy *et al.*, 2003, 2004 and 2005; Ravisankar *et al.*, 2006). The peaks in the range of 540–535 cm^{-1} and 630–625 cm^{-1} for orthoclase and weak shoulder at 410–405 cm^{-1} for albite were observed in the samples. The peak pertaining to the range 540–535 cm^{-1} is due to Si-O asymmetrical bending vibrations and 630–625 cm^{-1} is due to Al-O-coordination vibrations (Ramasamy *et al.*, 2004).

Carbonate Class

The carbonates consist of those minerals containing the anion $(CO_3)^{2-}$ and include calcite and aragonite (both calcium carbonate), dolomite (magnesium/calcium carbonate) and siderite (iron carbonate). Carbonates are commonly deposited in marine settings when the shells of dead planktonic life settle and accumulate on the sea floor. Carbonates are also found in evaporitic setting (*e.g.* the Great Salt Lake) and also in Karst regions, where the dissolution and reprecipitation of carbonates leads to the fonnation of caves, stalactites' and stalagmites. The carbonate class includes the nitrate and borate minerals.

Table 36.3: IR Absorption Frequencies cm^{-1} of Albtite and Orthoclase* Present in the Samples

T1	*T2*	*T3*	*T4*	*T5*	*T6*	*T7*	*T8*	*T9*	*T10*
408	408	–	410	–	407	408	–	410	–
535*	536*	535*	–	540*	–	535*	535*	–	540*
630*	–	630*	–	630*	–	628*	626*	625*	626*

It can be observed from the Table 36.4, that I.R. absorption peaks appear at 715–710, 875–870, 1440–1435, 1800–1795 and 2520–2515 cm^{-1} and are assigned to calcite. The I.R. absorption peaks of calcite are reported by many workers (Hunt and Turner, 1953; Adler and Kerr 1962, 63; Chester and Elderfield, 1967; Clarence Karr, 1975; Ghosh, 1978; Bottcher *et al.*, 1997; Lopez *et al.*, 2000; SenthilKumar *et al.*, 2001; Benedetto *et al.*, 2002; Reig *et al.*, 2002; Ramasamy *et al.*, 2003 and 2005).

Table 36.4: IR Absorption Frequencies cm^{-1} of Kaolinite, Montmorillonite* and Illite Present in the Samples

T1	*T2*	*T3*	*T4*	*T5*	*T6*	*T7*	*T8*	*T9*	*T10*
471	471	475	475	–	476	–	477	–	475
525*	525	523*	522*	523*	525*	525*	524*	525*	525*
925 * *	–	922**	923**	–	925**	920**	922**	–	920**
1025	102	1023	1025	1025	1025	1025	1025	1021	1025*
–	–	163222	1635*	1632	1635*	1635*	1635*	1635	1633*
–	–	3441**	3448*	3435	3447*	3447*	3439*	3448	3446*
3619	361	3623	3620		3619	3619	3620	3623	3621
–	–	–	3695	–	–	–	3690	–	–

According to the frequency assignments reported by Herzberg (1945) for carbonate minerals the different modes of vibrations are a symmetric stretching, V_1; an out-of-plane bending, V_2; a doubly degenerate asymmetric stretching, V_3; and a doubly degenerate planar bending V_4. But the symmetric oscillation represented by V_1 is reported to be infrared inactive and hence only three fundamentals are ordinarily encountered. These have been recorded for various calcite group minerals in the regions of absorption at approximately 1430 cm^{-1} (V_3), 909–833 cm^{-1} (V_2) and 769–666 cm^{-1} (V_4). In the present investigation, the calcite spectra show major proportion having the absorption bands at approximately 1435 cm^{-1}, 874 cm^{-1} and 713 cm^{-1} and minor bands at 1796 cm^{-1} and 2516 cm^{-1}.

The I.R. spectra of calcite and dolomite are characterized by bands at 711 cm^{-1} and 727 cm^{-1} respectively. These, bands appear unperturbed and sharp with varying intensity depending on the concentration of these minerals. But in the present study the 727 cm^{-1} peak is completely absent in all the samples showing the absence of dolomite. If dolomite were present, then phase change would have occurred at lower temperature.

Table 36.5: IR Absorption Frequencies cm^{-1} of Calcite Present in the Samples

T1	*T2*	*T3*	*T4*	*T5*	*T6*	*T7*	*T8*	*T9*	*T10*
713	714	713	712	713	713	713	713	713	713
875	875	874	874	874	874	874	874	874	874
1437	1438	1437	1436	1435	1436	–	1435	1438	1435
–	–	–	1796	–	1796	1795	1795	1796	1795
2516	–	–	2516	2516	2520	2518	2518	2518	2520

Clay Minerals

Kaolinite

Kaolinite is a mineral with the chemical composition $Al_2Si_2O_5 (OH)_4$. It is layered silicate mineral, with one tetrahedral sheet linked through oxygen molecules to one octahedral sheet of alumina octahedral. It is known as china clay and kaolin. It is softy, earthy, usually white mineral, produced by chemical weathering of feldspars. Due to extremely fine nature (finer than slit), it is mixed with water and transported in tanks as a liquid slurry. It is used in ceramics, medicines, bricks, paper, as a food additive, in toothpaste, and in cosmetics. A recent use is as a specially formulated spray to fruits, vegetables and other vegetation to repel or deter insect damage. A traditional use is .to soothe an upset stomach, similar to the way parrots (and later, humans) in South America originally used it.

The presence of the absorption bands at 3690, 3620, 1030 and 475 cm^{-1} indicate kaolinite (White, 1971; Clarence Karr, 1974; Hlavay *et al.*, 1977; Russell, 1987; Crowley and Vergo, 1988; Clark *et al.*, 1990; Summer, 1995; Gustafsson *et al.*, 1999; Madejova and Komadel, 2001; Xu *et al.*, 2001; Benedetto *et al.*, 2002; Ramasamy *et al.*, 2004, 2005; Ko and Chu, 2005; Ravisankar *et al.*, 2006).

Kaolinite shows a characteristic i.r. absorption sequence in the range of 3700–3600 cm^{-1}. These absorbances correspond to stretching vibrations of inner surface OHs 3696 cm^{-1} and inner OH 3620 cm^{-1}. Absorbance at 1105 and 1034 cm^{-1} are attributed to anti-symmetric Si-O-Si stretching, whereas those at 1018 cm^{-1} are attributed to stretching of an atom bound to both Si and tetrahedral Al (Summer, 1995).

According to Russell (1987), if four peaks are observed in the region of 3697–3620 cm^{-1}, the mineral is said to be ordered state. However, in the present study only two peaks are observed at 3690 and 3620 cm^{-1}. This suggests the mineral is in disordered state.

Montmorilinite

Montmorilinite is a very soft phyllosilicate mineral that typically forms in microscopic crystals, forming clay. Chemically it is hydrated sodium calcium aluminum magnesium silicate hydroxide $(Na, Ca) \times (Al, Mg)_2 (Si_4 O_{10})(OH)_2.nH_2O$. The exact ratio of cations varies with source. It is a member of the smectite family *i.e.*, 2 : 1 clay, meaning that it has 2 tetrahedral sheets sandwiching a central octahedral sheet. It is the main constituent of the volcanic ash weathering product, bentonite. It is used in the oil drilling industry as a component of drilling mud, making the water slurry viscous which helps in keeping the drill head cool and removing drill debris. It is also used as a soil additive to hold soil water in drought prone soils, to the construction of earthen dams and levees and to prevent the leakage of fluids. It is used as a component of foundry sand and as a desicant to remove moisture from air and gases. It is also used therapeutically on humans.

The presence of peaks at 3450–3445, 1635–1630 and 525–520 cm $^{-1}$ in all the samples indicates the presence of montmorillonite. The most distinguishing feature of the montmorillonite spectrum is the broad absorption band that ranges from 3300 to 3500 cm^{-1}. This band typically centered around 3400 cm^{-1} may be due to H-O-H stretching of water molecules present in the interlayer region of montmorillonite (Summer, 1995).The peak pertaining to the range 1635–1630 cm^{-1} due to OH deformation of water and 530–525 cm^{-1} is due to Al-O-Si-deformation (Madejova and Komadel, 2001).

Illite

Illite is a non-expanding, micaceous mineral. It is a phyllosilicate or layered silicate. Structurally illite is quite similar to muscovite or serticite with slightly more silicon, magnesium iron and water and slightly less tetrahedral aluminum and interlayer potassium. It occurs as an alteration product of muscovite and feldspar in weathering and hydrothermal environments. It is common in sediments, soils and argillanceous sedimentary rocks as well as in some low grade metamorptic rocks.

Infrared features of illite are rarely diagnostic because it is so variable in chemical composition. The presence of peak around 920 cm^{-1} in the samples is attributed to illite (Farmer, 1979; Balasubramanion, 1984; Summer, 1995; Clark *et al.*, 1990; Post and Noble, 1993; Xu *et al.*, 2001; Benedetto *et al.*, 2002).

Table 36.6: Band Assignments for Different Minerals of Sediment Samples

Minerals	*Frequency (cm^{-1})*	*Tentative Assignments*	*References*
Quartz	458 and 455	Si–O asymmetrical bending vibration	Hlavay *et al.*, (1978); Coates(1977)
	695– 693	Si–O symmetrical bending vibration	
	775	Si–O symmetrical stretching vibration	
	800	Si–O symmetrical stretching vibration	
Feldpar	540–533	Si–O asymmetrical bending vibration	Coates (1977); White (1971)
	589–586	O–Si (Al)–O bending vibration	Zhang *et al.* (1996); Coates (1977)
	644–640	Al–O–Co-ordination vibration	Hiavay *et al.* (1977); Coates (1977)
Kaolinte	908	Al (OH) vibrations	Ramasamy and Kamalakkanan (1997)
	1035–1030	Si–O Stretching	Ghosh, 1978; Ramasamy and
	3620	Inner OH stretching vibrations	Kamalakkanan (1997) Russell (1987); Summer (1995)
	3696–3695	Inner surface OH stretching vibrations	Russell (1987); Summer (l995)

XRD Analysis

Samples for X-ray powder diffraction (XRD) studies were packed in shallow cavities in glass slides to mipimize preferred orientation. The X-ray patterns of soil samples were recorded at room temperature by using X-ray diffractometer (D500, Siemens) having a curved graphite crystal diffracted monochromator, with a source of CuKα radiation and NaI(Tl) scintillation detector from Material Science Division (MSD), Indira Gandhi Centre for Atomic Research (IGCAR), Kalpakkam, Tamil Nadu, India. Qualitative mineralogy of the soil samples is determined with the standard interpretation procedures of XRD. Major mineral in the studied sample is quartz. Typical X-ray diffractrogram of soil samples are shown in Figure 36.3.

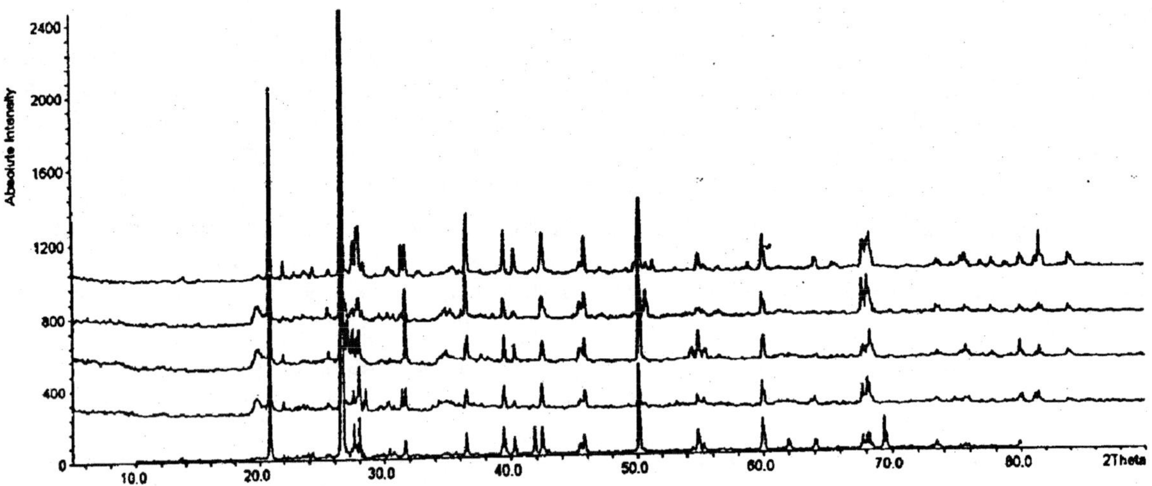

Figure 36.3: Typical X-ray Diffractrogram of Sediment Samples

Mineral-wise Analysis

Qualitative mineralogy of the coastal sediment samples is determined with the standard interpretation procedures of XRD. The mineral wise discussion is as follows:

Quartz

Quartz is ubiquitous and forms one of the most abundant mineral in all the samples. It is identified by the distinctive reflection at 4.254A°, 3.343A°, 2.456A°, 2.281A°, 2.127A°, 1.817A°, 1.541A°, 1.382A°, 1.374A° and 1.371A°.

Feldspar

The feldspar mineral is invariably present in all samples. Albite is identified by basal reflection at 4.030A°, 3.780A°, 3.684A°, 3.663A°, 3.196A° and 2.933A°. The basal reflection at 4.222A°, 3.777A°, 3.470A°, 3.310A°, 3.290A°, 3.240A°, 2.992A°, 2.901A° and 2.571A° indicated the presence of orthoclase.

Gypsum

It is identified by the basal reflections at 4.283A°, 3.065A° and 2.873A°.

Calcite

It is identified distinct reflections in the d spacing of 3.035A°, 2.495A°, 2.285A°, 2.095A° and 1.875A°.

Rutile

The d spacing of 4.283A°, 2.487A° and 1.687A° indicated the presence of rutile in the samples.

Garnet

Alamdine garnet is seen in the sample by the presence of the reflection with the d values of 2.881A°, 2.572A° and 1.540A°.

Conclusion

The qualitative identification of minerals in coastal sediment samples was carried out by FT-IR and XRD techniques. The infrared analyses of sediment samples from different locations of Tuna indicate the presence of quartz, orthoclase, albite; kaolinte, montmorillonite apd calcite. Among the different minerals quartz is present invariably in all the samples. Hence it is considered to be main or major constituent of the sample and feldspar is also present in different compositions *i.e.*, albite and orthoclase. The crystalline nature of the mineral is also attempted using quartz and discussed.

The XRD analyses of the selected samples show presence of quartz, albite, orthoclase, gypsum, calcite, rutile and garnet. Among these minerals quartz showed large number of peaks in the diffractrogram. This indicates that quartz is the most abundant mineral in the sediment samples and is supported by the i.r. studies.

The combined method of FT-IR and XRD showed that it is a powerful technique for mineral analysis in geological samples.

References

Adler, H.H. and Kerr, P.F., 1962. Infrared spectra symmetry and structure relations of some carbonate minerals. *Am. Min.*, 48: 839.

Adler, H.H. and Kerr, P.F., 1963. Infrared study of aragonite and calcite. *Am. Min.*, 47: 700.

Balasubramanian, K., Bheemshna and Gopinath, H., 1984. *Proc. Nat. Sci. Acad.*, 50: 491.

Bottcher, M.E., Gehlken, P.L. and Steele, D.F., 1997. Characterisation of inorganic and biogenic magnesian calcites by Fourier Transform Infrared Spectroscopy. *Solid State Ionics.*, 101: 1379.

Boruah, R.K., Neog, A.K., Borah, P.C., and Baruah, Gp., 1999. Petrographic study of silica bearing rocks from Assam, India. *Intl. J. Eng. Mat. Sci.*, 6: 338.

Benedetto, G.E.D., Laviano, R., Sabbatini, L. and Zambonin, P.G., 2002. Infrared spectroscopy in the mineralogical characterization of ancient pottery. *J. Cul. Her.*, 3: 177.

Chester, R. and Elderfield, H., 1967. The application of infrared absorption spectroscopy to carbonate mineralogy. *Sedimentology*, 9: 5.

Chester, R. and Green, R.N., 1968. The infrared determination of quartz in sediments and sedimentary rocks. *Chem. Geo.*, 3: 199.

Clarence Karr, Jr., 1974. *Infrared and Raman Spectroscopy of Lunar and Terrestrial Minerals.* Academic Press, New York.

Clark, R.N., King, T.V.V., Kiejwa, M., Swayze, G.A. and Verge, N., 1990. High spectral resolution reflectance spectra of minerals. *J. Geophys. Res.*, 95: 12653.(1990).

Coates, J.P., 1977. *The IR Analysis of Quartz and Asbestos.* Nelioth Offset Ltd., Chesham, England.

Crowley, J.K. and Vergo, N., 1988. Near-infrared reflectance of mixtures of kaolin group minerals; use in clay. *Clay and Clay Minerals,* 36: 310.

Fanner, V.C., 1974. *The IR Spectra of Minerals.* Mineralogical Society, London.

Fanner, V.C., 1979. *Infrared Spectroscopy: Data Handbook for Clay Materials and other Non-metallic Minerals,* (Ed.) Van Olphen and Fripit. Pergaman Press, Oxford, London.

Ganguly, J. and Banerjee, B.K., 1985. Appraisal of infrared spectra of isolated clay fractions from different soil profiles. *Trans. Ind. Cer. Soc.,* 26: 44.

Ghosh, S.N., 1978. Infrared spectra of some selected minerals, rocks and products. *J. Mat. Sci.,* 13: 1877.

Herzberg, G., 1945. *Molecular Spectra and Molecular Structure, II: Infrared and Raman Spectra of Polyatomic Molecules.* D. Van Nostrant Co., Inc., New York, p. 178.

Hlavay, J., Jonas, K., Elek, S. and Inczedy, J., 1978. Characterization of the particle size and the crystallinity of certain minerals by infrared spectrophotometry and other instrumental methods. II. Investigation on quartz and feldspar. *Clay and Clay Minerals,* 26: 139.

Hunt, J.M. and Turner, D.S., 1953. *Anal. Chem.,* 25: 1169.

Ko, T.H. and Chu, H., 2005. Spectroscopic study on sorption of hydrogen sulfide by means of red soil. *Spectro Chimica Acta,* Part–A, 61: 2253.

Lopez, M.C.B., Martinez-Alono and Tascon, J.M.D., 2000. Mineral matter characterization in olivne stones by joint use of LTA, XRD, FT-IR and SEM-EDX. *App. Spect.,* 54: 1712.

Lyon, R.I.P., 1967. *Infrared Absorption Spectroscopy, Physical Methods in Determinative Mineralogy,* (Ed.) J. Zussman. Academic Press, New York, p. 371.

Madejova and Komadel, 2001. Baseline studies of the clay minerals ociety source clays: Infrared methods. *Clay and Clay Minerals,* 49: 410.

Neog, A.K., Boruah, R.K., Sahu, A.P., Borah, P.C., Ahmed, W. and Boruah, G.D., 1999. XRD and IR of Deopani clay. *Asian. Chem. Letts.,* 3: 172.

Post, J.L. and Noble P.N., 1993. The near infrared combination band frequencies of dioctahedral hematites, micas and illites. *Clay Minerals,* 41: 639.

Ramasamy, V., Dheenathayalu, M.., Ponnusamy, V., Murgesan, S.and Mullainathan, S., 2003. Characterization of quartz and feldspars in White Granites. *J. Curr. Sci.,* 3(1): 181.

Ramasamy, V., Murgesan, S. and Mullainathan, S., 2004. Characterization of minerals and relative distribution of quartz in Cavery River sediments from Tamil Nadu, India: A FT-IR study. *Bulletin of Pure and Applied Sciences,* 23 F(No.1–2): 1–7.

Ramasamy, V., Murgesan, S. and Mullainathan, S., 2004. Fourier transform infrared analysis of some sediments from Palaru River, Tamil Nadu, India. *J. Curr. Sci.,* 5(2): 599.

Ramasamy, V., Murgesan, S. and Mullainathan, S., 2005. Distribution and characterization of minerals in Cavery river sediments by grain size analysis: A new approach by FT-IR study. *The Indian Mineralogist,* 39(2): 91.

Ravisankar, R., Rajalakshmi, A. and Manikandan, 2006. Mineral characterization of soil samples in and around saltfield area, Kelambakkam, Tamil Nadu, India. *Acta Ciencia Indica,* 32(3): 341.

Reig, F.B. Adelankndo, J.V.G. and Moreno, M.C.M., 2002. FT-IR quantitative analysis of calcium carbonate (calcite) and silica (quartz) mixtures using the constant ratio method: Application to geological samples. *Talanta*, 58: 811.

Russell, J.D., 1987. *Infrared Methods: A Handbook of Determinative Methods in Clay Mineralogy*, (Ed.) M.J. Wilson. Blackie and Son Ltd..

Senthil Kumar, P., Parthasarathy, G., Sharma, D.S., Srinivasan, R. and Krishnamurthy, P., 2001. Mineralogical and geochemical study on carbonate veins of Salem-Attur fault zone, Southern India: Evidence for carbonate affinity. *J. Geo. Soc. Ind.*, 58: 15.

Tuddenham, W.M., and Lyon, R.J.P., 1960. Infrared techniques in the identification and measurements of minerals. *Anal. Chem.*, 32: 1630.

Summer, M.E., 1995. *Hand Book of Soil Science*. University of Georgia, Boca Raton, Hondor Press, New York.

White, J.L., 1971. "Interpretation of infrared spectra of soil minerals. *Soil Science*, 112: 22.

Xu, Z., Comilsen, B.C., Popko, D.C., Penning, W.O., Wood, J.R. and Hwang, J.Y., 2001. Quantitative mineral analysis by FT-IR spectroscopy. *Int. J. Vib. Spec.*, 5: 4.

Yariv, S.H. and Mendelovici, E., 1979. The effects of degree of crystallinity on the infrared spectrum of hematite. *App. Spectroscopy*, 33: 410.

Chapter 37

Seasonal Analysis of Sanganer Handmade Paper and Cardboard Industries Effluent

Shweta Kulshreshtha, Nupur Mathur and Pradeep Bhatnagar
Environmental Molecular Microbiology Unit, Department of Zoology, University of Rajasthan, Jaipur – 302 004

ABSTRACT

Water bodies are constantly used as receptacles for the untreated wastewater or poorly treated effluents accrued from industrial activities, which has rendered many water bodies unsuitable for both primary and/or secondary usage. One such type of case was studied, in which Amani Shah Drainage of Sanganer (near Jaipur) received waste effluent from textile, blue pottery, handmade paper and cardboard industries and discharged it into the Dhund River. In this study seasonal changes in effluent quality parameters (physico-chemical) of handmade pulp and paper and cardboard industries, Amani Shah Drainage and connecting drain were investigated in three different season *i.e.* pre-monsoon, monsoon and post-monsoon to assess the suitability of effluent for discharge into the Amani Shah Drainage. Seasonal study of industrial effluent will be helpful in making strategies for preventing pollution. Different parameters like pH, COD, BOD, salinity, conductivity, TDS and turbidity were analyzed and compared. Variation in the above parameters will be helpful in deciding the dilution required for the treatment of effluent. Seasonal study emphasized that Amani Shah drainage has high pollution load during the monsoon season whereas handmade paper industry, cardboard industry drain and connecting drain has high pollution load during the pre-monsoon and post-monsoon season. It indicates that Amani Shah drainage receives the effluent of other industries, so its characteristics was very different from handmade paper and cardboard industries effluent and need different type of strategy for its treatment.

Keywords: *Handmade paper industry, Cardboard industry, Amani Shah Drainage, Seasonal changes, Physico-chemical characteristics, Organic content.*

Introduction

Pollution of natural waters with waste effluents arising from various industries has become a serious problem in India, as industrial growth and development have been on a very large scale. The State of Rajasthan has seen a phenomenal rise in the industrial sector in the last decade. When improperly handled and disposed, industrial wastes imperil both human health and the environment. Human exposures (occupational and non occupational) to industrial wastes have led to health effects ranging from headaches, nausea, lung, and skin irritations, to serious ailments like congenital malformations (Shinka *et al.*, 1991; Morikawa *et al.*, 1997).

Sanganer located about 16 km south of Jaipur, the state capital, has a population exceeding 2 million. The total area of Sanganer town is about 635.5 km^2. It is famous for handmade paper and cardboard industries, textile industries and blue pottery industries. However all these industries discharge their waste effluent into the Amani Shah Drainage (Gupta *et al.*, 2006).The ecological and toxicological problems resulting from the discharge of wastewaters from these industries into drainage (Amani Shah Ka Nallah) have been among the most important water pollution problems in this area. Usha (1989) reported eczema, contact dermatitis, asthma, chronic bronchitis, tuberculosis, bladder cancer, and irritation of the eyes among the workers of textiles industries in Sanganer. High concentrations of heavy metals such as Cu, Cd, Zn, Pb, and Ni have been reported in this area (Khan, 1995). This problem is further aggravated because the water from the drainage is used directly for irrigation purposes (Mathur *et al.*, 2006).

Paper-mill effluents are characterized by the presence of color and suspended solids, bad smell, high concentration of nutrients that cause eutrophication of receiving waters, and high toxicity overall (Lacorte *et al.*, 2003). Ali and Sreekrishnan (2001) reported that effluents from pulp and paper mills are highly toxic and are a major source of aquatic pollution. Seasonal changes in the physico-chemical characteristics of water bodies (Sachidanandamurthy *et al.*, 2006; Bhatt *et al.*, 1992; Smitha *et al.*, 2007) as well as in the effluent of pulp and paper industries (Malaviya *et al.*, 2007; Tiku *et al.*, 2007, Nanda *et al.*, 1999) have been reported earlier. Suriyanarayanan *et al.* (2005) analyzed the characteristics of wastewater from a small paper mill which is using wastepaper as a raw material. Malaviya *et al.* (2007) and Singh *et al.* (1996) reported significant changes in effluent quality parameters with the seasonal and spatial distribution. All physico-chemical parameters indicated the presence of pollutants in the effluent. Seasonal changes affect the pollution load of the water body because if there is large amount of water in the drainage during the rainy season there may be low pollution load or high pollution load (if water is flowing). During the summer season, there is very less amount of water in the drainage which may increase the pollution load. Assessment of pollution load is necessary for the successfully running of treatment plant and it is also helpful in making strategy for treating industrial waste.

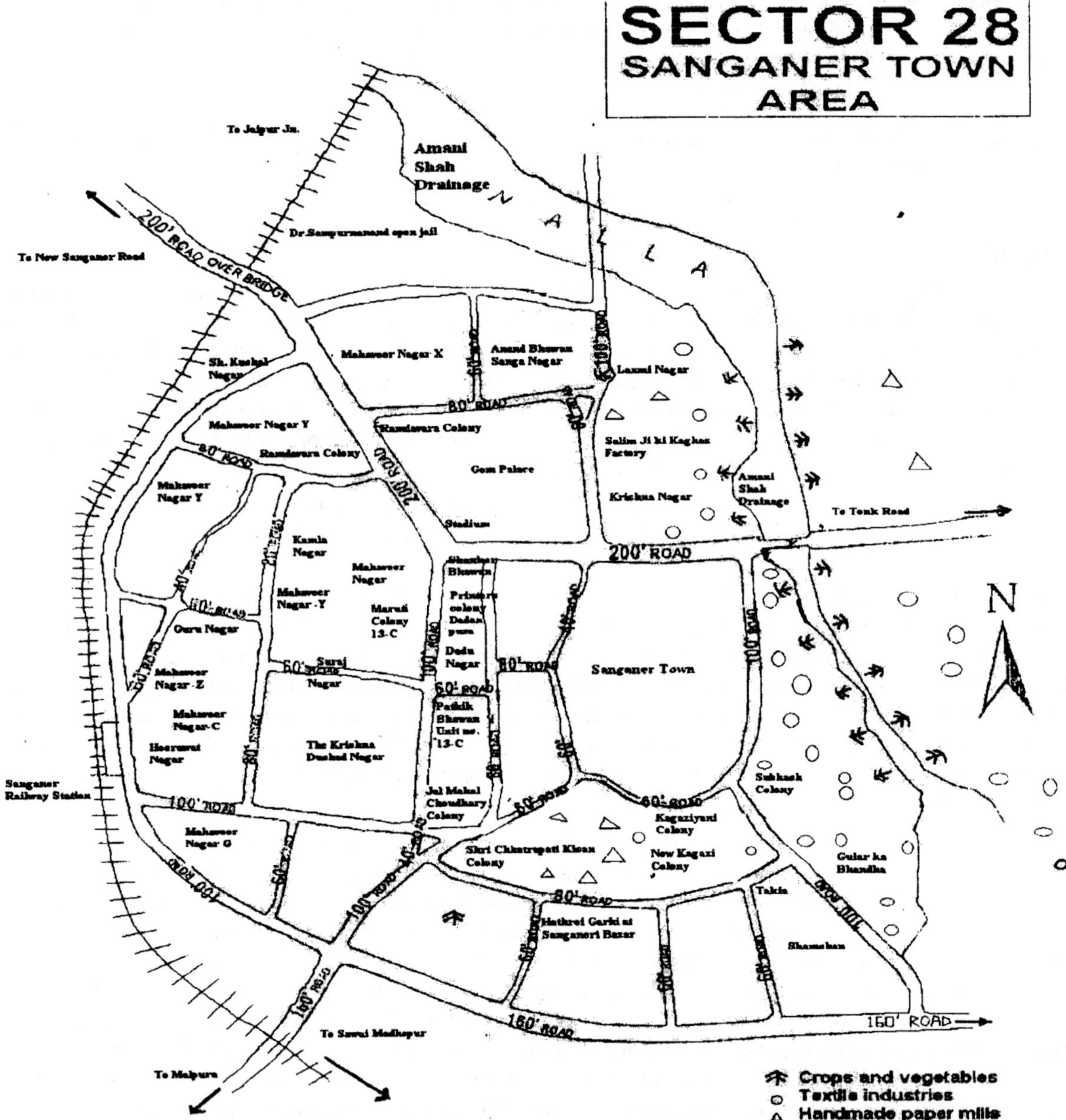

Figure 37.1: Shows the Area of Sanganer.
Handmade paper industries and cardboard industries are located near the Amani Shah Drainage.
It also shows the agricultural fields near the Amani Shah Drainage.

Although a lot of study has been done on physico-chemical study of various water bodies and pulp and paper industries, study on seasonal changes in physico-chemical characteristics of handmade pulp and paper and cardboard industries and their effect on Amani Shah Drainage of Sanganer, remains untouched because handmade paper and cardboard industries of Sanganer were thought to be non-polluting industries. They utilize waste hosiery cuttings, cotton rags and waste paper respectively which need very less amount of chemicals for processing due to high content of cellulose (Saakshi, 2005). Usage of chemical in these industries was limited to some steps of making paper such as sizing, dying and making of some special type of paper.

However, large amount of water is required in all steps of making paper except drying resulting in large amounts of effluents containing chemical, dyes and organic pulping residues. Assessment of seasonal changes in physicochemical characteristics of industrial effluent can be helpful in the application of pollution abatement techniques. Therefore in this study effluent quality parameters of handmade pulp and paper and cardboard industries, Amani Shah Drainage and connecting drain (connects all handmade paper and cardboard industries to the drainage) were investigated in three different season *i.e.* pre-monsoon, monsoon, post-monsoon within a period of two year to assess the suitability of effluent for discharge into the Amani Shah Drainage.

Materials and Methods

Site Selection

Two water samples were collected from six different locations in three different seasons:

1. Handmade paper industry (1) Samples were taken from 2 different small scale paper industry located one km away from the drainage.
2. Handmade paper industry (2) Samples were taken from 2 different large scale paper industry located half km away from the drainage.
3. Cardboard industry: It is also located in the area of Kagzi colony which is one km away from the drainage.
4. Cardboard drain: It is a drain which comes out from the industry. Sampling was done at 100 m away from the industry.
5. Connecting drain: This drain connected the above industries to the drainage. All discharges from all industry were collected by this drain and discharged into the drainage. Two samples collected from this drain 100 m and 200 m away from the discharging point.
6. Amani Shah Drainage: This drainage gets effluent from all the textile, blue pottery, paper and cardboard industries. Sampling was done from the point where connecting drain discharges effluent from handmade paper and cardboard industries effluent.

Collection of Samples

Samples were collected in different seasons *i.e.* pre-monsoon, monsoon and post-monsoon for a period of two year *i.e.* December 2005–November 2007. Five samples were collected during each season. The samples were stored in clean, sterile, screw capped glass containers, at 4°C.

Analysis of Samples

Samples were analyzed immediately for physico-chemical parameters by Systronic water analyzer kit model no. 371. This kit has electrodes of pH, temperature, salinity, conductivity, total dissolved solids (TDS). All these electrodes were maintained as given in manufacturer's instruction manual. All electrodes were calibrated with the standard solution before use. For calibrating pH standard buffer solution of pH 4 and pH 7 were used. Electrode of salinity was calibrated by 10ppt NaCl solution whereas conductivity and TDS electrode were calibrated by using 0.01 N KCl solutions. For measuring turbidity kit was provided with cuvettes. 100 NTU solutions were used as standard for turbidity. Samples were also analyzed for organic content such as biological oxygen demand (BOD) and chemical oxygen demand (COD) as given in the APHA manual. For BOD analysis, membrane electrode method was used because samples were highly colored.

Data Analysis

Data was analyzed statistically by calculating mean and standard deviation. In addition, the data were analyzed to evaluate the level of compliance with the effluent quality guideline for discharge into surface water (Table 36.3).

Result and Discussion

Effluents from the industries have considerable impact on water quality, primarily because of the large volume of wastewater they generate and the physical and chemical properties of their effluents. These effluents, if directly discharged into nearby water bodies, can cause health hazards and disturb the ecological balance. The wastes released from such industries cause soil, surface and groundwater pollution, besides causing a number of adverse effects on agricultural products, animals and health of people living in that area. The situation further aggravates as seasonal vegetables are grown in the surrounding area using this drainage water. Khan *et al.* (1995) reported effect of these textile industry effluents on physico-chemical characteristics of the water from Amani Shah Nallah drainage. Analyses of Sanganer tie and dye industry effluents also revealed heavy metal contamination (Ni, Zn, Pb, Cd, Fe), quite above permissible limits.

The picture becomes more alarming as these waters are directly being used for agriculture, in the land adjoining the drainage. In fact, during summer months, when very little amount of water is present in Amani Shah Ka Nallah drainage, the entire drainage area is converted into agricultural field. Toxic effluents from the Sanganer textile dyeing and printing industries being discharged directly, without any treatment, into Amani Shah Nallah drainage have been reported to contain highly mutagenic compounds. These compounds have also contaminated the surface and even undergroundwater, thereby, making it unfit for irrigation and drinking (Mathur *et al.*, 2005).

The seasonal change in physico-chemical characteristics of effluent of handmade paper and cardboard industries, connecting drain and Amani Shah Drainage of Sanganer showed considerable variations. Temperature of effluent affects physical chemical and biological processes in water bodies and therefore, the concentration of many variables (Chapman and Kimstach, 1992). Temperature values of the effluent samples showed that it depends on the season of collection of sample. Both the handmade paper industries showed similar minimum and maximum temperatures. A similar trend was obtained in the case of Cardboard industry 1 (minimum: 28.4°C, maximum: 35°C), Cardboard industry 2 (minimum: 28.7 °C, maximum: 33.2 °C), connecting drain (minimum: 28.45°C, maximum: 34°C) and Amani Shah Drainage (minimum: 27.2 °C, maximum: 33.9 °C). However, the temperatures were always higher in pre-monsoon season. Comparative analysis of the mean temperature of all effluent emphasized on the fact that handmade paper and cardboard industries and connecting drain discharged effluents of higher temperature than the Amani Shah Drainage. However, this did not increase the overall temperature of Drainage. Therefore temperature was found in the range describe by general standard for discharge of environmental pollutants GSR 801(E). Variation in temperature according to the season also reflects the fact that these industries do not use any process which generates heat.

pH is an important variable in water quality assessment as it influences many biological and chemical processes within a water body and all processes associated with water supply and treatment. In the case of handmade paper industry mean pH value was found to be 4.04, 4.27 and 4.27 during pre-monsoon, monsoon and post-monsoon season respectively. These low pH values indicate that some acidic substances are being used by these industries which comes out with the effluent. Only the pH of effluent of handmade paper industry 2 was found slightly basic (7.25) in the monsoon season, acidic (5.88) in pre-monsoon season, and highly acidic (4.9) in post-monsoon season (Table 37.1). These results are found in good agreement with Khan *et al.* (2006) who describe the use of rosin and alum by these industries which is probably contributing to the inherent acidity. Although pH of effluent of cardboard industries was found slightly acidic but it falls in the range as described by general standard for discharge of environmental pollutants GSR 801 (E) and Minimal National Standard for pulp and paper industry except in two cases effluent pH 6.45 (post-monsoon) and 6.39 (pre-monsoon) of cardboard industry 1 and 2 respectively. This revealed the use of waste paper in making cardboard because due to using of acidic compounds in making paper, papers are usually acidic in nature and thereby transfer this acidity to water during the pulp making process. Effluent of connecting drain was again found to be acidic in all seasons because this drain collected effluent from the handmade paper and cardboard industries and discharged it into the Amani Shah Drainage. A slightly different scenario was found in the effluent of Amani Shah Drainage. pH of this was found basic to slightly acidic in three seasons and lies in the range of both standards. Reason behind this is possibly discharge of alkaline effluents from some other industries in the drainage which neutralized the effect of acidic effluent of handmade paper and cardboard industries. During the pre- and post-monsoon season all industrial effluent was found to be highly acidic which is neutralized during the monsoon season when mixed with other industrial effluent.

Mean value of salinity of effluents were found higher in pre-monsoon and post-monsoon seasons than that of monsoon season except in the case of Amani Shah Drainage (Table 37.1). Salinity of Amani Shah Drainage was found higher (1.62 ppt) than that of pre-monsoon (1.60 ppt) and post-monsoon season (1.34 ppt). The reason for this alteration is probably the flow off water during the monsoon and mixing of effluent from other industries.

Conductivity of any effluent reveals the presence of compounds in it which can dissociate into ions. The compounds which cannot dissociate into ions cannot conduct electric current (Chapman and Kimstach, 1992). Such observations were found in the case of cardboard industry 1 during post-monsoon season where observed TDS value (1.76 ppt mS/cm) was high but conductivity of the sample was low. Amani Shah Drainage showed highest conductivity (13.35 mS/cm) in the monsoon season due to the dilution of effluent sample in the water which increase the mobility of ions and due to the mixing of discharges from other industries which increase the pollution load. Conductivity values were found to be very high when compared with the results of Malaviya *et al.* (2007) for pulp and paper industry. Conductivity of the effluent of Handmade paper industry 1, cardboard industry 1, cardboard industry 2 and connecting drain were found to be very high *i.e.* 7.02 mS/cm, 6.35 mS/cm, 4.8 mS/cm, 5.89 mS/cm respectively in pre-monsoon season. Lowest conductivity values were found in handmade paper industry 1 (3.41 mS/cm), handmade paper industry 2 (2.95 mS/cm) and cardboard industry 1 during the post monsoon season but conductivity was found to be lowest in monsoon season in effluent of cardboard industry 2 and connecting drain (Table 37.1). In the case of connecting drain and Amani Shah Drainage much lower values were observed due to the presence of excessive water and flowing of water with high speed. No values were observed above the limit during the monsoon season.

Table 37.1: Physico-chemical Characteristics (mean±standard deviation) of the Effluent of Handmade Paper Industries, Cardboard Industries and its Drain, Amani Shah Drainage and Connecting Drain which were Collected during December 2005–November 2007

Sl.No.	*Sample*	*Season*	*Temp. (°C)*	*pH*	*Salinity (ppt)*	*Conductivity (mS)*	*T.D.S. (ppt)*	*Turbidiy (NTU)*
1.	Handmade paper	Pre-monsoon	35.1±1.7	4.04±0.015	2.92±0.425	7.02±2.085	2.54±0.34	38–77
	industry 1	Monsoon	30.1±2.3	4.27±0.085	1.68±0.735	3.69±0.265	1.87±0.24	41–5.9
		Post-monsoon	28.2±1.1	4.27±0.015	2.08±0.29	3.41±0.855	1.70±0.50	12–49
2.	Handmade paper	Pre-monsoon	35±1.2	5.88±1.58	2.59±1.41	3.26±0.95	1.625±0.68	69–24
	industry 2	Monsoon	29.5±2.0	7.25±1.0	1.17±0.35	3.07±1.02	1.46±1.01	15–55
		Post-monsoon	28.2±1.3	4.9±0.87	1.60±0.51	2.95±0.875	1.425±0.35	120–59
3.	Cardboard industry 1	Pre-monsoon	35.0±0.95	6.74±0.66	2.29±0.01	6.35±0.09	2.39±0.41	0.38–7.6
		Monsoon	29.8±1.9	6.51±1.02	1.96±0.22	3.75±0.09	1.76±1.01	7.9–21
		Post-monsoon	28.4±1.6	6.45±0.135	2.25±0.39	2.96±0.315	1.45±0.22	16–18
4.	Cardboard industry 2	Pre-monsoon	33.2±1.2	6.39±1.01	2.25±1.34	4.80±0.125	2.09±0.06	1.1– 0.06
		Monsoon	29.8±1.7	7.15±0.02	1.69±0.02	2.495±0.345	1.27±0.26	1.4–98
		Post-monsoon	28.7±1.7	6.64±0.215	2.83±1.11	4.51±4.02	2.24±1.06	0.33–32
5.	Amani Shah Drainage	Pre-monsoon	33.9±2.1	8.23±1.125	1.60±0.235	4.20±0.02	1.39±0.21	22–20
		Monsoon	27.2±2.6	7.82±0.515	1.62±0.515	13.35±11.55	1.08±0.23	55–80
		Post-monsoon	28.4±1.4	6.54±0.01	1.34±0.37	2.985±0.125	1.465±0.02	12–22
6.	Connecting Drain	Pre-monsoon	34±2.1	6.85±0.52	2.19±0.125	5.89±0.985	1.91±0.02	43–55
		Monsoon	29.5±2.0	6.76±0.215	1.23±0.725	2.38±1.025	1.11±1.0	11 0–40
		Post-monsoon	28.45±1.6	6.32±0.34	1.69±0.365	3.21±0.46	1.56±0.125	43–119

Mean value of total dissolved solids (TDS) of handmade paper industry 2 were found below the BIS limit for industrial effluent discharge into inland surface waters in all seasons but it was found much above the limit in the effluent of handmade paper industry 1 (2.54 ppt) during the pre-monsoon season (Table 37.1). This may be due to the little amount of water in its drain because evaporation of water is high in pre-monsoon season. TDS values of small scale and large scale handmade paper industries also reflects the fact that small scale industries discharged more solid content in the effluent than that of large scale industries. Total dissolved solids were found to be high and above BIS limit in the effluent of cardboard industries 1 (2.39 ppt during pre-monsoon season) and cardboard industry 2 (2.24 ppt during post-monsoon season).

All effluent samples collected from handmade paper industry 1 and 2 showed much greater variation in turbidity in three different seasons than that of cardboard industry 1 and 2. The reason of this variation is the presence of colored organic pulping residues in the effluent which interferes with the light path. When effluent of connecting drain and Amani Shah Drainage were analyzed, high turbidity (40–110 and 55–80 respectively) was found during the monsoon season due to the mixing of other industrial effluent and fluctuations in the water body due to which setteled particles also float in the water and increase the turbidity (Table 37.1).

Table 37.2: Organic Content (Mean±Standard deviation) of the Effluent of Handmade Paper Industries, Cardboard Industries and its Drain, Amani Shah Drainage and Connecting Drain which were Collected During December 2005–November 2007

Sl.No.	*Sample*	*Season*	*BOD*	*COD*
1.	Handmade paper industry 1	Pre-monsoon	27.7±1.79	6406±876
		Monsoon	35.6±1.00	6988±301
		Post-monsoon	28.8±8.4	5396±532
2.	Handmade paper industry 2	Pre-monsoon	31.2±9.94	6676±333.6
		Monsoon	21.8±7.2	3785±140.8
		post-monsoon	20.4±1.2	4294±646
3.	Cardboard industry 1	Pre-monsoon	13.3±4.90	5174±293.5
		Monsoon	13.8±5.12	2133±197
		post-monsoon	29.2±2.2	2334±510
4.	Cardboard industry 2	Pre-monsoon	25.8±10.2	4994±236.6
		Monsoon	13.2±4.19	5241±484
		Post-monsoon	19.2±7.77	3040±225.6
5.	Amani Shah Drainage	Pre-monsoon	14.8±0.23	4076±270.1
		Monsoon	16.2±2.2	4224±510
		Post-monsoon	13.8±1.37	3040±158.1
6.	Connecting Drain	Pre-monsoon	12.8±1.2	5532±259.6
		Monsoon	16.2±1.25	6406±646
		Post-monsoon	11.8±0.23	7904±140.8

In the case of handmade paper industry's effluent BOD values decreases in three consequent season but COD of the effluent was found to be higher in post-monsoon season than that of monsoon season. General trend was observed in all effluents according to which BOD values are lower than the COD values. Similarly BOD of connecting drain was found to be lower than that of COD in post-monsoon season. On comparing BOD values with MINAS (Minimum National Standards, CPCB, 2000) for pulp and paper industry and General Standards for discharge of environmental pollutants, it was found that BOD values are in the range of discharge standards (Table 37.2). However, COD values were found much higher than the above standards (Table 37.2). High values of COD and low values of BOD reflect the reduced activity of indigenous micro flora present in the effluent during the post monsoon season, which are responsible for natural bioremediation.

Table 37.3: Standards for Discharge of Industrial Effluent into the Water Bodies Prescribed by Various Government Bodies

Sl.No.	*Parameter*	*Discharge Limits*	*Standards Prescribed by Various Bodies*
1.	pH	6.5 to 8.5	Minimum National Standards, CPCB, 2000; Indian Standard Drinking Water-Specification BIS 10500: 1991
2.	BOD (mg/L)	30	Minimum National Standards, CPCB, 2000; Part-A of General Standards for discharge of Environmental Pollutants, GSR 801 (E) EPA 1993, prescribed by CPCB;
3.	COD (mg/L)	350	Minimum National Standards, CPCB, 2000
4.	Temperature	Shall not exceed 5°C above the receiving water	Part-A of General Standards for discharge of Environmental Pollutants, GSR 801 (E) EPA 1993, prescribed by CPCB
5.	Colour, (Hazen units, Max)	25	Indian Standard Drinking Water-Specification BIS 10500 : 1991
6.	Odour	Unobjectionable	Indian Standard Drinking Water-Specification BIS 10500 : 1991
7.	Turbidity (NTU, Max)	10	Indian Standard Drinking Water-Specification BIS 10500 : 1991
8.	EC (S/cm)	2500	WHO, 1984
9	TDS (mg/L)	2000	Indian Standard Drinking Water-Specification BIS 10500 : 1991

The results of this study thus indicate that pH of all effluent was found to be acidic in all season. Industrial effluents as well as connecting drain effluent possess high salinity, TDS and COD during the pre-monsoon and post-monsoon season whereas conductivity was found to be high in monsoon season. BOD of all effluent was found to be in the range of discharged standards. A different type of trend was observed in Amani Shah drainage due to the mixing of effluent with the other industrial effluents and seasonal fluctuation which is responsible for mixing of deposited matter in the drainage. As a result of this pH was found to be slightly basic in the monsoon season. Also high salinity, TDS and COD and low conductivity was found in monsoon season.

All this data will be helpful in making strategies for the treatment of effluent during the different seasons. This analysis also emphasized on the fact that effluent of Amani Shah Drainage requires different type of treatment due to the mixing of effluent of other industries but connecting drain possess very similar characteristics to the handmade paper and cardboard industries effluent so it can be used to collect the effluent and transfer it to the effluent treatment plant.

References

Ali, M. and T.R., 2001. Aquatic toxicity from pulp and paper mill effluents: A review. *Adv. Environ. Res.*, 5: 175–196.

APHA (American Public Health Association, Washington DC), 1995. Standard methods for the examination of water and wastewater 14th edn. Part-A of General Standards for discharge of Environmental Pollutants, GSR 801 (E), EPA, 1993.

Bhatt, S.D. and Pathak, J.K., 1992. Assessment of water quality and aspects of pollution in a stretch of river Gosti (Kumaun: Lesser Himalaya). *J. Environ. Biol.*, 13(2): 113–120.

Chapman, D. and Kimstach, V., 1992. The selection of water quality variable. In: *Water Quality Assessments*, (Ed.) D. Chapman. Chapman and Hall Ltd., London pp. 51–119.

Gupta S., Kumar, A. and Seth, G., 2006. Seasonal variation of some heavy metals in surface water of Sanganer, Jaipur. *Indian J. Environ. Sci.*, 10(1): 75–77.

Khan, T.I., Kaur, N. and Vyas, P.C., 1995. Effects of industrial effluents on physico-chemical characteristics of Amani Shah Ka Nallah: A case study. *J. Environ. Pollut.*, 2(3): 147–150.

Lacorte, S., Latorre, A., Barceloa, D., Malmqvist, R. and Welander, T., 2003. Organic compounds in paper-mill process waters and effluents. *TrAC Trends Ana. Chem.*, 22(10): 725–737.

Malaviya, P. and Rathore, V.S., 2007. Seasonal variations in different physico-chemical parameters of the effluents of Century pulp and paper mill, Lal Kuan, Uttarakhand. *J. Environ. Biol.*, 28(2): 219–224.

Mathur, N., Bhatnagar, P., Nagar, P. and Bijamia, M.K., 2005. Mutagenicity assessment of effluents from textile/dye industries of Sanganer, Jaipur (India): A case study. *Ecotoxicol. Environ. Saf.*, 61(1): 105–113 IF: 1.2.

Mathur N., Bhatnagar, P. and Verma, H., 2006. Genotoxicity of vegetables irrigated by industrial wastewater. *J. Env. Sci.*, 18(5): 964–968.

Minimum National Standards (MINAS) 2000 for large pulp and paper mills with chemical recovery. CPCB.

Morikawa, Y., Shiomi, K., Ishihara, Y. and Matsuura, N., 1997. Triple primary cancers involving kidney, urinary bladder and liver in a dye workers. *Am. J. Ind. Med.*, 31: 44–49.

Nanda, P., Sudarshan, P. and Behera, K.M., 1999. Physico-chemical and microbial analysis of the Orient Paper Mill effluent. *Env. Eco.*, 17(4): 975–977.

Tiku, D.K., Kumar, A., Sawhney, S., Singh, V.P. and Kumar, R., 2007. Effectiveness of treatment technologies for wastewater pollution generated by Indian pulp mills. *Environ. Monit. Assess.*, 132: 1–3.

Sachidanandamurthy, K.L. and Yajurvedi, H.N., 2006. A study on physico-chemical parameters of an aquaculture body in Mysore city, Karnataka, India. *J. Environ Biol.*, 27(4): 615–618.

Saakshi, 2005 .Quality control in handmade paper industry. *Hath Kagaz*, 12(7–9): 6–14.

Shinka, T., Sawada, Y., Morimoto, S., Fujinaga, T., Nakamura, J. and Ohkana, T., 1991. Clinical study on urothelial tumors of dye workers in Wakayama City. *J. Urol.*, 146: 1504–1507.

Singh, R.S., Marwaha, S.S. and Khanna, P.K., 1996. Characteristics of pulp and paper mill effluent. *J. Indian Poll. Control*, 12(2): 163–172.

Smith, A.K. and Suthers, L.M., 1999. Effects of sewage effluent discharge.

Smitha, P.G., Byrappa, K. and Ramaswamy, S.N., 2007. Physico-chemical characteristics of water samples of Bantwal Taluk, South Western Karnataka, India. *J. Environ. Biol.,* 28(3): 591–595.

Suriyanarayanan, S., Jayakumar, D. and Balasubramanian, S., 2005. Physico-chemical characteristics of paper industry effluents: A case study. *J. Environ. Sci. Eng.,* 47(2): 155–160.

Usha, M., 1989. Impact analysis of industries in Sanganer. P.G. Diploma Field Study Report, Indira Gandhi Center for HEEPS, University of Rajasthan, Jaipur, India.

Chapter 38

Water Quality Assessment of Ayroor River (Kerala) with Special Reference to Zooplankton Diversity

C. Shaji[1], S. Suvidha[1] and L. Bindu*[2]
[1]Department of Botany, Sree Narayana College, Kollam – 691 001, Kerala
[2]Zoological Survey of India, M-Block, New Alipore, Kolkata – 700 053

ABSTRACT

The study was carried out in Ayroor river (Kerala) to assess the water quality for a period of six months from December, 2006 to May, 2007. Physico-chemical parameters of water and zooplankton diversity were estimated. High values of BOD (10.2–15.6 mg/l), COD (160–230 mg/l), TDS (200–1400 mg/l), free CO_2 (22–88 mg/l), nitrate (1.65–13.99 mg/l) and MPN coliform (5000/100 ml) coupled with low dissolved oxygen content (2–4.2 mg/l), low diversity index (1.170 –1.563) and presence of indicator species confirm high organic pollution in the river.

Keywords: *Water quality parameters, Zooplankton diversity, Ayroor river.*

Introduction

India is endowed with several hundreds of natural water bodies. Due to rapid industrialization, urbanization and modern agricultural activities majority of these water bodies are deteriorating, causing environmental hazards. Water quality evaluation has attained wide recognition in recent

* E-mail: bindulajapathi@yahoo.com

years due to increased degradation of the aquatic environment. Several workers have documented the water quality profile of the major river basins and estuarine systems of India (Datta *et al.*, 1988; Shaji and Patel, 1991, Unni *et al.*, 1992, Rajaram *et al.*, 2005; Bhadra *et al.*, 2005). The biological approach in monitoring environmental quality has become an accepted fact. Zooplankton has been used frequently to assess the water quality (Roy, 1982; Sinha, 2001; Shamla, 2001; Patil and Karikal, 2001; Saikia and Das, 2003). However in Kerala little attention has been paid till date to study the water quality based on the diversity of zooplankton. Ayoor river of Trivandrum district flows 30km from the beginning point to the end in Edava kayal. Quite a large number of people are using the river water for washing clothes, bathing and cattle wanting. The river receives household and municipal wastes, human and animal excreta and agricultural wastes. An effort has been made to find out the magnitude of pollution in Ayroor river based on physico-chemical characteristics and zooplankton diversity.

Materials and Methods

The present study was carried out in Ayroor river between Villikadavu and Ayroor bridge (located between 8°.46–8°.47′ N latitude and 76°.38–76°.44′ E longitude). Two sampling stations were selected for monthly collections of water and zooplankton samples during the period from December, 2006 to May, 2007. Sampling station-1 located at Villikadavu (upstream) and station-2 located at Ayroor bridge (down stream).

The water samples were collected from both the stations before 10.30 A.M. and analyzed for 15 parameters namely water temperature, pH, phenolphthalein alkalinity, total alkalinity, free CO_2, dissolved oxygen, BOD, COD, nitrate, phosphate, sulphate, TDS, chloride, total hardness and calcium following standard methods (APHA, 1995).

Coliform count was made only once *i.e.* from the samples collected in May, 2007 using multiple tube test (Mackie and Mc cartney, 1996).

Plankton samples were collected with the help of plankton net. The samples were preserved in 5 per cent formalin and analyzed as per method of Vollenweider (1969). Zooplankton have been identified with the help of standard publications (Needham and Needham, 1972; Corliss, 1979; Michael and Sharma, 1988 and Edmondson, 1992).

Shannon's diversity index (Shannon and Weaver, 1949) was calculated by the formula:

$$\overline{D} = -\sum pi \log_e pi$$

where,

$\overline{D}$: Represents Shannon's diversity index.

pi is the proportion of the species 'i' in the total *i.e.* number of individuals of species 'i' divided by the total number of individuals of all species in the sample.

Results and Discussion

The results of physico-chemical analysis of Ayroor river water are summarized in Table 38.1. The values of physico-chemical parameters did not show much difference between the two sampling stations. Hence, the average of the two stations are considered in the text. Water temperature varied from 29–32.4°C with an average of 31.1°C. The minimum temperature was noticed in December, 2006 and maximum in April, 2007. Water temperature has a close relation to the variations of atmospheric temperature. It has profound influence on the physico-chemical and biological characteristics of ecosystem. It also regulates species composition and multiplication of aquatic biota (Shaji and Patel, 1994).

Table 38.1: The Range and Average Values of Physico-chemical Parameters of the River Water During December, 2006 to May, 2007

Sl.No.	Parameters*	Range	Average
1.	Temperature (ºC)	29.0–32.4	31.1
2.	pH	5.75–7.47	6.5
3.	Phenolphthalein alkalinity	0–0	0
4.	Total alkalinity	20.0–140	57.4
5.	Free CO_2	22.0–88.0	45.08
6.	Dissolved oxygen	2.0–4.2	3.1
7.	BOD	10.2–15.6	13.41
8.	COD	160.0–230.0	172.8
9.	Nitrate	1.65–13.99	6.25
10.	Phosphate	0.08–0.46	0.34
11.	Sulphate	14.2–33.3	22.8
12.	TDS	200.0–1400	883.2
13.	Chloride	31.24–102.24	80.6
14.	Total hardness	12.0–124	64.0
15.	Calcium	9.61–48.9	23.46

* All parameters are expressed in mg/l except temperature and pH.

The pH of the water ranged from 5.75 and 7.47 with an average value of 6.5. The minimum was in January, 2007 and maximum in December, 2006. Wetzel (1972) stated that pH of Indian waters ranges from 8 to 9. Close monitoring of pH values for acidic and alkaline concentration helps to identify zones of pollution and other quality conditions for use of water (Clark *et al.*, 1977). According to Klein (1957) pH between 6.7 and 8.4 are suitable, while pH below 6.7 and above 8.5 are detrimental. The average value for the river (6.5) was below 6.7, indicating poor water quality. Total alkalinity varied from 20 to 140 mg/l with an average value of 57.4 mg/l. The lower value was recorded in February and higher in March. Alkalinity itself is not harmful to human beings, still the waters with less than 120 mg/l are desirable for domestic uses. Phenolphthalein alkalinity was totally absent in the river water. Free CO_2 was present in comparatively higher concentration. Its concentration varied from 22–88 mg/l with an average of 45.08 mg/l. The higher concentration of free CO_2 is attributed to the presence of decomposable organic matters as suggested by Unni (1972).

The dissolved oxygen (DO) concentration fluctuated between 2.0 and 4.2 mg/l with an average value of 3.1 mg/l, being minimum in April and maximum in March. Since DO is an index of physical and chemical processes going on in the water, the presence of oxygen demanding pollutants like organic wastes cause rapid depletion of DO from water (Jameel, 1998). Its depletion in the river water is the most critical manifestation of pollution.

Biological oxygen demand (BOD) values ranged from 10.2 to 15.6 mg/l with an average of 13.41 mg/l being maximum in May, indicating contamination by organic wastes (Gupta and Sharma,

2002). The river water has BOD values much higher than the permissible limit given by WHO (1992). The chemical oxygen demand (COD) values fluctuated between 160 and 230 mg/l with an average of 172.8 mg/l. The higher values of COD also indicate the presence of oxidizable organic matters (Garg, 1998).

Nitrate content of the river water ranged from 1.65 to 13.99 mg/l with an average of 6.25 mg/l. Palmer (1980) considered the nitrate value 0.3 mg/l and above as eutrophic nature of the water. The observed values are greater than 0.3 mg/l, indicating eutrophication in the river. Concentration of phosphate was very less and it fluctuated between 0.08 and 0.46 mg/l with an average of 0.34 mg/l. The low values of phosphate can be attributed to anabolic uptake by algae, as suggested by Curtis and Mara (1994). Sulphate varied from 14.2 to 33.3 mg/l with an average of 22.8 mg/l. The higher value indicates pollution (Shaji and Patel, 1991).

Total dissolved solids (TDS) ranged from 200 to 1400 mg/l with an average value of 883.2 mg/l. The average value is above the maximum desirable limit (WHO, 1992) and it may produce distress in aquatic biota. Chloride values varied from 31.24 to 102.24 mg/l with an average of 80.6 mg/l. High concentration of chloride reflects the presence of high amount of organic pollutants (Shanti *et al.*, 2002). Chloride content also increases with the degree of eutrophication (Sinha, 1986; Shaji and Patel, 1991). Total hardness of water varied from 12 to 124 mg/l being high during March and low during January. The average value is 64 mg/l. According to Kannan (1991) water with hardness value more than 180 mg/l is very hard. Hardness is governed by the content of calcium and magnesium, of which calcium is the most abundant substance of the natural water. The values of calcium fluctuated between 9.51 and 48.9 mg/l with an average of 23.46 mg/l. The higher concentration of Calcium can be sourced to sewage and agricultural wastes.

Maximum permissible limit of coliform in drinking water is 1–9/100 ml of sample. The observed MPN coliform is 5000/100 ml. This value indicates high level of bacterial contamination in the river.

The zooplankton was represented by 20 genera belonging to four groups *viz.* Cladocera (4 genera), Rotifera (7), Protozoa (6) and Copepoda (3). All the four groups were always present, but relative contribution to the total plankton by each group had changed dramatically. Rotifera was encounted as the most dominant group in its population. This group was mainly represented by some more or less constant species which in decreasing order of their abundance were *Anuraeopis* sp., *Brachionus calyciflorus, Trichocerea* sp., *Tesdinella* sp., *Filinia* sp., *Polyarthra* sp. and *Keratcila* sp. Protozoa was found to be the second dominant group. It comprised of 6 genera, among them *Paramecium caudatum* was found as the most abundant species followed by *Halteria grandinella, Tetrahymena pyriformis, Diflugia* sp., *Ceratium* sp. and *Arcella* sp. Four genera namely *Diaphanosona, Daphnia, Moina* and *Sinocephalus* constituted the group Cladocera. *Diaphanosona* and *Moina* were the frequent genera among them. Copepoda was represented by species of *Calanus, Heliodiaptomus* and *Diaptomus.* Although presented though out the study, this group never formed a conspicuous feature of the zooplankton. Khan and Rao (1981) identified *Paramecium caudatum, Halteria grandinella, Brachionus calyciflorus, Keratcila tropica* and *Filinia terminalis* as indicators of pollution.

The estimated DI values are given in Table 38.2. Wilhm and Dorris (1968) attempted to relate species diversity index of Shannon and Weaver (1949) to water quality and hypothesized that value greater than 3.0 indicates clean condition, 1.0 indicates severe pollution, values in between indicate moderate pollution. The observed values ranged from 1.170 to 1.563 with an average of 1.352. Considering the range of DI values it could be concluded that the river is moderately polluted.

Table 38.2: Shannon's Diversity Index of Zooplankton in the River from December 2006 to May, 2007

Dec	Jan	Feb	Mar	Apr	May	Average Value
1.364	1.563	1.442	1.307	1.170	1.268	1.352

High values of BOD, COD, TDS, free CO_2, nitrate and MPN coliform coupled with low DO content, low DI and presence of indicator species confirm high organic pollution in the river. Use of water for washing clothes bathing and drinking by cattle may cause health hazards to nearby inhabitants. It is therefore, advisable that authorities should take appropriate steps to make the river clean and free from contamination.

References

APHA, 1995. *Standard Methods for the Examination of Water and Wastewater,* 19th edn. American Public Health Association, New York.

Bhadra, B., Chakraborthy, B., Das, S. and Nanda, A.K., 2005. Investigation of some basic water quality parameters of the North Bengal Terai river, Kalyani, a tributary river Torsa and comparison there of with the main stream. *J. Environ. Biol.,* 26: 277–286.

Clark, JW., Viessman, W. and Hammer, M.J., 1977. *Water Supply and Pollution Control.* Dun Donnelly Publ., New York, pp. 257.

Corliss, J.O., 1979. *The Ciliated Protozoa: Characterization, Classification and Guide to the Literature,* 2nd edn. Pergamon Press, New York, pp 454.

Curtis, T.P. and Mara, D.D., 1994. The effect of sunlight on mechanisms for the die-off of faecal coliform bacteria in waste stabilization ponds. *Research Monograph* No. 1, (Ed.) D.D. Mara. University of Leeds, England.

Datta, N.C., Bandopadhyay, B.K., Majumder, A. and Abuja, D., 1988. Hydrobiological profile of Hoogly Sector (Bally to Bandel) of river Ganga. In: *Ecology and Pollution of Indian Rivers,* (Ed.) R.K. Trivedy. Ashish Publishing House, New Delhi, India, pp. 113–129.

Edmondson, W.T., 1992. *Freshwater Biology.* International Books and Periodical Supply Service, New Delhi.

Garg, S.K., 1988. *Sewage Disposal and Air Pollution Engineering, Vol 2: Environmental Engineering,* 11th edn. Khanna Publ., pp. 188–189.

Gupta, T.R.C. and Sharma, A., 2002. Impact of sewage on the hydrobiology of Tungabhadra river at Harihar, Karnataka. In: *Ecology of Polluted Waters, Vol. 2,* (Ed.) Arvind Kumar. A.P.H. Publ. Corp., New Delhi, pp. 791–807.

Jameel, A., 1998. Physico-chemical studies in Vyakondan Channel water of Cauvery. *J. Poll. Res.,* 17(2): 111–114.

Kannan, K., 1991. *Fundamentals of Environmental Pollution.* S. Chand and Co. Ltd. Publ., New Delhi.

Khan, M.A. and Rao, I.S., 1981. Zooplankton in the evaluation of pollution. In: *Biological Indicators and Indices of Environmental Pollution.* Proc. Workshop held at Hyderabad, CBPCWP, New Delhi.

Klein, L., 1957. *Aspects of River Pollution.* Butterworth Scientific Publication, London.

Mackie and Mc Carteny, 1996. *Practical Medical Biology,* 14th edn. (Ed.) J.G. Collee, A.G. Fraser, B.P. Marmion, and A. Simmons, pp. 883–918.

Michael, R.G. and Sharma, B.K., 1988. *Fauna of India and Adjacent Countries: Indian Cladocera.* Zoological Survey of India, Calcutta.

Needham, J.G. and Needham, P.R. 1972. *Guide to the Study of Freshwater Biology.* Holdenday Inc., San Francisco.

Palmer, C.M., 1980. *Algae and Water Pollution.* Castle House Publishers Ltd., USA.

Patil, H.S. and Karikal, S.M., 2001. Zooplankton diversity of Bhutnal Water reservoir at Bijapur, Karnataka State. In: *Water Quality Assessment and Zooplankton Diversity,* (Ed.) B.K. Sharma, pp. 236–249.

Rajaram, R., Srinivasan, M. and Rajasegar, M., 2005. Seasonal distribution of physico-chemical parameters in effluent discharge area of Uppanar estuary Cuddalore, South-East coast of India. *J Environ. Biol.,* 26(2): 291–197.

Roy, S.P., 1982. Seasonal variation and species diversity of aquatic Coleoptera in a freshwater pond at Bhagalpur, India. *Oriental Insects,* 16: 55–62.

Saikia, S.K. and Das, D.N., 2003. Diversity indices of zooplankton and water quality in two freshwater ponds of Arunachal Pradesh, India. *Geobios,* 30: 113–116.

Shaji, C. and Patel, R.J., 1991. Chemical and biological evaluation of pollution in the river Sabarmati at Ahmedabad, Gujarat. *Phykos,* 30: 91–100.

Shaji, C. and Patel, R.J., 1994. Phytoplankton ecology of a polluted pond at Anand, Gujarat. *Annals of Biology,* 10(2): 191–197.

Shannon, C.E. and Weaver, W., 1949. *Mathematical Theory of Communication.* Univ. Illinois Press, Urbana, III.

Shanthi, K., Ramaswamy, K. and Lakshmanaperumalasamy, P., 2002. Hydrobiological study of Singanallur lake at Coimbatore, India. *Nature Env. Poll. Tech.,* 1(2): 97–101.

Sharma, B.K., 2001. Zooplankton diversity: Freshwater plankton and semi planktonic rotifers. In: *Water Quality Assessment, Biomonitoring and Zooplankton Diversity,* (Ed.) B.K. Sharma, pp. 190–215.

Sinha, M.P., 1986. Limnobiotic study on trophic status of a polluted fresh water reservior of coal field area. *Poll. Res.,* 5(1): 13–17.

Sinha, R.K., 2001. Biomonitoring of fresh water ecosystems. In: *Water Quality Assessment, Biomonitoring and Zooplankton Diversity,* (Ed.) B.K. Sharma, pp. 111–117.

Unni, K.S., 1972. An ecological study of the macrophytic vegetation of Doodhari Lake Raipur (M.P). 3. Chemical factors. *Hydrobiology,* 40: 25–26.

Unni, K.S., Chauhan, A., Varghese, M. and Naik, L.P., 1992. Preliminary hydrobiological studies of river Narmada from Amarkantak to Jabalpur. In: *Aquatic Ecology,* (Ed.) Misra and Saksena. APH Publ. House, New Delhi, pp. 221–229.

Vollenweider, R.R., 1969. *A Manual on Methods for Measuring Primary Production in Aquatic Environment.* IBP, Handbook No. 12 Blackwell, Oxford.

Wetzel, P.S., 1972. The role of carbon in hard water of Marl Lake. In: *Nutrients and Eutrophication: A Limiting Nutrient Controversy,* (Ed.) G.B. Lichens. Amer. Soc. Limnol. Oceanogr., Allen Press, Kanasas, 449.

WHO, 1992. *Guideline for Drinking Water Quality,* 2nd edn. APHA, A WWA, WPCF, Washington D.C., USA.

Wilhm, J.L. and Dorris, T.C., 1968. Biological parameters for water quality criteria. *Bioscience,* 18: 477–481.

Chapter 39

Influence of Silicate Solubilizing Bacterial (SSB) Inoculation to Rice in Different Soils

P. Kalaiselvi[1]* and S. Anthoni Raj[2]
[1]PhD Scholar and Corresponding Author, [2]Former Dean
Department of Environmental Sciences, Tamil Nadu Agricultural University, Coimbatore – 641 003, Tamil Nadu

ABSTRACT

Rice is the staple food crop grown extensively all over the world and occupies a pivotal position in the food security system of India. It is a silicicolus plant and a known silicon accumulator and is a plant that benefits from silicon nutrition. The agricultural soils are largely composed of silicate minerals and many soils contain an inadequate supply of available silica (Si). There is a similarity for silica (Si) and potassium (K), not only in their co-existence in soil but also in their biochemical role in plants. It has been shown in a laboratory study that the silicate solubilizing bacteria (SSB) isolated from rice husk and soil of granite crusher yard identified as a member of the genus *Bacillus,* releases K in addition to Si from silicate bearing minerals in the agricultural soils. A pot culture experiment was conducted at Agriculture College and Research Institute, Madurai, and examined the influence of silicate solubilizing bacteria to rice with different soils (clay loam, red soil and sand). The application of SSB recorded higher levels of Si and K in field impounded water, soil and plant, markedly influenced the crop growth. It has also been proved that the application of SSB to rice reduces the use of commercial K fertilizer.

Keywords: *SSB, Silica, Potassium and Different soils.*

---

* E-mail: kalaiphd@gmail.com

Introduction

Rice (*Oryza sativa* L.) is the staple food crop grown extensively all over the world. The nutritional requirements of the crop increases with intensive cultivation and increased adoption of nitrogen responsive high yielding varieties. Increased nitrogen use makes the rice crop tender and succulent.

Rice is a silicicolens plant and a known silicon accumulator and requires high silicon for healthy growth and stable and high productivity (Ma and Takahashi, 2002). The plant benefits from sihcon nutrition; silicon is present in the chain from soil through plant to animal. Silicon does not form an integral part of any cellular constituent but primarily deposited on the wall of epidermis and vascular tissues. Silicon is considered essential for rice, maize, sugarcane and other graminaceous plants. Silicon at up to 10 per cent of dry weight is accumulated in the shoot, and more than 90 per cent of silicon is present in the form of silica gel (Ma and Takahashi, 2002). Silica gel is deposited on the cell wall of epidermal cells of leaves, stems, and hulls, forming a silica-cuticle double layer and a silica-cellulose double layer (Raven, 2003). Silicon is also deposited on the bulliform cells, dumbbell cells, and long and short cells on the surface of leaves and hulls. The deposition of silicon enhances the strength and rigidity of cell walls and thus increases the resistance of rice to diseases, pests, and lodging, improving light-receiving plant form in a community and decreasing transpiration (Ma and Takahashi, 2002; Ma, 2003). Thus, silicon plays an important role in enhancing the resistance of rice to multiple stresses, including biotic and abiotic stresses (Ma, 2004). High accumulation of silicon in rice has been attributed to the ability of the roots to take up silicon (Takahashi *et al.*, 1990; Richmond and Sussman, 2003).

Silicon application was found to significantly increase the grain yield by increasing the number of panicles per plant, spikelets per panicle, matured grain per panicle, test weight and harvest index. Therefore silicon nutrition and silica accumulation in rice is an important factor in improving productivity.

There is a similarity for Si and K, not only in their co-existence in soil but also in their role in plants. Both these ions do not form an integral part of any of the cellular constituent but they significantly influence the crop growth and yield. Rice contains 4–20 per cent silica and about 1.5 per cent K. Potassium is considered as a traffic policeman for the plant system which regulates all macro and micronutrients for their specific functions and activities. It is one of the most soluble elements in soil-plant system. Rice absorbs more K from non-exchangeable fractions. The mean K removal for production of every tonne of rice grain is around 26 kg. Potassium availability in the soil is influenced by many factors especially with microbial activity in the rhizosphere region. Microbial activity might release K from the non-exchangeable reserve. The silicate solubilizing bacteria were found to release K as well, in the process of liberating silicon (Xue-zhiyong, 1996).

Therefore identification of an efficient soil bacterium that might solubilize silicate could pave way for release of other nutrients in soil. The present investigation was undertaken in order to identify an efficient SSB and to test its influence on the release of Si and K and also its capacity to substitute the potash requirement of rice in different soils.

Materials and Methods

Isolation of Silicate Solubilizing Bacteria (SSB)

Silicate Solubilizing Bacteria (SSB) were isolated from rice husk and soil of granite crusher yard using silicate solubilizing bacterial medium (Bunt and Rovira, 1955) containing 0.25 per cent insoluble magnesium trisilicate.

Identification and Characterization of the Bacterial Isolates

The morphological, nutritional and biochemical characteristics of the bacterial isolates selected for the present investigation were determined as per the methods of Gerherdt *et al.* (1981). The Gram reaction, endospore staining acid and gas production, hydroloysis of starch, lipolytic activity, liquefaction of gelatin, catalase and nitrate reductase activities were carried out. Based on the characteristics of the bacterial cultures isolated from rice husk (SSB I) and from soil of granite crusher yard (SSB II) were identified employing the key described in Bergey's manual of determinative bacteriology (1974).

In vitro Determination of Silicate Solubilizing Ability of Bacterial Isolates

The extent of silicate solubilization by the selected isolate was assessed both qualitatively and quantitatively under *in vitro* condition using plate assay and broth assay by substituting the medium with 0.25 per cent insoluble $MgSiO_3$. Based on clear zone obtained in the plate assay and extent of silica content in the supernatant of broth assay the silicate solubilizing potential of the bacterial isolates were determined.

Effect of Silicate Solubilizing Bacteria on Rice

A pot culture experiment was taken up with three different soils *viz.*, clay loam, red soil and sand to assess the effect of silicate solubilizing bacteria on silicate and potassium release. Earthen pots of 30 cm diameter were taken and the pots were filled with respective soils (unsterilized) like sand (8.50 kg pot^{-1}), clay loam (Madukkur series, Typichaplustalf) (6.21 kg pot^{-1}) and red soil (Araiyur series, Udichaplustalf) (6.40 kg pot^{-1}). ADT 36 seedlings (25 days old) were planted @ 2 plants hill^{-1} with 4 hills pot^{-1}. The river sand was included in this study as it is the pure form of silica (quartz) available in nature. The sand washed free of organic matter with acid.

The SSB I isolated from rice husk and SSB II isolated from soil from granite crusher yard were inoculated in soil using peat based medium @ 5 g pot^{-1}. A separate uninoculated control was also maintained for each soil. The pots were irrigated uniformly with 900 ml of water at an interval of 2 days. The fertilizer required per pot was calculated from the quantity of soil taken in earthen pot and from the recommended fertilizer dose of 100 : 50 : 50 kg NPK ha^{-1}. The 'N' fertilizer as urea was applied @ 0.72 g, 0.54 and 0.50 g for sand, red soil and clay loam, respectively. Superphosphate was applied @ 1 g, 0.75 g and 0.69 g for sand, red soil and clay loam, respectively. As the experiment was intended to assess potassium release, potassic fertilizer was not applied. At periodical intervals the water and soil samples were withdrawn and the available silica and potassium in water as well as in soil and plant were determined.

Biometric Observation

Growth Characters

The plant height was recorded in 3 plants in each pot at 0, 10, 30, 60 and 90 days intervals and at harvest.

Estimation of Available Silica

Aliquots of 50 ml were taken in polyethylene beaker to which 1 ml of reagent A (Fifty ml of conc. HCl was mixed with 50 ml of distilled water) and 2 ml of reagent B (10 per cent Ammonium molybdate solution adjusted to pH 7 to 8) were added. After 10 min, 1.5 ml of reagent C (10 per cent Oxalic acid solution) was added and mixed thoroughly. The yellow colour developed was read at 420 nm in a colorimeter. The available silica in water was calculated from a standard curve and expressed as mg L^{-1} of silica (Saxena, 1989).

Standard Silica Solution

Sodium metasilicate 0.2 g, was dissolved in 100 ml of distilled water (0.566 mgL^{-1} SiO_2) which was further diluted to give a series of standards from 1 to 10 mg of silica (SiO_2) per 50 ml.

Estimation of Available Potassium

Culture filtrate samples was collected and filtered and fed to the flame photometer and the potassium content was determined (Jackson, 1973).

Estimation of Available Silica in Soil

The available silica in soil samples were estimated using the ammonium acetate extract of the soil employing the method described by Imaizumi and Yoshida (1958). The soil was extracted with 1 N ammonium acetate (1 : 10) adjusted to pH 4.0 and shaked for 5 h and filtered. Ten ml of the filtrate was taken in a dry polyethylene beaker and 5 ml of 0.6 N HCl and 10 per cent ammonium molybdate were added and mixed. After 3 minutes of incubation, 10 ml of 17 per cent sodium sulphite solution was added. The blue colour developed was read at 650 nm in a colorimeter. The available silica was calculated from the standard curve and expressed as mg kg^{-1} of silica (SiO_2).

Estimation of Available Potassium in Soil

Five gram of soil sample was transferred into a polyethylene shaking bottle. To this 25 ml of 1 N ammonium acetate extractant was added and the contents were shaken in a mechanical shaker for 5 minutes. The contents were filtered through Whatman No. 1 filter paper and the filtrate was fed into the flame photometer and the available K content was determined (Jackson, 1973).

Analysis of Plants

Representative plant samples collected at periodical intervals were dried at 60°C for three days in a hot air oven and ground in a Wiley mill to pass through a 20 mesh sieve. The silica and potassium content of plant samples were estimated from the triacid extracts.

Estimation of Silica

0.5 g of oven dried plant sample was taken and transferred to 50 ml Erlenmeyer flask. Ten ml of triple acid mixture was added and the contents were digested. The resultant solution from the digestion were carefully transferred with repeated washings into tall stainless steel beakers containing 1–1.5 g anhydrous sodium carbonate in suspension such that there was sufficient alkali in excess after neutralization of the acid. The alkali suspension was then boiled for 3–5 minutes to ensure complete dissolution of silica. The resultant solution after cooling was made upto 250 ml and stored in polyethylene bottles. From this 2 ml of aliquots was taken and treated with 2 ml of 1 : 1 HCl and agitated for a while to remove the CO_2 evolved. Then 2 ml of 10 per cent ammonium molybdate was added and the contents in the flask were allowed to stand for 5 minutes, the interference of iron was then removed by the addition of 0.5 ml of 5 per cent solution of hydroxylamine HCl and phosphorus by the addition of 1 ml of 10 per cent oxalic acid. The resultant silicomolybdate complex after dilution to volume was reduced by the addition of 2 ml of 0.5 per cent ascorbic acid and the volume made upto 50 ml. The blue colour developed was kept on standing for 15–20 minutes and read at 650 nm in a colorimeter. The silica content was calculated from the standard graph (Nayar *et al.*, 1975).

Estimation of Potassium

Potassium was estimated from the triple acid extracts of the plant sample using the flamephotometer (Jackson, 1973). The extract was directly fed into the flame photometer and the readings were recorded.

Statistical Analysis

The statistical analyses of the experimental data were carried out by adopting standard method (Gomez and Gomez, 1984).

Results and Discussion

Influence of Silicate Solubilizing Bacterial Inoculation in Different Soils

Silica Content of Impounded Water, Soil and Plant

Silica Content in Impounded Water and Soil

The results on the inoculation of SSB to rice on the solubilization of silica from different soils are summarized in Table 39.1. The water impounded in pots containing three types of soil exhibited silica which was however higher in the water from clay loam (Madukkur series, Typichaplustalf) than from red soil (Araiyur series, Udichaplustalf) and sand. In general the silica content was higher in water samples during active growth stage. Among the two isolates SSB I was found to release slightly higher amount of silica in water than SSB II. The maximum amount of silica released by SSB I was 20.8 mg L^{-1} on 90 DAT from the clay loam. On the other hand in red soil SSB II released higher quantity than the SSB I with 20.1 mg L^{-1} on 90 DAT compared to 19.3 mg L^{-1} by SSB I. SSB inoculation augmented the silica level in pot impounded water.

Table 39.1: Influence of SSB Inoculation on the Silica and Potassium Contents of Water

Sl.No.	Treatment	Silica (mg L^{-1})					Potassium (mg L^{-1})				
		0 DAT	10 DAT	30 DAT	60 DAT	90 DAT	0 DAT	10 DAT	30 DAT	60 DAT	90 DAT
1.	Clay loam–Control	12.0	12.9	13.0	14.0	16.0	2.0	2.0	5.0	4.0	3.0
	Clay loam–SSB I	12.4	13.8	16.0	18.0	20.8	3.0	4.0	6.0	8.0	7.0
	Clay loam–SSB II	12.2	13.6	15.6	16.9	17.6	2.0	3.0	5.0	7.0	6.0
2.	Red soil–Control	11.0	12.0	12.0	14.0	17.2	2.0	2.0	4.0	4.0	3.0
	Red soil–SSB I	11.6	12.6	13.0	16.8	19.3	2.0	3.0	3.0	5.0	4.0
	Red soil–SSB II	11.4	12.5	12.8	17.5	20.1	3.0	2.0	4.0	6.0	6.0
3.	Sand–Control	11.0	12.0	12.0	14.0	16.4	2.0	2.0	3.0	3.0	3.0
	Sand–SSB I	11.4	12.8	14.3	16.3	18.2	3.0	3.0	5.0	6.0	5.0
	Sand–SSB II	11.2	12.0	12.9	14.8	18.0	2.0	2.0	2.0	3.0	4.0
	SEd	0.48	0.51	0.55	0.64	0.75	0.10	0.11	0.18	0.22	0.20
	CD	1.01	1.07	1.15	1.34	1.57	0.21	0.23	0.37	0.46	0.41

DAT: Days After Transplanting.

The results on the influence of SSB inoculation on the available silica content of soil are presented in Table 39.2. The available silica was higher in clay loam followed by sand and red soil. Inoculation of SSB resulted in an enhancement of available silica in the soils and the enhancement was higher in clay loam followed by sand and red soil. The SSB I enhanced the silica level more than that by SSB II which however increased the silica content of water. The release was more in active growth and maturity stages of the crop (30–90 DAT). Among all the treatments a higher available silica was observed in clay loam treated with SSB I with 125.7 mg kg^{-1} on 90 DAT followed by clay loam treated with SSB II with 118.8 mg kg^{-1} on 90 DAT.

Table 39.2: Influence of SSB Inoculation on Available Silica and Available Potassium Contents of Soil

Sl.No.	*Treatment*	*Silica (mg L^{-1})*					*Potassium (mg L^{-1})*				
		0 DAT	*10 DAT*	*30 DAT*	*60 DAT*	*90 DAT*	*0 DAT*	*10 DAT*	*30 DAT*	*60 DAT*	*90 DAT*
1.	Clay loam–Control	57.8	62.5	75.8	89.5	95.6	325	350	358	316	269
	Clay loam–SSB I	78.2	92.5	99.8	113.7	125.7	358	388	392	341	285
	Clay loam–SSB II	73.3	89.3.	95.2	109.2	118.8	346	370	390	333	269
2.	Red soil–Control	33.6	40.2	48.3	59.5	70.0	202	202	213	191	182
	Red soil–SSB I	47.8	59.3	61.3	62.8	73.5	213	224	268	241	221
	Red soil–SSB II	45.9	55.7	59.2	69.5	78.3	269	290	311	278	261
3.	Sand–Control	47.4	55.7	68.3	72.5	88.7	22	22	36	31	27
	Sand–SSB I	68.7	78.3	91.5	97.8	103.0	22	34	56	46	36
	Sand–SSB II	65.8	73.5	89.5	100.3	100.3	22	45	48	42	36
	SEd	2.43	2.84	3.22	3.61	3.94	9.67	10.36	11.11	9.76	8.23
	CD	5.10	5.96	6.76	7.58	8.28	20.31	21.77	23.35	20.51	17.29

DAT: Days After Transplanting.

The silica content in water was found to be higher in clay loam followed by red soil and sand. The available silica was high in clay loam soil followed by sand and red soil. The clay loam taken for pot culture is a typical soil of traditional low land rice cultivation and this contained higher silica in water (20.8 mg L^{-1}) and soil (125.7 mg kg^{-1}). Inoculation soil with SSB I in this soil solubilized more silica than SSB II. SSB I has exhibited a higher silicate solubilizing potential under *in vitro* studies also. However SSB II was more effective in red soil. Jones and Handreck (1965) reported that in a wide range of soils of the same pH, the concentration of monosilicic acid in solution ranged from 3 ppm silicon (for a lateritic soil) to 37 ppm silicon (for a black clay vertisol). Chinnaswami and Chandrasekaran (1978) also found that available silica in clay soils varied from 155 to 282 ppm SiO_2 and the values for clay loam and sandy loam ranged from 130 to 201 and 57 to 153 ppm SiO_2 respectively. The results of the present study is in confirmation with the above findings. The available silica in soil was found to be higher in clay loam followed by sand. Although sand is totally a pure silica (quartz) the dissolution from it was lesser, as sand lacks nutrients for microbial proliferation.

In red soil the silica content was intermediate. Though the SSB was inoculated in this soil, solubilization of silica was observed to be lesser than the other soils. As red soils are lateritic that arose by continued leaching of silica and are inherently poor in silica a lesser content was observed. The lateritic soils contained greater amount of sesquioxides (Arogyaswamy, 1968 and Chinnaswami, 1974). Jones and Handreck (1963) and McKeague and Cline (1963) also reported that free sesquioxides removed silica from solution by adsorption. It was also likely that in red lateritic soils endowed with irrigation potential or groundwater, rice is cultivated season after season and the demand for silicon by the plant might increase in such soil.

Silica Content in Rice Plant

The results on the silica content of rice plants grown in different soils are presented in Table 39.3. The silica content in rice plant increased with age of the crop. Inoculation of SSB enhanced the silica levels in plants grown in all types of soils and the enhancement was relatively higher in clay loam. A higher silica content was observed in plants grown clay loam soil treated with SSB I with 197.5 mg g^{-1} at 90 DAT.

Table 39.3: Influence of SSB Inoculation of Silica and Potassium Contents of Plant

Sl.No.	Treatment	Silica (mg L^{-1})					Potassium (mg L^{-1})				
		0 DAT	10 DAT	30 DAT	60 DAT	90 DAT	0 DAT	10 DAT	30 DAT	60 DAT	90 DAT
1.	Clay loam–Control	48.8	68.8	97.2	157.9	160.3	0.45	0.58	0.78	0.99	0.84
	Clay loam–SSB I	68.9	88.2	132.8	187.3	197.5	0.67	0.96	1.03	1.44	1.18
	Clay loam–SSB II	67.8	86.2	127.3	184.5	195.3	0.78	0.87	1.99	1.36	1.12
2.	Red soil–Control	45.3	65.0	88.9	150.7	155.2	0.39	0.47	0.66	0.96	0.78
	Red soil–SSB I	62.5	78.9	118.0	166.3	176.3	0.52	0.88	0.98	1.27	1.09
	Red soil–SSB II	67.0	38.2	126.0	175.0	187.5	0.47	0.92	0.93	1.33	1.12
3.	Sand–Control	43.8	58.0	85.0	147.0	150.0	0.38	0.36	0.63	0.93	0.72
	Sand–SSB I	58.8	65.7	121.0	176.9	178.0	0.45	0.66	0.87	1.18	1.06
	Sand–SSB II	55.2	62.5	117.0	162.0	173.9	0.53	0.48	0.93	1.21	1.09
	SEd	2.37	2.82	4.65	6.85	7.18	0.02	0.03	0.04	0.05	0.04
	CD	4.99	5.93	9.76	14.40	15.08	0.05	0.06	0.09	0.10	0.09

DAT: Days After Transplanting.

The plants grown in the clay loam inoculated with SSB I showed higher silica content than the other. The silica content of the plant ranged from 48.8 mg g^{-1} to 197.5 mg g^{-1} in clay loam followed by 45.3 mg g^{-1} to 187.5 mg g^{-1} in red soil. The least was recorded in sand. An earlier report indicate that the amount of soil silicon available to plant is relatively very small and varies with different soils (du Pluessis and Burger, 1966). Therefore inoculation of SSB to such soils to augment the available silica status will be of immense benefit. The plants grown in the soils *viz.*, clay loam and sand which were inoculated with SSB I contained a higher silica content than SSB II, which might be due to its higher silicate solubilizing potential.

Potassium Content of Impounded Water, Soil and Plant

Potassium Content in Water and Soil

The results on the inoculation of SSB to rice on the solubilization of potassium from different soils are summarized in Table 39.1. The available potassium content increased in water as the crop growth advanced. Both the strains released the potassium to an almost same extent. Among the three soils a higher quantity of potassium was released in water from the clay loam treated with SSB I. The release was higher during 30–60 DAT followed by a decline afterwards. Inoculation of SSB resulted in higher levels of potassium in water and this was higher in clay loam followed by red soil and sand. Among the two isolates SSB II was found to increase the potassium level relatively greater than SSB I in red soil.

The results on the influence of SSB inoculation on the available potassium content of soil are presented in Table 39.2. The clay loam exhibited a higher potassium level followed by red soil and sand. Inoculation of SSB increased its level in general. The potassium level in general increased upto 30 DAT and declined afterwards. Among the two isolates SSB I released more amount of potassium compared to SSB II in clay loam and SSB II released more amount of potassium in red soil. The potassium content was higher with 392 kg ha^{-1} in clay loam treated with SSB I followed by 390 kg ha^{-1} in clay loam treated with SSB II at 30 DAT. The potassium content although increased in red soil and sand due to SSB inoculation.

The results on the potassium content of rice plants grown in different soils are presented in Table 39.3. A higher potassium content was observed in plants grown in clay loam soil treated with SSB I. The potassium content of plant increased from 0.45 per cent observed initially to 0.99 per cent after 60 DAT in uninoculated control in clay loam. SSB I inoculation to this soil augmented the potassium content in plants in clay loam. Although the increase in the potassium content of plants was observed in red soil and sand the increase was higher due to SSB II in these soils than SSB I.

The release of potassium from different soils were found to be higher due to SSB inoculation. A relatively higher release was observed in clay loam followed by red soil and sand. This might be due to variation in the content of K-bearing silicate minerals in these soils. In soil allophanes, chlorite, kaolinite, montmorillonite, vermiculite and illite are usually present and all these have absorbed K and this can be readily released by bacteria. The composition of different clay minerals in the soil play major role in the release of potassium (Tandon and Sekhon, 1988). It is also worth examining the type of silicate minerals in soils when one determines potash nutrition.

The available potassium content of water was high in clay loam treated with SSB I. The available potassium content of soil was high in clay loam followed by red soil and sand. The plants grown in clay loam inoculated with SSB I contained a higher level of potassium than the others. It is not surprising to note a variation in the K content of plants grown in different soils as the release of potassium varied with the composition of the clay minerals present in that soil and SSB inoculation also increased its level depending upon the type of soil.

The common substitution for Si^{++++} are Mg^{++}, Fe^{++}, Fe^{+++}, Al^{+++}. The negative lattice charges are compensated by the cations located between the three layer units (*e.g.*) Ca^{++}, Mg^{++}, K^{+}, Na^{+} and H^{+}. The K^{+} may be liberated to the immediately exchangeable form both by decomposition of K-bearing feldspars and by replacement of interlayer K^{+} ions of micas. Further bacterial cells are negatively charged quite similar to clay and the bacterial cells present in the immediate vicinity of soil solution might adsorb a part of it and release subsequently for plant absorption when these cells drift to rhizosphere in search of their own nutrient excreted in the exudates. This can be readily absorbed by plants and/or part of this may be refixed between the layers of particles of micaceous and montmorillonite minerals. Therefore any change in exchangeable ions by way of silicate solubilization might release K^{+} by (*i*) direct release of K due to breakdown of K-bearing minerals and (*ii*) Si^{++++} ions causing changes in exchangeable ion status of minerals. This can occur in any type of soil where SSB is introduced. The release of K observed in sand which is purely quartz might arise from the irrigation water used and its accumulation over the period. The irrigation water of this area contain 1 to 2 ppm (Muralikannan, 1996).

Biometric Observation

The results on the influence of SSB inoculation on plant height of rice plants grown in different soils are presented in Table 39.4. A higher plant height was observed in all the soils treated with SSB. The plants grown in clay loam was taller than those grown in other soils. Inoculation of both the isolates of SSB enhanced the plant height over that observed in control. The plants inoculated with strain SSB I showed marked increase in height followed by strain SSB II. In red soil and sand, although SSB I inoculation augmented plant height upto 30 DAT and SSB II treated plants showed increased height at advanced stages of crop growth.

Table 39.4: Influence of SSB Inoculation on Plant Height

Sl.No.	Treatment	Plant Height (cm)				
		0 DAT	10 DAT	30 DAT	60 DAT	90 DAT
1.	Clay loam–Control	11.3	17.0	27.8	69.0	78.1
	Clay loam–SSB I	14.0	21.0	34.8	83.0	88.2
	Clay loam–SSB II	10.0	20.0	32.4	81.0	86.1
2.	Red soil–Control	13.2	16.4	29.9	61.0	67.5
	Red soil–SSB I	11.5	17.5	38.4	67.0	72.4
	Red soil–SSB II	10.2	17.2	37.2	67.0	75.8
3.	Sand–Control	12.0	20.0	38.2	57.0	67.3
	Sand–SSB I	14.0	22.0	39.4	62.3	72.6
	Sand–SSB II	13.5	20.5	39.0	68.4	78.4
	SEd	0.51	0.78	1.45	2.81	3.12
	CD	1.08	1.65	3.04	5.90	6.56

DAT: Days After Transplanting.

The plant height was observed to be higher in all the SSB treated soils. The highest plant height was recorded in the clay loam followed by red soil and sand which might be due to release of more silica and potassium in the clay loam besides its inherent capacity to support plant growth compared to others. In addition to this the SSB I inoculated plants showed marked increase in height than that inoculated with SSB II which might be due to auxin production in the soil by SSB I besides its higher silicate solubilizing ability. The addition of 100 mg SiO_2 kg^{-1} to soil was found to increase plant height (Ashoka Rani *et al.*, 1997). The application of silicon generally increased the uptake of other nutrients like phosphorus, calcium and magnesium and the formation of carbohydrate (Islam and Saha, 1969). The increased plant growth observed due to SSB inoculation in the present study might therefore be due to (*i*) solubilization and enhanced levels of available silica in water and soil and (ii) the enhanced silica levels influencing the uptake of other nutrients like P, Ca and Mg. As SSB I was found more effective in clay loam soil. Further the plant height was enhanced by SSB I which might be likely due to a higher auxin production by this isolate.

Conclusion

The agricultural soils are largely composed of silicate minerals, many soils contain an inadequate supply or naturally low in available Si. Farmers are not replacing Si removed by rice. This will be a great limiting factor to sustainable rice production especially when they resort to high nitrogen fertilizer. So the results from this study clearly revealed that SSB can be used as a biofertilizer to rice not only to solubilize silica but also to make other nutrients like 'K' available from silicate minerals and organic soil fraction. Supplementary silica through silicate is expensive but a bacterium solubilizing polymerized silica in soil might provide the crop requirements at low cost. This might be useful in all rice growing soils and particularly soils like red laterite where silica is much less. The silicate solubilizing bacteria thus gains importance and further investigation in various soils in which rice is grown becomes an obvious necessity.

Acknowledgement

I thanks to Department of Microbiology, College of Agriculture and Research Institute, TNAU and ASPEE foundation, Mumbai, India for carried the research and financial assistance.

References

Ashoka Rani, Y., Narayanan, A., Saraswathi, Y., Devi and Subbaramamma, P., 1997. The effect of silicon application on growth and yield of rice plants. *Ann. Plant Physiol.*, 11(2): 123–128.

Bergey's, Manual of Determinative Bacteriology, 1974. (Eds.) R.E. Buchanan, N.E. Gibbons, S.T. Cowan, J.O. Holt, J. Liston, R.G.E. Murray, C.F. Niven, A.W. Ravin, R.Y. Stanier. 8th edn. The Williams and Wilkins Co., Baltimore, USA, p. 1247.

Bunt, J.S. and Rovira, A.D., 1955. Microbiological studies of some subantartic soils. *J. Soil Sci.*, 6: 119–128.

Gerherdt, P., Murrary, R.G.E., Costelow, R.N., Nexter, E.W., Wood, W.A, Krieg, N.B. and Phelleps, G.B., 1981. *Manual of Methods for General Bacteriology*. American Society for Microbiology, Washington DC, p. 524.

Gomez, K.A. and Gomez, A.A., 1984. *Statistical Procedures for Agricultural Research*. John Wiley and Sons, New York, p. 680.

Imaizumi, K. and Yoshida, S., 1958. Edaphological studies on silicon supplying power of paddy fields. *Bull. Natl. Inst. Agri. Sci., Japan Ser.*, 8: 261–304.

Raven, J.A., 2003. Cycling silicon: The role of accumulation in plants. *New Phytol.*, 158: 419–430.

Tandon, H.L.S. and Sekhon, U.S., 1988. *Potassium Research and Agricultural Production in India*. Fertilizer Development and Consultation Organisation, New Delhi, p. 144.

Muralikannan, N., 1996. Biodissolution of silicate, phosphate and potassium by silicate solubi using bacteria in rice ecosystem. *M.Sc.(Ag.). Thesis,* Submitted to Tamil Nadu Agricultural University, Coimbatore, p. 115.

Imaizumi, K. and Yoshida, S., 1958. Edaphological studies on silicon supplying power of paddy fields. Bull. *Natl. Inst. Agri. Sci., Japan Ser.*, 8: 261–304.

Jackson, M.L., 1973. *Soil Chemical Analysis.* Prentice Hall of India (Pvt.) Ltd., New Delhi, p. 114.

Nayar, P.K., Misra, A.K. and Patnaik, S., 1975. Rapid microdetermination of silicon in rice plant. *Plant and Soil,* 42: 491–494.

Ma, J.F. and Takahashi, E., 2002. *Soil, Fertilizer, and Plant Silicon Research in Japan*. Elsevier Science, Amsterdam.

Ma, J.F., 2003. Functions of silicon in higher plants. In: *Silicon Biomineralization*, (Ed.) W.E.G. Muller. Springer Verlag, Berlin, pp. 127–147.

Ma, J.F., 2004. Role of silicon in enhancing the resistance of plants to biotic and a biotic stresses. *Soil Sci. Plant Nutr.*, 50: 11–18.

Takahashi, E., Ma, J.F. and Miyake, Y., 1990. The possibility of silicon as an essential element for higher plants. *Comments Agri. Food Chemistry*, 2: 99–122.

Richmond, K.E. and Sussman, M., 2003. Got silicon? The non–essential beneficial plant nutrient. *Curr. Opin. Plant Biol.*, 6: 268–272.

Xue-Zhiyong, 1996. Effect of silicate bacterium on release of potassium in different soils and on increase yield of sweet potato. *Soil and Fetilizers*, pp. 28–26.

Chapter 40

Yield and Economics of Intercropping of Pulses in Pigeonpea

D.D. Dudhade, G.P. Deshmukh, S.S. Nawake and P.N. Harer
Pulses Improvement Project, Mahatma Phule Krishi Vidyapeeth, Rahuri – 413 722, Dist. Ahmednagar, Gujarat

ABSTRACT

Field experiments were conducted during the kharif seasons of 2006–07 and 2007–08 to explore the possibilities of growing intercrop of pulses with pigeonpea. Pigeonpea equivalent yield, net returns, LER and benefit cost ratio were higher with pigeonpea + mungbean inter cropping system (30.00 q/ha, Rs 33,553/ha, 1.34 and 3.23 respectively) over sole crop of pigeonpea (19.88 q/ha, Rs. 24,512/ha, 1.00 and 2.51 respectively). The monetary advantages was maximum with intercropping treatment of pigeonpea + mungbean (1 : 3).

***Keywords**: Pigeonpea, Intercropping, Mungbean, Urdbean, Rajmah.*

Introduction

Pigeonpea [(*Cajanus cajan* (L.) Millsp,] is an important crop of Maharashtra cultivated in the *kharif* season. The deep rooted system of this crop has made it more suitable for its cultivation under drought conditions and poor soils. Intercropping has been recognized as potentially valuable system of crop production in semiarid tropics. Evidence suggests that intercropping can provide substantial yield advantages compared to sole cropping. The intercropping increases yield and stabilizes production Pigeonpea is a long duration, widely spaced crop. The growth of the crop is quite slow during initial stage, sufficient wide space remains vacant in between two rows for this period. This space can be utilized for growing one or more short duration, quick growing, high yielding varieties of many crop without having any competitive effect on pigeonpea. Intercropping with short duration pulses in pigeonpea enhances the total productivity (Sarma *et al.,* 1995). Rao and Mishra (1989) stated that growing two legumes together helped in increasing productivity as well as achieving higher land equivalent ratio. Hence to explore the possibilities of growing intercrop of pulses with pigeonpea, the present investigation was planned.

Materials and Methods

A field experiment was conducted during *Kharif* 2007 and 2008 at the Pulses Improvement Project, of Mahatma Phule Krishi Vidyapeeth, Rahuri, Dist. Ahmednagar. The experimental soil was clayey loam with pH 8.47 and 8.51, EC-0.38 and 0.17 dS/m, available nitrogen 295 and 163 kg/ha, available phosphorus 13.64 and 5.0 kg/ha, and available potash 357 and 190 kg/ha during 2006–07 and 2007–08, respectively. The experiment consisted of 9 treatment combinations with three replications in Randomized Block Design. The treatments included T_1: Sole pigeonpea, T_2: Sole soybean, T_3: Sole mungbean, T_4: Sole urdbean, T_5: Sole rajmah, T_6: Pigeonpea + Soybean (1 : 3), T_7: Pigeonpea + mungbean (1 : 3), T_8: Pigeonpea + urdbean (1 : 3) and T_9: Pigeonpea + rajmah (1 : 3). Sowing of pigeonpea cv. Vipula, Soybean cv. Phule kalyani, Mungbean cv. Vaibhav, Urdbean cv.TPU-4, and Rajmah cv. Varun was done in the second fortnight of June in both the years. The gross plot size was 5.10 m × 3.60 m and net plot size was 3.90 × 3.20 m. Before sowing the seeds of all crops were treated with Thirum @ 2.5 g/kg of seed. Thinning was done 15 days after sowing in each crop to keep the plant to plant spacing of 10 cm in mungbean, urdbean, soybean and rajmah and 20 cm in pigeonpea. Recommended fertilizer was give through urea and single super phosphate. The different crops were harvested at their physiological maturity. Total productivity pigeonpea equivalents of intercropping systems were computed by converting yield of intercrops in to yield of pigeonpea on the basis of their existing market price. The economics of each system was also worked out as per current prices and cost of inputs prevailing during each year.

Results and Discussion

Yield Attributes

The mean yield attributes of pigeonpea are given in Table 40.1. Among the different intercropping systems, three rows of mungbean intercropped in pigeonpea recorded higher values of yield contributing characters *viz.*, number of branches/plant (10,0), number of pods/plant (186.0), number of grains/pod (4.1), and 100 grain weight (10.6 g) of pigeonpea than rest of treatments.

Table 40.1: Number of Branches/Plant, Number of Pods/Plant, Number of Grains/Pod and 100 Grain Weight of Pigeonpea as Influenced by Intercropping System (mean of two years)

Treatments	*Branches*	*Pods/Plant*	*No. of Grains/Pod*	*100 Grain Weight*
Pigeonpea + soybean	9.86	183.1	3.51	10.40
Pigeonpea + mungbean	10.00	186.0	4.14	10.6
Pigeonpea + urdbean	9.73	183.6	3.41	10.0
Pigeonpea + rajmah	10.0	184.3	3.50	10.1

Yield

Yield of Pigeonpea varied significantly in different cropping systems (Table 40.2). The highest yield of pigeonpea was obtained by with pigeonpea sole crop which was significantly superior to all other system. The lowest yield of pigeonpea obtained in pigeonpea + rajmah. The pigeonpea + mungbean gave significantly higher pigeonpea equivalent grain yield (30.00 q/ha). Sole pigeonpea gave 19.88 q/ha. This observations are in conformity with the finding of Karntz *et al.*, (1976) and Madhusudana Rao *et al.*, (1989). Rao and Willey (1980) also indicated that the slow establishing and late maturing pigeonpea combined well with early maturing of pulses to give very large advantage as measured by LER. The results are in conformity with Goyal *et al.* (1981), Gahlot *et al.* (1978), Rao *et al.* (2003).

Table 40.2: Pigeonpea Equivalent and Economics of Pigeonpea Intercropping System (Pooled Data)

Treatments	Pigeonpea Equivalent Yield (q/ha)			Gross Returns (Rs/ha)	Mean Expenditure (Rs/ha)	Net Returns (Rs/ha)	LER	B : C Ratio	Monetary Advantages (Rs/ha)
	2006–07	2007–08	Pooled						
Pigeonpea + soybean	26.84	29.35	28.10	45918	15489	30429	1.14	2.96	3736
Pigeonpea + mungbean	29.50	30.51	30.0	48616	15063	33553	1.34	3.23	8513
Pigeonpea + urdbean	26.25	26.11	26.18	42336	15063	27273	1.29	2.81	6131
Pigeonpea + rajmah	24.13	23.36	23.75	38714	16798	21916	1.32	2.30	5352
Sole pigeonpea	18.23	21.53	19.88	40754	16242	24512	1.00	2.51	
Sole soybean	19.35	19.84	19.60	41554	18240	23314	1.00	2.28	
Sole mungbean	15.53	16.39	15.98	22366	15404	6962	1.00	1.45	
Sole urdbean	11.27	12.04	11.66	20849	15404	5445	1.00	1.35	
Sole rajmah	9.43	14.34	11.89	21730	21010	720	1.00	1.03	
SE+	5.25	5.36	4.80	3901.5		3585.3	1.48	0.32	
CD at 5 per cent	7.87	8.04	7.20	5849.2		5375.3	0.22	0.49	

All the intercropping system except pigeonpea + rajmah gave significantly higher monetary returns than sole crop of pigeonpea and other sole crop. The highest net returns of Rs. 33,553/ha was obtained with pigeonpea + mungbean combination which was Rs. 9041/ha more than sole pigeonpea. The cost benefit ratio of this combination was 3 : 23 which was highest among all system. The monetary advantages was maximum with intercropping treatment of pigeonpea + mungbean (1 : 3).

In conclusion pigeonpea + mungbean cropping system was found to be most superior.

References

Gahlot, K.S., Ramvishal, N. and Vishwakarma, S.J., 1978. Urd T9 as in intercrop with *arhar* T+2. *Indian Fmg.*, 27(10): 7–8.

Goyal, S.N., Patel, N.L., Patel, N.M. and Ahlawat, R.P.S., 1991. Intercropping studies in pigeonpea under rainfed conditions. *Indian Journal of Agronomy*, 36(1): 49–51.

Krantz, B.A., Virmani, S.M., Singh Sardar and Rao, M.R., 1976. Intercropping for increased and more stable agricultural production in the semi arid tropics. In: *Report Presented at the Symposium on Intercropping in the Semi-arid Areas* at Morogoro, Tanzania.

Madhusudhan Rao, D.V., Ankaliah, R., Subba Rao, D.V. and Satyanarayana, G., 1989. Studies on the effect of green gram intercropping in pigeonpea under rainfed condition. *Andhra Agric. J.*, 36: 112–115.

Rao, L.J. and Mishra, S., 1989. It pays to intercrop groundnut and during winter in West Bengal. *Indian Fmg.*, 39: 25.

Rao, M.M., Ramalakshmi, D., Khan, M.M., Sree, S.P. and Reddy, M.V., 2003. Evaluation of genotypic compatibility of pigeonpea cultivars with urdbean and mungbean during post-rainy season in vertisols. *Indian Journal of Pulses*, 16(2): 108–111.

Rao, M.R. and Willey, R.W., 1980b. Preliminary studies of inter cropping combinations based on pigeonpea or sorghum. *Expt. Agric.*, 16(1): 29–30.

Sarma, N.N., Sarma, D. and Paul, R., 1995. Intercropping of greengram, blackgram and sesame in pigeonpea under rainfed different seeding method. *Indian J. Agrono.*, 40: 386–387.

Chapter 41

Observation on the Avifaunal Diversity in and Around Shringarbandh Lake, Bondgaon (Surban), District Gondia, Maharashtra

S.V. Bhandarkar[1] and R.N. Chavan[2]

[1]Centre for Higher Learning and Research in Zoology, N.H. College, Bramhapuri. Dist. Chandrapur, M.S.

[2]Mahatma Gandhi Arts, Science College, Armori, Dist. Gadchiroli, M.S.

ABSTRACT

Shringarbandh lake Bondgaon (Surban) is a man made 300 year old perennial freshwater ecosystem surrounded by rice field, situated at 3 km near the Itia Doh water reservoir across the Garvi river and 18 km away from the Navegaon National Park. This lake has biological potential and rich in flora and fauna. This lake is more productive because of the nutrients from adjoin watershed gets accumulated in the form of compost, sediments etc which are rich in organic constituents responsible for the growth of aquatic weeds phytoplankton and zooplankton therefore this is a good wet land habitat which make availability of food items to various organisms like the birds in ample proportion.

This lake is unique in the Vidarbha region of Maharashtra state for the habitat of world's largest flying bird Saras crane (Grus antigon) last from 50 years. It is important to know more about the abundance of species diversity from this region. The 41 birds species are detected, classified for their feeding habits have been discussed.

Keywords: *Shringarbandh Lake, Avifauna, Feeding habits.*

Introduction

The Shringarbandh lake Bondgaon (Surban), District Gondia, Maharashtra state has rich freshwater biodiversity and made unique in the Vidarbha region for the survival of World's largest flying bird Saras crane. This lake is situated 3 km near the Itia Doh water reservoir constructed across the Garvi River. The worldwide estimate shows that at least 2 out of every 3 bird species are in decline although only about 11 per cent of all birds are already officially threatened with extinction. 4 per cent are endangered or critically endangered, 7 per cent birds are in lightly better condition in terms of number but still remain highly to extinction. The most threatened major groups included rails and cranes, terrestrial game birds and pelagic sea birds. The causes responsible for the decline of birds are human interference, over hunting, stealing of eggs, exotic species invasion and chemical pollution (Khedkar, 2005). At least three quarters of all threatened birds species are in trouble because of intensive agriculture, heavy livestock grazing etc. In other words biodiversity is achieving a tremendous threat, where collection of base line data related to flora and fauna is important. Present work has been undertaken to study the avifaunal diversity of lake and its surrounding area. If such studies are not carried out, many of the existing species may go unnoticed or unrecorded.

Materials and Methods

The birds observation was carried out during the study period 2006–2007 using 7 × 50 binocular. The observation was done in a every week. The identification of species was done with. The book of Indian birds by Salim Ali (1996), Pakshikosh by Maruti Chitampalli (2002), Indian birds by Salim Ali and Lakih Fatehali (1969) and The Hamlyn photographic guide to Birds of the World by Andrew Gosler (1991).

Table 41.1: Avifauna in and Around Shringarbandh Lake, Bondgaon (Surban), Gondia District

Sl.No.	*Common Name*	*Zoological Name*	*Habit*
1.	Baya weaver bird	*Ploceus philippinus*	GR
2.	Black drango	*Dicrurus adsimilis*	IN
3.	Blue rock pigeon	*Columba livia*	GR
4.	Bronze winged jacana	*Metopidius indicus*	FR/NF
5.	Brown fish owl	*Bubo zeylonensis*	CA
6.	Common babbler	*Turdoides caudatus*	OM
7.	Common green pigeon	*Treron phoenicoptera*	FR
8.	Common myna	*Acridotheres tristis*	OM
9.	Common snipe	*Gallinago gallinago*	MI/A
10.	Crow pheasant	*Centrpus sinensis*	CA
11.	Golden oriole	*Oriolus oriolus kundoo*	IN
12.	Goldfronted chloropsis	*Chloropsis aurifrons foontalis*	IN/NF
13.	Hoopoe	*Upupa epops*	IN
14.	Indian jungle crow	*Corvus macrorhynchos culminatus*	OM
15.	Indian magpie robin	*Copsychus saularis*	IN/NF
16.	Indian night jar	*Caprimulgus asiaticus*	IN

Contd...

Table 41.1–Contd...

Sl.No.	*Common Name*	*Zoological Name*	*Habit*
17.	Indian robin	*Saxicoloides fulicata*	IN
18.	Indian roller	*Coracias benghalensis*	IN
19.	Koel	*Eudynamys scolopacea*	OM
20.	Large Egret	*Ardea alba*	CA/A
21.	Little cormorant	*Phalacrocoraz niger*	PI
22.	Open billed stork	*Anastomus oscitans*	CA/A
23.	Pied Kingfisher	*Ceryle rudis*	PI
24.	Pond heron	*Ardeola grayii*	CA/A
25.	Purple heron	*Ardea purpurea*	CA/A
26.	Purple sunbird	*Nectarinia asiatica*	NF
27.	Red munia	*Estrilda amandava*	GR/IN
28.	Red turtle dove	*Streptopelia tronquebarica*	GR
29.	Red vented bulbul	*Pycnonotus cafer*	OM
30.	Red wattled lapwing	*Vanellus indicus*	CA
31.	Rose ringed parakeet	*Psittacula krameri*	GR/FR
32.	Rufousbacked shrieked	*Lanius schach erythronotus*	IN
33.	Saras crane	*Grus antigone*	OM
34.	Shaheen falcon	*Falco peregrinator*	CA
35.	Shikra	*Accipiter badius*	CA
36.	Small blue Kingfisher	*Alcedo atthis*	PI
37.	Small egret	*Egretta garzetta*	CA/A
38.	Small green bee-eater	*Merops orientalis*	IN
39.	Spotted munia	*Lonchura punctulata*	GR
40.	White breasted waterhen	*Amauromis phoeniccurus*	IN/GR/A
41.	White necked Stork	*Ephippiorhynchos asiaticus*	CA/A

IN: Insectivorous, GR: Granivorous, OM: Omnivorous, CA: Carnirorous, PI: Piscivorous; FR: Frugivorous, NF: Nectar feeder, A: Aquatic.

Results and Discussion

In the regular survey for the observation of birds, 41 species were recorded including world's largest flying bird Saras crane, It is observed that the less number of birds were noticed during late summer, late rainy season and early winter, while more number of birds are recorded during spring, early monsoon and late winter (Kedar, 2005).

Among all the species the dominance were remained with carnivorous birds (28 per cent), followed by Insectivorous (20 per cent), Omnivorous (15 per cent), Granivorous (10 per cent), Piscivorous (7 per cent), IN/NF and GR/IN (5 per cent each), while Frugivorous, Nectar feeder, GR/FR, MI/A, and FR/NF (2 per cent each). Only 7 species are found to be aquatic birds.

Rathod and Sharma (2000) and Sivaperuman and Jayson (2000) also recorded similar findings in their study area. It is observed that the visiting and residential avifauna of Shringarbandh Lake depends upon the availability of food and concluded that they might be food specific (Kedar and Patil, 2005).

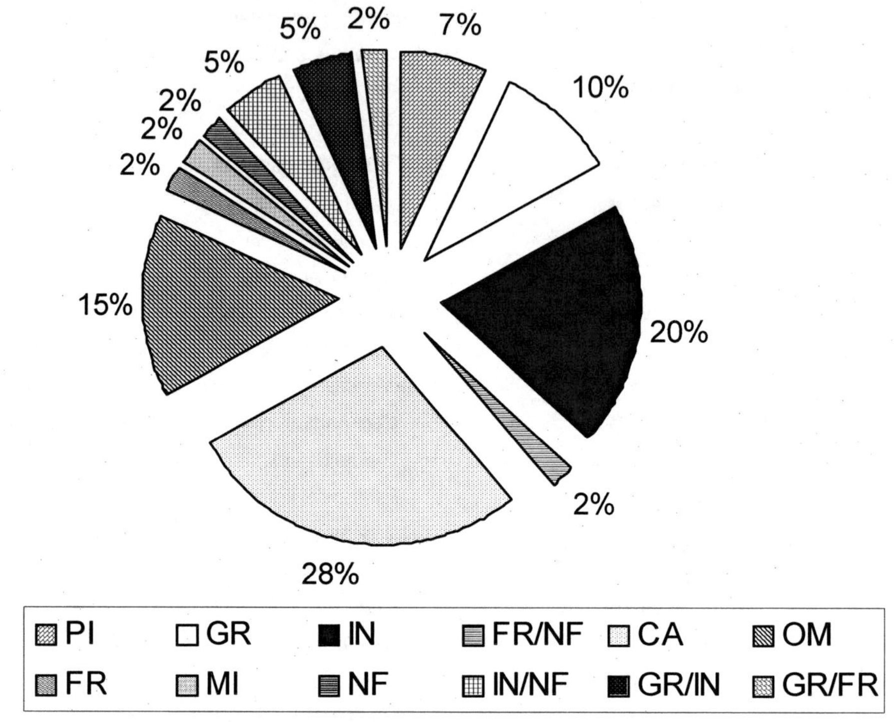

Figure 41.1: Percentage of Bird Species in Shringarbandh Lake

References

Ali, S., 1996. *The Book of Indian Birds*. Bombay Natural History Society, Mumbai.

Ali, S. and Fatehali, L., 1969. *Indian Birds*. National Book Trust of India.

Chitampalli, M., 2002. *Pakshikosh*. Sahitya Prasar Kendra, Nagpur, p. 405

Gosler, A., 1991. *The Hamlyn Photographic Guide to Birds of the World.* Reed International Book Limited.

Kedar, G.T. and Patil, G.P., 2005. Avifaunal diversity of Rishi lake, Karanja (Lad) Maharashtra with reference to food preference and feeding habits. *J. Aqua Biol.*, 20(1): 35–38.

Khedkar, G.D., 2005. Report on "The studies of Nathsagar reservoir bird sanctuary area in relation to the bird habitat." In: *IWP Sponsored Regional Workshop on Conservation and Management of Lakes in South Asia.* Hyderabad, Andhra Pradesh, India.

Rathore, V. and Sharma, R.K., 2000. Avifauna of a lake in district Etawah, Uttar Pradesh, India. *Zoos' Print Journal*, 15(6): 275–278.

Sivaperuman, C. and Jayson, E.A., 2000. Birds of Kole Wetlands, Trisur, Kerala. *Zoos Print Journal*, 15(10): 344–349.

Chapter 42

Correlation Coefficient of Physico-Chemical Properties with Zooplankton in Sager Lake Bhandara, Maharashtra

S.V. Bhandarkar[1] and G.N. Gopale[2]
Centre for Higher learning and Research in Zoology,
N.H. College, Bramhapuri, Dist. Chandrapur, M.S.
[2]Department of Fishery, J.M. Patel College, Bhandara, M.S.

ABSTRACT

The present investigation attempts to study correlation coefficient of physico-chemical property *viz.* Temperature. pH, DO, CO_2, Transparency, Total alkalinity, Total hardness, Silica, Phosphate, Nitrate and Total zooplankton from the Sager Lake, Bhandara. A study on water quality was conducted for one year and correlation coefficient was worked out for all possible properties, significant linear relation among Zooplankton population and physico-chemical properties was observed.

Keywords: *Correlation coefficient, Zooplankton, and Physico-chemical properties.*

Introduction

Sager lake is located near the national highway no. 6 towards Vainganga river, 1 km East of the Bhandara district, Maharashtra. The Lake receives the water from the catchment during monsoon period as well as from municipal drainage. The lake is extensively used for washing and bathing activities and exhibits characteristics of eutrophicated ecosystem. As a part of limnological studies on

Sager lake, the water quality in terms of different physico-chemical properties. Since a strong correlation exists among the physico-chemical properties hence a systematic calculation and interpretation of coefficient of correlation gives us an idea of rapid water quality monitoring methods (Aravinda *et al.*, 1998). According to Reid (1961) the plankton population on which the whole aquatic life depends directly or indirectly governed by the interaction of a number of physical, chemical and biological conditions and the tolerance of the organisms to variations in one or more of these conditions. In the following account, an attempt has been made to correlate certain physical and chemical properties with the plankton populations. In the r-value analysis if some of the correlation were found significant then they would be very useful in assessing the water quality parameters.

Materials and Methods

The surface water samples were collected for only one year (December 2006 to November 2007). Using glass bottles collected the samples. All samples were collected at monthly intervals during morning hours in first week of every month. The collected samples were brought to the laboratory for the estimation of various physicochemical parameters. The procedures were followed as per the standard methods (APHA, 1995). Zooplankton samples were collected with the help of plankton net made of bolting cloth No. 25 from the littoral zone. The quantitative estimation by using Sedgewick-Rafter Cell and expressed in numbers per liter. For all possible coefficient of correlation among the various properties, calculations were performed on computer in PF/Ex/W-XP.

Results and Discussion

The correlation coefficient among each possible pair of physico-chemical and biological properties is computed and listed in Table 41.3. The range of 'r' value for physico-chemical properties is –0.8323 to 0.9316 (pH–Free CO_2 and Total alka.–DO) and for Total zooplankton–Physico-chemical properties is –0.7188 to 0.06718 (Nitrate–Zooplankton and pH–Zooplankton) respectively.

Temperature is the determining factors in the seasonal distribution of organisms (Prasad, 1956). In the present study the maximum number of plankton were recorded during winter season. The zooplankton showed a negative correlation with temperature (r = –0.02183). George (1962) found a higher production during low temperature and low production during high temperature while Hynes (1970) reported that the plankton is maximum during summer while it is minimum in winter. The temperatures along with other environmental parameters are responsible for growth and abundance of different zooplankton communities (Ruttner-Kolisko, 1974).

The water temperature was positively correlated with the Free Carbon dioxide, Total alkalinity, Total hardness, Nitrate and Phosphate (r = 0.6811, 0.4348, 0.4093, 0.5184, 0.5395) respectively.

The pH is considered as an important ecological factor and is the result of the interactions of various substance in the water (Tamlurkar *et al.*, 2006). The pH was a positively correlated significantly with the zooplankton (r = 0.6718). Salaskar *et al.* (2003) also reported the positive correlation between pH and zooplankton in Powai Lake Mumbai. The highest positive correlation is observed between pH and Dissolved oxygen (r = 0.6179) while negative correlation is observed between pH and Free Carbon dioxide (r = –0.8323).

Dissolved oxygen content showed a positive correlation with Total zooplankton (r = 0.5650) while the strongest positive correlation is found with total alkalinity (r = 0.9316). On the other hand the weakest correlation found with water temperature (r = 0.1910). Salaskar and Yeragi, (2003) also shows the similar findings in Powai lake Mumbai.

In the present investigation the free Carbon dioxide did not show a specific variation (Table 41.1). It is found to be negatively correlated with Total zooplankton (r = 0.5598). The strong correlation found with Nitrate (r = 0.8390) while the weakest with Phosphate (r = 0.0080).

Table 41.1: Monthly Variations in Physico-chemical Properties of Sager Lake

Month	*Water Temp.*	*pH*	*DO*	*Free CO_2*	*Trans-parency*	*Total Alkalinity*	*Total Hard.*	*Silica*	*Nitrate*	*Phosphate*
Dec	20.4	7.30	5.54	3.00	210	120	50	25.65	0.080	0.036
Jan	18.8	7.38	5.40	3.30	190	112	48	17.15	0.090	0.039
Feb	23.2	7.40	5.65	3.80	170	125	48	11.58	0.105	0.054
Mar	24.7	7.32	5.84	3.95	132	132	57	12.65	0.115	0.088
Apr	29.0	7.20	6.25	4.40	95	146	54	8.37	0.090	0.079
May	30.3	7.25	6.48	4.80	140	154	56	14.76	0.123	0.076
Jun	26.2	7.58	7.50	3.60	120	172	60	18.36	0.106	0.098
Jul	23.0	7.64	8.00	2.80	135	175	63	22.85	0.088	0.102
Aug	25.6	7.62	7.64	2.10	175	165	65	26.16	0.064	0.089
Sep	23.0	7.68	7.32	2.60	198	170	62	25.25	0.046	0.089
Oct	20.4	7.74	7.06	2.45	222	158	57	21.15	0.028	0.076
Nov	19.6	8.00	6.80	2.05	240	132	48	16.24	0.015	0.058

Table 41.2: Monthly Variation of Zooplankton Population (µ/l) in Sager Lake

Month	*Total*	*Copepods*	*Rotifers*	*Cladocera*	*Protozoa*	*Misc.*
Dec	205	82	85	22	11	5
Jan	265	116	95	34	18	2
Feb	290	108	134	28	18	20
Mar	192	92	58	20	16	6
Apr	230	72	85	40	24	9
May	424	166	171	39	37	11
Jun	266	123	77	39	18	9
Jul	421	172	135	70	34	10
Aug	459	194	166	59	34	6
Sep	657	241	246	109	47	14
Oct	789	312	246	138	84	9
Nov	485	191	176	74	36	8

The transparency showed a positive correlation only with silica and pH (r = 0.4834 and 0.6005) respectively. It is also positively correlated with Total zooplankton (r = 0.5296).

The correlation of Total zooplankton with Total alkalinity and Total hardness is positive (r = 0.5084 and 0.3646) respectively. The Total alkalinity is significantly correlated with Total hardness (r = 0.8961) and Total hardness is significantly correlated with phosphate (r = 0.8836).

Table 41.3: Simple Correlation Between Properties of Sager Lake

Properties	*Water Temp.*	*pH*	*DO*	*Free CO_2*	*Transparency*	*Total Alkalinity*	*Total Hardness*	*Silica*	*Nitrate*	*Phosphate*	*Total Zooplankton*
Water temp.	1										
pH	–0.496	1									
DO	0.191	0.6179	1								
Free CO_2	0.6811	–0.8323	–0.4589	1							
Transparency	–0.7961	0.6005	–0.0863	–0.7281	1						
Total Alkalinity	0.4348	0.3533	0.9316	–0.1774	–0.3064	1					
Total Hardness	0.4093	0.193	0.8201	–0.2019	–0.0341	0.8961	1				
Silica	–0.3817	0.43	0.4913	–0.72	0.4834	0.3849	0.4922	1			
Nitrate	0.5814	–0.8136	–0.3381	0.839	–0.7706	–0.1118	0.005	–0.432	1		
Phosphate	0.5395	0.209	0.7996	0.008	–0.5413	0.8892	0.8836	0.1037	0.0967	1	
Total Zooplankton	–0.2183	0.6718	0.565	–0.5598	0.5296	0.5084	0.3646	0.4512	–07188	0.266	1

The nutrients are important for the biological activities. In the present study the zooplankton has weak positively correlated with the phosphate and negatively related with the nitrate (r = 0.2663 and –0.7188) respectively, while the silica has a positive correlation with zooplankton (r = 0.4512). Salaskar and Yeragi (2003) reported the similar findings in Powai Lake Mumbai.

The present findings shows that maximum correlation of physico-chemical properties with zooplankton population are found to be positive, it is indicated that majority of the physico-chemical factor are responsible to fluctuate the zooplankton population.

References

APHA, 1995. *Standard Methods for the Examination of Water and Wastewater,* 19th edn. American Public Health Association.

Aravinda, H.B., Manjappa, S. and Puttaiah, E.T., 1998. Correlation coefficients of some physico-chemical parameters of River Thunga Bhadra, Karnataka. *Poll. Res.,* 17(4): 371–375.

Das, S.M. and Shrivastava, V.K., 1956. Quantitative studies on fresh water plankton: Correlation between plankton and hydrobiological factors. *Proc. Nat. Acad. Sci., India,* 26B(4): 243–253.

Dhembare, A.J. and Pondhe, G.M., 1997. Correlation of groundwater parameters of Pravara Area, Maharashtra, India. *J. Aqua. Biol.,* 12(1 and 2): 32–33.

George, M.G., 1962. Diurnal variation in two shallow ponds in Delhi. *Hydrobiol.,* 18(3): 263–273.

Hynes, H.B.N., 1970. *The Ecology of Running Waters.* University of Toronto Press, Toronto, 555 p.

Prasad, R.R., 1956. Further studies on the plankton of the inshore waters off Mandapam. *Ind. J. Fish.,* 3(1): 1–45.

Reid, G.K., 1961. *Ecology of Inland Waters and Estuaries*. D. Van Nostrand Company, New York.

Ruttner-Kolisko, 1974. *Plankton Rotifera, Biology and Taxonomy.* Stuttgart, E. Schweizerbear-tschi Verlagsluchhanlung (nageleuo-besmiekes).

Salaskar, P.B. and Yeragi, S.G., 2003. Seasonal fluctuations of plankton population correlated with physico-chemical factors in Powai Lake, Mumbai, Maharashtra. *J. Aqua. Biol.,* 18(1): 19–22.

Sunkad, B.N. and Patil, H.S., 2004. Water quality assessment of Fort lake of Belgaum (Karnataka) with special reference to zooplankton. *J. Environ. Biol.,* 25(1): 99–102.

Tamlurkar, H.L. and Ambhore, N.E., 2006. Correlation coefficients of some physico-chemical characteristics of Alisagar Dam water, District Nizamabad (A.P.) India. *J. Aqua. Biol.,* 21(2): 115–118.

Chapter 43

Comment on Rotifer Diversity in Two Water Bodies of Brahmapuri, Maharashtra

S.V. Bhandarkar, A.B. Jagne and W.R. Bhandarkar
Centre for Higher Learning and Research in Zoology, N.H. College, Bramhapuri – 441 206, Maharashtra

ABSTRACT

The freshwater bodies in the Bramhapuri are enriched with the diversity of Zooplankton, a present study had been undertaken on the Rotifer fauna of two bodies *viz.* Kalikar lake and Lalbodi pond located at different places in Bramhapuri town in District Chandrapur, Maharashtra. First lake situated near the market area and latter is situated near Municipal sewage dumping station, regular municipal sewage were dumped, the study had disclosed the presence of 19 species of rotifers belonging 10 genera, spread over 7 families and 3 orders. Kalikar lake has shown more diversity than the latter one.

Keywords: *Rotifer, Bioindicator, Lake and Eutrophication.*

Introduction

The Rotifers are regarded as valuable bio-indicator to depict the trophic status of water quality of their environments within limnosaprobity (Sladecek, 1983; Berzins and Pejler, 1989). In the wide and extensive studies on the biodiversity of rotifer fauna from central India relatively less work has been done on Maharashtrian aquatic habitats. The rotifers from the collection sites are undergoing rapid environmental degradation with a very fast rate of eutrophication due to direct introduction of raw domestic sewage in the catchment area.

Among both these water bodies rotifers were present even though Copepods and Cladocerans have shown its representation in the Lalbodi pond. In highly eutrophic waters, Copepod, small shrimp and fish, the predators of rotifers, can not survive longer and ultimately eliminated from the waters. But rotifers have higher ability to tolerate eutrophication than the predators. The rotifers can survive, grow and reproduce very well in high eutrophication waters (Liu Fengqi, 1996).

As no attempts was made on the study on this groups of these two water bodies, among them the Kalikar lake is highly eutrophic one. The author has studied its Rotifer composition.

Materials and Methods

The plankton samples were taken by hauling the 40 µm meshed fine net horizontally from the littoral region of four spots of two water bodies, the samples were collected seasonally from Feb 06 to April 07 and preserved in 4 per cent formalin. Observation and photography were takes place with the computerized Digipro 4.0, labomed camera light microscope.

Results and Discussions

Many rotifer species exhibit a wide range of tolerance to varying environmental conditions and an amazing diversity of this group is observed in all types of aquatic habitats (Hanazata and Yasuno, 1985; Bhatia and Rana, 1987; Takamura *et al.*, 1989; Chandrasekhar, 1996 and Dhanapathi, 1997).

We observed that the great population of the rotifers appeared in waters with high eutrophication, in these waters there is no copepods which is main predator of rotifers. If the water includes copepods, the rotifer population is very small or there is no rotifers in the waters. (Liu Fengqi, 1996), latter is the same situation which is found in the Lalbodi where small population of rotifers are observed.

Instead of using simple presence or absence criterion of individual species as indicators; community composition or group of species, their actual and relative abundance could be estimated in the biomonitoring of aquatic systems (Michael, 1985). Rotifers serve as a valuable indicator of water quality Arora (1962, 1963, 1965, 1966).

A study on the collections of the Kalikar lake with high eutrophication has 16 sp. while Lalbodi pond with no eutrophication has only 3 species. All these species are listed below in phylogenetic sequence.

Class: **Rotifera;** *Sub-class*: **Monogononta;** *Order*: **Ploimida;** *Family*: **Brachionidae**

Brachionus calyciflorus, Pallas, 1776

Lorica flexible, smooth. Anterior stout spines broad at the base and with rounded tips. Median spines slightly longer than laterals. Posterior spines absent.

Size: length–350 µm, width–180 µm.

Brachionus calyciflorus famphiceros, Ehrenberg, 1838

Median occipital spines slightly longer than laterals; posterior lateral spines very long and posterior spines also present near foot opening.

Size: length–210 µm, width–160 µm

***Brachionus calyciflorus* var. *Hymani*, Dhanapathi, 1974**

Lorica flexible. anterior stout spines with rounded tips. Saw-tooth like prominent short spiny extensions present at the base of median spines. Posterior latetal spines and short posterior spines flanking the foot opening are present.

Size: length–320 µm, width–190 µm.

***Brachionus bidentata*, Anderson, 1889**

Six occipital spines present of which the laterals and medians are of same length. Lightly stippled. Forms with single lateral posterior spine were also present.

Size: length–170 µm, width–130 µm.

***Brachionus durgae*, Dhanapathi, 1974**

Lorica smooth, nearly round. Anterior dorsal margin with six saw-tooth like spines of nearly equals length. Foot opening situated below the centre of the ventral plate, pear shaped and flanked by small anchor shaped projections.

Size: length–240 µm, width–190 µm.

***Brachionus falcatus*, Zacharias 1898**

Anterior dorsal margin with six unequal spines, the medians longs and curved ventrally at the end. Posterior spines very long, bent inwards and in some forms almost touch each other at their tips.

Size: length–180 µm, width–90 µm.

***Brachionus quadridentatus*, Hermann, 1783**

Lorica broader than long, with six occipital spines of which the median is longs and curved outward. Two posterior spines present

Size: length–220 µm, width–160 µm.

***Brachionus urceolaris*, Multer, 1773**

Lorica broadly oval, thin, with dorsal and ventral plates. Anterior dorsal margin with ventral plates. Anterior dorsal margin with six spines of equal length Ventral margin nearly straight. Foot opening posterior.

Size: length–250 µm, width–190 µm.

***Brachionus rubens*, Ehrb. 1838**

Lorica oval, Occipital margins with six spines, medians longest and intermediates longer than laterals and with peculiar asymmetric shape, each spine with a narrow anterior part, then rounding outwards and forming a broad base. Four inner occipital with short ridges, posterior spines absent.

Size: length–190 µm, width–120 µm.

***Keratella tropica*, Apstein, 1907**

Lorica elongate, Dorsal plate with three hexagonal plaques and a small squarish area between the last median plaque and the posterior margin. Six anterior occipital spines, medians longest, pointed and curved outwardly, intermediates shortest. Right posterior spines longer than left.

Size: length–110 µm, width–70 µm.

***Platyas quadricornis*, Ehremberg, 1832**

Lorica firm stippled, Occipital margin with two stout spines having truncated ends. Posterior spines equal in length.

Size: length–190 µm, width–140 µm.

Family: Colurellidae

***Colurella adriatica*, Ehrenberg, 1837**

The anterior margin is convex. The lorical plate wider in the middle and covers half of the foot posteriorly. Toes slender.

Size: length–90 µm, width–50 µm.

***Lepadella ovalis*, Muller, 1786**

Lorica more or less circular in out line and compressed dorsoventrally. Dorsal plate convex and ventral plate nearly flat. Anterodorsal margins somewhat straight, anteroventral margins with V shaped notch. Toes relatively short and acutely pointed.

Size: length–110 µm, width–70 µm.

Family: Lecanidae

Lecane arculata

Harring, 1914 Lorica ovate, anterior margin straight and coincident with spines at external angles. Markings on dorsal plate not clear. Posterior segment broad rounded and extends beyond the dorsal plate. Toes slender eodina in pointed claws.

Size: length–130 µm, width–70 µm.

***Lecane donnerianus*, Dhanapathi, 1976**

Lorica is broadly ovate. Anterior dorsal margin is straight and ventral slightly concave and with two small spiny projections at the external angles. Dorsal plate smaller than ventral, ventral plate with transverse ridge in front of the first foot joint. The second foot joint is large and subsquare. Toes long, claws very long and acutely pointed.

Size: length–80 µm, width–50 µm.

Family: Notomatidae

***Notommata pseudocereberus*, de Beauchamp, 1908**

Body transparent. Corona extends ventrally with a post-oral chin. Auricles moderate with continuous ciliation along corona. Head and neck separated by transverse folds. Abdominal part tapers into a tail. Foot with two joints. Foot glands elongate with accessory glands. Toes slender and conical ending in points.

Size: length–230 µm, width–110 µm.

Family: Asplanchnidae

***Asplanchna brightwelli*, Gosse, 1850**

Body large transparent, thin, greatly contractile without lorica. Corona reduced to a thin course of cilia around the head. Vitellarium horse shoe-shaped.

Size: length–420 µm, width–280 µm.

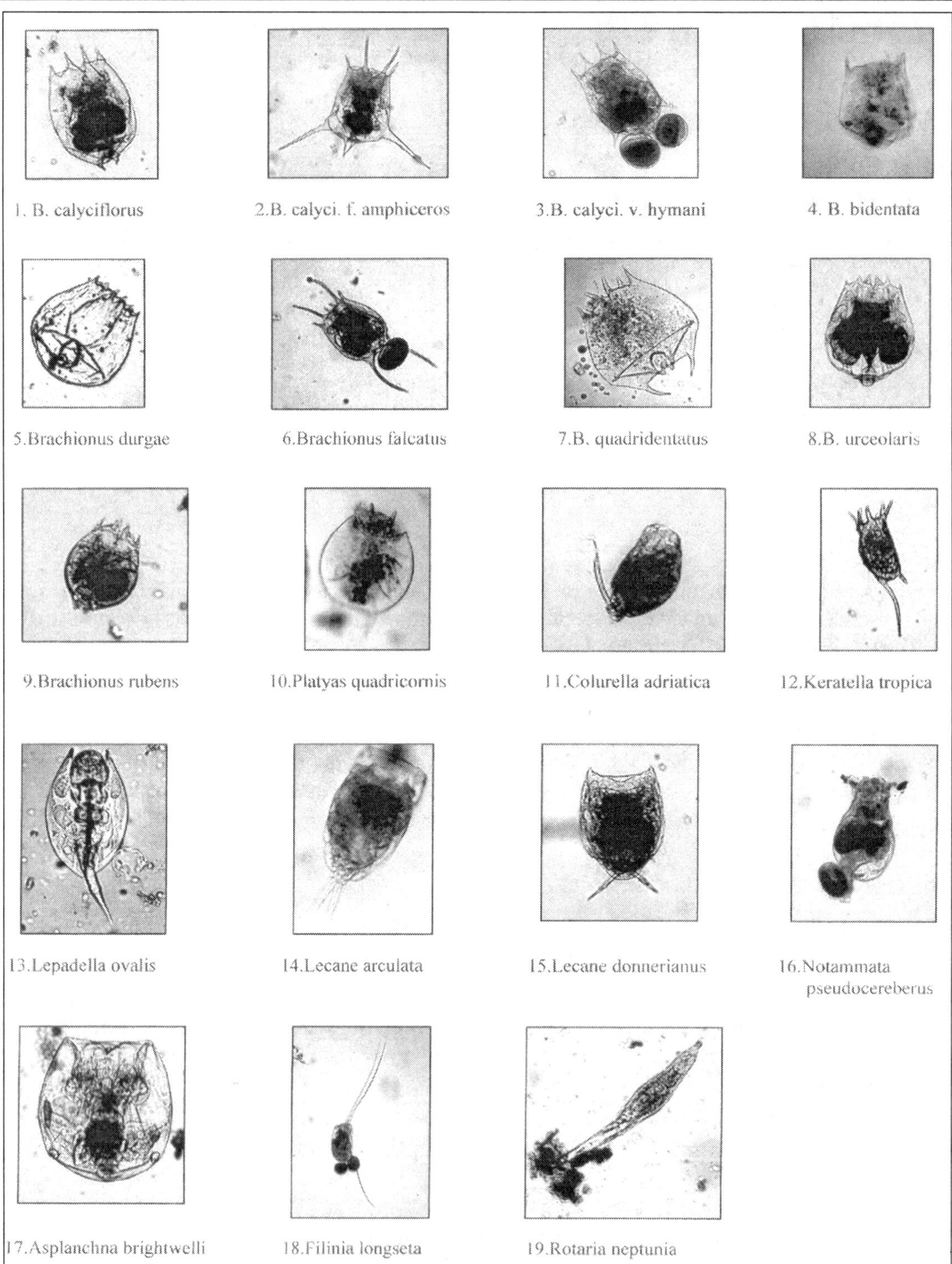

Plate 43.1: 19 Species of Rotifers

Order: Gnesiotrocha, Family: Filinidae

***Filinia longiseta*, Ehrenberg, 1834**

Body thin, without lorica, barrel-shaped and with two long equal, movab1e antero-lateral spines and one immobile posterior spine. Anterior setae usually folded ventrally. Posterior seta inserted almost one-fourth body length away from caudal end. Corona in the form of circumapical girdles. Foot absents.

Size: length–140 µm, width–85 µm

Order: Bdelloida; Family: Philodinidae

***Rotaria neptunia*, Ehrenberg, 1832**

Elongate rotifer known with two trochal discs. Foot slender, very long, five jointed and telescopically projected. Third foot joint has a pair of equal and jointed spurs. Last foot joint with three slender, long, equal and pointed toes.

Size: length–130 µm, width–50 µm

Table 43.1: Showing the Presence (+) and Absence (–) of Species in Two Different Water Bodies, *viz.*, Kalikar Lake (KL) and Lalbodi Pond (LP)

Sl.No.	*Species*	*KL*	*LP*
1.	*Brachionus calicyflorus*	+	–
2.	*Brachionus calicyflorus f. amphecerus*	+	–
3.	*Brachionus calicyflorus ver. hymani*	+	–
4.	*Brachionus bidentata*	+	–
5.	*Brachionus durgae*	+	–
6.	*Brachionus falcatus*	+	–
7.	*Brachionus quadridentatus*	+	–
8.	*Brachionus urceolaris*	+	–
9.	*Brachionus rubens*	+	–
10.	*Platyas quadricornis*	+	–
11.	*Colurella adriatica*	–	+
12.	*Keratella tropica*	–	+
13.	*Lepadella ovalis*	+	–
14.	*Lecane arculata*	–	+
15.	*Lecane donerrianus*	+	–
16.	*Notomata psuedocereberus*	+	–
17.	*Asplanchna brightwelfi*	+	–
18.	*Filinia longiseta*	+	–
19.	*Rotaria neptunia*	+	–

References

Arora, H.C. 1966b. Studies on Indian Rotifera–V. *Arch. Hydrobiol.*, 61: 482–483.

Arora, H.C., 1962. Studies on Indian Rotifera–I. *J. Zool. Soc. India*, 14: 33–44.

Arora, H.C., 1963a. Studies on Indian Rotifera–II. *J. Zool. Soc. India,* 15: 112–121.

Arora, H.C., 1965. Studies on Indian Rotifera– VI. *Hydrobiologia,* 26: 444–456

Arora, H.C., 1966a. Studies on Indian Rotifera–III. *J. Zool. Soc. India,* 16: 1–6.

Berzins, B. and Pejler, B., 1987. Rotifer occurrence in relation to pH. *Hydrobiologia,* 182: 171–182

Bhati, D.P.S. and Rana, K.S., 1987. Zooplankton in relation to abiotic components in the fort moat of Bharatpur. *Proc. Nat. Acad. Sci. India,* 57(B): 237–242.

Chandrashekhar, S.V.A. and Kodarkar, M.S., 1995. Studies on Brachionus from Sarrornagar Lake, Hyderabad, Andhra Pradesh, India. *J. Aqua. Biol.,* 10(1 and 2): 48–52.

Dhanapathi, M.V.S.S.S., 1997. Variations in some rotifers of the family Brachionidae. *J. Aqua. Biol.,* 12(1 and 2): 35–38.

Hanazata, T. and Yasuno, M., 1985. Population dynamics and production of Caldoceran zooplankton in the highly eutrophic lake, Kasumigaura. *Hydrobiologia.* 124: 13–2.

Liu Fengqi, 1996. Production and application of rotifers in aquaculture. *Aquaculture Magazine,* 22(3): 16–22

Michael, R.G., 1985. Use of rotifers and cladocerans as potential bioindicators of Indian freshwater ecosystems. In: *Proc International Symposium on Biological Monitoring of the State of the Environment (Bioindicators),* New Delhi, October 11–13, 1984. Indian National Science Academy, pp. 82–83.

Siadecek, V., 1983. Rotifers as indicators of water quality. *Ibid,* 100: 169–201.

Takimura, K.Y, Sugaya, N., Takamura, T.M. Yosuno and Iwakuma, T., 1989. Primary production of phytoplankton and standing crop of zooplankton and zoobenthos in hyper-eutrophic lake, Teganuma. *Hydrobiologia,* 173: 173–184.

Chapter 44

Studies on Population Fluctuations and Abundance of Predators Fauna in Rice Ecosystem

K. Verma* and M.A. Khan
Biological Control Laboratory, Department of Entomology, G.B. Pant University of Agriculture and Technology, Pantnagar – 263 145, Uttarakhand

ABSTRACT

On studying the occurrence of predator population in rice ecosystem through sweep net collections at weekly intervals, it was found that population of predators was more in earlier to middle part of the season, and followed a decreasing trend during the later part *i.e.* in October before harvesting. However, population of spiders and mirids remained more or less constant through out the season.

Keywords: *Predators, Spiders, Rice ecosystem.*

Introduction

Predators are the most important biocontrol agents found most conspicuously in rice ecosystem. Each predator preys upon and consumes many pests in its life time. It is reported that 80–90 per cent

* E-mail: kulverma@rediffmail.com

of the eggs of certain pests are consumed by the predators. Important predators found in rice ecosystem are mirid bug, *Cyrtorrhinus lividipennis* (Reuter); coccinellid beetle, *Coccinella acuata* (Fabricius); staphyllinid beetle, *Paederus* sp.; vallid bug, *Microvella* sp.; spiders, *Murpisa pomatix* (Tikader), *Lycosa pseudoannulata* (Bosenberg and Strand); water strider, *Limnogonus niditus* (Mager) and carabid, *Casnoida indica* (Thunberg).

The spiders have been rated to be the most important group of predators in the nature and agricultural ecosystem. The biotic and abiotic factors of an area often play an important role on the population build up of spiders. (Bhathal and Dhaliwal, 1991). In most of the studies, predators are mentioned as to thrive on sucking insect pests in rice ecosystem, but ants, lady beetles, staphyllinid beetles, gryllid, green meadow grasshopper and mirid bug are reported to also eat eggs of lepidopteran pests (Yasumatsu and Torii, 1968).

In a view to quantify the abundance of natural predators and effect of the environmental factors on their population in a crop season, the present study was conducted in Pantnagar region in the crop year *Kharif* 2002–03.

Materials and Methods

The study on abundance of predators in rice ecosystem was made under natural control *i.e.* where no management practice to control any pest was applied. Sweep net collections were used to estimate the population of predators. A total of 20 sweeps were made in each of the four replication plots at weekly interval starting from 42 DAT. The collected predators were separated into groups and counted. An average of the four replications was calculated for each group of predators. In addition to the actual count, square root transformations were also calculated as $\sqrt{(X+0.5)}$, X being the actual number of the predator.

Results and Discussion

The present study was aimed at to estimate the population of various predators associated with rice ecosystem exerting natural control on the various rice insect pests.

Population of the predators in terms of average of their actual count and the square root transformations, as found at each week starting from 42 DAT is given in the Table 44.1 and is illustrated in the Figure 44.1.

Perusal of the Table 44.1 and the Figure 44.1 revealed that the population of coccinellids was initially high (3.57) in the first week of August; it decreased in the second week (3.04) and again increased in the third week (4.21) of the same month. Thereafter it decreased constantly up to mid September (2.06). Population was again high in the start of October (2.44) and decreased in the third week.

The population of staphyllinids was quite variable ranging uom 1.00 to 2.39. Initially the population was low (1.87) and at its peak in the third week of August and thereafter it maintained a declining trend. In the later part of the season, the population was quite less except in the second week of October (1.93).

Damsel flies were found more in numbers during early to middle part of the season. From the first week of August (2.82) the population increased till the third week of the same month (3.70), then after the population remained stable with little variations up to the last of September (3.46). The population declined to 1.93 in the third week of October touching the lowest (1.32) in the previous week.

Table 44.1: Population Fluctuations of the Predators of Rice Insect Pests in Pantnagar Region (*Kharif* 2002–03)

Average Number per 20 sweeps (*)

Week	*Coccinellids*	*Staphylinids*	*Damsel Flies*	*Dragon Flies*	*Spiders*	*Mirids*
Aug I	12.25	3.00	7.50	4.25	7.50	5.25
	(3.57)	(1.87)	(2.82)	(2.17)	(2.82)	(2.39)
Aug II	8.75	1.25	11.75	5.75	11.25	11.50
	(3.04)	(1.32)	(3.50)	(2.50)	(3.42)	(3.46)
Aug III	17.25	5.25	13.25	4.50	12.75	12.25
	(4.21)	(2.39)	(3.70)	(2.23)	(3.64)	(3.57)
Aug IV	7.00	1.00	9.00	5.25	18.25	16.50
	(2.73)	(1.32)	(3.08)	(2.39)	(4.33)	(4.12)
Sep I	4.50	3.00	5.25	3.50	13.00	1.75
	(2.23)	(1.87)	(2.39)	(4.00)	(3.67)	(1.50)
Sep II	2.75	2.50	8.75	0.75	12.50	4.50
	(2.06)	(1.73)	(3.04)	(1.11)	(3.60)	(2.23)
Sep III	4.25	1.25	9.25	6.75	14.25	2.25
	(2.17)	(1.32)	(3.12)	2.69)	(3.84)	(1.65)
Sep IV	1.75	0.75	10.50	11.00	9.75	8.75
	(1.50)	(1.11)	(3.31)	(3.39)	(3.20)	(3.04)
Sep V	4.75	0.50	11.50	6.25	10.25	7.50
	(2.29)	(1.00)	(3.46)	(2.59)	(3.27)	(2.82)
Oct I	5.50	0.50	11.50	6.25	10.25	7.50
	(2.44)	(1.00)	(3.46)	(2.59)	(3.27)	(2.82)
Oct II	4.25	3.25	1.25	2.75	10.75	5.25
	(2.17)	(1.93)	(1.32)	(1.80)	(3.35)	(2.39)
Oct III	3.00	0.75	3.25	1.75	10.25	4.00
	(1.87)	(1.11)	(1.93)	(1.50)	(3.27)	(2.12)

*: Value in parentheses are the square root transformations.

Dragon fly population was more in the period from the first week of August (2.17) to the fifth week of September (2.59). The population peaked during the first week of September (4.00). In the end of the season, the population continuously decreased to 1.50.

Spiders were more abundant than any other group of predators and the population remained more or less the same throughout the season. In the first week of August, the population was 2.82, then it increased to touch the peak in the fourth week of August (4.33). Then after the population declined and remained stable around 3.27 as in the third week of October.

The population of mirids also followed similar trend as that of spiders but in lesser abundance

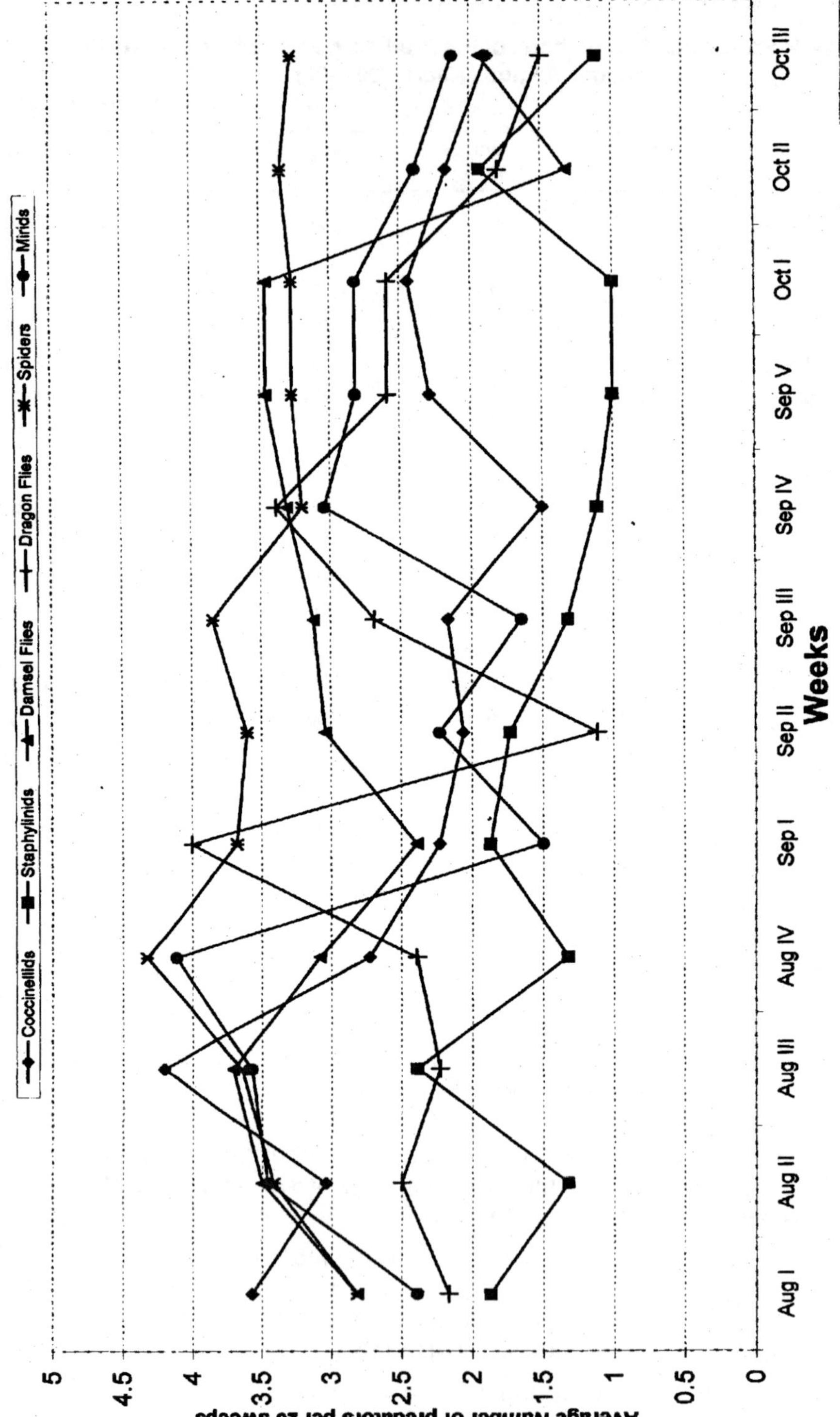

Figure 44.1: Population Fluctuations of the Predators of Rice Insect Pests in Pantnagar Region

particularly in the later half of the season. The population continuously increased from 2.39 in the first week of August to 4.12 in the fourth week of August. It decreased drastically in the following week (1.50) but regained to 3.04 in the fourth week of September. After this a stable trend was observed till end of the season (2.12).

These observations revealed that the population of predators was more in earlier to middle part of the season, sufficiently high during the later middle part *i.e.* in September, but declined in the end before harvesting. But the population of spiders remained more or less constant throughout the season.

The findings were agreeing to earlier reports that spiders and coccinellids were dominant among the predators (Anonymous, 1999). However, in present study dominance of coccinellids and mirids was observed in the beginning of the season, dragon flies peaked twice in early and later phases of mid season, while population of damsel flies and spiders remained constant for most of the season except in end. Another study by Kumar *et al.* (1996) reported that in early crop growth stages, the percentage of infested leaves gave comparatively lower number of live caterpillars in comparison to August, yet it declined sharply in the month of October. This was due to predation of the pests by a large number of natural enemies, especially spiders, which were prevalent during booting and flowering stages of the crop.

Population of predators also depends upon the population of their food *i.e.* insect pests, which in turn depends upon crop age and other abiotic factors. As reported by Singh and Singh (2000), crop age and insect pest population had positive correlation in rice ecosystem. Thus predator population also increases with the age of rice crop, however, up to the stage prone to be damaged by the insect pests.

References

Anonymous, 1999. *Research Highlights (1998–1999).* Project Directorate of Biological Control, Bangalore, India, pp. 6.

Bhathal, J.S. and Dhaliwal, G.S., 1991. Studies on the population dynamics of white backed planhopper and its natural enemies on rice. *Indian J. Ent.,* 53(1): 1–17.

Kumar, P., Singh, R., Pandey, S.K. and Kumar, P., 1996. Population dynamics of rice leaf folder *Cnaphalocrocis medinalis* (Guenee) in relation to stage of the crop, weather factors and predatory spiders. *J. Entomological Res.,* 20(3): 205–210.

Singh, R.B. and Singh, R., 2000. Influence in biotic and abiotic parameters on the population build up of spiders under rice agrosystem. *Shashpa,* 7(2): 117–123.

Verma, K., 2003. Studies on BIPM of major lepidopterous pests in rice ecosystem with special reference to tritrophic interaction. *M.Sc. Thesis,* G.B. Pant University of Agriculture and Technology, Pantnagar, Uttarakhand, India, 136 pp.

Yasumatsu, K. and Torii, T., 1968. Impact of parasites, predators and diseases on rice pests. *Ann. Rev. Entomol.,* 13: 295–324.

Chapter 45

Biosuppression of *Scirpophaga incertulas* (Walker)

K. Verma* and M.A. Khan
Biological Control Laboratory, Department of Entomology,
G.B. Pant University of Agriculture and Technology,
Pantnagar – 263 145, Uttarakhand

ABSTRACT

The present investigations were carried out with a view to assess the feasibility of bio-intensive pest management under different doses of the egg parasitoids for the control of *Scirpophaga incertulas* (Walker) in rice ecosystem at Pantnagar. The population trends represented by the damage percentage of yellow stem borer were studied in different treatments *viz.*, BCP 1 (where *T. chilonis* and *T. japonicum* parasitized eggs were released @ 1,00,000 per hectare per week) and BCP 2 (where *T. chilonis* and *T. japonicum* eggs were released @ 1,50,000 per hectare per week) and the natural control plot. Yellow stem borer infestation was high at early tillering and late reproductive stages, peaked at flowering stage of the crop. The overall reduction in damage ranged from 33.77 to 80.65 per cent over control. Thus in overall biocontrol practices were successful in suppression of the pest; BCP 1 and BCP 2 treatments were at par.

Keywords: *Trichogramma, Scirpophaga, Biocontrol.*

---

* Corresponding Author: E-mail: kulverma@rediffmail.com.

Introduction

The rice stem borer, *Scirpophaga incertulas* (Walker) is the major insect pest of rice, ranked at second position in India after rice blast disease by Geddes and Iles (1991).

The borer infests the crop continuously from seedling to maturity stage. About a week after hatching, the larvae from the leaf sheaths bore into the stem and staying in the pith feed on the inner surface of walls. Such feeding often results in severing of the apical parts of the plant from the base. When this kind of damage occurs during the vegetative stage of the plants, the central leaf does not unfold, turns brown and dries off, although the lower leaves remain green and healthy. This condition is known as 'Dead Heart (DH)'. The affected tillers dry out without bearing panicles. After panicle initiation, severing of the growing plant parts from the base results in drying of the panicle which may not emerge at all, and there that have already emerged do not produce grains, this condition is known as 'White Ear Head (WH)' stage (Pathak, 1968).

The eggs of *Scirpophaga incertulas* are mostly parasititzed by *Trichogramma* sp., *Telenomus* sp. and *Tetrastichus* sp. (Singh and Dhaliwal, 1994). Also it has been reported that use of *Triehogramma japonieum* @ 50,000 to 1,00,000 per hectare per week during egg laying period of *Scirpophaga incertulas* has given promising results (Singh, 2001).

In a view of the above, this study was conducted at the Crop Research Centre (CRC), Pantnagar on behalf of the Biological Control Laboratory, Pantnagar in *Kharif* 2002–03 to ascertain the effectiveness of inundative releases of *Trichogramma chilonis* and *T. japonicum* at different doses for the control of *Scirpophaga incertulas* (Walker).

Materials and Methods

The experiment was conducted at the Crop Research Centre, Pantnagar. Summary of the treatment is given below:

A. Experimental Setup Details

Variety: Swarna, A long duration and high yielding variety.

Plot Size: 25 m × 20 m= 500 m^2

Treatments: 3 (including control)

Replications: 4

Design: Randomized Block Design (RBD)

B. Treatment Details

1. Biocontrol Plot 1 (BCP 1): *Trichogramma japonicum* and *T. chilonis* in that manner alternatively @1,00,000 parasitized eggs per hectare at weekly intervals starting from 35 DAT up to 112 DAT.
2. Biocontrol Plot 2 (BCP 2): *Trichogramma japonicum* and *T. chilonis* in that manner alternatively @1,50,000 parastized eggs per hectare at weekly intervals starting from 35 DAT up to 112 DAT.
3. Natural Control Plot: No control measures were adopted during all stages.

C. Surveillance Through Pheromone Traps

A pheromone trap was installed on the front side of the experimental field to keep watch over the onset of infestation of *Scirpophaga incertulas.*

D. Method of Treatment

For application of the tricho-cards and to save theses from rains and predators, the cards were stapled inside a thermocol cup, whose mouth was then covered with a muslin cloth fastened by a rubber band. This cup was tied in an inverted position to the stick installed in the plot so that rain drops would not enter in the cup. Each dose of the parasitoids was divided into four parts in this fashion and applied in each plot under treatment, at four corners of a rectangle 15 m × 10 m.

E. Method of Monitoring the Damage

Following parameters were adopted for monitoring the damage caused by the target pests:

1. To make observations in each plot at weekly intervals.
2. Random selection of 10 hills in each plot to record the total number of tillers and the damaged ones and to calculate the per cent damage.
3. Tying a piece of red thread to the counted hills to avoid recounting.
4. Not removing any damaged part or eggs etc. from any of the experimental plots, in order to keep their balance undisturbed.

F. Nature of Observations Recorded

Stage of the Crop	*Observations per 10 Hills*
i. Vegetative stage	(*a*) Total number of tillers
	(*b*) Number of damaged tillers
ii. Ear Head stage	(*a*) Total number of panicle bearing tillers
	(*b*) Number of white earheads

G. Estimation of Extent of Damage

To estimate the damage percentage, following formula was used:

$$\text{Per cent Infestation} = \frac{\text{Number of damaged tillers}}{\text{Total number of tillers}} \times 100$$

The bio-efficacy of different treatments was evaluated by calculating the per cent reduction in infestation with respect to different treatments over control, as follows:

$$\text{Per cent reduction in infestation} = \frac{\text{Per cent infestation in control} - \text{Percent infestation in the treatment}}{\text{Per cent infestation in control}} \times 100$$

Results and Discussion

To investigate the level of damage in different treatments, weekly observations on infested tillers in form of dead hearts (DH) and white ear heads (WE) were conducted during the vegetative and reproductive stages respectively, as given in the Table 45.1 and illustrated in the Figure 45.1.

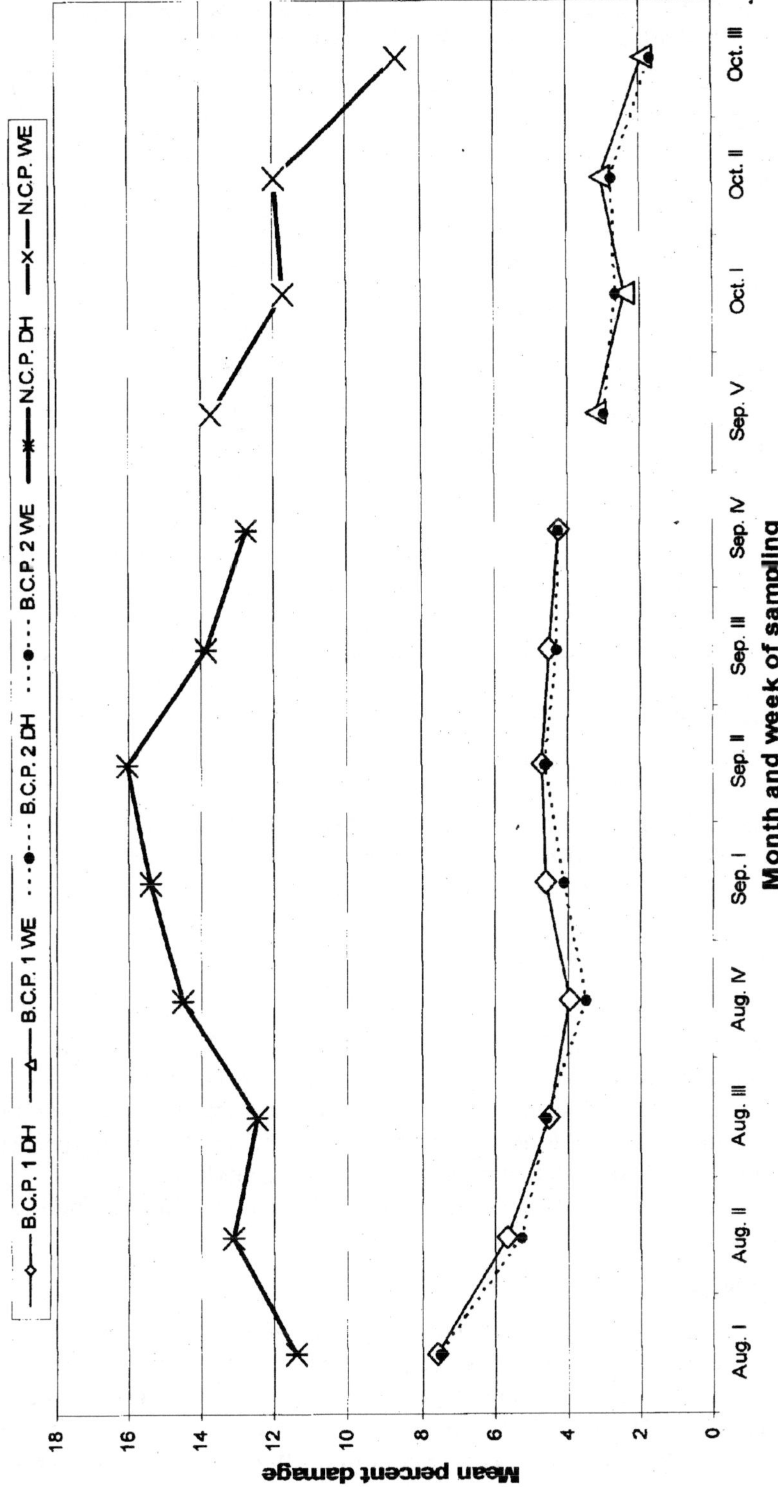

Figure 45.1: Mean Per cent Tiller Damage Caused by Yellow Stem Borer Under Different Protection Treatments

Table 45.1: Mean Per cent Tiller Damage Caused by Yellow Stem Borer, *Scirpophaga incertulas* (Walker) Under Different Protection Treatments

Month and Week of Sampling	*B.C.P.I*		*B.C.P.2*		*N.C.P.*		*Mean*
	DH	*WE*	*DH*	*WE*	*DH*	*WE*	
Aug I	7.55	–	7.49	–	11.40	–	8.81
Aug II	5.68	–	5.26	–	13.14	–	8.03
Aug III	4.48	–	4.56	–	12.50	–	7.18
Aug IV	3.93	–	3.49	–	14.51	–	7.31
Sep I	4.58	–	4.09	–	15.39	–	8.02
Sep II	4.68	–	4.61	–	16.06	–	8.45
Sep III	4.51	–	4.26	–	13.86	–	7.54
Sep IV	4.22	–	4.20	–	12.77	–	7.06
Sep V	–	3.20	–	2.95	–	13.74	6.63
Oct I	–	2.42	–	2.63	–	11.73	5.59
Oct II	–	3.06	–	2.74	–	11.99	5.93
Oct III	–	1.89	–	1.67	–	8.63	4.06
Mean	4.95	2.64	4.74	2.49	13.70	11.52	7.05

Biological Control Plot 1 (BCP 1)

In the treatment, weekly inundative releases of *Trichogramma japonicum* and *T. chilonis* alternatively, were made @ 1,00,000 per hectare. The tiller damage ranged from 1.89 per cent to 7.55 per cent. Maximum infestation 7.55 per cent was observed in the first week of August which declined to 3.93 per cent in the last week of the same month. The infestation of yellow stem borer again increased to 4.68 per cent in the second week of September and went on decreasing up to 2.42 per cent in the first week of October, with a trifle increase (3.06 per cent) in the second week. It further drastically decreased to 1.89 per cent in the third week of October.

Biological Control Plot 2 (BCP 2)

A higher dose of the parasitoids @ 1,50,000 per hectare was inundatively released under this treatment to study effectiveness of the parasitoids at higher density. The tiller damage under this treatment ranged from 1.67 per cent to 7.49 per cent. The maximum damage to tillers (7.49 per cent) was observed in the first week of August, which decreased to 3.49 per cent in the end of August and again rose to 4.61 per cent in the second week of September. After again decreasing to 4.26 per cent in the third week of September and to 2.63 per cent in early October, the infestation reached 1.67 per cent in the third week of October.

Natural Control Plot (NCP)

In this treatment, no management practice was followed. The yellow stem borer infestation ranged from 8.63 per cent (in the third week of October) to 16.06 per cent (in the second week of September). In the fIrst week of August, the tiller damage was 11.40 per cent, which increased to 13.14 per cent in the second week but again decreased (12.50 per cent) in the third week of the same month. The damage increased to its highest in the second week of September and declined thereafter with little increase

(13.71 per cent) in the fifth week of September. After this, the infestation decreased steadily with little fluctuation. From analyzing the damage variations under this treatment, population fluctuations could be revealed. The incidence of yellow stem borer was high at tillering stage, which kept on increasing and showed its peak in the second week of September *i.e.* at the flowering stage. Incidence of yellow stem borer was comparatively more at the start of the season than at the end of the season excluding its peak during September.

From the Table 45.2 and Figure 45.2, it is revealed that in BCP 1 treatment, the dead hearts varied from 3.93 to 7.55 per cent (Mean = 4.95 per cent), while in control these were 11.40 to 16.06 per cent (Mean = 13.70 per cent). The reduction in the dead hearts was in the range of 33.77 to 72.92 per cent (Mean 63.87 per cent) over control. In case of white ear heads, which were 1.89 to 3.20 per cent (Mean= 2.64 per cent), the reduction varied from 74.48 to 79.37 per cent (Mean = 77.08 per cent).

Table 45.2: Bioefficacy of *Trichogramma chilonis* and *T. japonicum* Against Yellow Stem Borer, *Scirpophaga incertulas* (Walker)

Month and Week of Sampling	*BCP I*				*BCP 2*				*NCP*	
	DH	*Per cent Reduction*	*WE*	*Per cent Reduction*	*DH*	*Per cent Reduction*	*WE*	*Per cent Reduction*	*DH*	*WE*
Aug I	7.55	33.77	–	–	7.49	34.3	–	–	11.40	–
Aug II	5.68	56.77	–	–	5.26	59.97	–	–	13.14	–
Aug III	4.48	64.16	–	–	4.56	63.52	–	–	12.50	–
Aug IV	3.93	72.92	–	–	3.49	75.95	–	–	14.51	–
Sep I	4.58	70.24	–	–	4.09	73.42	–	–	15.39	–
Sep II	4.68	70.86	–	–	4.61	71.3	–	–	16.06	–
Sep III	4.51	67.46	–	–	4.26	69.26	–	–	13.86	–
Sep IV	4.22	66.95	–	–	4.20	67.11	–	–	12.77	–
Sep V	–	–	3.20	76.71	–	–	2.95	78.53	–	13.74
Oct I	–	–	2.42	79.37	–	–	2.63	77.58	–	11.73
Oct II	–	–	3.06	74.48	–	–	2.74	77.15	–	11.99
Oct III	–	–	1.89	78.1	–	–	1.67	80.65	–	8.63
Mean	4.95	63.87	2.64	77.08	4.75	65.33	2.50	78.30	13.70	11.52

In the other treatment plot (BCP 2) where the egg parasitoids were released inundatively @ 1,50,000 parasitized eggs per hectare per week, the reduction in damage was not significantly different from that of BCP 1. The dead hearts varied from 3.49 to 7.49 per cent (Mean= 4.75 per cent) and reduction in dead hearts over control ranged from 34.30 to 75.95 per cent (Mean= 65.33 per cent). In case of white ear heads, the damage was from 1.67 to 2.95 per cent (Mean = 2.50 per cent) and the reduction in this case varied from 77.15 to 80.65 per cent (Mean= 78.30 per cent) over control.

The results obtained were in accordance with that reported by Brar (2000) that 7–9 releases of *T. chilonis* and *T. japonicum* @ 1,00,000 per hectare starting from 30 DAT proved as effective as standard insecticide treatments for the control of stem borer.

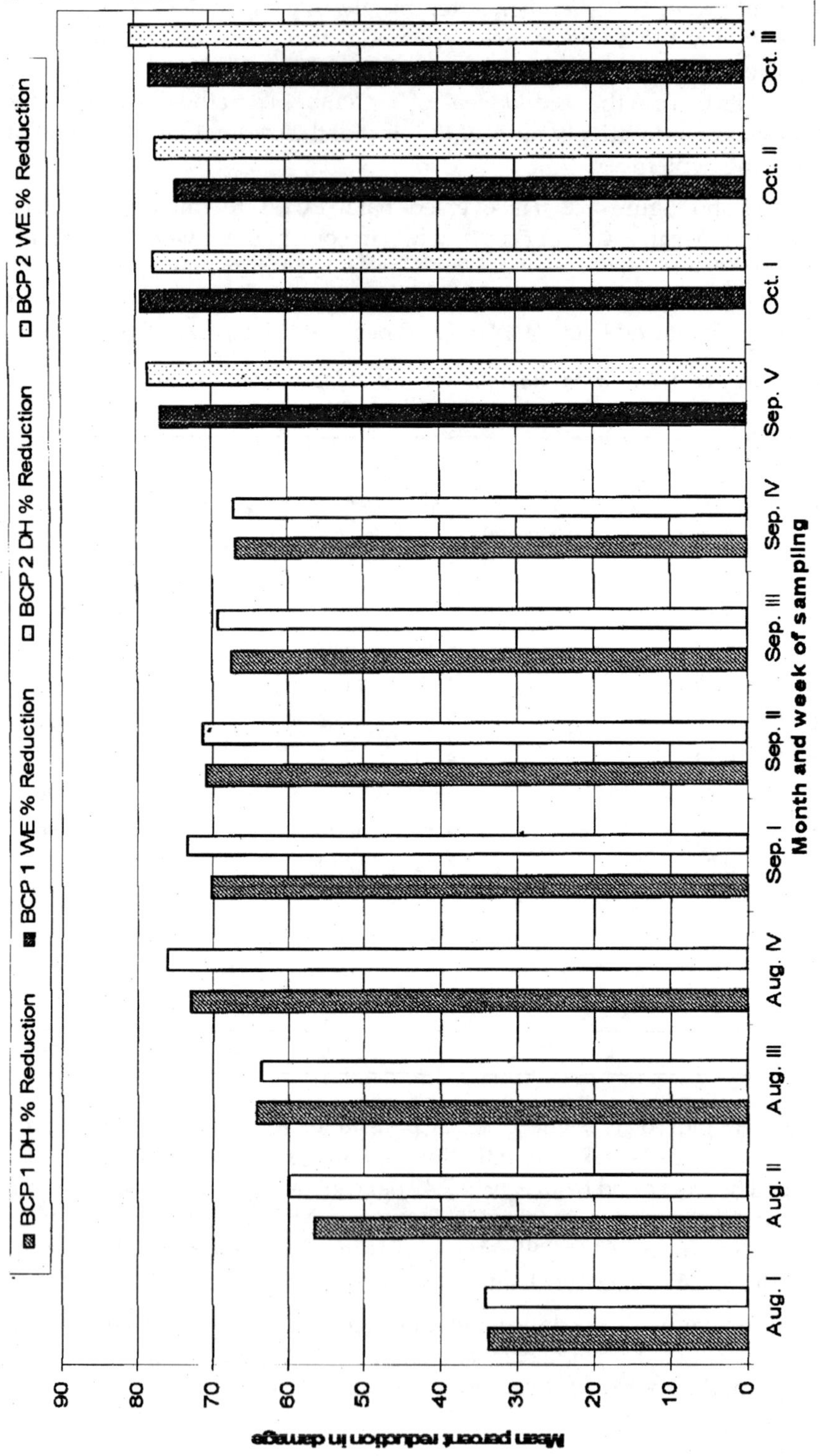

Figure 45.2: Bioefficacy of *Trichogramma* spp. Against Yellow Stem Borer

The results showed that overall reduction in tiller damage in the parasitoids released plots ranged from 33.77 to 80.65 per cent over control. On the basis of reduction in damage it can be concluded that inundative releases of *Trichogramma japonicum* and *T. chilonis* were effective against yellow stem borer in rice ecosystem. It helps to suppress the damage at the time of increased activity by the pest in untreated plots. The continuous releases also help to maintain the effective population density of the principal egg parasitoids which leads to a sustainable and effective management of yellow stem borer in the season.

Increasing population density of the parasitoids may not result in increased effectiveness as revealed in the studies. Since the reduction in damaged tillers over control did not differ significantly in the two treatments, which revealed that increased doses of egg parasitoids did not give any additional benefit. A dose of 1,00,000 parasitized eggs per hectare is sufficient enough for successful management of the yellow stem borer and 1,50,000 parasitized eggs per hectare only result in increased cost of protection without any significant additional benefit.

References

Brar, K.S., 2000. Biological control of insect pest of field crops. In: *One Day Seminar on Biocontrol of Pests*. Indian Society for the Advancement of Insect Science, Punjab Agricultural University, Ludhiana, pp. 4–6.

Geddes, A.M.W. and Iles, M., 1991. The relative importance of crop pests in south Asia. In: *Nat. Resour. Inst. Bull.* No. 39, Natural Resources Institute, Chatham, Maritime, Kent, U.K.

Katti, G., Pasalu, L.C., Verma, N.R.G. and Dhandapani, N., 2001. Quantification of natural biological control in rice ecosystem for possible exploitation in rice IPM. *Indian J. Ent.*, 63(4): 439–448.

Parez, L.A. and Cadapan, E.P., 1986. The efficacy of *Trichogramma* spp. as biological control agent against some rice insect pests. *Phillipp. Entomol.*, 6: 463–470.

Pathak, M.D., 1968. Ecology of common insect pests of rice. *Ann. Rev. Ent.*, 13: 257–273.

Singh, J. and Dhaliwal, G.S., 1994. Insect pest management in rice: A perspective. In: *Trends in Agricultural Insect Pest Management*, (Eds.) G.S. Dhaliwal and R. Arora. Commonwealth Publishers, New Delhi, India, pp 56–112.

Singh, S.P., 2001. Advances in biological control in India. In: *Proceedings of National Symposium on Biocontrol Based Pest Management for Quality Crop Protection in the Current Millennium*, July 18–19, PAU, Ludhiana, India, pp 12.

Verma, K., 2001. Studies on BIPM of major lepidopterous pests in rice ecosystem with special reference to tritrophic interaction. *M.Sc. Thesis*, G.B. Pant University of Agriculture and Technology, Pantnagar, Uttarakhand, India, 136 pp.

Chapter 46

Effect of Soil Solarization on Damping Off Disease Management in Vegetable Nursery

V. Bali Reddy, G. Bindu Madhavi, D. Venkata Reddy, V. Chenga Reddy, K. Gurava Reddy and M. Chandra Sekhar Reddy
Regional Agricultural Research Station, Lam, Guntur – 522 034, A.P.

ABSTRACT

Vegetable nursery management through soil solarization was investigated in relation to polythene thickness moisture status of nursery beds, temperature, microbial population, seedling stand, temperature difference between atmosphere and nursery beds was recorded 6–10°C under different moisture regime. Fungal population of pathogenic and non-pathogenic was changed after solarization with and without polythene. The extent of damping off reduction was 25.8 to 57.6 per cent over control treatment.

Introduction

Most of the vegetables crops are first raised in nurseries and later transplanted in the main field. Successful nursery rising is very difficult in *kharif* season due to damping off and weed population. Several soil borne pathogens cause heavy losses not only in nursery but also in main field after transplanting. Repeated sowing of seeds in the same plot results in a high inoculum build up in the soil through seeds and other susceptible planting materials. Tropical and subtropical part of the India is very hot during April to June. In these months solar radiation can be easily trapped by white polythene to increase heat of the soil for partial sterilization. This approach for disease and weed management has been studied thoroughly in Israel (Horowitz, 1980; Katan *et al.*, 1976). In India, soil solarization has been studied for controlling chick pea and pigeonpea wilt (Arya and Mathew, 1993). The results from the study as effect of solarization on vegetable crops are reported here.

Materials and Methods

The experiment was conducted during 2005–07 at Krishi Vigyan Kendra, Darsi, Prakasam District, A.P. Raised nursery beds were prepared in with 3 meter length, 80 cm width and 8" height. Dry, moist, wet and flood soil moisture levels were created by adding water in nursery beds during polythene mulching. White polythene of two different thickness *i.e.*, 80 µm and 130 µm were used for experimentation. Moisture percentage of nursery bed was calculated by weight (w/w) basis. Polythene was mulched between the last week of April to 2nd week of June. Polythene was tightly sealed from all sides through soil. The fungal population in soil was recorded colony forming units/g (Martin 1950) prior to polythene mulching and just after removal to determine microbial status of pre and post solarized soil. Total number of mature seedling stand was recorded in per square meter area basis after twenty-five days of sowing of seeds. Germination percentage of tomato chilli and brinjal was counted after 10 days of sowing. Each laboratory experiment was replicated six times while field experiments three times.

Results and Discussion

Polythene Thickness

Effect of polythene thickness revealed that 130 µm thick polythene is better than 80 µm for reducing the total fungal population and increasing the seedling stand of tomato and chilli in per unit area (Table 46.1). The total weed population was slightly increased from 42 to 64 under thick polythene treatment. The total fungal population, total seedling stand of tomato and chilli was lower under thin polythene. In the present finding, decrease of fungal population was recorded from thin to thick polythene mulching. Sharma and Sharma (2002) reported increase in total microbial population from 25 µm to 50 µm thickness. The difference in soil type and agro-climatic conditions might have interfered with microbial diversity. In this region 130 µm thick polythene is found better than 80 µm for reducing total fungal population. 300 gauge polythene sheets mulching resulted isolation on only six fungi from rhizosphere of pigeon pea (Arya and Mathew, 1993).

Table 46.1: Effect of Polythene Thickness on Vegetable Nursery Management in Soil Solarization

Treatments	*Solarized Beds*		*Unsolarized Beds (Control)*
	Thin Polythene (0.08 mm)	*Thick Polythene (0. 13 mm)*	
Total fungal population (cfu/g soil)	2.66×10^4	2.2×10^4	3.2×10^4
Total chilli seedling stand (area/m^2)	225.33	278.33	195.3
Total tomato seedling stand (area/m^2)	42268.3	292.3	193.0
Total weed population after seven days (area/m^2)	42	64	580.0

*: Average of three replications; **: Average of six replications.

Moisture Status of Nursery Beds

Four different moisture regime of nursery bed was calculated as 0.42, 2.9, 11.1 and 19.5 per cent respectively in dry, moist, wet and flood beds. The moisture content of control (unmulched) bed was 0.81 per cent w/w basis. The minimum temperature was recorded in morning while maximum temperature at 2:30 p.m. The study revealed that maximum temperature 52°C was recorded in moist nursery bed holding 2.9 per cent water content in w/w basis. The next higher temperature 51°C was observed in dry beds having 0.42 per cent water only. The wet and flooded solarized beds showed slightly lower temperature of 49°C and 48°C respectively. The temperature was 46°C in unsolarized control beds while 42°C at atmosphere. The minimum temperature varied from 35° to 37°C in the treatments. It indicated that minimum thermal regime in solarized beds in relation to moisture status was almost stabilized during whole period of solarization maximum temperature was variable. Tu *et al.* (1991) reported that highest temperature 57°C of the mulched plots and lowest 27°C in comparison to unmulched plots (39°C and 29°C). Dry and moist beds recorded 5°C and 6°C increase in soil temperature over control beds. Katan *et al.* (1976) reported that maximum 49.5°C temperature in mulched soil with 0.03 mm thick polythene. However, in present study the thickness of polythene was 0.13mm. Temperature increase was reported 10.3°C and 10.9°C respectively in unirrigated and irrigated soil by Sharma and Sharma (2002) under hilly region of Himachal Pradesh. The moist and dry nursery beds were equally effective for increase in temperature during solarization.

Effect of Nursery Diseases

Soil solarization of nursery beds revealed that it was very effective in reducing the damping off in tomato, chilli and brinjal. The seed germination was comparatively higher in solarized beds in comparison to unsolarized beds. Solarization has greater impact on these vegetable crops. The germination percentage of these crops varies from 74 to 82 in polythene mulched solarized beds (Table 46.2).

Table 46.2: Effect of Soil Solarization on Damping Off and Bacterial Leaf Spot Incidence in Nursery Beds

Crop	*Solarized Seedling Stand (Per cent)*	*Unsolarized (Control) Seedling Stand (Per cent)*
Tomato	78	62
Chilli	74	50
Brinjal	82	52

Effect of Soil Fungi

It is evident from the Table 46.3 that total soil fungi population was reduced in polythene mulched solarized beds. The change in total fungi was also recorded in pre-solarized beds to post solarized beds without polythene mulch. The fungi was also categorized in two groups *viz.*, pathogenic and non-pathogenic. The pathogenic fungal population was 11.8 × 10^3 colony forming units/g in pre-solarized bed and 10.8 × 10^3 cfu/g in solarized beds without polythene. It was drastically reduced with pathogen-solarized beds under three moisture regimes except flooded beds. It is clear that most of the pathogenic fungi reduced in maximum amount in moist beds and hence moist conditions of soil are effective for solarization. When non-pathogenic fungi were compared, the dry beds showed least population followed by moist beds. Interesting result observed that the total non-pathogenic fungi

Table 46.3: Effect of Moisture Status of Nursery Beds on Fungal Population (10^3 cfu/g soil) in Relation to Solarization

	Post-Solarization					
	Before Solarization	*Without Polythene*	*With Polythene*			
			Dry	*Moist*	*Wet*	*Flooded*
Non-pathogenic fungi						
A. terreus	8.0	10.0	8.2	6.0	5.3	3.5
A. quadrilinatus	10.0	11.5	2.0	1.9	1.5	4.5
A. niger	8.0	4.2	1.2	3.2	3.0	4.3
A. flavus	3.5	2.6	1.0	0.8	1.2	2.3
Pencillium sp.	1.0	1.5	3.0	8.7	12.5	2.5
Unidentified	4.0	3.5	2.5	1.5	3.3	2.5
Sub total non-pathogenic	34.5	33.3	17.9	22.1	26.8	19.6
Pathogenic fungi						
S. rolfsii	3.0	2.0	0.0	0.3	1.5	1.9
Fusarium spp.	2.5	3.0	1.2	0.9	1.0	2.0
Rhizoctonia solani	2.0	2.0	2.0	0.4	0.5	0.9
Pythium spp.	3.0	2.8	2.5	2.2	2.0	2.1
Curvularia lunata	0.5	0.3	0.0	0.0	0.0	0.0
Macrophomina phaseolima	0.8	0.7	0.4	0.0	0.0	0.0
Sub total pathogenic	11.8	10.8	5.3	3.9	5.4	8.3
Total fungi	46.3	44.1	23.2	26.0	32.2	27.9

were more in post-solarized non-mulched beds. Because thermophilic nature of the fungi resting spores of fungi are activated resulting in increasing fungal population. Similar results were obtained with *Fusarium* sp. also where chalymodospore have activated after solarization and gave more count after planting. However, after polythene mulching, these fungi were killed and reduced. It was due to increase of latent temperature. However, total non-pathogenic fungi also decreased in polythene mulched solarized beds.

References

Arya, Arun and Mathew, D.S., 1993. Soil solarization of *Fusarium udam*. *Indian Phytopathology*, 46(2): 151–154.

Horowitz, M., 1980. Weed research in Israel. *Weed Sci.*, 28: 457–460.

Katan, J.A. Greenberger, Alon, H. and Grinstein, A., 1976. Solar heating by polythene mulching for the control of diseases caused by soil borne pathogens. *Phytopathology*, 66: 683–688.

Martin, J.P., 1950. Use of acid, rose Bengal and streptomycin in the plate method for estimating soil fungi. *Soil Science*, 69: 215–233.

Pandey, K.K., 2001. Important diseases of vegetable crops and their integrated management. In: *National Training Course on Vegetable Production Technology*, Directorate of Extension, Ministry of Agriculture, New Delhi, pp. 177–186.

Sharma, M. and Sharma, S.K., 2002. Effect of soil solarization on soil microflora with special reference to *Dematophora necatrix* in apple nurseries. *Indian Phytopathology,* 55(2): 158–162.

Tu, C.C., Cheny, Y.H. and Hwang, S.C., 1991. Effect of solar energy and green manures on the control of southern blight of tomato. *Plant Protection Bulletin (Taipei),* 33(1): 95–102.

Chapter 47

Screening of Eight Plants for their Repellent Action Against Adult Khapra Beetle, *Trogoderma granarium* (Everls) Infesting Wheat Grain

S.C. Dwivedi and Santosh Yadav
Ecotoxicology Laboratory, Department of Zoology, University of Rajasthan, Jaipur – 302 004, Rajasthan

ABSTRACT

Acetone and pet-ether extract of *Emblica officinalis, Prosopis spicegera, Ricinus communis, Tagetes erecta, Cassia fistula, Gossypium arboreum, Zinnia angustifolia, Nyctanthes arbortristis* were screened for their repellent action against khapra beetle. The khapra beetle showed strong repellent behaviour against *Gossypium arboreum* in acetone extract and minimum repellency was recorded in *Tagetes erecta* pet ether extract. However, repellent property has been confirmed in all the plant species.

Keywords: *Plant extracts, Repellent action, Trogoderma granarium.*

Introduction

Trogoderma granarium is a devastating and formidable pest of stored wheat and wheat products, which belongs to order coleoptera and family dermestideae. The distribution of this pest is extensive, more common in hot arid countries like India, west to the mediterranean and north east Africa (Burges 1959).

Frequent use of insecticides and fumigants has led to resistant to insect and environmental pollution due to bio-accumulation of residues in food grains.

This inspired to undertake this study on some biodegradable plant origin insecticides for the management of this pest which cause irritability, growth retardation and reduced reproduction abilities in the insects.

In comparison to the chemical insecticides, plant origin insecticides are not only easily biodegradable, less pollutant to the environment but are economical also and readily available to the farmer.

Many indigenous plants have been reported to posses repellent property against Khapra beetle. Abrol and Chopra (1963) had reported several indigenous plants for their repellent action against Khapra beetle. Dwivedi and Bajaj (2000) assessed *cassia* leaf extract and later Dwivedi and Sharma (2002) investigated repellency of 5 plant extracts *viz. Chenopodium album, Achras Sapota, Trigonella Foenum graecum, Quisqualis indica* and *Tamarindus indica* against Khapra beetle. Though substantial work has been carried out on the effect of plants and their products on various parameters against stored product pest (Gundu Rao and Majumdar 1962, Nawrot et al. 1982, Ahmed and Eapen 1986 and Quadri 1973) however, very scanty work has been carried out on aboriginal flora for their efficacy as repellent as such the present stufly is aimed to explore the possibility of using plant extracts as repellent which may prove effective wheat grain protectants and reduce population of *Trogoderma granarium.*

Materials and Methods

Rearing of *Trogoderma granarium* (Everts) was done in laboratory in presterilized jars containing disinfected wheat grain by providing optimum conditions, of 35±2°C temperature of 60±10 per cent RH. 30 gms of powdered plant material was extracted in soxhlet apparatus for 8 hrs. in 300 ml of solvent as per Deshmukh and Borle, 1975 method using acetone and pet ether as solvent.

The extract was filtered and after vacuum evaporation it was kept in refrigerator as stock solution (100 per cent Conc.).

The repellency was estimated by 'Y' shaped alfactometer following Read et al (1970). The olfactometer has three arms *viz.* Base arm, Control arm, and Experimental arm. Each arm of olfactometer measures 6 cm in length and 1 cm in diameter. A piece of sponge soaked in I ml of plant extract (100 per cent Conc.) was placed in experimental arms, whereas control arm contained soaked sponge piece in the same amount of solvent. 10 newly emerged adults were released in the center of the olfactometer through the base arm and after 30 min. the number of individuals in control, base and experimental arms were counted. Three replica were run for each plant extract. The repellency percentage was calculated using the formula suggested by Granett *et al.* (1949).

$$\text{Per cent Repellency} = \frac{\text{Insects in cont. arm} - \text{Insects in expt. arm}}{\text{Insect in cont. arm}} \times 100$$

Results and Discussion

Trogoderma granarium exhibited repellent behaviour against plant extracts. The data given in Table 47.1 reveal that *Gossypium arboreum* acetone extract exhibited maximum repellency (92.26 per cent) whereas *Tagetes erecta* pet ether extract registered minimum repellency (59.97 per cent). *Gossypium's* pet ether extract recorded 76.14 per cent repellent action. These result are supported by the findings of Ahmed and Eapen (1986) and Giga and Munestsi (1990). Repellent properties of *Gossypium* can be attributed to its Gossypol contents (Bottger, and Patana, 1966).

Table 47.1: Response of *T. granarium* (Everts) to Eight Plant Acetone and Pet Ether Extracts

Sl.No.	*Name of Plant*	*Family*	*Mean Per cent Repellency in Acetone*	*Mean Per cent Repellency in Pet Ether*
1.	*Gossypium arboreum* (seeds)	*Malvaceae*	92.26	76.142
2.	*Ricinus communis* (seeds)	*Euphoarbiaceae*	87.5	84.033
3.	*Cassiafistula* (leaf)	*Leguminaceae*	78.328	71.428
4.	*Zinnia angustifolia* (leaf)	*Asteraceae*	87.5	75.046
5.	*Nyctanthes arbortristis* (seeds)	*Oleaceae*	78.22	65.067
6.	*Tagetes erecta* (floral head)	*Asteraceae*	62.476	59.97
7.	*Prosopis spicegera* (leaf)	*Leguminaceae*	76.501	68.40
8.	*Emblica officinalis* (fruit)	*Euphorbiaceae*	76.501	62.476

Ricinus communis and *Zinnia angustifolia* exhibited significant results against Khapra beetle. *R. Communis* exhibited 87.5 per cent repellency in acetone and 84.03 per cent in pet ether extract, which could be due to ricin ingredients present in *R. Communis* which was isolated and identified by Kwon *et al.* (1991). Further result are in conformation with the findings of Kachare (1994) and Babu *et al.* (1989).

Zinnia angustifolia registered 87.5 per cent repellency in acetone and 75.046 per cent in pet ether extract. *Cassia fistula* exhibited 78.328 per cent 'and 71.428 per cent in acetone and pet ether respectively. The results are in accordance with the observations made on various stored product pest using different botanical products by Abrol and Chopra (1963).

Rich glycosidal contents of *cassia* are the suggested chemical factor behind its repellent potential. Similar, repellent action has also been noticed in other glycoside rich plants such as *Ageratum conyzoides* against *C. chinensis* (Pandey *et al.* 1986).

Nyctanthes arbortristes and *Prosopis spicegera* registered 78.22 per cent and 76.501 per cent in acetone and 65.067 per cent and 68.40 per cent repellency in pet ether extract respectively. These findings find support from observation-recorded by Ramamurthy and Venugopal (1997). They found that Neem oil at 1,2 and 3 per cent; NSKE at 2, 3 per cent; leaf extract of *Prosopis, Ipomoea, Nochi* and *Ocimum* at 10 per cent and ether extract of vitex at 1 per cent when applied reduced *Sitotroga cerealella* infestation in field.

Emblica officenalis and *Prosopis spicegera* acetone extract registered 76.501 per cent repellency whereas *E. officenalis* pet ether results are at par with *T. erecta* having 62.476 per cent repellency these results are in accordance with the observations recorded by Dwivedi and Shekhawat (2004).

These results are also find support from the observation of Hethelyi *et al.* (1988) who determined the insecticidal properties of oils of *T. erecta, T. minuta* and *T. patula.*

From the results recorded, it can be concluded that acetone extract out last in repellent properties pet ether extract. This presumption gets support from the observation made by Nawrot *et al.* (1982).

Thus, it can be inferred that the aforementioned plant extract act as good repellent which may be suggestive of alternative to chemical/synthetic pesticides.

Acknowledgement

The authors feel great pleasure to thank Professor and Head of Zoology Department, Dr. P.P. Bakre for providing necessary facilities.

References

Abrol, B.K. and Chopra, I.C., 1963. Development of indigenous vegetable insecticides and insect repellent. *Bull RRL, Jammu*, 1(2): 156–158.

Ahmed, S.M. and Eapen, M., 1986. Vapour toxicity and repellency of some essential oils to insects pest. *India Perfume*, 30(1): 273–278.

Babu, T.R., Reddy, V.S. and Hussaini, S.H., 1989. Effect of edible and non-edible oils on the development of the pulse beetle (*C. chinensis)* and on viability and yield of Mungbean. *Tropical Sci.*, 29: 215–220.

Bottger, G.T. and Patana, R., 1966. Growth development and survival of certain Lepidoptera fed *gossypol* in the diet. *J. Econ. Entomol.*, 59(5): 1166–1168.

Burges, 1959. Studies on the Dermestid beetle, *T. granarium* Everts. III. Ecology in mott storage. *Ann. Appl. Biology,* 47: 445–462.

Deshmukh, S.D. and Borle, M.N., 1975. Studies on insecticidal properties of indigenous plant products. *Ind. J. Ent.*, 35(1): 111–118.

Durront J.H., 1921. Insects associated with grain and C. Repts, grain pests (war) committee. *Royal Society. London*, 9: 33–52.

Dwivedi and Bajaj, 2000. Repellent action of seven plant extract against *T. granarium* (Caleoptera : Dermestidae). *Uttar Pradesh J. Zool.*, 20(1): 97–99.

Dwivedi and Sharma, Y., 2002. Investigation on repellent responses of Khapra beetle *T. granarium* (Coleoptera : Dermestidae) to five plant species. *Indian Biologists,* 34(2): 55–58.

Dwivedi, S.C. and Shekhawat, N.B., 2004. Repellent effect of some indigenous plant extracts against *T. granarium* (Everts). *Asian J. Exp. Sci.*, 18(1–2): 47–51.

Giga, D.P. and Munestis, M.M., 1990. The effectiveness of vegetable and citrus oil as protectants of cowpea against infestation by *Callosobruchus rhodesianus* (Pic). *Plant Protection Quarterly,* 5(4): 148–151.

Granett, P., Haynes, H.L., Cannola, D.P., Bowery, T.G. and Barker, G.W., 1949. Two butoxypolyene glycol compound as fly repellents for live stock. *J. Econ. Ent.*, 42(2): 281–286.

Gundo Rao, H.R. and Majumdar, S.K., 1962. Repellency of species, aromatic plant material and essential oil to adults of *T. Castaneum.* In: *Proceedings Symposium on Utilization of Medicinal Plants,* Jammu (India).

Hethelyi, E., Tetenyi, P., Kaposi, P., Danos, B., Kerocz, Z. and Bubi, G., 1988. GC/MS investigation of antimicrobial and repellent compounds. *Herba Hungarica*, 27(2–3): 89–105.

Kachare, B.V., Khaire, V.M. and Mote, U.N., 1994. Efficacy of different vegetable oil as seed treatment in increasing storage ability of pigeonpea seed against pulse beetle. *Ind. J. Ent.*, 56(1): 58–62.

Kwon, O.K., Song, B.H., Park, H.M. and Jeong, Y.R., 1991. Identification of insecticidal ingredient in castor oil plant against the brown plant hopper. Research reports of the rural development administration. *Crop Protection*, 33(2): 74–79.

Nawrot, J., Bloszyk, E., Grabraczyk and Drozbz, B., 1982. Feeding detterent activity of the compositae plant extract for the selected storage pests. *Prace-Naukowe-Instytutu-Ochrony Roslin*, 4(1): 37–44.

Pandey, N.D., Mathur, K., Pandey, S. and Tripathi, R.A., 1986. Effect of some plant extract against pulse beetle, *C. chinensis* Linn. *Ind. J. Ent.*, 48(1): 85–90.

Qadri, S.S.H., 1973. Some new indigenous plants repellents for storage pests. *Pesticides*, 7(12): 18–19 and 22.

Rood, D.P., Feency, P.P. and Root, R.B., 1970. Habitat selection by the aphid parasite, *Diaerchtiella rapae* (Hymenoptera : Braconidae) and Hyper parasite *Charpis brasicae* (Hymenoptera : Cypinidae). *Can. J. Ent.*, 102: 1567–1578.

Chapter 48

Some Aspects on Vision in Relation to Light in Birds

P. Jasmine, M. Surendran and D.J. Prakash
Neurophysiology Laboratory, Postgraduate and Research Department of Zoology, Voorhees College, Vellore – 632 001, Tamil Nadu

ABSTRACT

While the bimodality is an universal phenomenon in skeletal muscles of vertebrates, it was of interest to find out the influence of presence or absence of light on the nature of caliber spectra of optic nervers in avian species particularly Pigeon and Owl. The fibre diameter analysis shows and unimodal pattern in both the species. It is not able to identify what type of never fibre is connected to what type of retinal cell. Therefore it is suggested that a direct correlation is essential to establish this fact. The results have been appropriately discussed.

Keywords: *Influence of light on vision, Pigeon and owl, Optic nerve fibre, Diameter pattern-myelin thickness, Fascicles.*

Introduction

Of all the nerves, the optic nerve has always been of particular interest, because despite its appearing to leave the brain in the manner of peripheral nerve, it is a central tract of the nerve fibres.

As such several electro physiological properties have been studied in considerable detail. However, the histological studies of the optic nerves of vertebrates particularly diurnal and nocturnal birds are few, and not always in agreement. Earlier studies of this central tract Bishop and O'Leary, 1942; Bishop, Jeremy and Lance, 1953, Chang, 1956; Lennox, 1958; Sefton and Surnburn, 1964) have shown that is contains groups of fibres differing from each other in there should, conduction velocity and destination in the brain.

It might be expected that those fibre groups differing in conduction velocity would also form distinct groups when arranged according to the fibre diameter (Hursh, 1939; Rushton, 1962).

The fibre diameter analysis of the optic nerve has been the attempted by various investigators (Bishop *et al.*, 1953; Chacko, 1954). Such an analysis describes a unimodal distribution of fibre diameter and different peaks. There appears to be some discrepancy between the results of various workers with regards to the number of peaks.

All the above the studies were used histological methods which stained the myelinated fibres only. It is therefore the present study was undertaken to find out the pattern fibre diameter distribution; to resolve the discrepancies as regards the fibre caliber spectra of optic nerves of pigeon (Diurnal) and owl (Nocturnal).

This would give an insight in understanding the physiological aspects about the diurnal and nocturnal vision.

Materials and Methods

Animals

All the experiments were conducted on adult Pigeon (Columbia livia; Class: Aves; Order: Coulmbiformes) and Owl (Bubo bubo: Class: Aves; Order; Strigiformes). The pigeon weighed about 260 grams. While the owl weighed 250 grams. Only adult male animals were used.

Dissection of the Optic Nerve

Following the opening of the skull and the location of optic chiasma, a short-length of optic nerves from both left and right sides were traced to the base of the eye balls and removed intracranially. The removed nerve was placed in Ringer's solution. The removed nerves were fixed in 1 per cent unbuffered Osmium tetroxide for 7 days. Following the fixation, the tissues were washed is distilled water for three hours to remove the fixative. The tissues were then treated with graded series of alcohol from 10 per cent to 100 per cent. After dehydration, the materials were cleared in xylene and embedded in way. Sections were cut at 8 μm thickness.

Dissection of the Eye Ball

The eye balls are spherical situated one on either side of head in orbits. The exposed eye ball was dissected out to free it from the surrounding connective tissues. The removed eye balls were then fixed in aqueous Bouin's fixative. The dehydration was done by passing the tissue through the ascending series of alcohol that is ilel, 10 per cent to 70 per cent. The impregnated tissues were then embedded 8 μm thick sections were cut using Rotary microtome. The slides were stained with heamatoxyline and counterstained using eosin.

Analysis of Nerve Fibres

Number of Fascicles

The number of fascicles occupied in the cross sectional area in the entire trunk of each nerve was counted.

Measurement and Counting of Fibres

A sampling method was used in measuring the fibres. No problem was encountered with the circular nerve fibres while the maximum and minimum diameters were averaged out to mean diameters in non-circular fibres.

Thickness of Myelin Sheath

The difference between the external and internal myelin sheath diameters was calculated and converted to microns.

Results

From the study some interesting observations have emerged.

Number of Fascicles

Pigeon

The number of fascicles in the left optic nerve of pigeon was 31. The right optic nerve contain about 37 fascicles. The number of nerve fibres vary in these fascicles. The area of the fascicles is comparatively smaller than that of owl. The size of the nerve fibres varies from smaller to larger diameters.

Owl

In owl the number of fascicles vary both in left and right optic nerves. The left optic nerve show about 34 fascicles while the right optic nerve 42 fascicles. Each fascicles consists of nerve fibres of varying diameters; the smaller, medium and larger nerve fibres which are randomly distributed.

Diameter of Afferent Fibres

Pigeon

In Pigeon the caliber spectra of the nerve fibres of optic nerves in pigeon show the distribution of fibres in range between 1–2 μm and 6–7 μm. The fibres less than the diameters of 2 μm are absent.

The left optic nerve shows an unimodal pattern showing a single peak in the range 3–4 μm (13787). The next maximum number of fibres is found in the range 2–3 μm (11813). The fibre diameter with a range of 4–5 μm contain 11482 fibres varies the tapering diameters in the range between 5–6 (4253) to 6–7 (3255) μm.

Owl

The calibre spectra of the nerve fibres of optic nerves in owl show the distribution of fibres in the range between 1–2 μm and 6–7 μm. The fibres less than the diameters of 2 μm are absent. The left optic nerve shows and unimodal pattern showing a single peak in the range 3–4 μm (19956). The next maximum number of fibres is found in the range 2–3 μm (14744). The fibre diameter with a range of 4-5 μm contain 11650 fibres varies the tapering diameters in the range between 5–6 to 6–7 μm.

Myelin Thickness

Pigeon

The range of myelin thickness in left optic nerve fibre is from 1–3 μm. 1–2 μm is larger than the other larger fibres with thicker myelin sheath (17527 and 20012 fibres). Accordingly the fibres with the thickness of myelin in the range 3 –4 are a fewer in the number (15125 fibres).

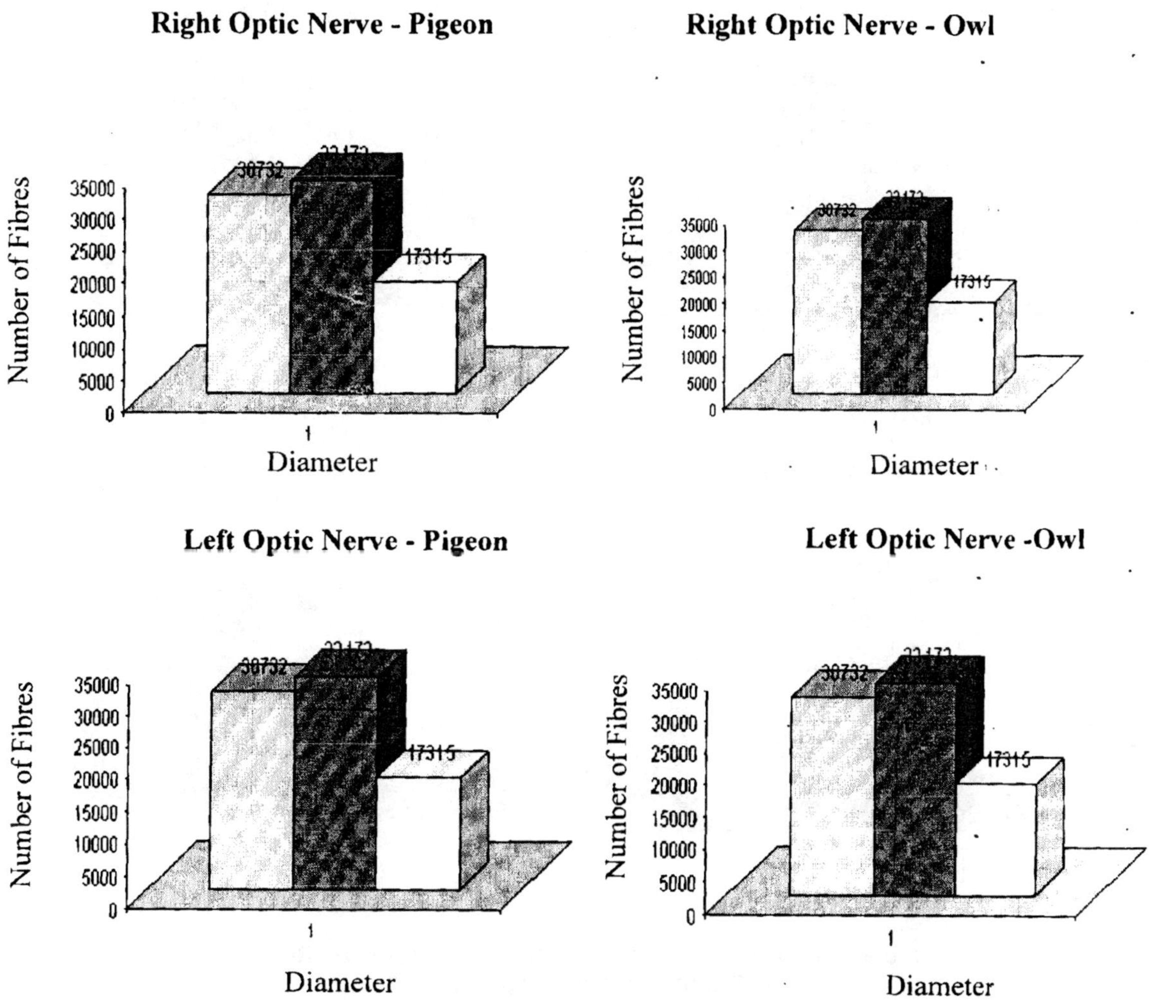

Figure 48.1: Calibre Spectra of Right and Left Optic Nerves in Pigeon and Owl

The myelin thickness of right optic nerve fibre in pigeon ranges from 1–4 µm. The fibres with 2 to 3 µm thickness of myelin contain more number of fibres. Where as fibre diameter in 1–2 have the next larger number of fibres (14744). The larger fibres are fewer in number bur contain 11650 fibres.

Owl

The range of myelin thickness in left optic nerve fibre is from 1–3 µm. From the observations it is clear that the number of fibres with myelin thickness 1–2 µm is larger than the other larger fibres with thicker myelin sheath (30732 and 33173 fibre). Accordingly the fibres with the thickness of myelin in the range 3–4 are a fewer in number (17315 fibres).

The myelin thickness of right optic nerve fibre in owl ranges from 1–4 µm. The fibres with 2–3 µm thickness of myelin contain more number of fibres. Where as fibre diameter in 1–2 have the next larger number of fibres (25755). The larger fibres are fewer in number but contain 13562 fibres.

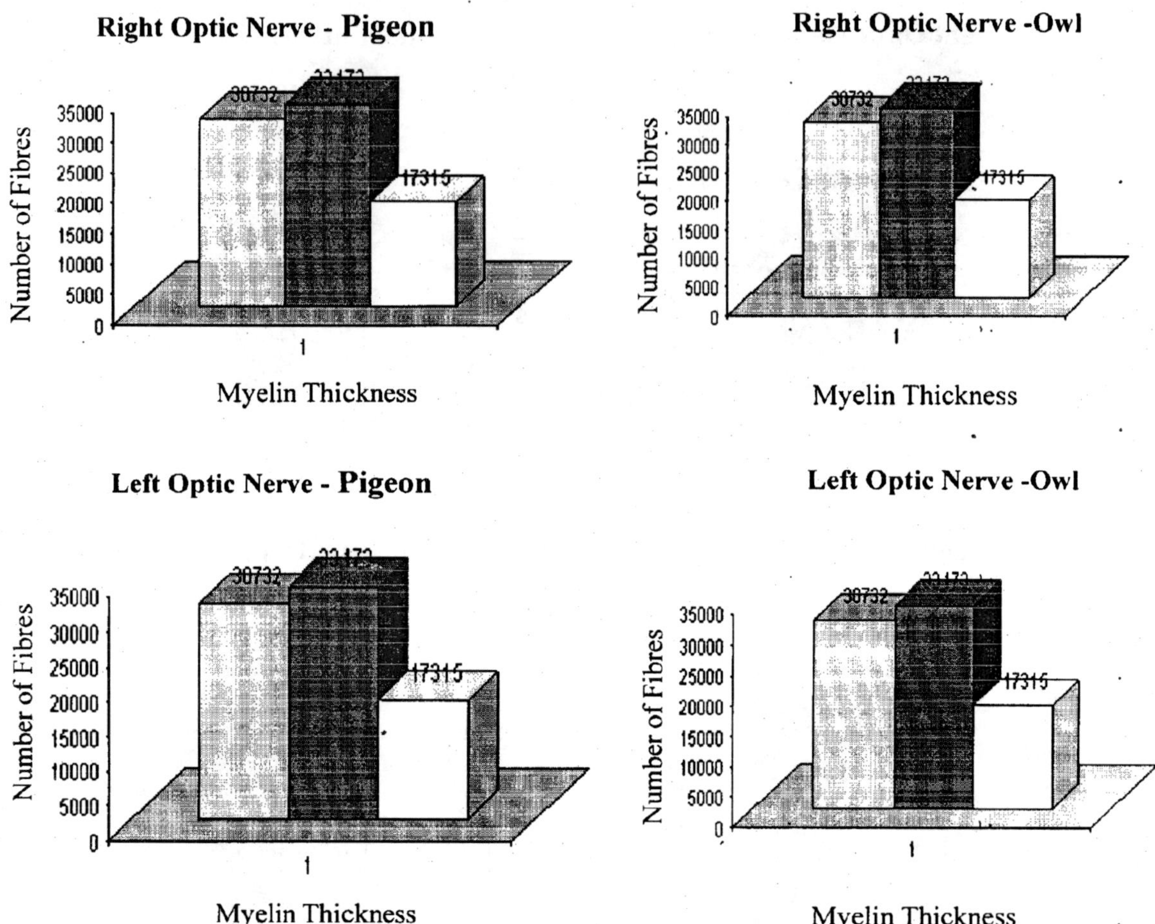

Figure 48.2: Myelin Thickness and Number of Fibres in Right and Left Optic Nerves in Pigeon and Owl

Discussion

Histological study can provide functionally specific spectra of nerve fibres (Boyd and Davey, 1968) close of their termination, as fibres can be divided into functional groups on the basis of being seen to supply different structures. For the main nerve trunks, histology can usually provide only a single over all spectrum, as at this site the fibres cannot be divided by function since if is then usually impossible to trace an individual fibre to its termination.

Fascicles

The fascicles contain nerve fibres of mixed diameter (all sizes) even though a large number of small fibres distributed peripherally while large fibres centrally. The larger and smaller fibres are not separately distributed. This clearly shows that the nerve axons with the a particular function are most grouped in to a separate fascicle (Chacko, 1954).

Secondly, with the increased number of fascicles, the number of fibres the also increased (Bishop *et al.*, 1953). However, number of fibres in one fascicle is not the same in the other. Each Fascicle shows not only variations in size but also the number of fibres.

Percentage of Fibre Distribution

The nerve fibre distribution shows what type of fibres seen and how they are distributed plus the diameter of the same. Accordingly, it has already been observed that an unimodal pattern is prevailing in the optic nerves of Pigeon and Owl. The percentage of fibres represents the available number of fibres in the respective diameters of fibers among the total fibres. Hence, the pattern should necessarily reflect the histogram of the fibre diameter. Therefore, the unimodal pattern shows the dominance of a particular group of fibres while others do not. This suggests that the fibres with the diameter 3–4 μm of unknown origin play a significant role in the vision of the birds (Donvan, 1967).

Myelin Thickness

Myelin thickness appears to be an important issue in the conduction of incormation from photoreceptors to the brain (Lennox, 1958). The axons of nerve fibres are encased in a proteolipid sheath extending from just distal to the cell body to within 2 μm of the terminal synapses.

Microchemical studies have shown that the wavy network of an axon is referred to as a neurokeratin. The chemical nature of myelin differs in the peripheral nerves from that observed in the central nervous system (Hursh, 1939). The myelin sheaths are also appearently formed by processes derived from glial cells. Luse (1956) also suggests that myelin is formed centrally by glial cell process.

Conduction Velocity

The results of electro physiological investigations of the optic nerves of Pigeon and Owl indicate that there are many groups of nerve fibres with respect to conduction velocity. The important major groups appear to categorize about those fibres with respect to conduction velocities of 12–18 m/sec; 6–12 m/sec; 12–18 m/sec; 18–24 m/sec; 24–30 m/sec; 30–36 m/sec and 36–42/Sec, using the conversion factor of Hursh (1939). In this study, fibres were seen as larger as 6–7 μm, the smallest being 1.0 μm. But only a single size grouping revealed with peak at 3–4 μm. Thus it is quite possible that there are definite functional groups of nerve fibres with in the pigeon and owl optic nerves without such groupings appearing as distinct elevations in the total fibre caliber spectrum. Increase in thickness and complexing is accompanied by waves of retinal cell death's which are probably related in some way to the orderly establishment of the precise neuro sensory network (Sefton and Surbum, 1964). The intricate and interrelated process lead to the architecture that apparently comprise a self-directing and self-sustaining complex.

References

Boyd, I.A. and Davey, M.R., 1968. *Composition of Pheripheral Nerves.* Edinburgh: Living Stone.

Bishop, G.H. and O'Leary, J.L., 1942. Factors determining the form of the potential record in the vicinity of the synapses of the dorsal nucleus of the lateral geniculate body. *J. Cell. Comp. Physiol.*, 19: 315–331.

Bishop, P.O., Jeremy, D. and Lance, J.W., 1953. The optic nerve properties of a central tract. *J. Physiol. Land.*, 121: 415–432.

Chacko, L.W., 1954. Fibre groups in the primary optic pathway of the eat. *J. Neurophysiol.*, 19: 224–231.

Chang, H.T., 1956. Fibre groups in the primary optic pathway of the cat. *J. Neurophysiol.*, 19: 224–231.

Donvan, A., 1967. The nerve fibre comporition of the cat optic nerve. *J. Anat.*, 101: 1–11.

Hursh, J.B., 1939. Conduction velocity and diameter of nerve fibres. *Am. J. Physiol.*, 127: 131–139.

Lennox, M.A.C., 1958. Single fibre responses to electrical stimulation in cat's optic tract. *J. Neurophysiol.*, 21: 62–69.

Lase, S.A., 1956. Formation of Myelin in the central nervous system of mice and rats, as studied in the electron microscope. *J. Biophys. Biochem. Cytol.*, 2: 777–784.

Rushton, W.A.H., 1962. *Visual Pigments in Man.* Liverpool University Press, Liverpool.

Sefton, A.J. and Surburn, M., 1964. Electrical activity of lateral geniculate nucleus and optic tract of the rat. *Vision Res.*, 4: 315–328.

Chapter 49

On Some Ciliates (Protozoa) of Bor Wildlife Sanctuary, Maharashtra, India

L. Bindu*

Zoological Survey of India, M-Block, New Alipore, Kolkata – 700 053

ABSTRACT

8 species of free living ciliates (protozoa) were identified from the water bodies of Bor wildlife sanctuary in Maharashtra of which 6 species are new records from Maharashtra and 2 species are reported for the first time from the Bor wildlife sanctuary.

Keywords: *Protozoa, Ciliates, Bor wildlife sanctuary, Maharashtra.*

Introduction

Bor wildlife sanctuary covers an area of 61 sq. km. in Nagpur district in the North-eastern part of the Maharashtra state and is covered by mixed deciduous and teak forests and bamboo brakes. This is the home of many interesting species of animals and birds. The protozoan fauna of this area is completely unattended and as part of the faunistic survey of the Maharashtra State of the Zoological Survey of India, Kolkata collections were made from the sanctuary and examined for the occurrence of the protozoa. It is herewith reporting 8 species of ciliates from the water bodies spread over 6 families and 6 orders.

---

* E-mail: bindulajapathi@yahoo.com.

Systematic Account

Phylum: CILIOPHORA

Class: KINETOFRAGMINOPHORA

Subclass: GYMNOSTOMATIA

Order: PROSTOMATIDA

Family: PRORODONTIDAE

1. *Prorodon discolor* (Ehrenberg)

1835. *Holophyra discolor* Ehrenberg, Abh. Preuss. Acad. Wiss.. Berlin, p. 251.

1930. *Prorodon* (Holophrya) *discolor* Kahl. Urtiere oder protozoa, (in: Dahl's Tuerwelt Dtsch.), Jena, pt. 18, p. 76, Figure 8(4).

Material Examined

4exs.. Bor WLS, 25-09-2007, freshwater, Coll. Bindu L.

Diagnosis

Body oval, narrower posteriorly, mouth terminocentral, pharyngeal tube with trichites extending above one-fourth of the body; macronucleus single, rounded; contractile vacuole single and situated near posterior end.

Distribution

West Bengal, India.

Remarks

New record from Maharashtra.

Order: PLEUROSTOMATIDA

Family: AMPHILEPTIDAE

2. *Loxophyllum levigatum* Sauerbey

1928. *Loxophyllum levigatum* Sauerbrey, *Arch Protistenk.*, Jena, 62, p. 377.

Material Examined

2exs., Bor WLS, 25-09-2007, freshwater, Coll. Bindu L.

Diagnosis

Body leaf-like, anterior end with distinct beak, posterior end rounded; body margin smooth; macronuclei 7–15 in number, globular and scattered; contractile vacuole single and located at posterior end on dorsal side.

Distribution

West Bengal, India.

Remarks

New record from Maharashtra.

Order: KARYORELICTIDA

Family: LOXODIDAE

3. *Loxodes vorax* Stokes

1885. *Loxodes vorax* Stokes, Amer. Mon. micro J., 6, pp. 121–127.

Material Examined

3exs., Bor WLS., 25-09-2007, freshwater, Coll. Bindu L.

Diagnosis

Body leaf-like, flattened, anterior end beak-like and curved ventrally; posterior end rounded; macronuclei two in number and lying very close to each other.

Distribution

West Bengal, India.

Remarks

New record from Maharashtra.

Order: TRICHOSTOMATIDA

Family: PLAGIOPYLIDAE

4. *Plagiopyla nasuta* Stein

1860. *Plagiopyla nasuta* Stein, S.B. Ges Wiss., Prague, p. 58.

Material Examined

3exs., 25-09-2007, freshwater, Coll. Bindu L.

Diagnosis

Body reniform or ovoid, narrower anteriorly; cytostome situated near median line at the end of peristome; cytopharynx long; single and terminal contractile vacuole; macro-nucleus round.

Distribution

West Bengal, Orissa, India.

Remarks

First record from Maharashtra.

Order: COLPODIDA

Family: COLPODIDAE

5. *Colpoda cucullus* Muller

1773. *Colpoda cucullus* O.F. Muller, *Verminum terrest et. Fluviatil S. animal infusor. etc. historia., Hafniae et Lipsiae,* I and II.

Material Examined

2exs., 25-09-2007, freshwater, Coll. Bindu L.

Diagnosis

Body typically kidney shaped, cytostome located about middle of the body, frontal dentations 8–10 and meridians 29–34; macronucleus oval; contractile vacuole single and posterior.

Distribution

West Bengal, Andhra Pradesh, Assam, Karnataka, Jammu and Kashmir, Maharashtra, Madhya Pradesh, Orissa, Punjab, Tamil Nadu and Uttar Pradesh.

Remarks

First record from Bor WLS.

6. *Colpoda maupasi* Enriques

1908. *Colpoda maupasi* Enriques, Arch. Zool. Exp. Gen., Notes et Rev. (4) 8, pp.7–11, Figs. 3–5,9.

Material Examined

3exs., 25-09-2007, freshwater, Coll. Bindu L.

Diagnosis

Body oval and cylindrical, cytostome located at about one-fourth of the body from the anterior end, frontal dentations 6–7 and meridians 16–18, macronuclcus spherical.

Distribution

Bengal, India (exact locality not cited) (see Sandon, 1927).

Remarks

First record from Maharashtra.

Order: HYMENOSTOMATIDA

Family: FRONTONIIDAE

7. *Frontonia depressa* (Stokes)

1886. Colpoda depressa Stokes, *Ann. Mag. Nat. Hist.,* (5)17.

Material Examined

4exs., 25-09-2007, freshwater. Coll. Bindu L.

Diagnosis

Body ovoid, flattened, body size small, contractile vacuole single with strong radiating canals and located below the middle, macronucleus short, sausage-shaped with single micronucleus.

Distribution

West Bengal, India.

Remarks

New record from Maharashtra.

8. *Frontonia leucas* (Ehrenberg)

1838. *Bursaria leucas* Ehrenberg. *Die. Infusionsthierchen als Volkommene organisimen, Leipzig,* p. 329. pl. 34. Fig. 8.

Material Examined

6exs., 25-09-2007, freshwater, Coll. Bindu L.

Diagnosis

Body elongated or ovoid, rounded at both ends, body size larger, contractile vacuole single with long radiating canals and located at the middle of the body; macronucleus ellipsoid with several micronuclei.

Distribution

West Bengal, Jammu and Kashmir, Rajasthan and Maharashtra, India.

Remarks

First record from Bor WLS.

Acknowledgement

The author is grateful to Dr. Ramakrishna, Director, Zoological Survey of India, Kolkata for the facilities provided. The thanks are also due to Dr. A.K. Das, Emeritus Scientist, Zoological Survey of India, Kolkata for the help rendered during the study of the specimens.

References

Bhatia, B.L., 1936. *The Fauna of British India including Ceylon and Burma, Protozoa: Ciliophora,* xxii+493 pp. Pergamon Press, Oxford.

Enriques, P., 1980. Sulla morfologie E sistcmica del Gcnera Colpoda. *Arch. Zool. Exp. Gen.,* Notes et Rev., (4)8, 1–15.

Kahl, A., 1930. Urtiere oder Protozoa. I. Wimpertierc odcr Ciliate (Infusoria), Fine Bearbeitung der freilebenden und ectocommensalen Infusoricn der Erde. unter Ausschluss der marinen Tintinnidae. In: Dahl, F. Tierwelt Dtsch. (Pts. 18, 21, 25 and 30), 886 pp., Jena (G. Fischer).

Sauerbrey, E., 1928. Beobachtungen uber einige neue oder wenig bekannte marine Ciliaten. *Arch. Protistenk.,* LXII, pp. 335-407.

Stokes, A., 1885. Notices of new freshwater Infusoria. *Amer. Mon. Micr. J.,* 6: 121–127.

Chapter 50

Effect of Silicate Solubilizing Bacterial (SSB) Application Along with Rice Residues on Growth and Yield Attributes of Rice

P. Kalaiselvi[1]* and S. Anthoni Raj[2]

[1]Ph.D. Scholar, [2]Former Dean

Department of Environmental Sciences, Tamil Nadu Agricultural University, Coimbatore – 641 003, Tamil Nadu

ABSTRACT

Rice is the staple food crop grown extensively all over the world and occupies a pivotal position in the food security system of India. It is a silicicolus plant and a known silicon accumulator and is a plant that benefits from silicon nutrition. The agricultural soils are largely composed of silicate minerals and many soils contain an inadequate supply of available silica (Si). There is a similarity for silica (Si) and potassium (K), not only in their co-existence in soil but also in their biochemical role in plants. It has been shown in a laboratory study that a silicate solubilizing bacteria (SSB) isolated from rice husk and identified as a member of the genus Bacillus, releases K in addition to Si from silicate bearing minerals in the agricultural soils. A field experiment was conducted at Agriculture College and Research Institute, Madurai, and examined the influence of silicate solubilizing bacteria to rice with rice residues like straw, rice husk and black ash (rice husk ash). The application of SSB recorded higher levels of Si and K in field impounded water,

* Corresponding Author, E-mail: kalaiphd@gmail.com

soil and plant, markedly influenced the crop growth and grain yield (4.1 t ha-I) of rice. It has been concluded that the application of SSB increased the grain yield of rice, mainly due to the supply of Si and K which might have resulted due to solubilization of soil minerals and decomposition of rice residues. It has also been proved that the application of SSB to rice reduces the use of commercial K fertilizer.

Keywords: *SSB, Bacillus sp., Silica, Potassium, Rice yield.*

Introduction

Sustainability in agriculture and crop production is not merely achieved by cutting down the off farm inputs like inorganic fertilizers and chemicals but also by replenishing the soil with what has been removed from it by the crop, through on farm inputs like organic manures particularly by recycling crop residues. In rice based cropping system there is a good lot of residues available for recycling. Rice stubbles (inclusive of basal culm and root) constitute 3–5 t ha^{-1}. Rice straw although use as cattle feed when harvested dry, abandoned in farm site when it is wet and spoiled by rain or when harvested by harvester. Black ash (rice husk ash) accumulates in rice mills as a waste at throwaway prices waranting safe disposal as it poses environmental problems in the rice mills located amidst human habitation. These rice residues contain a good lot of nutrients like Si, K and others. But so far their use in rice cultivation was avoided for various reasons. In the present study these residues were utilized to supplement the crop requirement and as to rapidly degrade these siliceous materials and solubilize the nutrients SSB was combined and investigated. The results showed that application of rice straw, rice husk and black ash @ 5 t ha^{-1} augmented the nutrients in soil and plant, enhanced the growth and yield of rice. When they were applied in combination with SSB their favourable effect were further enhanced.

The black ash rich in carbon (15–30 per cent), silica (60–65 per cent), potassium (0.95 per cent) and several nutrients available in rice mills are applied occasionally to garden land, but application to rice is unknown. Application of ash to soil increases soil temperature particularly in winter, there by providing a congenial environment for nutrients uptake. An increase in nitrogen fixing micro flora due to application of charred rice husk was also reported (Beagle, 1978). Hence the present study was conducted with a view to study the effect of application of rice residues along with SSB on rice crop and soil.

Materials and Methods

Isolation of Silicate Solubilizing Bacteria (SSB)

Silicate Solubilizing Bacteria (SSB) were isolated from rice husk and soil of granite crusher yard using silicate solubilizing bacterial medium (Bunt and Rovira, 1955) containing 0.25 per cent insoluble magnesium trisilicate.

Identification and Characterization of the Bacterial Isolates

The morphological, nutritional and biochemical characteristics of the bacterial isolates selected for the present investigation were determined as per the methods of Gerherdt *et al.* (1981). The Gram reaction, endospore staining acid and gas production, hydroloysis of starch, lipolytic activity, liquefaction of gelatin, catalase and nitrate reductase activities were carried out. Based on the characteristics of the bacterial cultures isolated from rice husk (SSB I) and from soil of granite crusher yard (SSB II) (Plate 50.1) were identified employing the key described in Bergey's manual of determinative bacteriology (1974).

Plate 50.1: Silicate Solubilizing Bacterial Isolates SSB I and SSB II, SSB I: From Rice Husk, SSB II: From Granite Crusher Yard

In vitro Determination of Silicate Solubilizing Ability of Bacterial Isolates

The extent of silicate solubilization by the selected isolate was assessed both qualitatively and quantitatively under in vitro condition using plate assay and broth assay by substituting the medium with 0.25 per cent insoluble $MgSiO_3$. Based on clear zone obtained in the plate assay and extent of silica content in the supernatant of broth assay the silicate solubilizing potential of the bacterial isolates were determined. In this the SSBI was found to be superior in solubilizing silica (Plate 50.2).

Field Testing of Silicate Solubilizing Bacteria I (SSB I) on Rice

A field trial was laid out in the Central Farm of Agricultural College and Research Institute, Madurai in a randomized block design in the *rabi* season (December–March). ADT 36 rice variety seedlings (24 days old) were planted in plots of 4 × 3 m size with 15 × 10 cm spacing at the rate of 2 seedlings $hill^{-1}$. The rice residues like straw, husk and black ashes were trampled @ 5 t ha^{-1} prior to

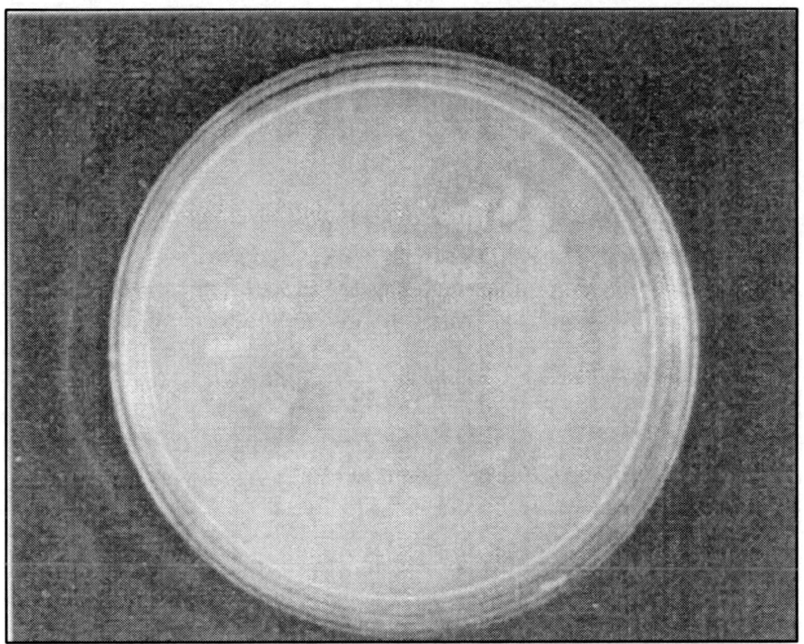

Plate 50.2: Cleaning Zone of Silicate Solubilzing Bacteria I (SSB I) *Bacillus* sp.

planting. The crop was supplied with Nitrogen (N), Phosphorus (P) and Potassium (K) fertilizers, as the treatments at the recommended dose of 75, 50 and 25 kg of NPK ha^{-1} respectively. The nitrogen, phosphorus and potassium were supplied as urea, super phosphate and muriate of potash respectively. The plots were inoculated with silicate solubilizing bacteria I @ 2 kg ha^{-1} (10 g $plot^{-1}$).

The bacterial isolate SSB I from rice husk was used in this field experiment. The treatment details given below.

T_1: Control

T_2: SSB I alone

T_3: Rice straw @ 5t ha^{-1}

T_4: Rice straw @ 5t ha^{-1} + SSB I

T_5: Rice husk @ 5t ha^{-1}

T_6: Rice husk @ 5t ha^{-1} + SSB I

T_7: Black ash @ 5t ha^{-1}

T_8: Black ash @ 5t ha^{-1} + SSB I

The biometric observations like growth characters, dry matter production, yield components, biomass at harvest, grain yield and 1000 grain weight were also recorded adopting the methods described below.

Biometric Observation

Growth Characters

The plant height was recorded in 5 plants in each plot at 0, 15, 30, 60 and 90 Days After Planting (DAP).

Yield Attributes

The number of total and productive tillers and the number of panicles were counted in 5 hills plot^{-1} and the mean was arrived. The number of fully matured grain were counted in 5 panicles from each plot and the fully matured grain percentage was calculated. Thousand grain weight was recorded from the representative samples collected from the threshed paddy.

Biomass at Harvest and Grain Yield

The biomass was calculated from the weight of the harvested sheaves. The biomass was expressed as t ha^{-1}. Immediately after threshing and cleaning, grains from each net plot was dried and weighed. The grain yield was then computed and expressed in t ha^{-1}.

Statistical Analysis

The statistical analyses of the experimental data were carried out by adopting standard method (Gomez and Gomez, 1984).

Results and Discussion

Effect of Application of SSB I on Biometrics of Rice

The results on the influence of applications of rice residues and SSB I inoculation on plant height, root and shoot length of rice are presented in Tables 50.1 and 50.2. The addition of straw, husk and black ash @ 5 t ha^{-1} augmented the plant height and inoculation of SSB I along with these organic siliceous materials further enhanced it. The highest plant height was observed in straw along with SSB I inoculation up to 97.0 cm on 90 DAT. This was followed by husk combined with SSB I inoculation with 95.5 cm. The lowest plant height was recorded in uninoculated control and its was 76.5 cm at 90 DAT.

The root length of the plants ranged from 13.0 cm to 19.0 cm. The root length was augmented by application of rice residues straw, husk and black ash either alone or along with SSB I inoculation. The root length was found to be higher in all the SSB I treated plots. The lowest root length was recorded in control uninoculated. The root length was generally higher in rice straw @ 5 t ha^{-1} + SSB I at all sampling period than other treatments.

The shoot length also increased in a similar way. The augmentation of shoot height was higher initially upto 30 DAT in black ash @ 5 t ha^{-1} + SSB I treatment. However, at later stages rice straw @ 5 t ha^{-1} + SSB I recorded a higher shoot length than other treatments. At 90 DAT the highest shoot length was recorded in rice straw @ 5 t ha^{-1} + SSB I treatment followed by rice husk @ 5 t ha^{-1} + SSB I with 78.0 cm and 77.5 cm, respectively. The shoot length was relatively lesser in uninoculated control.

The results revealed that the plants grown in the treatment rice straw with SSB I showed superiority over all other treatments.

The plant height, root length and shoot length were found to be higher in SSB I treated plots. Straw and husk application showed a higher plant height. The organic matter present in these materials as well as silica rendered soluble and other nutrients made available might have exerted a favourable

Table 50.1: Influence of Application of Rice Residues and SSB I Inoculation on the Plant Height, Root and Shoot Length of Rice

Sl.No.	Treatment	Plant Height (cm)					Root Length (cm)					Shoot Length (cm)				
		0 DAT	15 DAT	30 DAT	60 DAT	90 DAT	0 DAT	15 DAT	30 DAT	60 DAT	90 DAT	0 DAT	15 DAT	30 DAT	60 DAT	90 DAT
1.	Control	31.3	35.4	52.6	67.8	76.5	8.0	9.0	11.0	12.0	13.0	23.3	26.4	41.6	55.8	63.5
2.	SSB I alone	32.2	38.2	56.5	76.2	90.0	8.2	10.2	12.5	14.0	15.0	24.0	28.0	44.0	62.2	75.0
3.	Rice straw @ 5 t ha^{-1}	34.0	41.5	63.9	74.1	82.0	10.2	11.0	13.0	16.0	16.0	23.8	30.5	50.9	58.1	66.0
4.	Rice straw @ 5 t ha^{-1} + SSB I	39.5	45.3	67.6	86.7	97.0	10.8	12.0	16.0	18.0	19.0	28.7	33.3	51.6	68.7	78.0
5.	Rice husk @ 5 t ha^{-1}	35.4	39.3	60.8	72.0	85.0	8.4	10.2	14.0	14.0	15.0	25.2	29.1	46.8	58.0	70.0
6.	Rice husk @ 5 t ha^{-1} + SSB I	34.0	41.6	63.8	85.0	95.5	9.5	11.0	15.0	16.0	18.0	25.9	30.6	48.3	69.0	77.5
7.	Black ash @ 5 t ha^{-1}	38.6	43.7	60.4	76.5	88.0	9.0	10.0	13.0	16.0	17.0	29.6	33.7	47.4	60.5	71.0
8.	Black ash @ 5 t ha^{-1} + SSB I	39.0	45.4	65.6	88.1	93.0	9.5	11.0	15.0	17.0	19.0	30.2	34.4	50.6	71.1	74.0
	SEd	1.47	1.72	2.52	3.20	3.64	0.37	0.43	0.56	0.63	0.68	1.09	1.27	1.95	2.57	2.97
	CD	4.37**	5.11**	7.51 **	9.52**	10.82**	1.11**	1.29**	1.68**	1.88**	2.02**	3.25**	3.78**	5.82**	7.66**	8.83 **

DAT: Days After Transplanting

*: Significant at 5 per cent level.

**: Significant at 1 per cent level

NS: Non significant.

effect on the growth of rice. The yield parameters like total and productive tillers hill^{-1}, number of grains panicle^{-1}, 1000 grain weight were found to be higher in SSB I treated plots. These parameters were relatively higher in rice straw combined with SSB I followed by black ash with SSB I. The dry matter production was also higher in SSB I treated plots and a higher DMP was recorded in straw with SSB I followed by black ash with SSB I. Silica nutrition has direct and indirect beneficial effects on the rice growth largely due to its unique physiological role (Okuda and Takahashi, 1965; Yoshida, 1975). Ashoka Rani *et al.* (1997) also supported that the application of silica to rice was found to increase the grain yield under both upland and water logged conditions. They also reported that the silicon application was found to significantly increase the grain yield by increasing the number of panicles plant^{-1}, spikelet panicle^{-1}, fully ripened grains panicle^{-1}, test weight and harvest index.

Table 50.2: Influence of Application of Rice Residues and SSB I Inoculation on Yield Parameters and Grain Yield of Rice

Sl.No.	*Treatments*	*No. of Tillers Hill^{-1}*	*No. of Productive Tillers*	*No. of Grains Panicle^{-1}*	*Fully Matured Grain (per cent)*	*1000 Grain Weight (g)*	*Biomass (t ha^{-1})*	*Grain Yield (t ha^{-1})*
1.	Control	8.6	6.6	90.3	70.2	21.4	7.3	3.1
2.	SSB I alone	10.2	7.5	97.6	82.0	23.1	7.5	3.3
3.	Rice straw @ 5 t ha^{-1}	10.7	9.9	106.2	76.1	22.2	7.4	3.7
4.	Rice straw @ 5 t ha^{-1} + SSB I	13.0	10.9	112.1	86.0	22.2	9.2	4.1
5.	Rice husk @ 5 t ha^{-1}	9.6	8.7	98.3	72.0	22.1	7.3	3.6
6.	Rice husk @ 5 t ha^{-1} + SSB I	10.4	10.3	100.3	84.6	23.4	8.1	3.9
7.	Black ash @ 5 t ha^{-1}	10.8	9.2	95.1	74.2	22.0	7.0	3.1
8.	Black ash @ 5 t ha^{-1} + SSB I	11.6	9.6	105.3	83.4	23.1	9.3	3.4
	SEd	1.14	1.22	2.23	1.93	1.0	0.7	0.29
	CD	2.45*	2.62*	4.80**	4.14**	NS	1.44 *	0.63*

*: Significant at 5 per cent level.

**: Significant at 1 per cent level

NS: Non significant.

The application of black ash of rice hull made rice seedlings healthy and strong and increased their biomass (Sawant *et al.*, 1994; Savant and Sawant, 1995). Talashilkar and Chavan (1996) reported that application of rice hull ash as a source of silicon resulted in significant increase in straw and grain yield of rice. Yield increase in rice due to siliceous slag application @ 1.5 to 3 t ha^{-1} continuously to each crop was also reported in field and pot experiments (Lian and Wang, 1983; Masaoka and Takawo, 1987).

Effect of Application of SSB I on Yield Attributes of Rice

The results on influence of application of rice residues and SSB I inoculation on the total and productive tillers, number of grains per panicle, fully matured grain percentage, 1000 grain weight, biomass and grain yield are summarized in Table 50.2.

In general SSB I treated plots showed increased number of tillers hill^{-1}. The mean total number of tillers hill^{-1} ranged from 8.6 to 13.0. The higher number of tillers recorded from the plots treated with rice straw @ 5 t ha^{-1} along with SSB I with 13.0 tillers hill^{-1} followed by 11.7 tillers hill^{-1} in rice straw alone. The lowest number of tillers hill^{-1} was recorded in uninoculated plot with 8.6. The treatment rice straw @ 5 t ha^{-1} + SSB I was superior to other treatments.

The mean total number of productive tillers ranged from 6.6 to 10.6 hill^{-1}. The higher number of productive tillers was recorded in the rice straw @ 5 t ha^{-1} + SSB I treatment and it was 10.9 hill^{-1} and the lowest number of productive tillers was recorded in uninoculated control with 6.6 hill^{-1}.

The number of grains panicle^{-1} was found to increase due to SSB I inoculation. The highest number of grains panicle^{-1} (112.1) was observed in rice straw @ 5 t ha^{-1} + SSB I treatment and the lowest number of grains panicle^{-1} (90.3) was recorded in uninoculated control.

The percentage of fully matured grains was found to be higher when organic siliceous materials were applied either alone or with SSB I. However, SSB I inoculation with rice residues has augmented the filled grains more. The highest filled grains was recorded in the treatment rice straw @ 5 t ha^{-1} + SSB I with 86.0 per cent.

The 1000 grain weight was found to be higher in all the SSB I treated plots. The maximum grains weight was observed in the plots treated with rice husk @ 5 t ha^{-1} along with SSB I with 23.4 g.

The biomass (straw and grain) at harvest was the highest with 9.3 t ha^{-1} in black ash along with SSB I application. The addition of rice residues either alone or with SSB I enhanced the biomass. This was further enhanced when combined with SSB I application. The increase was almost similar in straw and black ash and between husk and control.

The grain yield in this experiment ranged from 3.1 to 4.1 t ha^{-1}. Mere SSB inoculation recorded 3.3 t ha^{-1} compared to 3.1 t ha^{-1} in uninoculated control. Application of rice residues further increased the grain yield which was further enhanced when combined with SSB I. The highest grain yield of 4.1 t ha^{-1} that was recorded in straw along with SSB I followed by rice husk along with SSB I with 3.9 t ha^{-1}. The treatments were on par with each other. The results showed that application of rice residues with SSB I has enhanced the grain yield. This was supported by (Raghupathy and Chandrasekaran, 1980) who reported that in Tamil Nadu soils application of lignite fly ash was found to increase the rice grain and straw yield. Soil inoculation of silicate bacteria *Bacillus* circulans had a significant effect on the release of silicon and potassium which was reflected in the plant uptake (Zahro *et al.*, 1984).

Xue-Zhiyong (1996) also reported the increased yield of sweet potato by the silicate bacterium on release of potassium. Muralikannan (1996) also reported that the *Bacillus* sp. (SSB) that solubilized silicate also solubilized the potassium which effectively increased the grain yield of rice.

The addition of these siliceous materials along with SSB I might have increased the yield by physiological effects possibly through (*i*) erectness of leaves due to higher silica resulting in efficient photoreception from sunlight and carbon assimilation as suggested by Gaussmann (1962) and (*ii*) promotion of new root growth. (*iii*) improvement of oxygen supply to roots by increasing the volume and rigidity of the gas channel (Ponnamperuma, 1964), (*iv*) better translocation of phosphorus from source to sink (Cheng *et al.*, 1973) and (*v*) higher sugar, starch and protein content and their greater translocation due to higher silica (Okomato, 1959) and/or (*vi*) increased K content in plants offering other beneficial effects inherent to K.

Conclusion

Although the application of raw undecomposed straw, rice husk and black ash to rice soils are not practiced widely there is a scope for their utility particularly when the straw is damaged by monsoon rains and becomes unfit for animal consumption. Further the use of combine harvester cuts the sheaves into bits and leaves in band allover the field which is mostly unfit for animal consumption. It is also uneconomical for collection and composting due to high labour cost. Burning it in situ is also of environmental concern. Therefore, it is left in the field unutilized. The application of SSB I hastened the straw decomposition in situ at the time of preparatory cultivation and liberating its nutrients to the succeeding crop.

The agricultural soils are largely composed of silicate minerals, many soils contain an inadequate supply or naturally low in available Si. Farmers are not replacing Si removed by rice. This will be a great limiting factor to sustainable rice production especially when they resort to high nitrogen fertilizer. So this problem can be alleviated by application of the siliceous rice residues like rice straw, rice husk and black ash (rice husk ash). By substitution of organic siliceous materials to cultivable soil the recommended level of potassium can be reduced by 25 to 50 per cent. The silicate solubilizing bacteria thus gains importance and further investigation in various soils in which rice is grown becomes an obvious necessity.

Acknowledgement

I thanks to Department of Microbiology, Agricultural College and Research Institute, TNAU and ASPEE foundation, Mumbai, India for carried the research and financial assistance.

References

Ashoka Rani, Y., Narayanan, A., Saraswathi Devi, V. and Subbaramamma, P., 1997. The effect of silicon application on growth and yield of rice plants. *Ann. Plant Physiol.,* 11(2): 123–128.

Beagle, E.C., 1978. Rice husk conversion to energy. *FAO Agrl. Series Bulletin,* FAO, Rome, p. 175.

Bergey's, Manual of Determinative Bacteriology, 1974. (Eds.) Buchanan, RE., Gibbons, N.E., Cowan, S.T., Holt, J.O., Liston, J., Murray, R.G.E., Niven, C.F., Ravin, A.W., Stanier, R.Y. 8th edn. The Williams and Wilkins Co., Baltimore, USA, p. 1247.

Bunt, J.S. and Rovira, A.D., 1955. Microbiological studies of some subantartic soils. *J. Soil Sci.,* 6: 119–128.

Cheng, W.Y., Heitz, A. and Devilie, U., 1973. The effect of silica on sugarcane growth in pure nutrient solution. *J. Sci. Food Agric.,* 24: 13–15.

Gaussmann, W., 1962. Investigations on the effect of silica on uptake of phosphorus and other nutrients. *Phosphorsaure,* 22: 223– 241.

Gerherdt, P., Murrary, R.G.E., Costelow, R.N., Nexter, E.W., Wood, W.A., Krieg, N.B. and Phelleps, G.B., 1981. *Manual of Methods for General Bacteriology.* American Society for Microbiology. Washington DC, p. 524.

Gomez, K.A. and Gomez, A.A., 1984. *Statistical Procedures for Agricultural Research.* John Wiley and Sons, New York, p. 680.

Lian, S. and Wang, C.H., 1983. Effect of long term applications of siliceous slag on paddy yield and chemical properties of soil. *J. Agrl. Res. China.,* 32: 187–197.

Masaoka, Y. and Takawan, N., 1987. Effects of silicon in warm season forage crops on cell wall digestion by cellulose. *J. Japanese Soc. Grass land Sci.*, 32: 417–418.

Muralikannan, N., 1996. Biodissolution of silicate, phosphate and potassium by silicate solubilising bacteria in rice ecosystem. *M.Sc.(Ag.). Thesis*, Submitted to Tamil Nadu Agricultural University, Coimbatore, p. 115.

Nayar, P.K., Misra, A.K. and Patnaik, S., 1975. Rapid microdetermination of silicon in rice plant. *Plant and Soil*, 42: 491– 494.

Okamato, Y., 1959. Physiological studies on the effect of silicic acid upon flee plants. *Proc. Crop Sci. Soc., Japan.*, 28:35–40.

Okuda, A. and Takahashi, E., 1965. The role of silicon. In: *The Mineral Nutrition of the Rice Plant.* John Hopkins Press, Baltimore, Maryland, pp. 126–146.

Ponnamperuma, F.N., 1964. The mineral nutrition of rice plant. *Symp. Intern. Rice. Res. Inst.*, Johns Hopkins Press, Baltimore, Maryland, pp. 395–398.

Raghupathy, B. and Chandrasekaran, S., 1980. Effect of lignite fly ash on the release of phosphorus and its uptake by maize in lateritic soil. *Annamalai Univ. Agri. Res. Ann.*, 10: 73–77.

Savant, N.K. and Sawant, A.S., 1995. Nutrient composition of rice seedlings as influenced by rice hull ash application to seed bed. *Oryza*, 17: 188–194.

Sawant, A.S., Patil, V.H. and. Savant, N.K., 1994. Rice hull ash applied to seed bed reduces dead hearts in transplanted rice. *Int. Rice Res. Notes*, 19(4): 21–22.

Talashilkar, S.C. and Chavan, A.S., 1996. Effect of rice hull ash on yield and uptake of silicon and phosphorus by rice cultivars at different growth stages. *J. Indian Soc. Soil Sci.*, 44(2): 340–342.

Xue-Zhiyong, 1996. Effect of silicate bacterium on release of potassium in different soils and on increase yield of sweet potato. *Soil and Fetilizers*, pp. 28–26.

Yoshida, S., 1975. The physiolo of silicon in rice. *Food Fertilizer Tech. Centre Technical Bull.*, No. 25.

Zahro, M.K., Mohib, M., Abdel-Al, S.I. and Heggo, H., 1984. Significance of inoculation with silicate bacteria. *Zentral. Mikrobiol.*, 139: 349–357.

Chapter 51

Effect of Wheat Straw and Sugarcane Trash on Yield and Nutrient Status of *Agaricus bitorquis*

D. Venkata Reddy, V. Bali Reddy, G. Bindu Madhavi, V. Chenga Reddy, K. Gurava Reddy and M. Chandra Sekhar Reddy
Regional Agricultural Research Station, Acharya N.G. Ranga Agricultural University, Lam, Guntur – 522 034

ABSTRACT

The present investigation was undertaken with a view to see the possibility of utilising sugarcane trash as a base material in composting for cultivation of *Agaricus bitorquis*. Mushrooms grown on wheat straw yielded more fruit bodies as compared to sugarcane trash. Maximum number of buttons (142/bag), button weight of 9.0 g and average volume of 13.62 ml were observed on wheat straw compost. The maximum biological efficiency (11.5 per cent) and protein content (33.80 per cent) obtained on wheat straw substrate compost.

Keywords: *Agaricus bitorquis, Button yield, Protein content, Biological efficiency.*

Introduction

Mushroom is a most popular food delicacy in the diet of the world. Increasing population is creating an alarming situation in food problem in India. Malnutrition in terms of protein deficiency is becoming a major hazard in developing countries. Exploiting non-traditional food resources can make a substantial break through to meet serious food deficit. In these circumstances, popularizing mushroom as part and parcel of every day food is of paramount importance. Mushrooms have been identified as efficient means for conversion of worthless agro wastes into valuable protein. Common substrates used for mushroom production are wheat straw, paddy straw, sugarcane trash and chicken manure. Over dependence on wheat straw as the basal material has started its toll as its demand for cattle feed for paper industry has pushed up its price. Therefore, it becomes necessary to find the substitutes of wheat and paddy straw along with the evaluation of the better strains of mushroom species. Hence the present investigation was carried out with the objective of evaluation of the performance of *Agaricus bitorquis* on compost prepared from wheat straw and sugarcane trash.

Materials and Methods

Agaricus bitorquis isolate NCB-6 procured from The Mycologist, All India Coordinated Mushroom Improvement Project, College of Agriculture, Pune was used in the present investigation. Spawn was prepared by standard method (Gupta and Sharma, 1994)

Compost Preparation

For preparing compost as substrate for mushroom growth, long method of compost preparation was followed as per the procedure recommended by Indian Institute of Horticulture Research, Bangalore. Ingredients used for preparing compost were, Wheat straw or sugarcane trash–300 kg, Ammonium sulphate–9 kg, Super phosphate–9 kg, Urea–4 kg, Sulphate of potash–3 kg, Wheat bran–15 kg and Gypsum–30 kg.

Spawning

Double layer method of spawning was followed. These filled bags were kept on the racks in the cultivating room and the temperature in this room was maintained at 25±1°C and relative humidity of 85 to 90 per cent in order to accelerate spawn run in the bags. The spawn run in the bag was completed in 15–18 days depending on the strains and compost substrates used. The fully spread mycelium in the bags appeared white in colour and these bags were applied with casing soil. Watering was done by spraying carefully on the bags till the pin heads attain the button size (harvesting size). The growing mushroom need plenty of fresh air. For this, provision of free air movement inside the mushroom house was provided. Malathion (0.1 per cent) was sprayed in order to avoid insect infestation. Harvesting of mushrooms was done alternate day when buttons attained size of 2–3 cm in diameter. Harvested mushrooms were gently cleaned and packed in polythene bags.

Observations

Data on days required for complete spawn run on the compost, for appearance of pin heads, for first picking and other subsequent pickings, total cropping period, Number of buttons per bag, Yield of buttons per bag, Weight of buttons, Volume of buttons. The yield of fruit bodies on fresh weight basis was recorded and converted in to biological efficiency following standard formula. Protein content of *Agaricus bitorquis* was evaluated by analysing nitrogen content of mushroom by Microkjeldhal's digestion method and multiplying it by factor 6.5 (Jackson, 1973).

Nutritional Analysis of *Agaricus bitorquis*

Freshly harvested mushrooms were cleaned. Oven dried and powdered before analysis. Moisture and ash contents were estimated by following the method outlined by Association of Official Analytical Chemists (AOAC, 1980). Minerals were determined by Shimadzu AA-670 Atomic Absorption

Spectrophotometer. Protein (N × 6.25) was determined by the Microkjeldahl method. Amino acid contents were determined by the LKB-4151 Alpha plus automatic amino acid analyser. Fat was extracted in a soxhelt apparatus with petroleum ether for 8 hours. The free fatty acids in fats were converted to methyl esters by the acetyl chloride methanol procedure of Stoffel *et al.*, (1959) prior to gas chromatographic analysis to determine their molar ratios as reported by Al-Kaisey. Total carbohydrates were calculated by the difference.

Results and Discussion

Spawn Run of *Agaricus bitorquis* on the Compost Prepared from Sugarcane Trash and Wheat Straw

The complete spawn run of *A. bitorquis* on the compost prepared from sugarcane trash it takes only 17–18 days while on wheat strains 18–21 days were required. The difference in the days required for completion of spawn run may be due to differential rate of availability of materials in the compost prepared from different base materials. Similar results was reported by Shandilya *et al.* (1974) as, the double layer spawned trays of *Agaricus bisporus* needed 14 days, whereas, the trays of the remaining treatments required 18 days for completion of spawn run. Overstijns and Bockstacle (1975) reported that 14 days were required for completion of spawn run for *Agaricus bisporus* on compost prepared from wheat straw as a base material.

Days Required for Pin Head Formation

Pin heads appeared with in 12 days in wheat straw where as in sugarcane trash it took about 14 days. This result is in corroboration with the report of Flegg (1981). All or most of the fruiting bodies initials for a crop of *Agaricus bisporus* were formed 7–14 days after casing. Days required for picking were almost the same for both the substrate composts prepared from sugarcane trash and wheat straw *i.e.* first picking after 18 days, second picking after 46 days, third picking after 52 days, fourth after 61 days and fifth picking after 69 days.

Number of Buttons per Bag

The compost prepared from wheat straw recorded significantly a maximum number of buttons *i.e.* 142/bag. While the compost prepared from sugarcane trash recorded minimum number of buttons *i.e.* 86/per bag. The results (Table 51.1) also indicated that in all the treatments there was a gradual decrease in number of buttons per bag with the advancement in picking. Decrease in formation of buttons in subsequent pickings which could be attributed to decrease in the nutrient supply from the compost. Bahl (1984) also reported that after first few flushes the mushrooms become lighter and there is a decrease in number of buttons, with which the present results are in close conformity.

Button Yield

The compost prepared from wheat straw recorded a maximum yield of 1380 g/bag of 12 kg compost with button weight of 9.0 g and average volume is 13.62 ml. respectively. While the compost prepared from sugarcane trash recorded a minimum yield of 810 g/bag of 12 kg compost after all 5 picking at of 9.4 gms and volume 13.12 ml respectively. The results (Table 51.2) indicate that in all the treatments there was a gradual decrease in the yield of buttons/bag with the advancement in picking. This indicates decrease in the yield of buttons/bag in subsequent pickings which could be attributed to decrease in the nutrients supply from the compost substrate.

Table 51.1: Number of Buttons and Yield at First and Subsequent Pickings of *Agaricus bitorquis* on Composts Prepared from Sugarcane Trash and Wheat Straw

Sl.No.	Number of Pickings	No. of Buttons/Bag on Composts of		Yield of Buttons (g) on Composts of	
		Sugarcane Trash	Wheat Straw	Sugarcane Trash	Wheat Straw
1.	1st	24	39	234	377
2.	2nd	20	34	189	312
3.	3rd	17	28	171	298
4.	4th	14	23	117	225
5.	5th	11	18	99	168
	Total	86	142	810	1380
	SEM±	1.794	1.673	2.225	2.712
	CD	5.848	5.456	7.254	8.842
	CV@5 per cent	18.1	10.2	2.4	1.7

Table 51.2: Weight, Average Volume and Biological Efficiency of *A. bitorquis* Grown on Compost Prepared with Sugarcane Trash and Wheat Straw

Type of Compost	Weight of Individual Button of A. bitorquis (grams)	Average Volume of Button of Agaricus bitorquis (ml)	Biological Efficiency of Composts (Per cent)
Sugarcane Trash	9.4	13.12	67.5
Wheat straw	9.0	13.62	115

Anonymous (1999a) recorded that compost prepared from sugarcane bagasse yielded 945 g of *Agaricus bisporus* buttons per 10 kg of compost. Strain U-3 of *Agaricus bisporus* yielded 1639 g/10 kg of compost prepared from wheat straw Anonymous 1999The results of present investigation are in close conformity. Jain and Singh (1982) however, recorded the greater yield of 3335 g/tray where wheat straw was used as base material for composting. Suman (1993) reported that the average weight of fruiting body of *Agaricus bisporus* ranged between 8.5 and 10.09 g. In the present investigation the results were in close conformity.

Biological Efficiency of Composts Prepared from Sugarcane Trash and Wheat Straw

The compost prepared from wheat straw recorded significantly high biological efficiency of (11.5 per cent), while the compost prepared from sugarcane trash recorded 6.75 per cent biological efficiency (Table 51.2). Kachroo *et al.* (1979) reported that yield of *Agaricus bisporus* when converted to gram per kilogram dry weight of compost of filling shows an efficiency of more than 50 per cent for all the treatments. This indicates that the sugarcane trash could be used as a base material in composting for cultivation of *A. bitorquis* as its biological efficiency is 67.5 per cent.

Protein Content of *Agaricus bitorquis*

The compost prepared from wheat straw produced button having maximum protein content of 33.80 per cent while the compost prepared from sugarcane trash produced button having minimum protein content of 22.46 per cent. These observations in general indicates that mushroom fungus varies in protein content when grown on compost prepared from different base material, it also depends on the absorption capacity of nutrients by *A. bitorquis* fungus. Kumar *et al.* (1991) and Eicker (1993) reported protein content of *Agaricus bisporus* in the range of 20 to 40 per cent. However Hadwan *et al.* (1993) reported 36–40 per cent of protein in *Agaricus bisporus* on dry weight basis with which the present results are in close agreement.

Other Nutrients Present in the *Agaricus bitorquis* (On Fresh Weight Basis)

The mushrooms have a definite nutritive value. They contains the proteins, vitamins, minerals, carbohydrates, fats, crude fibre and also energy value. The moisture content on fresh weight basis is 89–91 per cent and carbohydrates are 5.3–6.28 per cent and the energy value is 24.4–34.4 k.cal. Ash, 0.97–1.26 per cent, Fat 0.25–0.65 per cent, Crude fibre 0.09–1.67 per cent, Glucose 0.04 per cent, Glycogen 0.91 per cent (Table 51.3). It was found that, the *Agaricus bitorquis* contains vitamins like Riboflavin, Niacin, Pantothenic acid, Thiamine and appreciable amounts of folic acid and ascorbic acid and minerals like potassium, phosphorus, iron and calcium. The nutritive value of *A. bitorquis* was similar these reported by Singer (1961).

Table 51.3: Nutritive Value of *Agaricus bitorquis* on Dry Fresh Weight Basis

Sl.No.	*Content*	*Nutritive Value in Per cent*
1.	Moisture	89–91
2.	Ash	0.97–1.26
3.	Fat	0.25–0.65
4.	Crude fibre	0.09–1.67
5.	Carbohydrates	5.3–6.28
6.	Energy	24.4–34.4 Kcal.
7.	Glucose	0.04
8.	Glycogen	0.91

Table 51.4: Nutritive Value of *Agaricus bitorquis* on Dry Weight Basis

Sl.No.	*Vitamins*	*Minerals*	*Nutrients*			*Fatty Acid*		*Mineral*	
			Content	*Per cent*	*Dry Weight (mg/100 g)*	*Name*	*Proportion*	*Name*	*Dry Weight (mg/100 g)*
1.	Riboflavin	Potassium	Protein		33.80	Myristic acid (C_{14})	4.56	Na	180
2.	Nicotinic acid (Niacin)	Phosphorus	Carbohy-drate		56.00	Palmitic acid (C_{16})	11.27	K	3800
3.	Pantothenic acid	Iron	Fat		1.2	Stearic acid ($C_{18:0}$)	6.00	Mg	140
4.	Thiamine	Calcium (trace amount)	Ash		9.0	Oleic acid ($C_{18:1}$)	1.41	Ca	150
5.	Folic acid (Appreciable amount)					Linoleic acid ($C_{18:2}$)	73.94	Fe	5.6
6.	Ascorbic acid (Appreciable amount)					Arachidic acid (C_{20})	2.84	Cu	4.3

Nutritional Analysis of *Agaricus bitorquis*

The white button mushroom on dry weight basis contains the maximum carbohydrates (56.00 per cent) which was followed by protein (33.80 per cent), 9 per cent mineral matter and minimum percentage of fat (1.2 per cent) was estimated (Table 51.4). Similar to those reported for *A. bisporus* by Huang *et al.* (1989); Hadwan *et al.* (1993).The unsaturated fatty acids accounted for 75.35 of the total fatty acids and linoleic acid was most abundant Huang *et al.* (1989) also recorded similar in distribution of fatty acids. On dry weight basis, potassium concentration was found maximum 3800 mg/100 g dry weight followed by Sodium, Calcium and Magnesium. Iron and Copper were found in very low concentration. The results are in confirmation with those observed by Hadwan *et al.* (1993).

References

Anonymous, 1999a. Testing of different strains of *Agaricus bisporus.* Report on Research Review committee. All India Co-ordinated Mushroom Improvement Project, College of Agriculture, Pune, pp. 6–7.

Anonymous, 1999b. Evaluation of different base material for composting. Report on Research Review Committee. All India Co-ordinated Mushroom Improvement Project College of Agriculture, Pune, pp. 3–4.

AOAC, 1980. *Official Methods of Analysis*, 13th edn. Washington, D.C.

Bahl, N., 1984. *Handbook on Mushrooms*. Oxford and IBH Publishing Co., New Delhi, pp. 29.

Eicker, A., 1993. Mushroom: A source of protein for Africa. *African J. Mycol. and Biotech*., 1(1): 12–13.

Flegg, P.B., 1981. How do your mushroom grow. *Mush. J.*, 108: 401–409.

Gupta, Y. and Sharma, 1994. *Mushroom Spawn Production*. Technical Bull. No. 5, NCRR&T Publication.

Hadwan, H.A., Alkaisey, M.T., Al-Tikriti, M.N., Alani, S.R. and Dhar, B.L., 1993. Evaluation of strains of *Agaricus bisporus* for yield and chemical composition. *Mush. Res.*, 2(2): 83–85.

Huang, B.H., Hung, K.H. and Chang, S.T., 1989. *Mushroom Sci.*, 12(2): 533–540.

Jain, V.B. and Singh, S.P., 1982. Effect of spawning method on the yield of *Agaricus brunnescens* Peck. *Progressive Hort.*, 14(4): 246–248.

Jakson, M.L., 1973. *Soil Chemical Analysis.* Prentice Hall of India (Ltd.), New Delhi, p. 159–161.

Kachroo, J.L., Ahmad, N. and Amanullah, Khan, T.N., 1979. Experiments in the improvements of compost formulae and procedure in *Agaricus bisporus* cultivation. *Mush. J.*, 82: 433, 435, 437, 441.

Kumar, S., Sharma, T.R., Kumar, S. and Goswami, A.K., 1991. Comparison of protein in six strains of *Agaricus bisporus. Plant Path. Biochem., New Delhi,* 18(2): 71–74.

Overstijns, A. and Bockstaele, L., 1975. The optimal time of casing for the commonly cultivated mushroom (*Agaricus bisporus*). *Champigon Culture*, 19(2): 55, 57, 59 [Cited from *Hort. Abst.* 45(12): 1975].

Shandilya, T.R., Seth, P.K., Kumar, S. and Munjal, R.L., 1974. Effect of different spawning methods on the productivity of *Agaricus bisporus. Indian J. Mycol. Plant Path.*, 4: 129–131.

Singer, R., 1961. *Mushrooms and Truffles: Botany, Cultivation and Utilization.* Leonard Hill, London, pp. 272.

Stoffel, W., Chu, F. and Ahrens, E.H. Jr., 1959. *Anal Chem.*, 31: 307–308.

Suman, B.C., 1993. Evolving a new strains of *Agaricus bisporus* by single spore isolation technique. *Indian J. Mycol. Plant Pathol.*, 23(3): 314–315.

Chapter 52

Sustainable Development of Environmental Conservation on Natural Land Resources and Population through Carrying Capacity and Footprints

S.R. Singh[1], Th. Manimala Devi[1]* and N. Sangbanbi Devi[2]

[1]Post Graduate Studies Centre, HRDRI, Canchipur – 795 003

[2]Imphal College, Imphal – 795 001

ABSTRACT

"Sustainable development of a particular place through environment and population carrying capacity by computation on ecologically productive land per capita in hectares, ecologically deficit land per capita in hectares and in percent available in decades and in annum expedite judicious as 0.17, 1.82, 1043.83 and 104.38 in 1991 and 0.14, 1.85, 1316.93, 131.69 in 2001 respectively for Thoubal District, Manipur. Sustainability on population increase of the district scored to destructive criteria consequently reflects the felt need of consciousness to planners and decision makers. Ecological footprint analysis of land resources and population on sustainable development highlighted the needs of sustainability guidance. Exhaustive analysis on sustainable development on total land accentuated to covet the expedite management. The sustainability on total area for developmental planning of Thoubal district and Manipur state have intensely established and manifest the enormous wealth of nature, for meeting the needs of the present

* Corresponding Author; E-mail: hrdri_1990@yahoo.com; (M) (O) 9863058178.

without compromising the ability of future generation to meet their own needs for human kind with a hope of conservation. Right planning for right action based on conservation of natural resources evince the way to solve the ever advancing exploitation of resources.

Keywords: *Sustainable development, Natural resources, Footprint, Environmental conservation and Carrying capacity.*

Introduction

Sustainable development encapsulates the idea of economic development with due care for the environment having wide currency following the World Commission on Environment and Development, the Brundtland Commission. The Commission in 1987 recommended sustainable development is the way to achieve global environmental security. The term itself had been used before the Brundtland Commission endorsed it, notably in the "World Conservation Strategy", jointly prepared by the World Conservation Union (IUCN), the United Nations Environment Programme (UNEP), the World Wide Fund for Nature (WWF) in 1980. (Ruth and William, 2000).

The World Commission on Environment and Development was created in 1993 and published "Our Common Future Report" in which it stated that sustainable development is a process of change as "Development that meet the needs of the present without compromising the ability of future generations to meet their own needs". Further the report stated that "Sustainable development is a process of change in which the exploitation of resource, the direction of investments, the orientation of technological and institutional change are all in harmony and enhance both current and future potential to meet human needs and aspirations".

Further, population and consumption are two sides of the same coin. The impact of population is very important consequently the consumption is clearly a serious issue. Truly a reduction in population growth in developing countries and reduction of consumption in the so called developed countries are both needed to achieve sustainability.

Sustainability in simplest idea is the status recognization about resources that consumed faster than they are produced, or renewed. The resource in common is depleted and eventually used up. If humanity's ecological resource demands exceed that nature can continually supply, the condition is termed ecological overshoot. In a sustainable world, society's demand on nature is in balance with nature's capacity to meet that demand.

Thoubal, the tiny district in the valley area of the Manipur state have to examine with criteria of wide angle video scope of sustainable development with respect to population increase as human activity since 1991. Thus, it is worthwhile to execute in debt scientific approach to the components of sustainable development with the undermentioned objective:

1. To determine the sustainability development of total area based on footprint.
2. To detect the sustainability development on population growth.

Materials and Methods

For the present investigation probability sampling of simple and stratified random sampling, systematic and cluster sampling and non probability sampling of convenience, snow ball, quote and focus groups sampling were adopted. Both the documentary *i.e.* books, report of surveys, memoirs, account of travels, 'etc. and field sources *i.e.* interview with living persons, scholars, scientists, research workers etc. were adopted for analysis of available data. (Arlene, 1995; Yadava, 1995)

For intransigence of sustainable development of the Thoubal district, the undermentioned formulae have adopted following Rees, 1996.

1. *Ecologically productive land per capita, EPLC (in Ha)*

$$\text{EPLC (in Ha)} = \frac{\text{TA (in ha)}}{\text{TP (in nos.)}}$$

where,

EPLC: Ecologically productive land per capita in hectares.
TA: Total area in hectares.
TP: Total population in numbers.

2. *Ecologically deficit land per capita (EDLC) in hectares*

$$\text{EDLC} = \text{FP} - \text{EPLC in hectares}$$

where,

EDLC: Ecologically deficit land per capita in hectares.
FP: 2 Hectares since India belongs to (2–3) Ha foot print.
EPLC: Ecologically productive land per capita in hectares.

3. *Ecologically deficit land per capita (EDLC) in percent available in decades*

$$\text{EDLC in per cent available in decades} = \frac{\text{(EDLC in Ha)}}{\text{(EPLC in Ha)}} \times 100$$

4. *Ecologically deficit land per capita (EDLC) in percent available in annum*

$$\text{EDLC in per cent available in annual} = \frac{\text{(EDLC in Ha)}}{\text{(EPLC in Ha)}} \times 10$$

Results and Discussion

The population of Thoubal district in 1991 recorded 2,93,958 persons and 51,400 hectares of total land area. The calculated value of ecologically productive land per capita accord 0.17 hectares and ecologically deficit per capita as 1.82 hectares. The ecologically deficit per capita in percent available in decades and annum compute as 1043.83 and 104.38 respectively. Thoubal district recorded a population of 3,64,140 persons and 51,400 hectares in total land area in 2001. The calculated value of ecologically productive land per capita accord 0.14 hectares and ecologically deficit per capita as 1.85 hectares. The ecologically deficit per capita in percent available in decades and per annum compute as 1316.93 and 131.69 respectively (Table 52.1). The illucidated data were portrayaled in Figure 52.1

Manipur state recorded a population of 18,37,149 persons and 22,32,700 hectares of total land area in 1991. The calculated value of ecologically productive land per capita in hectares accord 1.21

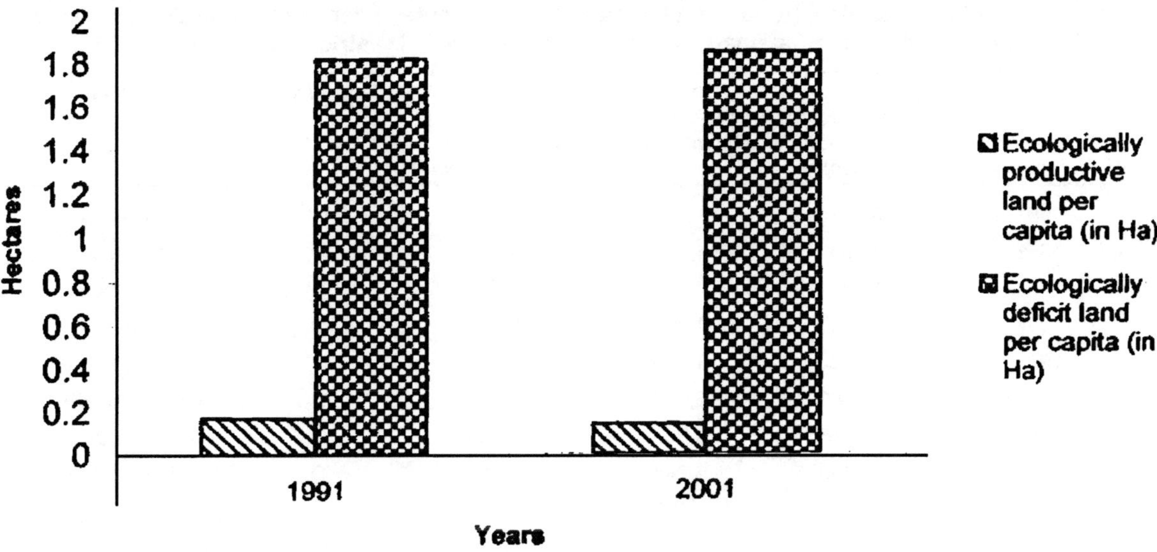

Figure 52.1: Sustainability Development on Total Land Area of Thoubal District

and ecologically deficit per capita as 0.78 hectares and the percent available in decades and annual compute as 64.56 and 6.45 respectively. The population in 2001 recorded 22,93,896 persons and 22,32,700 hectares in total land area. The calculated value of ecologically productive land per capita accord 0.97 hectares and ecologically deficit per capita as 1.02668 hectares. The percentage available in decades and annual compute as 105.48 and 10.54 respectively (Table 52.1). The enlisted data were plotted in Figure 52.2.

Table 52.1: Determination of Sustainability on Total Land Area for Developmental Planning of Thoubal District and Manipur State

District/State	*Year*	*Total Population (TP) in nos.*	*Total Area (TA) in Ha*	*Ecologically Productive Land Per Capita (EPLC) in Ha.*	*Ecologically Deficit Land Per Capita (EDLC)*			*Remarks*
					In Ha	*In per cent Available in Decades*	*In per cent Available in Annum*	
Thoubal	1991	2,93,958	51,400	0.17	1.82	1043.83	104.38	Assuming 2 Ha
	2001	3,64,140	51,400	0.14	1.85	1316.93	131.69	as Footprint
Manipur	1991	18,37,149	22,32,700	1.21	0.78	64.56	6.45	Assuming 2 Ha
	2001	22,93,896	22,32,700	0.97	1.02	105.48	10.54	as footprint

Sources: 1. Statistical abstract Manipur 200l. 2. Statistical abstract Manipur 2006. 3. Statistical abstract of Manipur 1982.

Thoubal district recorded the total population with a range from 76,782 to 3,64,140 persons from 1951 to 2001. During these 50 years *viz.*, 1951 to 2001 the population increase per annum, ranges from 4,452 to 7,835 and corresponding percentage assigned from 2.19 per cent to 5.79 per cent (Table 52.2). The exemplified data were graphically plotted and presented in Figure 52.3.

Table 52.2: Total Population, Population Increase, their Percentage per Annum and Sustainability Status as per Criteria of Sustainable Development of Thoubal District, Manipur (1951–2001)

Year	*Population Computed Based on Decennial "r" (in nos)*	*Population Increase per Annum (in nos)*	*Percentage of Population Increase per Annum (per cent)*	*Sustainability*	*Remarks*
1951	76,782	–	–		Criteria of
1952	81,234	4,452	5.79	D	Sustainable
1953	85,945	4,711	5.79	D	Development
1954	90,930	4,985	5.80	D	SD,
1955	96,203	5,273	5.79	D	Sustainable
1956	1,01,782	5,579	5.79	D	<0.5 per cent p.a.
1957	1,07,685	5,903	5.79	D	C, Critical
1958	1,13,930	6,245	5.79	D	0.1–1.5 per cent p.a.
1959	1,20,537	6,607	5.79	D	D, Destructive
1960	1,27,528	6,991	5.79	D	>2.0 per cent p.a.
1961	1,34,924	7,396	5.79	D	
1962	1,39,005	4,081	3.02	D	
1963	1,43,210	4,205	3.02	D	
1964	1,47,543	4,333	3.02	D	
1965	1,52,006	4,463	3.02	D	
1966	1,56,605	4,599	3.02	D	
1967	1,61,343	4,738	3.02	D	
1968	1,66,224	4,881	3.02	D	
1969	1,71,252	5,028	3.02	D	
1970	1,76,433	5,181	3.02	D	
1971	1,81,771	5,338	3.02	D	
1972	1,86,242	4,471	2.45	D	
1973	1,90,823	4,581	2.45	D	
1974	1,95,518	4,695	2.46	D	
1975	2,00,327	4,809	2.45	D	
1976	2,05,255	4,928	2.45	D	
1977	2,10,305	5,050	2.46	D	
1978	2,15,478	5,173	2.45	D	
1979	2,20,779	5,301	2.45	D	
1980	2,26,210	5,431	2.45	D	
1981	2,31,781	5,571	2.46	D	
1982	2,37,274	5,493	2.36	D	
1983	2,42,897	5,623	2.36	D	

Contd...

Table 52.2–Contd...

Year	*Population Computed Based on Decennial "r" (in nos)*	*Population Increase per Annum (in nos)*	*Percentage of Population Increase per Annum (per cent)*	*Sustainability*	*Remarks*
1984	2,48,654	5,757	2.37	D	Criteria of
1985	2,54,547	5,893	2.36	D	Sustainable
1986	2,60,580	6,033	2.37	D	Development
1987	2,66,758	6,178	2.37	D	SD,
1988	2,73,080	6,322	2.36	D	Sustainable
1989	2,79,552	6,4 72	2.37	D	<0.5 per cent p.a.
1990	2,86,178	6,626	2.37	D	C, Critical
1991	2,92,960	6,782	2.36	D	0.1–1.5 per cent p.a.
1992	2,99,399	6,439	2.15	D	D,
1993	3,05,983	6,584	2.19	D	Destructive
1994	3,12,711	6,728	2.19	D	>2.0 per cent p.a.
1995	3,19,588	6,877	2.19	D	
1996	3,26,615	7,027	2.19	D	
1997	3,33,797	7,182	2.19	D	
1998	3,41,137	7,340	2.19	D	
1999	3,48,638	7,501	2.19	D	
2000	3,56,305	7,667	2.19	D	
2001	3,64,140	7,835	2.19	D	

Sources: 1. Statistical abstract Manipur 2001. 2. Statistical abstract Manipur 2006. 3. Statistical abstract of Manipur 1982.

$$\text{Rate constant, } r = \frac{ln\, N_t - ln\, N_0}{T} \quad \text{based on decennial}$$

where,

Nt: Final population; N_0: Initial population; T: Time period

The sustainability of the district under "Sustainable Development" was computed following the criteria for sustainable development by Murai (2005) with respect to Global Environment and Population Carrying Capacity. As per criteria, human activity of population increases in lesser than 0.5 per cent p.a. between 1.0 to 1.5 per cent p.a. and greater gramped than 2 per cent p.a. were categorized as sustainable, critical and destructive respectively.

Table 52.1 revealed the population, total area, ecologically productive land per capita and different components of ecologically deficit per capita etc. of the Thoubal district and Manipur state during 1991 and 2001. The total area of Thoubal district during 1991 observed 51,400 hectares with a population of 2,93,958 persons. During 2001, the district accord the population of 3,64,140 persons in the same area. The ecologically productive land per capita was 0.17 hectares and 0.14 hectares in 1991 and 2001 respectively and ecologically deficit per capita was 1.82 hectares in 1991 and 1.85 hectares in 2001. Ecologically deficit land per capita in percent available in decades was 1043.83 in 1991 and

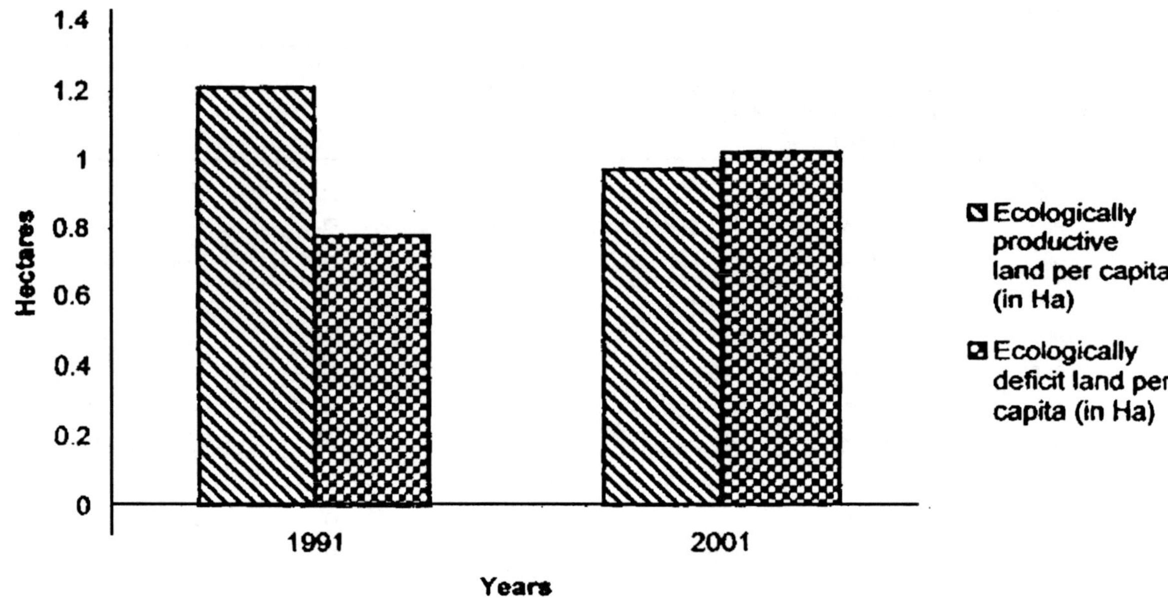

Figure 52.2. Sustainability Development on Total Land Area of Manipur State

1316.93 in 2001 and the per cent available in annual was 104.38 in 1991 and 131.69 in 2001. The finding explored the ecological crisis imposed by the requirements and needs of the population increase in the area. The similar impact of population increase prevailed in the district extending to different districts of the state then to the nation then upto the global level. Manipur state in 1991 recorded 18,37,149 persons of population in 22,32,700 hectares of total land area. The state have accord 1.21 hectares of ecologically productive land per capita, 0.78 hectares of ecologically deficit per capita, 64.56 percentage available in decades and 6.45 percentage available in annual. The state also recorded 22,93,896 persons of population, 22,32,700 hectares of total area, 0.97 hectares ecologically productive land per capita, 1.02 ecologically deficit land per capita in hectares, 105.48 percentage available in decades and 10.54 percentage available in annual during 2001. It is evident from the present finding that EDLC exceed than 100 and raised to 10.5 per cent per annum. The finding reflect the impact of population increase to the natural land resources. The trend of overshoot and collapse has coincidence with the population increase of present studying district and highlighted the situation to the planners and developmental programmers.

Further, the present work clarified the ecological footprint as a resource accounting tool inexorably used to address underlying sustainability questions. It measures the extent to which people of the district is using nature's resources faster than they can regenerate. It further illustrates ecological resources depleted by populations both geographically and socially. Furthermore, the ecological footprint clarifies the relationship of resources use to equality by explicitly tying individual's and groups activities to ecological demands. These connections help decision makers more accurately and equitably shape policy in support of social and environmental justice (Schmidt-Bleek, 1992a,b; Pearce, 1994; Rees, 1994a; Rees, 1992).

As long as we do not know how much of nature's capacity we use or how resource use compares to existing stocks, overshoots may go under use increasing the ecological deficit and reducing natures

Figure 52.3(*a*): Graph Showing Computed Annual Population Based on Decennial "r" of Thoubal District, Manipur

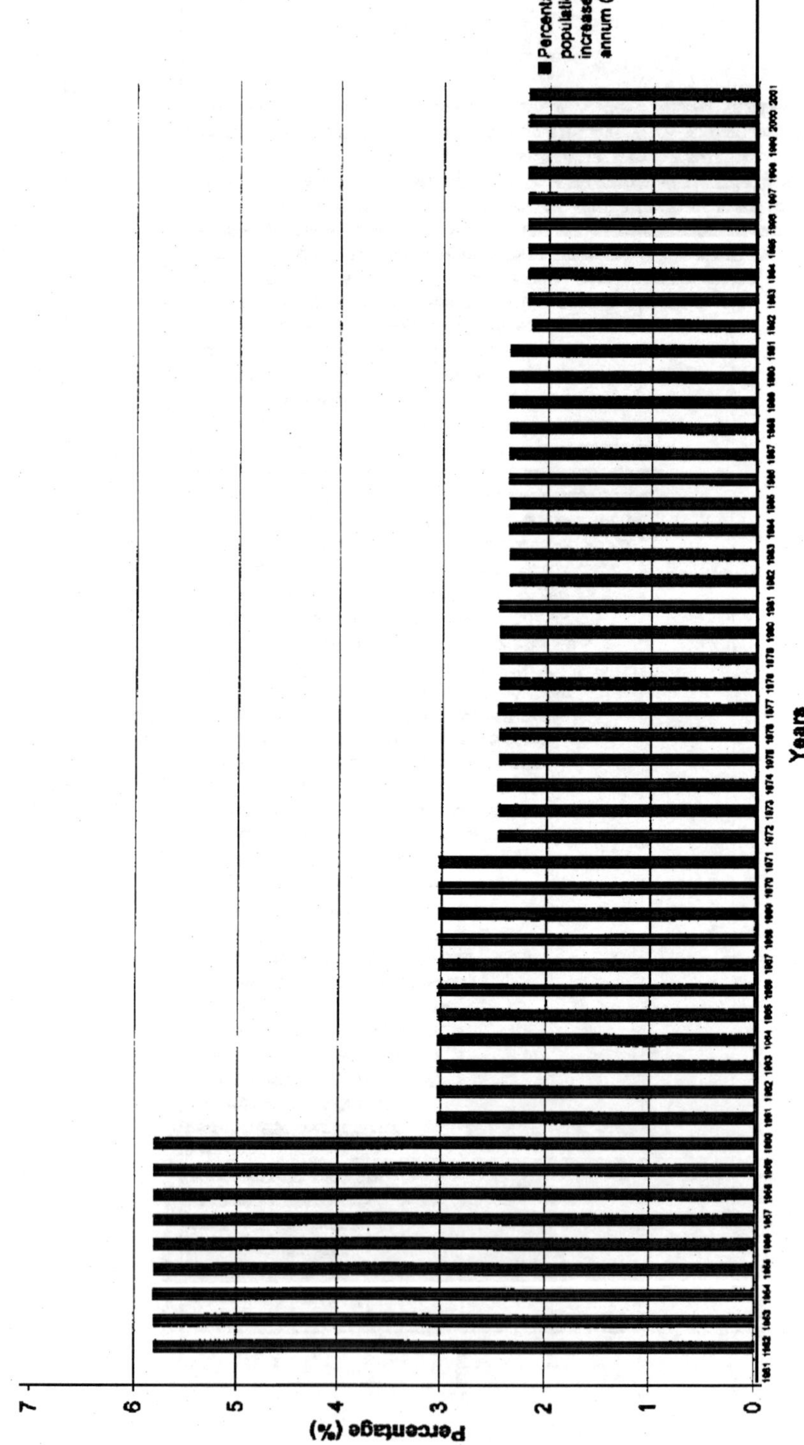

Figure 52.3(*b*): Graph Showing the Percentage of Population Increase per annum of Thoubal District, Manipur

capacity to meet society's need. Continued overshoot is not inevitable. The ecological footprint provides a systematic resource accounting tool that can help us plan for a place in which we all live well, within the means of our one planet.

Table 52.2 depicts the total population, population increase, their percentage per annum and sustainability status of the Thoubal district. During 1951 to 2001, the district recorded the total population ranges from 76,782 to 3,64,140 persons. The district accord the increase of population per annum ranges from 4452 to 7835 and 12.19 per cent to 5.79 per cent as the percentage of population increase per annum. Based on the criteria of sustainable development on the percentage of population increase per annum, the district crossed the critical and belongs to the destructive level as its value being greater than 2 per cent per annum. It is evident from the present finding that the district needs an additional area to accommodate the increased ecological load of people alive today. If the population pressure increase at the present trend than in the next decade nearly 5 additional area would be needed for all else demanding resources due increase of population. In other words just to maintain the present rate of ecological decline. The finding was in corroborative with that of other workers at different polaces (Rees and Wackernagel, 1994).

The present finding highlight a cautionary signal in certain commodities and resources and suggestions to a variety of concrete sustainability guidelines and supports to broadly based programme of reform that could redirect us in the direction of developmental progress by right utilization of resources and conservation. The sustainability on total area for developmental planning of Thoubal district and Manipur state need intensely analysis and established formulation to manifest the enormous wealth of nature, the land covering area of the district and state.

Acknowledgement

The authors are thankful to the Director, Department of Economics and Statistics, Government of Manipur for providing needful data, Deputy Secretary UGCNERO, for MRP to Miss Th. Manimala Devi.

References

Arlene, F., 1995. *How to Ask Survey Questions?* Sage Publications India Pvt. Ltd.

Meadows, D.L., 1995. It is too late to achieve sustainable development, now let us strive for survivable development. In: *Towards Global Planning of Sustainable Use of the Earth: Development of Global Eco-engineering,* (Ed.) S. Murai. Elsevier, Amsterdam, p. 359–374.

Murai, S., 2005. *Global Environment and Population Carrying Capacity.* http://www.unu.edu/unupress/unupbooks/uu03pe/uu03pe0c.htm.

Pearce, D., 1994. Sustainable consumption through economic instruments. In: *Paper prepared for the Government of Norway Symposium on Sustainable Consumption*, Oslo, 19–20 January.

Rees, W., 1992. Ecological footprints and appropriated carrying capacity: What urban economics leaves out. *Environment and Urbanization*, 4(2): 121–130.

Rees, W., 1994a. Sustainability, growth and employment: Towards an ecologically stable, economically secure and socially satisfying future. In: *Paper Prepared for the IISD Employment and Sustainable Development Project*. Winnpeg, Manitoba: International Institute for Sustainable Development, (Revised version in Alternatives 21:4) [October/November 1995].

Rees, W. and Wackemagel, M., 1994. Ecological footprints and appropriated carrying capacity; Measuring the natural capita requirements of the human economy. In: *Investing in Natural Capital: The Ecological Economics Approach to Sustainability*, (Eds.) A.M. Jansson, M. Hammer, C. Folk, and R. Costanza. Island Press, Washington, pp. 362–390.

Rees, W.E., 1996. Revising carrying capacity: Area based indicators of sustainability. *Population and Environment: A Journal of Interdisciplinary Studies*, 17: 3.

Ruth, A.E. and William R.E., 2000. *The Environment Encyclopedia, Vol. 9*. Salmonella Tropical Dry Forest.

Schmidt-Bleek, F., 1992a. MIPS: A universal ecological measure. *Fresenius Environmental Bulletin*, 1: 306–31.

Schmidt-Bleek, F., 1992b. MIPS: Revisited. *Fresenius Environmental Bulletin*, 2: 407–412.

Yadava, S.S. and Yadava, K.N.S., 1995. *Statistical Analysis for Social Sciences*. Manak Publications.

Chapter 53

Biomedical Waste Management: A Prophalactic Measure for Future Health Hazards–A Case Study of Warangal District of Andhra Pradesh

G. Benarjee*, K. Srikanth, G. Laxman Prasad, G. Ramu and K. Narasimha Ramulu

Department of Zoology, Kakatiya University, Warangal – 506 009, Andhra Pradesh, India

ABSTRACT

The Biomedical waste is generated from healthcare establishments, blood banks, laboratories and research institutes. Biomedical waste which the hazards and risks multiply not just due to the generators and operators but the general community is get effected with these substances and it is the source of environmental pollution of the present day. Hospitals are the one of the major source of biomedical waste material and its quantity is increasing day-to-day in geometrical ratio. However, the disposal of this material is a big challenging task due to their hazardous nature. Therefore, it is endeavored to make new initiatives for the clean and eco-friendly environment in Warangal town and its surroundings, because this region is known for good medical services and medical expertise in the entire Northern Telangana region of Andhra Pradesh. Hence an attempt has been made to carryout an extensive survey on existing healthcare establishments to estimate the quantum of Biomedical waste produced in this area. The present study emphasized on the quantity and the quality of Biomedical waste produced in this area. A

* Corresponding Author; E-mail: gbgsss@yahoo.co.in.

detailed assessment has been carried out on the techno-economical viability of the biomedical waste management and the disposal of biomedical waste generated in this area in an economical and eco-friendly manner, thus a pollution free atmosphere could be established. Further the aim of the present work is for the safe disposal of highly infectious solid waste and thus it helps in improving environmental quality in this area, which will have positive impact on the general health conditions of the community of Warangal district.

Keywords: *Biomedical waste, Health hazards, Prophalactic measures, Warangal district.*

Introduction

The biomedical waste may cause environmental problems. Biomedical waste generates primarily from healthcare establishments. The handling, management and the disposal-off this material is relatively a recent issue and however there is a lot of information gap and apprehentions existing in this area of research. The management of biomedical waste is still in its infancy all over the world. The improper management of Biomedical waste causes serious environmental problems in terms of air, water and land. Biomedical waste management wherein the hazards and risks multiply not just to the generators and operators but also affect the general community as a whole. Biomedical waste handling and disposal is a serious concern to the public. Additional fuel for public concern about the medical waste is that this waste may transmit the human immuno-deficiency virus (HIV), hepatitis-B or C virus and other agents associated with bloodborne diseases (Kosha Shah *et al.*, 2006). This public concern has resulted promalgamation of certain laws and enforced rules and regulations by many states and local governments to regulate the collection, transport, and disposal of the biomedical waste. Although there is a paucity of information on this issue, still some authors (Brown, 1993; Bineesha and Mahendra, 1999; Pruthvish *et al.*, 1996, 1998; Rao and Garg, 1994; Acharya and Singh Meeta, 2000 and Srivastava, 2000) have made their attempts to probe into this area of research.

The present study made an attempt on to estimate the number of institutions, their capacity (no of beds) and the quantity of Biomedical waste that these institutions generate every day and its safe disposal to avoid the health risk. In the present study an attempt has also been made on how to reduce the health hazards caused by this material. The Warangal district of Andhra Pradesh was the place selected for this study.

Materials and Methods

Warangal is a cosmopolitan town located at 17°19′ and 18°36′ North latitude and 78°49′ and 80°43′ East longitude (Figure 53.1) with population of more than 3.2 millions. In it's radius it is the first major town in the North Telangana districts of Andhra Pradesh. In this area initially a survey has been conducted on the number of medical care institutions like hospitals, nursing homes, dental clinics, diagnostic centers and blood banks where the Biomedical waste is generated were listed out. The biomedical waste material produced in these institutions every day were collected in a separate containers following the methods as prescribed by Singh and Sarma, (1996) and Srivastava, (2000). This biomedical waste material was later on segregated in accordance with CPCB norms (CPCB rules 1998) which suggests the management and handling of Biomedical waste as presented in Table 53.1. Finally the incinerator available at M/s Kakatiya Mediclean Services, Warangal was used to dispose off the waste. The incinerator's capacity and it's specifications were as presented in Table 53.2.

Figure 53.1: Map of India shows Location of Warangal District in Andhra Pradesh

Table 53.1: Handling of Biomedical Waste in Colour Bags (Rule 1998 of Biomedical Waste Management)

Bag Colour Code	Material Segregated	Treatment
Yellow bags	Human body parts, placenta, human tissues, histopathology specimens, organs, any pathological surgical waste, microbiological waste, biotechnology waste, contaminated cotton, bandages, bedding, other linen material dressings, Wabs, MOPs, solid plaster, casts	Incineration/deep burial
Black bags/ puncture proof containers	Needles, cut glasses, blades, scalpels, slides, nails, lancets, anything that can cut or puncture.	Disposal into secured land fill
Blue bags	Disposable syringes, I.V. tubes, catheters, drains and any other tubings, blood bags, urine bags, stents, gloves, aprons.	Autoclaving/microwaving/ chemical treatment and destruction/shredding.

Table 53.2: The Capacity and Technical Specifications of Incinerator Used for the Final Disposal of the Biomedical Waste

Sl.No.	Specifications	Capacity
1.	Burning capacity	100 kg/hr
2.	Combustion efficiency	99.0 per cent
3.	Population control equipment	Cyclone separator, venture scrubber, droplet separator
4.	Burners	1 no for primary chamber
		1 no for secondary chamber
5.	Mode of waste handling	Waste feeding by conveyor
6.	Mode of ash handling	Manual
7.	Doors	1 waste charging door in primary chamber
		2 ash removal door in primary chamber
		1 ash removal door in cyclone separator
8.	Operating hours	(Total quantity)/(Capacity/hr)+1–1½ hrs for preheat and cooling of furnace
9.	I.D. Fan	3400 m^3/hr
	F.F. Fan	1100 m^3/hr
	Re-circulation pump	1 HP
	Burner motor	240 W each
10.	Types of fuel	HSD/Diesel
11.	Type of burner operation	Automatic on-off
12.	Connected electrical load	Approx. 12 KW (AC3 phase, 400/440v, 50 cycles, 4 wire)
13.	Operating temperature	
	Primary chamber	800°C±50°C
	Secondary chamber	1050°C±50°C

Results and Discussion

The Biomedical waste generated in the health care facilities and their disposal is of great concern to the public. Although sporadic incidents of public illness caused by exposure of Biomedical waste occurred else where, which show the potential risk that the Biomedical waste creates and despite of the existence of stipulate norms (CPCB rules 1998) as prescribed by the government. However, people are not paying much attention to handle this issue to make the society free from the potential risk that the Biomedical waste creates. The quality and quantity of the Biomedical waste generated from various healthcare institutions and allied establishments in the Warangal district was presented in Table 53.3. The data indicates that such institutions are more in private sector (111) than the public sector (007). It is estimated that in the urban area of Warangal district there is a total of 214 kg Biomedical waste is generated every day which was @ 0.14 kgs per bed per a day. However in the rural areas of the district the scenario is quite different, where the total quantity of the waste generated was 62 kg @ 0.12 kgs per bed per a day. Hospital is one of the complex institutions which is frequented by the people from all walks of life in the society without any distinction between age, sex, race and religion. The produce of the Biomedical waste is increasing in its quantity and nature due to advances in scientific knowledge and at the same time it is also creating its impact (Rao and Garg, 1994). The most appropriate way of identifying the various categories in Biomedical waste is may be by sorting out the waste into colour coded plastic bags/containers. Accordingly in the present study the Biomedical waste was collected in the separate colour coded bags, where the yellow bags were used to collect the material such as Human body parts, placenta, human tissues, histopathological specimens, human organs, pathological surgical waste, microbiological waste, contaminated cotton, bandages, bedding, other linen material dressings, swabs, MOPs, solid plaster, casts. Whereas the black bags and puncture proof containers were used to collect needles, cut glasses, blades, scalpels, slides, nails, lancets, any thing that was used for cut or puncture and the blue bags were used to collect disposables such as syringes, IV tubes, catheters, drains, tubings, blood bags, urine bags, stents, gloves etc. Thus the Biomedical waste was procured at source where it was graded and it is the mandatory as notified by CPCB during the year 1998. Indoor air pollution due to biological waste can be reduced by covering the waste material properly, routing the waste in such a way that shortest distance is used to transport and sensitive areas are must be avoided.

Table 53.3: Biomedical Waste Generated in the Urban and Rural Areas of Warangal District (Per a day during the year from June, 2006–May, 2007)

Sl.No.	*Area*	*No. of Institutions*			*No. of Beds*			*Quantity of Waste Generated per Day*		*Nature of Waste*	
		Govt.	*Priv.*	*Total*	*Govt.*	*Priv.*	*Total*	*Per Bed (kgs)*	*Per Day Total (kgs)*	*Incinerable Waste (kgs)*	*Autoclave Waste (kgs)*
1.	Urban	7		7	961		961	0.06	64	39	25
			111	111		1850	1850	0.08	150	88	62
								0.14	214		
2.	Rural	5		5	278		278	0.02	8	5	3
			45	45		530	530	0.10	54	33	21
								0.12	62		
3.	Total	12	156	168	1239	2380	3619	0.26	276	165	111

*Govt: Government; Priv.: Private.

Blood, body fluids and body secretions are the constituents of biomedical waste which harbour most of the viruses, bacteria and parasites that cause infection. This passes via a number of human contacts, all of whom are potential 'recipients' of the infection. Tuberculosis, pneumonia, diarrhoea, tetanus, whooping cough etc., are other common diseases spread due to improper biomedical waste management. The hospital waste is an additional risk for patients and personnel who handle this waste and it poses a serious threat to public health by polluting the environment (Singh and Sarma, 1996). Although pollution load does not migrate completely and its load and impact may be reduced. Air pollution may also caused in both indoors and outdoors. Pathogens present in the waste can enter and remain in the air in an institution for a long period in the form of spores or as pathogens itself. This can result in hospital acquired infections (Noscomial infections) or occupational health hazards. The patients and their attendants also have a chance of contracting infections caused due to pathogens or spores which are air born. However, there are very limited statistics available in this field. The quantity of Biomedical waste produced in the Warangal district gives an exaggerative picture (Table 53.4). During the present study (The period which started from June 2006 to May 2007) a total of 1,20,805 kg waste was produced in which incinerable waste is 99,915 kg and autoclave waste is 20,890 kg. However the Biomedical waste is produced in a year vary from month to month. The wastage thus generated in the health care establishments showed a gradual increased from July to December where as it was reached to its peak during the October month. This rise indicates that the epidemics and the patients entry was more during October.

Table 53.4: Quantity of Biomedical Waste Produced in Warangal district (From June'2006–May'2007)

Month	*Incinerable Waste (kgs)*	*Autoclave (kgs)*	*Total Quantity (kgs)*
June'06	8200	1790	9990
July'06	8255	2000	10225
August'05	9325	1890	11215
September'06	9860	1810	11670
October'06	10205	1930	12135
November'06	8485	1680	10165
December'06	8395	1760	10155
January'07	7925	1545	9470
February'07	6790	1455	8245
March'07	7570	1680	9250
April'07	7650	1660	9476
May'07	7255	1690	8945
Total	99915	20890	120941

One of the causes for the sharp increase of occurrence of the infectious diseases is the improper waste management. The health hazards due to improper waste management can not only affect the occupants in institutions, but also spread in the vicinity of the institutions. The problem of occupational health hazards due to Biomedical waste is not publicized, as there is lack of information. Proper management of waste can solve the problem of occupational hazards to a large extent. Biomedical waste can cause health hazards to animals and birds. Harmful chemicals such as dioxins and furans can cause serious health hazards to animals and birds. Certain heavy metals can effect the reproductive

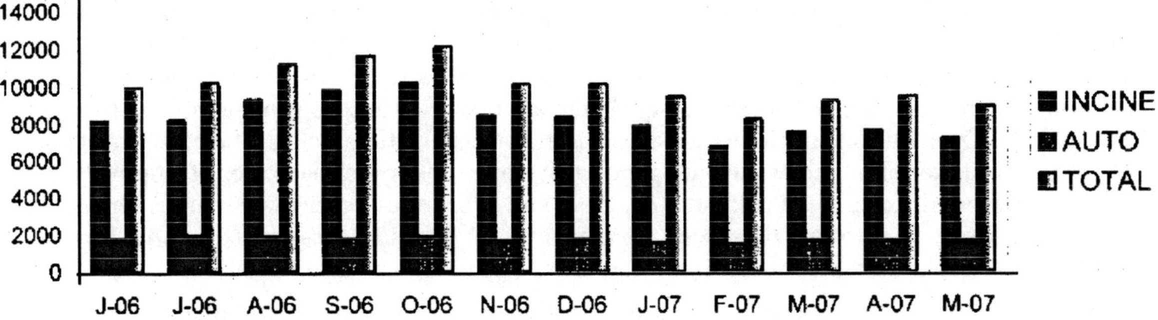

Figure 3.2: Monthly Biomedical Waste Generated in Warangal District
***INCINE: Incinerable waste; AUTO: Autoclave waste; TOTAL: Total Waste.**

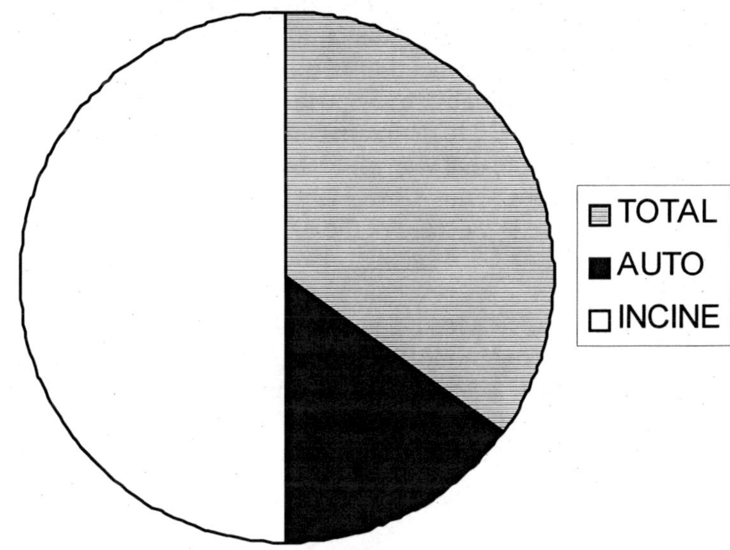

Figure 3.3: Total Quantity of Biomedical Waste in Warangal District
***INCINE: Incinerable waste; AUTO: Autoclave waste; TOTAL: Total Waste.**

nature of the animals. Biomedical waste can cause water pollution, if the waste is dumped in low lying areas, or in the vicinity of lakes and water bodies where it causes water pollution (Brown, 1993; Sriniwas, 1992). The liquid waste generated from healthcare institutions may also cause water pollution if not treated them properly. The harmful chemicals present in biomedical waste such as heavy metals can also cause water pollution. Water pollution can either be caused due to biologicals, chemicals or radioactive substances. The pathogens present in the waste can leach out and contaminate the ground water or surface water. Poor land filling technology may cause water pollution in the form of leachates. Exess nutrient leachate nitrates and phosphates from land fills which causes eutrophication.

According to the WHO, the global life expectancy is increasing year after year. However, deaths due to infectious disease are increasing. Open dumping of biomedical waste is the greatest cause for land pollution. Land filling is also harmful to a limited extent. Soil pollution from biomedical waste is caused due to infectious waste, discarded medicines, chemicals used in treatment and ash and other waste generated during treatment processes (Tumberg and Wayne, 1996). Heavy metals such as cadmium, lead, mercury etc., which are present in the waste will get absorbed by plants and then enter the food chain. Open burning of biomedical waste is the most harmful practice. The presence of plants and hazardous materials in the waste with generate harmful gases such as oxides of sulphur, oxides of nitrogen, carbondioxide etc. (Pruthish *et al.*, 1996) and suspended particulate matter. These when inhaled can cause respiratory diseases. Certain organic gases such as dioxins and furans or carcinogenic. Open burning of biomedical waste should be strictly avoided.

The hospitals, nursing homes, clinics, dispensaries, animal houses, pathological labs etc., are therefore required to be accessible to the Biomedical waste treatment plants or they must setup their own plants. Because the CPCB rules also envisage that common facility or any other such facilities must be used for waste treatment. Handling, segregation, mutilation, disinfection, storage, transportation and final disposal are vital steps for safe and scientific management of biomedical waste in any establishment (Acharya and Singh Meeta, 2000). Hence in view of the public interest and also to adhere to the CPCB rules it is suggested that all the healthcare establishments in the Warangal district of Andhra Pradesh must take initiative to dispose-off their biomedical waste in an appropriate scientific methods as envisaged by CPCB to maintain clean and hygienic environment in the district. Because the incredible amount of waste that was generated in the district might become instrumental in the outbreak of infections diseases at any time. Hence in view of public health and hygiene the proper way of disposal of this waste is the need of the day.

Acknowledgements

The authors thank M/S Kakatiya Mediclean Services for providing necessary technical and laboratory facilities to take up this programme. We profusely thank the medical care establishments in Warangal district for extending their co-operation during entire period of study. We express our sincere thanks to all medical doctors for their co-operation. We are also greatful to the Head, Department of Zoology, Kakatiya University for providing computer aided laboratory facilities during the period of this study.

References

Acharya, D.B. and Singh, Meeta, 2000. *The book of Hospital Waste Management.* Minerva Press, New Delhi, p. 15,47.

Armin, Roscncranz and Diwan, Shyam N, 2001. *Environmental Law and Policy in India,* 2nd edn. OUP, New Delhi.

Bineesha, P. and Mahendra, B.J., 1999. *Biomedical Waste dings of Intemation Conference on Biomedical Waste Management.* Tamil Nadu Pollution Control Board, Chennai.

Bineesha, P., Dayal, H.V. and Ramesh, S., 1998. Energy management in hospitals. In: *Proceedings of Technology Management in Healthcare System Seminar*, Defence Bio-Engineering and Electromedical Laboratory (DEBEL).

Brown, J., 1993. Hospital waste management that saves money and helps the environment and improves safety. *Med Waste, Regul. Anal,* 1(10): 1–8., Central Pollution Control Board, Vol. 4(iv), 1998.

Connet, Paul, 1996. *Medical Waste Incineration: A Mismatch Between Problem and Solution*. A publication of work on waste, USA, Inc., 82 Judson, Canton, NY 13617.

Pruthvish, S., Gopinath, D., Jayachandra Rao, M., Girish, N., Bineesha, P and Shivaram, C., 1996. *Exploratory Study of Hospital Waste Management in Bangalore.* M.S.R.M.C., Bangalore.

Pruthvish, S., Gopinath, D., Jayachandra Rao, M., Girish, N., Bineesha, P and Shivaram, C., 1998. *Information Learning Units for Safe Management of Healthcare Waste: A Manual on Healthcare Waste Management.* Healthcare Waste Management Cell, Department of Community Medicine, M.S. Ramaiah Medical College, Bangalore.

Rao, S.K.M and Garg, R.K., 1994. A study of hospital waste disposal system in service hospital. *Journal of Academy of Hospital Administration,* 6(2): 27–31.

Shah, Kosha A., Shah, Komal A. and Shah, A.R., 2006. Biomedical waste management practices: A Case study. *Env. poll. Cont. Jot.,* 9(4): 46–50.

Singh, I.B. and Sarma, R.K., 1996. Hospital waste disposal system and technology. *Journal of Academy of Hospital Administration*, 8(2): 4–8.

Sriniwas, 1992. *Hospital Acquired Infections: Guidelines for Control*. Ministry of Health and Family Welfare, Department of Health, Government of India, Nirman Bhavan, New Delhi.

Srivastava, J.N., 2000. Hospital waste management project at Command hospital, Air Force, Bangalore. In: *National Seminar on Hospital Waste Management: A Report.*

Turnberg, Wayne L., 1996. *Bio Hazardous Waste: Risk Assessment, Policy and Management.* John Wiley and Sons, Inc.

Chapter 54

Assessment of Irrigation Water Quality Parameters of Certain Tank Located Around Tumkur

C. Vijaya Bhaskar, Mahadevaiah and G. Nagendrappa*

Department of Chemistry, University of Mysore, Mysore

ABSTRACT

Quality of water is essential both for domestic and agricultural purposes. Water quality refers to the characteristics of a water supply that will influence its suitability for a specific use. Common problems for poor quality of irrigation water were cited and the water quality parameters determined were precisely evaluated and interpreted for the suitability of irrigation. For evaluation of irrigation water, emphasis is placed on the chemical and physical characteristics of the water and only rarely any other factors are considered important. In this background, an effort was made to assess the water quality of seven tanks located around Tumkur, Karnataka. The water samples from these tanks were collected and analyzed for pH, EC, Ca^{+2}, Mg^{+2}, Na^{+}, Cl, TDS, TH, CO_3^{-2}, HCO_3^{-} and boron. The SAR, RSC, RSBC values were also determined. The results have shown that all the important parameter values required for irrigation were found to be within the acceptable range. However, the SAR value of sample 4 was high (5.40) that together with EC value of 0.72 mS/cm, has moderate affect on infiltration rate of water into the soil. Except for these minor discrepancies, the qualities of all water samples collected from these tanks were good to excellent quality for irrigation. The chloride concentration of sample 4 was also slightly higher (4.6 m eq/l) than the permissible level (< 2 m eq/l) for irrigation.

Keywords: *Quality of water, Sodium adsorption ratio and Suitability for irrigation.*

* Corresponding Author; E-mail: gnagendrappa@yahoo.co.in.

Introduction

Water is one of the abundantly available substance in nature. It is vital for the existence of human beings and all other forms of life. It is available in nature as ground water, surface water and ice water. Water resources are harnessed for various purposes like drinking, irrigation, fish farming, recreations, power generations and other uses (Sarkar and Hassan, 2006). Although the natural supply of water is adequate in absolute terms but not available at right time, place in expected quantity and quality (Sood *et al.*, 1998). In addition, pollution of water is a wide spread phenomenon. Soil, water and air pollutions are proportional to population, urbanization change in physical, chemical and biological characteristics of water that affect life of man, animals and plants (Misra and Dinesh, 1991).

There are water quality guidelines for its satisfactory use in agriculture. It can vary in quality depending upon type and quality of dissolved salts that are present in significant amounts. The solubility of water is the main source for irrigation, surface waters have their own contribution for this purpose. Water pollution is regarded as a result of human activities and natural contamination. In agriculture, it is quite common to concentrate more on quantity of water. Therefore, the present study is an attempt made to measure the quality of surface water for the suitability of agriculture through variables, pH, electrical conductivity (EC), total dissolved solids (TDS), the sodium adsorption ratio (SAR), boron, bicarbonates (HCO_3^-), carbonates (CO_3^{-2}) and chlorides (Cl^-).

Materials and Methods

For the present study seven tanks are selected:

1. Kallambella
2. Bugudanahally
3. Maidala
4. Honnudike
5. Kunigal
6. Kadaba and
7. Hebbur.

Water samples from them were collected periodically for the evaluation of water quality and their suitability for irrigation. The tanks are situated around Tumkur town. These are mainly rain fed. However, three tanks 1, 5 and 6 have water resources from Hemavathi reservoir.

The water samples were collected in dry and clean polythene cans of three liter capacity from each tank in an interval of 30 days. The collected samples were brought to the laboratory for analysis to evaluate the parameters, pH, EC, soluble cations-calcium, magnesium, sodium, iron, boron, the anions-carbonate, bicarbonate and chloride through the procedures given in APHA (1995), Trivedy and Goel (1986) and Kudesia (2003).

According to Follett and Soltanpour (1999), Michael (1978) and Raghunath (1987) the irrigation water quality is judged by the values of four basic parameters namely,

1. Total soluble salt content (salinity hazard)
2. Relative proportion of sodium cations to other cations (sodium hazard)
3. Excessive concentration of elements that cause ionic imbalance in plants or toxicity and
4. Bicarbonate concentration as related to calcium plus magnesium (RSC).

Other factors, soil calcium carbonate, nitrate, potassium and phosphates also influence indirectly irrigation water quality (Michael, 1978).

Salinity Hazard

Salts in water reduce water availability to the crop such an extent that yield is affected. In irrigated areas, these salts often originate from a saline, high water table or from salts in the water. Salts that contribute to salinity of soil are water soluble and transported readily by water. The soluble salt content in irrigation water generally is measured either by determining its electrical conductivity (EC) or the actual salt content.

Sodium Hazard

The sodium hazard of irrigation water is usually expressed as the sodium adsorption ratio (SAR). It is calculated by the following formula (Rowell, 1996 and Richard, 1954):

$$SAR = \frac{NA^{+}}{\sqrt{(Ca^{+2} + Mg^{+2})/2}}$$

The infiltration of water into the soil is influenced by the quality of irrigation water, that includes salinity of the water and SAR value. High salinity water will increase infiltration whereas low salinity water will decrease the infiltration into the soil.

Toxicity

Toxicity problems occur if certain constituents in the soil or water are taken up by the plants and accumulate to concentrations that are high enough either to cause crop damage or to reduce yield. The ions of primary concern for such a toxicity are chloride, sodium and boron. Direct toxicity to crops may result from specific chemical element in irrigation water. The actual concentration of an element in water that causes toxic symptoms depends on the kind of crop.

Residual Sodium Carbonate

Water with high concentration of bicarbonate (HCO_3^-) will tend to precipitate calcium carbonate and magnesium carbonate when the soil solution concentrate through evapotranspiration increasing the SAR value (Michael, 1978) causing more sodium hazard. Eaton (1950) suggested the following formula to obtain RSC.

$$RSC = (CO_3^{-2} + HCO_3^-) - (Ca^{+2} + Mg^{+2})$$

Gupta and Gupta (1987) defined RSBC (Residual Sodium Bicarbonate) as:

$$RSBC = HCO_3^- - Ca^{+2}$$

Results and Discussion

Water used for irrigation with an EC less than 0.7 mS cm^{-1} and/or TDS less than 450 mg/l is considered to be safe but when it is in the range of 0.7–3 mS cm^{-1} may cause little to moderate salinity problems (Table 54.1). All irrigation waters contain potentially injurious salts and nearly all the dissolved salts are left in the soil after the applied water is lost by evaporation from the soil or through transpiration by plants (Rhoades *et al.*, 1992). The difference in the EC and TDS of the different tanks could be due to the variations in salt concentration, rainfall and temperature. The analysis have shown that EC values of water samples from all the tanks are in the range of 0.19 to 0.72 mS cm^{-1}

(Table 54.2) and are well within the permissible limits set by (Ayers and Westcott, 1989) for irrigation (Table 54.1). The higher SAR value of Ronnudike tank compared to other tanks could be due to its location, nature of soil, the smaller dimension and a small variation in rainfall and temperature of the area. Presence of sodium in irrigation water reacts with soil to reduce its permeability and its continuous use makes the soil impermeable. Frequent irrigation with high sodium water for considerable time makes the soil plastic and sticky in wet condition and form clods and crust in drying condition. In contrast, presence of calcium and/or magnesium in irrigation water retards the evil effects of sodium by increasing the permeability of the soil (Punmia and Lal, 1981; Asaduzzaman, 1985). High SAR values cause clay dispersion leading to slaking of soil aggregates and hence the collapse of the soil structure (Barzegar *et al.*, 1994). A combination of EC value > 0.7 mS cm^{-1} and SAR of 0–3 is ideal for irrigation while EC of 0.7–0.2 mS cm^{-1} and SAR of 0–3 is moderate and EC of 0.2 mS cm^{-1} and SAR of 0–3 is not an ideal for irrigation. Following this classification, six water samples, under study (Table 54.3) fulfill the ideal condition but the sample 4 deviates since its EC and SAR values were found to be 0.72 mS cm^{-1} and 5.40 respectively.

Table 54.1: Limits of Some Important Parameter Indices for Water Quality Rating and its Suitability for Irrigation

Category	*EC mS cm^{-1}*	*RSC m eq/l*	*SAR*	*Restriction*
I	<0.7	<1.25	<3	None
II	0.7–3.0	1.25–2.5	3–9	Medium
III	>3.0	>2.5	>9.0	Severe

Source: Ayers and Westcott, 1989; Eaton, 1950.

Table 54.2: Values of Different Parameters of the Samples

Samples	*Parameters*								
	pH	*EC mScm^{-1}*	*Ca^{+2}*	*Mg^{+2}*	*Na^{+}*	*CO_3^{-2}*	*HCO_3^{-}*	*Cl^{-}*	*B*
S_1	7.8–8.2	0.26	0.75	0.42	1.43	–	3.50	1.02	0.01
S_2	7.6–8.3	0.19	0.46	0.28	1.16	0.33	0.81	0.81	0.23
S_3	7.7–8.1	0.43	1.31	0.70	2.30	–	1.64	1.64	0.22
S_4	7.8–8.0	0.72	1.38	0.55	5.27	–	4.58	4.68	0.25
S_5	7.7–8.1	0.20	1.02	0.39	0.60	–	1.01	1.01	0.25
S_6	7.9–8.3	0.23	0.97	0.54	0.74	–	1.04	1.04	0.25
S_7	7.7–8.2	0.22	0.66	0.34	1.20	–	0.76	0.76	0.01

It would be clear from the data as in the Table 54.2, that the pH range is between 7.6–8.3 for all the samples and fulfill the suggestion, (James, 1971), in which water having pH 7–9 is safe for irrigation. The concentration of boron and chlorides of water samples were found to be in the range of 0.01–0.25 ppm and 0.76–4.68 (Table 54.2) indicating again the suitability of water for all crops (Table 54.4) except for the samples 4 whose chloride concentration is 3.56 m eq/l exceeding the permissible limit less than 2 m eq/l. Therefore, the sample 4 water may cause some injury to sensitive plants and reduce the phosphorous availability to plants and also reduce the concentration of organic acids in plants to sub optimal levels (Follett and Soltanpour, 1999).

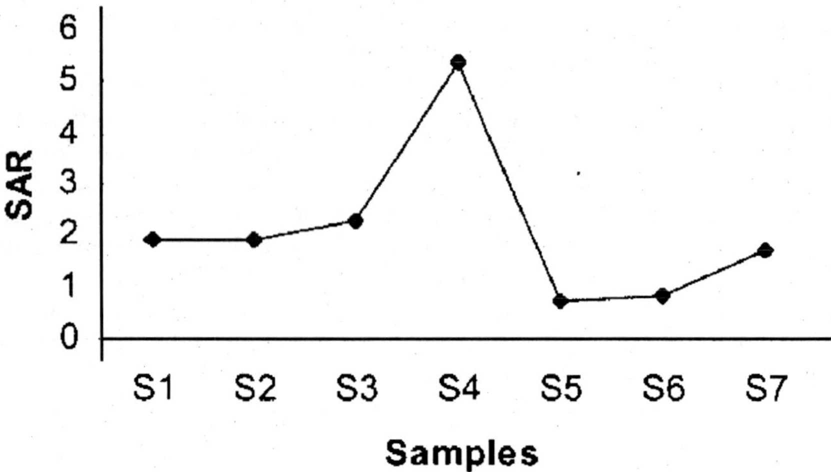

Figure 54.1: SAR Values for Different Samples

Table 54.3: SAR, RSC, RSBC Values

Samples	*SAR*	*RSC (m eq/l)*	*RSBC (m eq/l)*
S_1	1.90	2.27	2.60
S_2	1.90	0.59	1.43
S_3	2.30	2.56	3.01
S_4	5.40	1.12	1.63
S_5	0.71	1.31	1.65
S_6	0.85	1.94	2.38
S_7	1.70	1.32	1.58

Table 54.4: Boron and Chlorides

(a) Chloride Conc. (m eq/l) Effect on crops	*(b) Boron Conc. Effect on Crops*
<2 Generally safe for all plants	<0.5 ppm Satisfactory for all crops
2–4 Sensitive plants show injury	0.5–1.0 ppm Satisfactory for most crops
4–10 Moderately tolerant plants show injury	1.0–2.0 ppm Satisfactory for semi tolerant crops
> 10 Can cause severe problems	2–4 ppm Satisfactory for tolerant crops

Source: Follett and Soltanpour, 1999.

The water samples collected in the study area contained almost zero to trace amount of CO_3^{-2} whereas concentration of HCO_3^- upon analyses found to be varied from 1.91–4.57 m eq/l. Although bicarbonate is not a known toxic ion but known to cause zinc deficiency in rice and becomes severe when it exceeds 2 m eq/l for growing paddy rice (Ayers and Westcott, 1985). The RSC value of the water samples was found to be in the range of 1.59–2.57 m eq/l indicating almost the required limit for RSC value of 2.5 m eq/l (Eaton, 1950). The positive RSC values indicate that dissolved Ca^{+2} and Mg^{+2} ions were less than that of CO_3^{-2} and HCO^-_3 contents, The samples RSBC values varied from 1.43–3.01 m eq/l, which is satisfactory (5 mg/l) according to the criteria set by (Gupta and Gupta, 1987). Average TH values of the samples were below 100 mg/l which indicate that the waters are moderately hard.

Conclusion

The water quality investigation results for pH, EC, TDS, SAR, RSC, RSBC, chloride and boron revealed that Kallambella, Bugudanahally, Maidala, Kunigal, Kadaba and Hebbur tanks are found to be excellent to good quality on the scale of Ayers and Westcott and Easton for crop production. But Honnudike tank water has higher values for SAR and chloride that may cause problems of permeability and toxicity. Location and smaller dimension of the Honnudike tank may be responsible for such a behavior of its water.

Acknowledgement

The first author is highly grateful to the Management of Sri Siddaratha Education Society, Tumkur for their encouragement and H.N. Vijayendra, M.B. Nadoni, K.S. Kumar and M.S. Jayaprakash for their co-operation. We also thanks to M.K. Veeraiah and G. Manjunath for extending laboratory facility.

References

Ayers, R.S. and Westcott, D.W., 1989. *Water Quality for Agriculture*. FAO Irrigation and Drainage, Paper No. 29 (Rev. 1), pp. 1–109.

Asaduzzaman, M., 1985. *Hand Book of Ground Water and Wells*. BRAC Prakashana, Dhaka, pp. 196–207.

APHA, AWWA, WPCF, 1995. *Standard Methods for the Examination of Water and Wastewater*, 19th edition. Washington, D.C.

Barzegar, A.R. *et al.*, 1994. The effect of salinity and sodicity on disaggregation and tensile strength of an alfisol under different cropping system. *Soil Tillage Res.*, 32: 329–345.

Eaton, F.M., 1950. Significance of carbonate in irrigation waters. *Soil Sci.*, 67: 12–133.

Follett, R.H. and Soltanpour, P.N., 1999. Irrigation water quality criteria. *Crop Series.*

Gupta, S.K and Gupta, I.C., 1987. *Management of Saline Soils and Water*. Oxford and IBH Publ. Co., New Delhi, India, pp. 399.

James, G.V., 1971. *Water Treatment*, 4th edn. Tech Press, Scotland, pp. 175–181.

Kudesia, V.P. and Ritu, 2003. *Environmental Chemistry*. Pragati Prakashan, New Market, Meerut.

Michael, A.M., 1978. *Irrigation: Theory and Practice*. Vikas Publishing House Pvt. Ltd., New Delhi, pp. 713.

Misra, S.G. and Dinesh, M., 1991. *Soil Pollution*. Ashish Publishing House, New Delhi, pp. 7.

Punmia, B.C. and Lal, P.B.B., 1981. *Irrigation and Water Power Engineering*. Standard Publishers and Distributors, Delhi, pp. 28–32.

Pule-Meulenberg, F. and Emongor, V.E., 2005. Suitability of water for irrigated agriculture in some parts of Botswana. *J. of Applied Sciences*, 5(8): 1420–1423.

Raghunath, H.M., 1987. *Groundwater*, 2nd edition. Wiley Eastern Limited, New Delhi, India, pp. 344–369.

Rowell, D.L., 1996. *Soil Science Methods and Applications.* Longman, U.K., pp. 284–286.

Richard, L.A., 1954. *Diagnosis and Improvement of Saline and Alkali Soils*. USSL Staff Handbook 60.

Rhoades, J.D. *et al.*, 1992. *The Use of Saline Waters for Crop Production*. Irrigation and Drainage Paper 48, FAO, Rome.

Sarkar, A.A. and Hassan, A.A., 2006. Water quality assessment of a groundwater basin in Bangladesh for irrigation use. *Pakistan J. Bio. Sciences*, 9(9): 1677–1684.

Sood *et al.*, 1998. Assessment and management of underground water quality in Talwandi Sabo Tehsil of Bathinda District (Punjab). *J. Indian Soc. Soil Sci.*, 46: 421–426.

Todd, D.K., 1990. *Groundwater Hydrology*. Wiley International Edn, John Wiley and Sons, Inc., New York.

Trivedy, R.K. and Goel, P.K., 1986. *Chemical and Biological Methods for Water Pollution Studies*. Environmental Publications, Karad.

Talukdar, M.S.U. *et al.*, 1998. Suitability of groundwater for irrigation at Karimganj thana, Kishoreganj. *Progress, Agri.*, 9: 107–112.

Chapter 55

Assessment of Impact of Human Activity on Mangrove Ecosystem South East Coast of India

C. Sivasubramanian[1], S. Raveendran[1] and P. Kennedy[2]*

[1]P.G. and Research Department of Zoology, Khadir Mohideen College, Adirampattinam, Thanjavur District

[2]Department of Zoology, Government Arts College, Ariyalur

ABSTRACT

Mangroves are unique forests, representing an intermediate vegetation between the land and sea that grow in oxygen deficient water logged soils. Mangroves are both ecologically and economically impartment. Mangroves areas are though to be highly productive communities. Sometimes being the important primary producers in the estuarine food chain, leading to commercial fish and shellfish. The ecological role of mangrove is varied. Mangroves are important nursery, breeding, feeding and spawning grounds for some economically important bivalves, shrimps, crabs and fishes. Their role as nursery and feeding grounds for finfish and shellfish. The mangrove ecosystem of Muthupet in South East Coast of India is integral ecosystem of forest and fisheries. Increased pressure like indiscriminate wood falling, over grazing, land conversion and siltation on the natural resources of Muthupet has led mangroves being exploited beyond their sustainable potential continuous siltation has caused land emergence and micro delta formations. This paper emphasizes the significance of mangrove habitat and deals with the impact of human activity along the Muthupet estuarine complex and on the direct and indirect effect on mangroves and society and suggests remedial measures.

* Corresponding Author.

Introduction

Mangroves are the characteristic littoral plant formations of tropical and sub-tropical sheltered coastlines. They have been variously described as "Coastal woodland", ' Tidal forest" and Mangrove forest". Mangroves play an important role in moderating coastal storms at the interface between the land and sea. The coastal mangroves are vital shelterbelts, which afford protection to inland homesteads. Agricultural crops, livestock and aquaculture. The mangrove ecosystem is also known to support fishery by developing a detritus based food-web and be regulating a regional climate. It is the natural abode for many groups of animals and a large number of species are known to feed, breed, and take shelter in this ecosystem.

The mangrove wetland of Muthupet in Southeast coast of India is integral to the ecosystem of forest and fisheries. Increased pressure like indiscriminate wood felling, over grazing, land conversion and siltation of the natural resources of Muthupet has led mangroves being exploited beyond their sustainable potential. Continuous siltation has caused land emergence and micro delta formations.

This paper deals with the significance of mangrove habitat and is impact of human activity on mangroves and society along the Muthupet estuarine complex and suggests remedial measurers.

Study Area

The Muthupet mangrove wetland is located in the southern end of the cauvery delta (Let 10°46N, Long 79°51E) and occupies an area of approximately 12,000 ha, including a 1,700 ha lagoon. The river cauvery divides into five branches of Grand Anicut in Thanjavur district and one of the branches, vennar, further divides into five channels, namely Paminiyar, Koraiyar, Killaithankuyar, Marakakoraiyar and Valavanar. Paminiyar and Koraiyar join together and flow for about 7 kilo meters and drain into Muthupet lagoon. 61 plant species have been recorded (Oswin and Rahaman 1997) Only 5 mangrove species namely *Acanthus illicifolius, Agiceras corniculatum, Avicennia marina, Excoecaria agallocha* and *Luminitzera recemosa* are present in the Muthupet mangrove wetland. Among them *Avicennia marina* is dominant, Constituting more than 95 per cent of the tree population. The commercial and traditional products obtained from Muthupet mangrove-range from construction material to medicine and honey. The honey collected from the pure strands of *Avicennia marina* have a good medicinal value Mangroves are a potentially sustainable source of wood and domestic fuel. Hence the calorific value of wood and mangrove used is thrice that of the other wood. It is observed both men and women during all seasons carry head loads of *Avicennia logs*.

In Muthupet mangroves form the sale source of fodder for cattle. Other natural products harvested from Muthupet mangroves include crustaceans mollusks and finfishes. Valuable fishes like Mullets, Milkfish and crustaceans utilize mangrove as a nursery breeding ground and a source of food. About 73 species of fin fishes have been reported (Cecilia Pandian, 1996). But the local fishers consider that only 30 species are commonly occurring in the Muthupet mangrove water.

In the present study an attempt is made to find out the human induced stresses of Muthupet mangrove wetland. Efforts were taken to identify the impact of effluent discharge from the shrimp ponds and saltpan. In order to make a detailed study of the causes for degradation of mangrove forests, a survey was conducted and an interview conducted with the village people living in the area adjoining the Muthupet mangrove wetland from Adirampattinam to Thondiyakkadu.

Cattle Grazing

The local farming community that lives around the Muthupet mangrove wetland rears livestock, for various purposes such as milking, manuring, pulling carts and ploughing. Cattle growers utilize the mangrove forests mainly for livestock grazing. Flocks of ferrel cattle and domesticated cattle were observed in the mangrove. The owners let there cattle to their fate and do not bother about them. Skulls and bones of dead cattle are a common sight in the mangroves. The leaves and fruits of *Avicennia marina* serve as fodder for the cattle. Severe grazing paved way for enormous destruction of seedling and budding branches.

Clear Felling

Fire wood collection from the Muthupet mangrove forest has been declared illegal. But some of the locals continue to collect firewood from the mangroves, both for their own use and for sale. Some of the fishers collect small bundles of dry wood for their homes when they go fishing. Some others collect firewood from the mangrove and sell it usually to local tea shops and small hotels. The trees along the banks are the sole source of domestic fuel, the trees are felled indiscriminately. This has not only destroyed the productivity of these areas, but also has made the land useless for future use. Fortunately, clean felling under coupe system of management was abandoned a long time ago and now these degraded areas are being restored.

Dumping of Garbage

Waste garbage is simply dumped in to the streams, canals and ditches in mangrove areas and some of the effluents go directly to the estuaries every day. The wastes include soil, glass, paper, plastics, leather, dead animals and a profusion of plastic bags. The run off sewage from the village nearby drains into the river. The presence of the garbage, leads to serious health hazards. An abnoxious smell pervades around the place were the cows ruminating on the remnants left after partial combustion of the wastes and the eagle and kites hover around.

Aquaculture

Mangrove areas in many developing countries are increasingly being converted into large aquacultural ponds used mainly for rearing shrimps rather than fish due to high expert demand and the prices. Twenty-seven prawn farms occupy an area 1000 acre around the Muthupet mangrove wetland. These are located very close to the mangrove forest. These farms follow a modified extension system. They draw water either from the sea through canals or from the mangrove wetland. For aquacultural operations large quantities of water is pumped indirectly from the rivers. The pumping and releasing of water has disturbed the population of juvenile fish and crustaceans beneath the mangrove. The schooling of fish fry and juveniles of crustaceans have vanished around the aqua farming areas.

Waste water disposed from aquaculture ponds contains suspended material, excess feed and faecal matter and is high in ammonia, low oxygen content and pH value which affect the natural water bodies (Oswin and Rahaman, 1997).

Siltation

The poor flow of seawater into the estuarine mangroves is due to the silted narrow river mouths. This interferes with the entry of fish juveniles from the Bay of Bengal, which results in decline of fishery resources in the mangrove water bodies (Kathiresan, 2000). This problem exists in the Muthupet areas. The local people have felt the seriousness of the problem.

Most fishers said that the migration of Fish, Prawns and Crabs and their juveniles to the mangrove wetland has gone down in the last 20 years since the width and depth of the Lagoon mouths are shrinking every year.

The Muthupet mangrove wetland is get water for agriculture during the monsoon season from the Mattur dam. This water is supplied through a network of larger canals. These canals are silted up. It also created rigorous siltation on the riverbanks and lagoon.

The mouth of the mangrove estuary due to siltation affects the free flow of tidal water in and out of the mangrove wetland. This has many negative implications. Due to poor exchange of fresh water with the sea water and the mangrove environment becomes stagnant and evaporation of the stagnant water increases the salinity to a very high in a short period of time giving acute salinity problem both to the aquatic animal and to the plants (Selvam, 2001).

In the monsoon when the run-off from the land is excessive and the level of the rivers rises, it is the lagoons and the mangrove forest that come to the rescue, acting as water storage and allying the threat of inundation of low lying areas. The lagoon mouth and the lagoon as a whole are getting narrower every year. In some regions of the lagoon, particularly in the eastern part, the depth of the water is not even 30 cm during high tide. An interview with the fisherman revealed the fact that the fish potential of the river, lagoon and also especially the sea got affected. They reported at 50 per cent of their lagoon catch are loss and a drastic decrease in the potential fish catch at sea.

The decrease in the water level due to siltation and eutrophication has affected the nursery habitat of several valuable fin and shellfishes. The siltation process has also affected the filter feeding mechanism of the bivalves and Oysters, *Crassostrea madrasensis,* hence a rich population of dead Oysters was found deposited on the uplifted land. (Oswin and Rhaaman, 1996). A number of new Oyster beds have been formed in the lagoon, particularly in the region when the river Koriyar meets the lagoon. This prevents the free movement of fishing boats. It also damages nets during fishing and causes injuries to fishermen. The Oyster beds in vast areas are also responsible for the shallowness of the lagoon.

Anthropogenic Factors

The fishermen living near the mangrove forest use the mangrove resources for their livelihood and for firewood and fencing. The uncontrolled growth of population, lack of civic amenities, unplanted slum development; create health problems and social problems in Muthupet mangroves. Innumerable tourists visit Muthupet mangroves but the flow is not regulated due to lack of adequate staff strength and well-trained ecoguide etc. Consequently, the huge load of sewage and domestic garbage dumped in rivers affects the biota. The non-bio degradable wastes such as plastic bags and bottles, cans choke the estuary. The oil spills of the mechanized boats too do a considerable damage to the productivity of the ecosystem. It was a common practice to defecate in fields, riverbanks, bushes and near human habitation.

Conclusions

The Muthupet mangrove areas are an important integral part of the environment in Tamil Nadu. It is an important feeding ground for many coastal fishes and serve as good nurseries for juvenile fishes and crustaceans. They attract and shelter fishes in the shade of mangrove forests. To maintain and enhance fishery resources in mangrove areas, the natural environmental mangrove communities must be protected from human disturbances.

With the increase of urbanization and modernization of agriculture, the Muthupet mangrove ecosystem is subjected to environmental stresses due to aqua cultural effluents, domestic sewage, pesticides, herbicides and oil which playa major role in polluting this sensitive ecosystem. It is high time to think about a management scheme for this ecotonal zone, acting as cushion of buffer in between the mighty sea and the human habitation. Any imbalance in nature is reciprocative and irrational utilization without proper management for short-term benefit will ultimately drastically affect the long-term benefit. It is not only for the animals and their associates but also for our own mortal survival. The mangrove forest of Muthupet needs immediate attention for protection and conservation.

References

Cecelia-Pandian, 1996. Icthyofuna of Muthupet estuary. *Ph.D. Thesis,* Bharathidasan University, Tiruchirappalli, India.

Kathiresan, K., 2000. A review of studies on Pichavaram mangrove, South east India. *Hydrobiologica,* 430: 185–205.

Oswin, S.D. and Rahaman, A.A., 1996. An observation of the anthropogenic stress on the mangrove of Muthupet, Tamil Nadu. *Fourth Indian Fisheries Form,* School of Marine Science, Cochin University.

Oswin, S.D. and Rahaman, A.A., 1997. Floral diversity at Muthupet mangrove forests. In: *State Level Seminar on Biodiversity* at K.N.G. Arts College (W), Thanjavur, India.

Oswin, S.D. and Rahaman, A.A., 1997. Impact of aquaculture effluents on mangroves. In: *Proceeding of the National Seminar on Water Quality Issues in Aquaculture Systems,* (Eds.) R. Santhanam, V. Ramadhas and P. Gopala Krishnan, pp. 15–22.

Selvam, V., 2001. Anthropogenic and ecological basis of mangrove restoration. In: *Workshop on Rehabitilation of Degraded Mangrove Forests*. M.S. Swaminathan Research Foundation, Chennai.

Chapter 56

Impact of Weeds on Some of the Morphological Attributes of a Little Millet Crop

K.L. Barik
Department of Botany, North Orissa University,
Takatpur, Baripada – 757003, Orissa

ABSTRACT

Impact of weeds on some of the morphological attributes of a little millet crop (*Panicum sumatrense,* Roth ex. Roem and Schult) variety. SS. 81-1 was studied *in vivo* at Agriculture Research Station, Ankuspur, Orissa. The plant height, number of green leaves per plant, root length, number of roots, number of effective tillers per plant and the length of panicle were found to be high in weed free beds compared to those infested with weed. Most of the attributes in weed free and weed associated beds exhibited high significant differences ($p \leq 0.001$) in the later part of the growth period. The number of basal tillers however did not show any significant differences throughout the sampling period in all beds.

Keywords: *Weed, Impact, Morphological attributes, Little millet crop.*

Introduction

Millet–a staple food of tribals and also of the labour classes. Little millet (*Panicum sumatrense*) is one of the typical minor millet crops basically grown on the hilltops, hill slopes and also in the hill bases. Recently, the rice grower of the plains has taken up cultivation of this crop. It can withstand heavy rain and drought condition to a considerable extent. It has also ability to protect against various pathogenic activities. Much information on the crop-weed competition, weed management and their control measure are available with reference to rice and wheat. A lot work has also been carried out by Indian Council of Agricultural Research, through all India Co-ordinated Small Millets Improvement Project on Millets (ICAR, 1992–93, 1993–94, 1994–95, 1995–96, 1996–97). However very little work has been done on the impact of weed growth on morphological attributes of little millet crop. Taking all these aspects, an attempt was made to study the impact of weed growth on same of the morphological attributes of a little millet crop.

Study Site, Environment and Soil

The experiment was conducted at the Agriculture Research Station, Ankuspur (19°46′N, 94°21′E) situated at a distance of 22 km away from Berhampur University in the state of Orissa. The climate of the locality is monsoonal with 3 distinct seasons *i.e.* rainy (July to October), winter (November to February) and summer (March to June). The mean minimum and mean maximum atmospheric temperature ranged from 15.4°C to 26.13°C and 27.6°C to 37.81°C respectively during December and May. Annual rainfall was found to be 836.65 mm, out of which a maximum or 248.79 mm was observed in June. The soil was found to be sandy (75 per cent) with a pH of 6.58. The phosphorus and potassium contents of the soil were high whereas the amount of organic carbon (per cent) was very low.

Materials and Methods

ICAR Technology as proposed by Seetharam (1994) was employed for cropping with little modification depending upon the soil condition and climate of the locality. The 1st sampling was made on the 30th days after sowing (DAS) followed by 45DAS, 60DAS, 75DAS and also on the harvesting day (87DAS). Morphological characteristics *i.e.* plant height, number of basal tillers, number of green leaves, root length, number of rootlets, number of effective tillers and length of panicle were measured/ counted and averaged on different days after sowing. The plant height (cm) was measured from the root-stem transition region to the leaf tip or the panicle top which was the maximum. The root length (cm) was measured from the transition region to the tip of the longest root. Productive tiller number per plant was counted at harvest. The length of effective panicle (cm) per plant was also measured at harvest.

Results and Discussion

The height of the crop plant gradually increased with the increase in the days after sowing (DAS) and attained maximum levels on the 87DAS in both weed free arid weed associated beds. It was also observed that the crop height was maximum in weed free bed compared to that of weed associated bed with respect to different days after sowing. However, no appreciable difference was marked in the number of basal tiller during 30DAS and onwards in both weed free and weed associated beds. Number of green leaves gradually increased from 30DAS to 75DAS. Thereafter, the number got reduced to some extent in both the beds. Comparatively less number of green leaves were observed in weed associated bed. The length of root followed similar trend to that of the plant height during successive sampling period in both the beds. Number of rootlet of the crop plant followed almost similar trend to that observed in number of green leaves. The effective tiller of crop plant was found to be comparatively less in weed associated bed to that observed in weed free bed. In case of effective panicle length, comparatively longer panicles were seen in the weed free beds as against shorter panicle in the weed associated crop beds.

Statistical analysis (Student 't' test) reveals that the plant height and the root length in the weed free and weed associated beds were significantly different with level of significance $p \leq 0.001$. Number of basal tiller per plant did not, however, show any significant difference in the two beds. The number of green leaves in the early growth period showed some differences, significant at $p \leq 0.05$ level. The number of green leaves was found to be strictly different in the two beds ($p \leq 0.001$) during 60 DAS, 75 DAS and 87 DAS. The difference in the number of rootlet per plant in weed free and weed associated beds during later part of growth was highly significant ($p \leq 0.001$). Differences in the number of effective tilters of the two beds were significant at a level of $P \leq 0.05$. Variation in the length of the panicles of the two beds were found to be highly significant ($p \leq 0.001$). This indicates that the differences marked in the morphological attributes *i.e.* plant height, number of green leaves, length of root, number of rootlet, number of effective tiller and the length of effective panicle of the crop were mostly due to the interference of weed growing along with the crop (Table 56.1).

Table 56.1: Morphological Attributes of Crop Plant Showing the Mean ($\bar{x}$), Standard Deviation (SD), Coefficient of Variation (CV per cent), 't' Values and Level of Significance (p) in Weed Free and Weed Associated Beds at Different Days After Sowing (DAS)

Parameters	*Beds*	*Variables*	*30DAS*	*45DAS*	*60DAS*	*75DAS*	*87DAS*
Plant height in cm	Weed free	$\bar{x}$	55.4	77.5	106.8	117.3	122.5
		SD	1.16	1.89	1.15	1.89	1.41
		cv (%)	2.094	2.439	1.077	1.611	1.151
		n	10	10	10	10	10
	Weed associated	$\bar{x}$	50.0	67.6	95.5	106.1	110.3
		SD	235	2.65	2.27	2.22	2.18
		cv (%)	4.7	3.92	2.377	2.092	1.976
		n	10	10	10	10	10
		t	6.515	9.618	14.042	12.148	14.860
		p (≤)	0.001	0.001	0.001	0.001	0.001
No. of basal tillers	Weed free	$\bar{x}$	2.8	2.9	2.9	2.9	2.9
		SD	0.422	0.316	0.316	0.316	0.316
		cv (%)	15.071	10.896	10.896	10.896	10.896
		n	10	10	10	10	10
	Weed associated	$\bar{x}$	2.6	2.6	2.6	2.6	2.6
		SD	0.516	0.516	0.516	0.516	0.516
		cv (%)	19.846	19.846	19.846	19.846	19.846
		n	10	10	10	10	10
		t	0.949	1.568	1.568	1.568	1.568
		p (≤)	NS	NS	NS	NS	NS

Contd...

Table 56.1–Contd...

Parameters	Beds	Variables	30DAS	45DAS	60DAS	75DA S	87DAS
No. of green leaves	Weed free	$\bar{x}$	133	19.8	23.1	20.9	16.8
		SD	0.675	0.789	0.876	0.876	0.789
		cv (%)	5.075	3.985	3.792	4.191	4.696
		n	10	10	10	10	10
	Weed associated	$\bar{x}$	12.5	18.2	20.6	166	12.7
		SD	0.707	1.751	1713	1.838	1.636
		cv (%)	5.656	9.621	8.315	11.072	12.882
		n	10	10	10	10	10
		t	2.588	2.634	4.109	6.678	7.138
		P (≤)	0.05	0.05	0.001	0.001	0.001
Root length in cm	Weed free	$\bar{x}$	9.1	11.4	13.2	15.1	15.8
		SD	0.422	0.307	0.309	0.393	0.366
		cv (%)	4.637	2.693	2.341	2.603	2.316
		n	10	10	10	10	10
	Weed associated	$\bar{x}$	8.1	10.2	12.2	14.1	14.3
		SD	0.677	0.478	0.427	0.543	0.482
		cv (%)	8.358	4.686	3.500	3.851	3.371
		n	10	10	10	10	10
		t	3.964	6.680	6.000	4.718	7.838
		P (≤)	0.001	0.001	0.001	0.001	0.001
Number of roots	Weed free	$\bar{x}$	26.1	28.4	30.5	31.6	29.2
		SD	1.524	1.578	1.58	1.075	1.398
		cv (%)	5.839	5.556	5.180	3.402	4.788
		n	10	10	10	10	10
	Weed associated	$\bar{x}$	23.2	25.6	27.2	27.8	25.1
		SD	1.813	1.776	1.690	1.316	2.234
		cv (%)	7.815	6.937	6.213	4.734	8.900
		n	10	10	10	10	10
		t	3.872	3.727	4.511	7.072	4.920
		p (≤)	0.01	0.01	0.001	0.001	0.001
No. of effective tillers	Weed free	$\bar{x}$	–	–	–	–	3.2
		SD	–	–	–	–	0.422
		cv (%)	–	–	–	–	13.187
		n	–	–	–	–	10

Contd...

Table 56.1–Contd...

Parameters	Beds	Variables	30DAS	45DAS	60DAS	75DA S	87DAS
	Weed associated	$\bar{x}$	–	–	–	–	2.8
		SD	–	–	–	–	0.422
		cv (%)	–	–	–	–	15.071
		n	–	–	–	–	10
		t	–	–	–	–	2.119
		p (≤)	–	–	–	–	0.05
Panicle length in cm	Weed free	$\bar{x}$	–	–	–	–	28.0
		SD	–	–	–	–	0.654
		cv (%)	–	–	–	–	2.327
		n	–	–	–	–	10
	Weed associated	$\bar{x}$	–	–	–	–	26.2
		SD	–	–	–	–	1.195
		cv (%)	–	–	–	–	4.561
		n	–	–	–	–	1
		t	–	–	–	–	4.178
		p (≤)	–	–	–	–	0.001

Acknowledgement

The author gratefully acknowledge the financial assistance extended by University Grants Commission (UGC), New Delhi. Thanks are expressed to Professor B.N. Misra (Retired Professor of Botany), Professor A.K. Panigrahi (Head of the Department) and Dr. M.K. Misra (Associate Professor), Department of Botany, Berhampur University, Berhampur, Orissa for their co-operation throughout the progress of this investigation. The author is also indebted to Dr. R. C. Misra (Sr. Breeder and Officer in-charge), Dr. H.K. Mohapatra (Entomologist), Dr. S. Panda (Pathologist), Dr. B.K. Jena (Agronomist) and Mr. S.N. Biswal (Field Assistant) of Agriculture Research Station, Ankuspur for providing necessary help throughout the cropping.

References

ICAR, 1992–93. *Annual Report*. All India Coordinated Small Millet Improvement Project. Indian Council of Agricultural Research, GKVK Campus, Bangalore.

ICAR, 1993–94. *Annual Report*. All India Coordinated Small Millet Improvement Project. Indian Council of Agricultural Research, GKVK Campus, Bangalore.

ICAR, 1994–95. *Annual Report*. All India Coordinated Small Millet Improvement Project. Indian Council of Agricultural Research, GKVK Campus, Bangalore.

ICAR, 1995–96. *Annual Report*. All India Coordinated Small Millet Improvement Project. Indian Council of Agricultural Research, GKVK Campus, Bangalore.

ICAR, 1996–97. *Annual Report*. All India Coordinated Small Millet Improvement Project. Indian Council of Agricultural Research, GKVK Campus, Bangalore.

Seetharam, A., 1994. *Technology for Increasing Finger Millet and Other Small Millet Production in India*. Project Co-ordination Cell, AICSMIP, ICAR, Bangalore.

Chapter 57

Nutrient Dynamics in an Annual Fire Influencing Humid Subtropical Grassland of Indra Gandhi Wildlife Sanctuary, the Western Ghats, India

S. Paulsamy[1], S. Manian[2] and D. Suresh[1]

P.G. and Research Department of Botany,
Kongunadu Arts and Science College, Coimbatore – 641 029
[2]Department of Botany, Bharathiar University, Coimbatore – 641 046

ABSTRACT

Accumulation and transfer rates of nutrients *viz.*, nitrogen, phosphorus and potassium on annual basis were studied in a humid subtropical annual fire influencing dis-climax grassland located in Indra Gandhi Wildlife Sanctuary, the Western Ghats, India. The results of the study revealed that all the three nutrients were stored and transferred mainly by the dominant perennial grass, *Chrysopogon zeylanicus* Thw. than that of the other two species categories studied *viz.*, 'other grasses' and 'remaining species'. Of the plant parts analysed (aboveground standing live and standing dead, belowground and litter), the belowground parts of the community stored higher content of nitrogen and phosphorus. Among the three nutrients, nitrogen was taken up by the community in higher quantity (157.6 kg/ha/yr) followed by potassium (95.1 kg/ha/yr) and phosphorus (15.7 kg/ha/yr). The retention and release rates were also higher for nitrogen. The rapid rate of uptake, and accumulation of nutrients in standing live aboveground parts showed the availability of nutritious herbage to wild herbivores. Hence, the influence of annual fire for the greater transfer rates of nutrients is considered as a beneficial factor with respect to the welfare of wild herbivores.

Keywords: *Annual fire, Nutrient dynamics, Humid grassland, Chrysopogon zeylanicus, Grass Hills, Western Ghats, India.*

Introduction

Cycling of nutrients between the components of ecosystem is one of the useful functional attributes to find out the level of community development and the degree of influence by the physical factors (Odum, 1971). Occurrence of fire in vegetations particularly in grasslands where the complete aboveground parts are destroyed by burning, not only modifies the biotic composition but also the soil properties and hence the accumulation and transfer rate of nutrients (Schacht *et al.*, 1996; Bijker *et al.*, 2001). It is reported that the montane grasslands of south India are frequently experienced with fire during every summer (February–May) and this factor plays major role in controlling the community development and other functions of ecosystems (Champion, 1939; Bor, 1938; Singh and Krishnamurthy, 1981; Paulsamy, 1992).

After the postulation of the theory of uptake and complete release of nutrients (Liebig, 1940), many works have been done on the nutrient structure and its cycling in cropland and forest ecosystems (Nye, 1961; Ovington, 1965; Rodin and Bazilevich, 1967; Woodwell and Whittaker, 1968; Turvey, 1974; Pemadasa, 1981; Middleton, 1982; Gower *et al.*, 1992; Reich *et al.*, 1997). But, little number of works only carried out in grasslands on this aspect (Laudelent and Germain, 1954; Yamane and Sato, 1961; Allison, 1965; Dahlman *et al.*, 1967; Bokhari and Singh, 1974; Agrawa!, 1988; Scholes and Archer, 1997; Reich *et al.*, 2001) and no works have been carried out in the grassland communities of Indra Gandhi Wildlife Sanctuary, the Western Ghats, India. Hence, in the present study, an attempt has been made in the *Chrysopogon zeylanicus* dominated grassland of Grass Hills ecosystem, Indra Gandhi Wildlife Sanctuary, Western Ghats where annual summer fire is an integral part of the ecosystem to investigate the dynamics and cycling of certain nutrients *viz.*, nitrogen, phosphorus and potassium.

Study Area

The Grass Hills ecosystem in Anaimalais, the Western Ghats, India, comes under Indra Gandhi Wildlife Sanctuary spreads over an area of 19000 ha is located at latitude 10°32′8″ N and longitude 77°4′23″ E at the elevation range between 1650 to 2100 m above msl. The grassland in Grass Hills, dominated by a perennial grass, *Chrysopogon zeylanicus* Thw. is humid subtropical type (Champion, 1936) and at dis-climax stage in community succession (Paulsamy, 1992). The annual rainfall in Grass Hills for the past 20 years is as much as 3000 mm and the temperature ranges between 15 and 32°C. Annual summer fire, an integral part of this ecosystem is generally occurred during the months between February and May every year.

The grassland is interspersed with pockets of shola forests according to topography, and comprises a total number of 35 species in which 15 are grasses, 2 are sedges and 18 are forbs and among these species the perennial grass, *Chrysopogon zeylanicus* is dominant and contributes more than 70 per cent of density and bagal cover to the community (Paulsamy, 1992). Further, it is recognized as highly palatable species for the wild animals particularly for Indian gaur (Chinnamani, 1981). A rich variety of wild animals like sambar, Nilgiri tahr, great Indian gaur, elephant, wild boar, spotted deer, barking deer, lion-tailed macaque, Nilgiri langur, Malabar giant squirrel, porcupine, jackal, tiger etc. are inhabiting the Grass Hills ecosystem (Wilson, 1964). Since, it is a protected area for biodiversity conservation, little or no human influences are there in this ecosystem.

Materials and Methods

An one hectare experimental plot with homogeneous floristic composition was established in the grassland of Grass Hills to estimate the aboveground standing live and standing dead, belowground and litter biomass components. The short term harvest method of Odum (1960) was followed for biomass estimation at monthly intervals from February, 2004 to February, 2005. At every sampling time, 5 quadrats, each with the size of 1 × 1 m were laid randomly in the plot established for harvesting the aboveground parts. The harvested aboveground live and standing dead materials were sorted out quadrat-wise and separated into three species categories *viz.*, dominant grass, *Chrysopogon zeylanicus*, 'other grasses' and 'remaining species'. Litter component was also separated out into these three categories. To determine the belowground biomass, 5 monoliths each with the size of 25 × 25 × 30 cm were sampled by excavating the soil cores. The belowground parts were soaked in water for few hours and then washed in running water to remove the adhered soil particles. The biomass values of respective parts were determined after oven drying at 80°C for 48 h. Triplicate samples were maintained to estimate the biomass. At every month, species category-wise powdered materials of aboveground standing live and dead, belowground and litter components were analysed for total nitrogen (Piper, 1944), phosphorus (Jackson, 1962) and potassium (Peach and Tracey, 1956).

Month-wise soil samples were collected during the study period by digging 25 × 25 × 30 cm pits in the experimental plot and brought to the laboratory for analysis. Then they were air dried and analysed for nitrogen and phosphorus (Jackson, 1958) and potassium (Peach and Tracey, 1956). The N, P and K contents per gram dry weight of soil was multiplied of the bulk density and the results were expressed as $g/m^2/30$ cm.

Net uptake was calculated (Table 57.3) from the respective nutrient equivalent and the net community production which was estimated by summing up the positive increatments of the species categories from the field data (Paulsamy *et al.*, 2001). The outputs of litter and belowground disappearance of nutrients were calculated as per the methods of Sims and Singh (1971) and Mall and Billore (1974). The differences between the uptake and disappearance are assumed to be the amount of nutrient retained by the plant components.

Results and Discussion

Nutrient Concentration

Generally, variation has been noted in the mineral contents of plant parts across the three categories of plant species analyzed (Table 57.1). In all the cases, the belowground parts registered high concentration of nutrients followed by aboveground standing live, standing dead and litter components. Singh and Krishnamurthy (1981) pointed out that the grassland communities dominated by perennials usually store high concentration of nutrients in the rhizomes of their constituent species. The decline of nutrient concentration in standing dead and subsequently in litter parts may be due to weathering or to leaching by rain, or by translocation away from the dying tissue (Agrawal, 1988).

Standing State of Nutrients in Community and Soil

The standing state of nutrients in the average (annual) biomass of aboveground standing live and dead, belowground and litter parts of three categories of species is given in Table 57.2. The data showed that all the three nutrients *viz.*, nitrogen, phosphorus and potassium were stored in higher amount in various parts of the dominant grass, *Chrysopogon zeyianicus* than that of the other two categories, the 'other grasses' and the 'remaining species'.

Table 57.1: Mean Per cent Nutrients in Three Categories of Plant Species Present in the Annual Fire Influenced Grassland at Grass Hills, Indra Gandhi Wildlife Sanctuary, the Western Ghats

Plant Parts	*Nutrient (per cent)*								
	Nitrogen			*Phosphorus*			*Potassium*		
	Dominant Grass	*Other Grasses*	*Remaining Species*	*Dominant Grass*	*Other Grasses*	*Remaining Species*	*Dominant Grass*	*Other Grasses*	*Remaining Species*
Aboveground standing live	0.79[a]	0.79[a]	0.81[b]	0.08[ab]	0.07[ac]	0.09[bd]	0.48[a]	0.47[a]	0.49[a]
Aboveground standing dead	0.78[a]	0.77[a]	0.78[a]	0.06[a]	0.06[a]	0.08[b]	0.46[a]	0.46[a]	0.049[b]
Belowground	1.10[a]	1.08[a]	1.18[b]	0.11[a]	0.10[a]	0.11[a]	0.53[a]	0.51[a]	0.56[b]
Litter	0.76[a]	0.77[a]	0.78[a]	0.05[a]	0.05[a]	0.07[b]	0.46[ab]	0.44[ac]	0.047[bd]
Average	0.86	0.85	0.89	0.08[ab]	0.07[ac]	0.09[bd]	0.48[a]	0.047[a]	0.50[b]

Means followed by same letter(s) in horizontal rows are not significant to each other according to Duncan Multiple Range Test at 5 per cent level.

Table 57.2: Accumulation of Nutrients Based on Average Standing Crop in the Annual Fire Influenced Grassland at Grass Hills, Indra Gandhi Wildlife Sanctuary, the Western Ghats

Plant Parts	*Nutrient (g/m²)*								
	Nitrogen			*Phosphorus*			*Potassium*		
	Dominant Grass	*Other Grasses*	*Remaining Species*	*Dominant Grass*	*Other Grasses*	*Remaining Species*	*Dominant Grass*	*Other Grasses*	*Remaining Species*
Above ground standing live	3.67	1.83	2.13	0.35	0.17	0.25	2.23	1.08	1.29
Above ground standing dead	5.30	2.03	3.35	0.41	0.15	0.13	3.17	1.23	2.05
Below ground	7.80	2.61	2.37	0.75	0.25	0.23	3.46	1.13	0.94
Litter	1.08	0.45	0.40	0.08	0.03	0.03	0.66	0.25	0.23
Total	17.85	6.92	8.25	1.59	0.60	0.64	9.52	3.69	4.51

It may be explained that the magnitude of nutrients cycled through these categories of species corresponds, in general to the magnitude of the standing crop of the dry matter of the respective categories in Grass Hills (Paulsamy, 1995). Generally, the nitrogen and phosphorus storage were higher in belowground biomass followed by aboveground standing live and standing dead parts. However, in the species category, 'remaining species', the nitrogen accumulation was greater in above ground standing dead parts than in the other components. Similarly, for phosphorus, next to below ground biomass, the above ground standing live part of the categories, 'other grasses' and 'remaining species' have stored more materials. For potassium, the aboveground standing dead parts of the species categories, 'other grasses' and remaining species', accumulated higher element than that in the belowground biomass. In all cases, the litter parts stored lower quantity of all the three nutrients. These differences of nutrient storage may be explained due to their requirements in the metabolic processes, or to their relative availability in the ecosystem. The data on the accumulation of nutrients showed that the dominant grass, *Chrysopogon zeylanicus* alone accounted for more than half the quantity of this cycling pool. Thus, with respect to nutrient flux, it is known that the community metabolism of grassland in Grass Hills is being operated mainly by this single dominant grass. The higher annual accumulation of phosphorus (231.4 g/m^2/30cm) and potassium (443.5 g/m^2/30cm) in the soil reservoir pool in comparison to other tropical grasslands of India (Billore and Mall, 1976; Agrawal, 1988) indicates the least leaching losses of these nutrients despite the higher rainfall in this region.

Nutrient Budget

The species categories and their parts were arranged in block and arrow model (Billore and Mall, 1976), with each part considered as an independent compartment with an input and output (Figure 57.1). The values in the rectangular boxes are the average standing states of nutrients and were obtained by multiplying the average compartmental biomass with appropriate nutrient equivalents. The values on arrows are the annual transfer rates. The standing state of nutrients are only the measure of the storage capacity, of the compartments and therefore they are as not informative as the annual transfer rates. The uptake and losses of nutrients through disappearance were mainly contributed by the dominant grass, *C. zeylanicus* by its all parts than in the counter parts of the other two species categories *viz.*, 'other grasses' and 'remaining species' (Table 57.3). It indicates the major functional role played by the perennial grass, C. *zeylanicus* in the studied grassland community.

Table 57.3: Uptake and Losses of Nutrients in the Annual Fire Influenced Grassland at Grass Hills, Indra Gandhi Wildlife Sanctuary, the Western Ghats

Compartment	*Nutrient (g/m²/yr)*								
	Nitrogen			*Phosphorus*			*Potassium*		
	Dominant Grass	*Other Grasses*	*Remaining Species*	*Dominant Grass*	*Other Grasses*	*Remaining Species*	*Dominant Grass*	*Other Grasses*	*Remaining Species*
Net uptake in:									
Above ground standing live	7.78	4.19	3.79	0.78	0.37	0.42	4.73	2.49	2.29
Above ground standing dead	12.12	4.67	2.06	0.93	0.36	0.34	7.15	2.79	2.05
Below ground	3.60	2.38	1 18	0.36	0.22	0.19	1.74	1.12	0.98
Litter	3.09	1.34	1.18	0.20	0.09	0.11	1.87	0.76	0.71
Net loss from:	2.97	2.48	2.71	0.30	0.23	0.25	1.43	1.17	1.28
Below ground	3.86	3.24	3.34	0.39	0.30	0.31	1.86	1.53	1.59
Litter	0.61	0.53	0.43	0.04	0.03	0.04	0.37	0.30	0.26
Soil level		591.77			231.42			443.48	

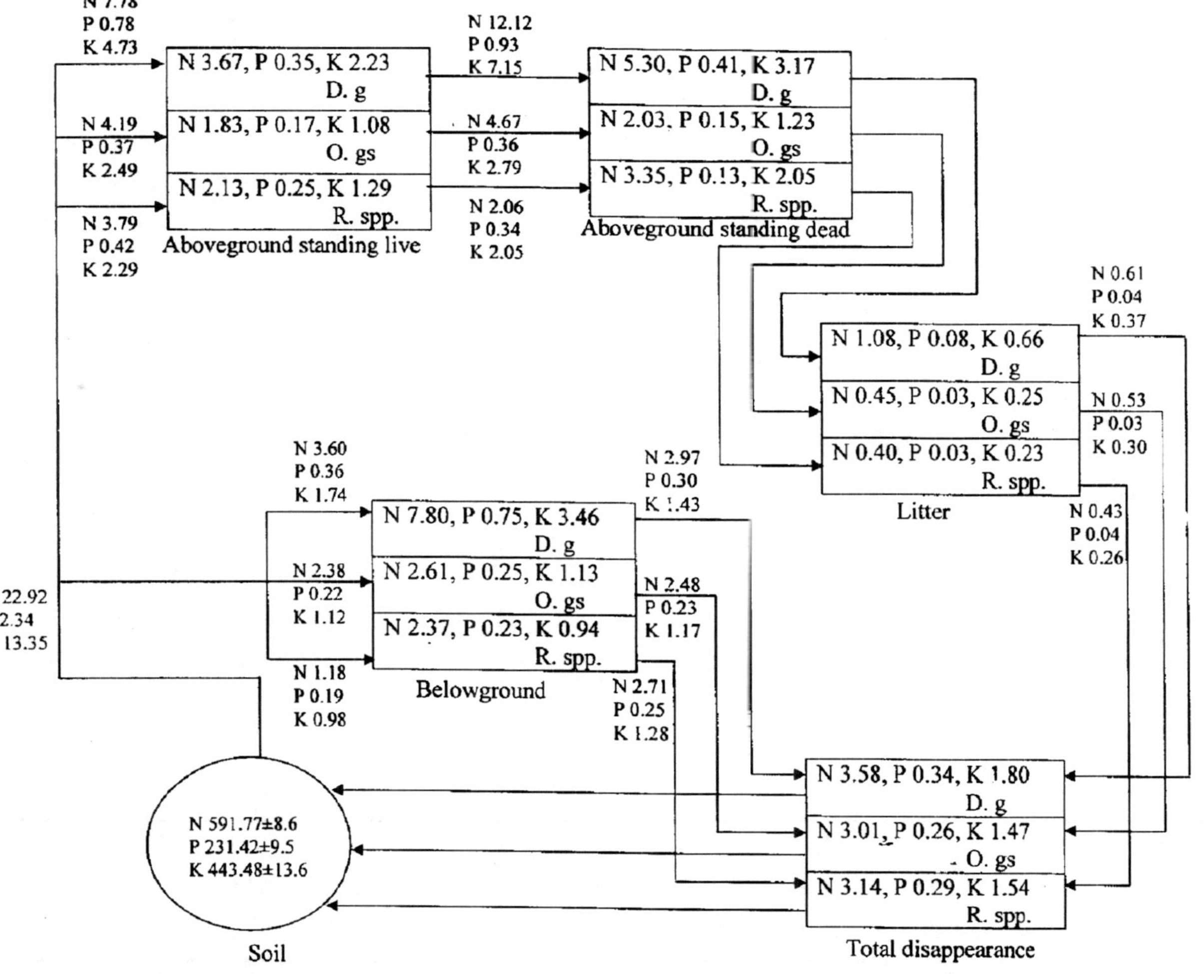

Figure 57.1: Net Accumulation and Transfer Rates of Nutrients ($g/m^2/yr$) in the Grassland Community of Grass Hills, the Anaimalais, Western Ghats

Values in rectangular boxes are accumulation of nutrients on average standing crop of nutrients and the valuation arrows are the transfer rate of nutrients on annual basis.

D.g: The dominant grass, *Chrysopogon zeylanicus;* O.gs: 'Other grasses'; R.spp: The 'remaining species'.

Table 57.4: Net Uptake, Release and Retention Rates of Nutrients in the Annual Fire Influenced Grassland at Grass Hills, Indra Gandhi Wildlife Sanctuary, the Western Ghats

Nutrient (kg/ha/yr)	*Uptake*				*Release*				*Retention*			
	Dominant Grass	*Other Grasses*	*Remaining Species*	*Total Content*	*Dominant Grass*	*Other Grasses*	*Remaining Species*	*Total Content*	*Dominant Grass*	*Other Grasses*	*Remaining Species*	*Total Content*
Nitrogen	77.80	41.9	37.9	157.60	35.80	30.10	31.40	97.30	42.00	11.80	6.50	60.30
Phosphorus	7.80	3.70	4.20	15.70	3.40	2.60	2.90	8.90	4.40	1.10	1.30	0.80
Potassium	47.30	24.90	22.90	95.10	18.00	14.70	15.40	48.10	29.30	10.20	7.50	47.00

The annual uptake, release and retention rates of nitrogen, phosphorus and potassium are given in Table 57.4. The data revealed that on annual basis, among the three nutrients, nitrogen was taken up by the plants in higher quantity (157.6 kg/ha/yr) followed by potassium (95.10 kg/ha/yr) and phosphorus (15.7 kg/ha/yr). Similarly, the retention and release rates on annual basis were also higher for nitrogen. Of the species of three categories, the dominant grass, *C. zeylanicus* contributed about 50 per cent of nutrient uptake and about 30 per cent each of retention and release rate of nutrients. It may be attributed to greater functional niche of the grass, *C. zeylanicus* in the grassland of Grass Hills in addition to its higher structural niche (Paulsamy, 1992).

The nutrient contents and transfer rates determined in the present investigation seem to be relatively higher in comparison to those reported in other grassland communities of India (Billore and Mall, 1976; Agrawal, 1988). It may be attributed to the presence of more available nutrients in the soil of Grass Hills owing to high decomposition co-efficient (Paulsamy, 1990). The suitable microclimate in soil may also facilitate the rapid rate of mineralization of dead matter and litter in humid vegetations (Witkamp and Vander Drift, 1961). Jenkinson *et al.* (2004) while working in grassland at Rothamsteads reported that the sustained availability of nutrients in favourable conditions after disturbances like fire enhances the uptake and other flux rates. Further, Misra (1980) pointed out on succession basis that nutrient transfer rates are rapid in young or disturbed communities and it is slow in the matured or climax communities. The dis climax stage of succession attained by the present study grassland due to annual fire, performs the rapid exchange of nutrients than the other climax grasslands of India. The complete destruction of aboveground parts of this grassland during summer (February–May) is subsequently followed by luxuriant growth of the constituent species from their bulk rhizomes (Paulsamy *et al.*, 1995) and this active regeneration facilitates the greater uptake of nutrients. This fact of rapid uptake of nutrients after fire in the subsequent rainy months may also decrease their leaching losses in Grass Hills. Further, the leaching loss of nutrients may also be influenced by the type of species growing in the community and other edaphic conditions (Personeni and Loiseau, 2005). Hence, it is known that annual summer fire in the grassland of Grass Hills induces the uptake of nutrients from soil through the greater net primary production which in turn supplies nutritious herbage to the wild herbivores. Therefore, annual summer fire occurs in Grass Hills is considered to be a beneficial factor with respect to nutrient dynamics and wildlife welfare.

References

Agrawal, A.K., 1988. Nutrient structure and dynamics in a temperate grassland community of Western Himalaya (Garhwal), India. *Tropical Grasslands*, 22(1): 33–39.

Allison, F.E., 1965. The enigma of soil nitrogen balance sheet. *Advance Agronomic Journal*, 7: 213–247.

Bijker, H.J., Sumner, P.D., Meiklejohn, K.I. and Bredenkamp, G.J., 2001. Documenting the effect of veld burning on soil and vegetation characteristics in giant's castle game reserve, Kwa-Zulu-Natal, Drakensberg. *South African Geographical Journal*, 83: 28–33.

Billore, S.K. and Mall, L.P., 1976. Nutrient composition and inventory in a tropical grassland. *Plant and Nature*, 45: 509–520.

Bokhari, U.G. and Singh, J.S., 1974. *Standing State and Cycling of N in Soil Vegetation Components of Prairie Ecosystems*. Preprint No. 108, NREL, Colorado State Univ.

Bor, N.L., 1938. The vegetation of the Nilgiris. *Ind For.*, 64: 600–609.

Champion, H.G., 1936. A preliminary survey of the forest types of India and Burma. *Indian For. Rec. (N.S.) Silv.*, pp. 236.

Champion, H.G., 1936. The relative stability of Indian vegetational types. *J. Indian Bot. Soc.*, 18: 1.

Chinnamani, S., 1981. Grasslands of South India and its ecology, floristic composition and utility. *Bull. Bot. Surv. India,* 23: 129–134.

Dahlman, R.C., Olson, J.S. and Doxtader, K., 1967. The nitrogen economy of grassland and dune soils. In: *Mimographed Proceedings of Biology and Ecology of Nitrogen*, Conference held at the University of California, Davis (National Academy of Science, Washington D.C.), pp. 54–82.

Gower, S.T. and Son, Y., 1992. Differences in soil and leaf litterfall nitrogen. I. Dynamics for five forest plantations. *Soil Science Society of American Journal,* 56: 1959–1966.

Jackson, M.L., 1962. *Soil Chemical Analysis.* Asia Publishing House, Bombay.

Jackson, M.L., 1958. *Soil Chemical Analysis.* Prentice Hall of India Private Ltd., New Delhi.

Jenkinson, D.S., Poulton, P.R., Johnston, A.E. and Powlson, D.S., 2004. Turnover of nitrogen 15-labeled fertilizer in old grassland. *Soil Science Society of American Journal*, 68: 865–875.

Laudeleut, H. and Germain, R., 1954. Primer resultants sur la dynamique chemique de' jacheres herbacees et des pastures a yangambi. International Congress on Soil Sciences, Leopold Ville.

Liebig, S.L., 1940. Die organische chemie in ihrer anwendung vor agricultur und physiologie. Braunschweig.

Mall, L.P. and Billore, S.K., 1974. An indirect estimation of litter disappearance in grassland study. *Current Science*, 43: 506–507.

Middleton, C.H., 1982. Dry matter and nitrogen changes in five tropical grasses as influenced by cutting height and frequency. *Tropical Grasslands,* 16: 112–117.

Misra, K.C., 1980. *Manual of Plant Ecology.* Oxford and IBH Publishing Co., New Delhi, Bombay and Calcutta, pp. 457.

Nye, P.H., 1961. Organic matter and nutrient cycles under moist tropical forests. *Plant and Soil,* 13: 333–346.

Odum, E.P., 1960. Organic, production and turnover in the old field succession. *Ecology,* 41: 34 49.

Odum, E.P., 1971. *Fundamentals of Ecology,* 3rd edn. W.B. Saunders Co., Philadelphia and London, pp. 574.

Ovington, J.D., 1965. Organic production, turnover and mineral cycling in woodlands. *Biol. Rev.,* Cambridge Phil. Soc., 40: 295–336.

Paulsamy, S., Maryhelan, G. and Manian, S., 1990. Studies on the litter decomposition in a tropical mixed community: Trends in ecotoxicology. In: *Proceedings of 10th Annual Session of AEB and Symposium on "Environmental Pollution on Biosystems"* held at Loyola College, Madras, pp. 263–266.

Paulsamy, S., 1992. Role of fire in the maintenance of *Chrysopogon zeylanicus* (Nees ex Steud.) Thw. dominated grasslands of Grass Hills, Western Ghats, India. *Ph.D. Thesis,* Bharathiar University, Coimbatore, India.

Paulsamy, S., Manian, S. and Udaiyan, K., 1995. Relationship in the response of grass, *Chrysopogon zeylanicus* and a fern, *Pteris aquilina* to annual summer fire on Grass Hills ecosystem, Western Ghats. *Indian Journal of Forestry,* 18(3): 221–225.

Paulsamy, S., Manian, S. and Udaiyan, K., 1995. A strategy to evaluate community dynamics through uptake/release rate of nutrients. *Range Mgmt. and Agroforestry,* 16(1): 15–20.

Paulsamy, S., Manian, S., Udaiyan, K., Arumugasamy, K., Nagarajan, N. and Kil, R.S., 2001. Biomass structure and dry matter dynamics in a fine influencing montane subtropical humid grassland, Western Ghats, Southern India. *Korean J. Ecol.,* 24(4): 227–232.

Peach, K. and Tracey, M.V., 1956. *Modern Methods in Plant Analysis.* Springer Verleg, Vol. IV.

Pemadasa, M.A., 1981. The mineral nutrition of the vegetation of a montane grassland in Sri Lanka. *Journal of Ecology,* 69: 125–134.

Personeni, E. and Loiseau, P., 2005. Species strategy and N fluxes in grassland soil a question of root litter quality or rhizosphere activity? *European Journal of Agronomy,* 22(2): 217–229.

Piper, C.S., 1944. *Plant Soil Analysis.* International Scientific Publication, New York.

Reich, P.R, Grigal, D.F., Aber, J.D. and Gower, S.T., 1997. Nitrogen mineralization and productivity in 50 hardwood and conifer stands on diverse soils. *Ecology,* 78: 335–347.

Reich, P.B., Peterson, D.W., Wedin, D.A. and Wrage, K., 2001. Fire and vegetation effects on productivity and nitrogen cycling across a forest-grassland continuum. *Ecology,* 82(6): 1703–1719.

Rodin, L.E. and Bazilevich, N.I., 1967. *Production and Mineral Cycling in a Terrestrial Vegetation.* Oliver and Boyd, Edinburgh, London.

Schacht, W.H., Stubbendieck, J., Bragg, T.R, Smart, A.I. and Doran, J.W., 1996. Soil quality response of reestablished grasslands to mowing and burning. *Journal of Range Management,* 49: 458–463.

Scholes, R.I. and Archer, S.R., 1997. Tree-grass interactions in savannas. *Annual Review of Ecology and Systematics,* 28: 517–544.

Sims, P.L. and Singh, J.S., 1971. Herbage dynamics and net primary production in certain ungrazed and grazed grassland in North America. In: *Preliminary Analysis of Structure and Function in Grasslands,* (Ed.) N.R. French. Range Science Department, Science Series No. 10, Colorado State University, pp. 59–124.

Singh, J.S. and Krishnamurthy, L., 1981. Analysis of structure and function of tropical grassland vegetation of India. *Indian Rev. Life Sci.,* 1: 225–270.

Turvey, N.D., 1974. Water in the nutrient cycle of a Papuan rainforest. *Nature,* 251: 414–415.

Wilson, J., 1964. Working plan for the Coimbatore south division. In: *Proc. Chief Conservator of Forests,* Madras. Proc. Mis. No. 186, pp. 317.

Witkamp, M. and Vander Drift J., 1961. Breakdown of forest litter in relation to environmental factors. *Plant and Soil,* pp. 295–311.

Woodwell, G.M. and Whittaker R.I-L., 1968. Primary production in terrestrial ecosystems. *American Zoologist,* 8: 19–30.

Yamane, I. and Sato, K., 1961. Soil plant relationship in the mountainous grassland of the Kawatabi farm. *Tohoku University Report,* 1957–1960.

Chapter 58

Prospect of Biomonitoring of Pollution in Aquatic System by Algal (Chlorophyceae) Assessment

Sushma Das Guru
Department of Botany, Ranchi Women's College, Ranchi – 834 001

ABSTRACT

The species richness and algal diversity of two freshwater bodies (Ranchi Lake and Hatania Talab) has been investigated for biomonitoring the water quality. During the investigation 56 species belonging to 26 genera of Chlorophyceae has been encountered. The predominance and species richness of four genera *Scenedesmus, Oedogonium, Closterium* and *Cosmarium* has been noticed. Out of 26 genera of the present investigation only 14 genera has been identified as pollution tolerant algal genera in the list of Palmer (1969), the author proposes that the rest of the genera can be tentatively designated as pollution tolerant genera till they are evaluated. The present communication also reflects the present status of algal diversity for future limnologists.

Keywords: *Freshwater, Biomonitoring, Chlorophyceae, Species richness.*

Introduction

In recent days, as a result of pollution, most of the water bodies have turned into unproductive and biological deserts. The extent of pollution stress and level of degradation of water bodies are now quantified by biomonitoring technology besides conventional methods (Kennish, 1992). Though this method of surveillance and assessment for the chemical analysis of water had originated abroad, but this approach has now been practiced in our country also (Rai and Kumar, 1980; Mohanty, 1983; Tripathi, 1989; Mohapatra and Mohanty, 1992; Venkatraman *et al.*, 1994; Ramkrishnan, 2003; Guru and Sahu, 2006; Guru, 2007) to monitor the quality for the water resources. In early days of this technology benthic macrofauna, fish and even aquatic plants were used as indicator species for pollution (Batiuk *et al.*, 1992; Phillips and Rainbow, 1993) but later it was noticed (Cairns *et al.*, 1994) that algae are more sensitive to pollution than other organisms. De Lang (1994) also found that algae to be more convenient organism as algal growth are linked with sunlight and nutrient contents of water directly.

The objective of this chapter is two fold. first, to make a biosurvey of various Chlorophycean algae present in the two water bodies namely Ranchi Lake and Hatania Talab in order to assess the biological integration of these aquatic resources; and second, to prepare the present status report of algal diversity for the future use of limnologists.

Materials and Methods

The present investigation was carried out in Ranchi Lake and Hatania Talab situated at the middle of the city. They harbor a large number of living organisms apart from the rich assemblage of chlorophycean members. The tanks are also used for many other purposes by the people living around, *e.g.*. for their daily domestic chorus, hence receive detergents, domestic wastes, sewage and affluent from various sources, run off water from surrounding agricultural fields containing insecticides and pesticides etc.

To study the biodiversity, algae (Chlorophyceae) were collected every month by filtering the pond water through bolting silk. The filtered samples were fixed in 4 per cent formalin. The mixed samples were identified and separated under the binocular microscope using the standard taxonomic keys.

Results

During the present investigation the author has recorded 56 species of chlorophycean members collectively from Ranchi Lake and Hatania Talab which are listed in the Table 58.1.

Discussion

In natural aquatic water bodies the algae are responsible for the biological production of energy which they transfer to the organisms placed on higher rung of trophic level (Walsh, 1975). It is also well established that phytocommunities exhibit seasonal stability as they reflect the changes in light, temperature and status of the nutrients in their habitat (De Lange, 1994). However, studies have demonstrated that various types of aquatic pollution are responsible for change in community structure (Palmer, 1969, 1980; Patrick, 1971; Trainor, 1984; Plafkin *et al.*, 1989; Phillips and Rainbow, 1993; De Lang, 1994). In our country also this subject has attracted a number of investigators due to ever increasing rate of aquatic pollution. Rai and Kumar (1980) studied the effect of various types of pollution on algae since they constitute a strategic component of aquatic environment and concluded that any disruption or damage of this will disturb the ecological food chain. Later, a number of workers have contributed valuable information in the field and even identified a large number of bioindicators of pollution tolerating algal genera (Mohapatra and Mohanty, 1992; Parvateesam and Mishra, 1993; Kanishk *et al.*, 1993; Siddiqui *et al.*, 1994; Hosmani, 2002; Ramakrishnan, 2003). The Ranchi Lake and Hatania Talab resemble tropical counterparts in showing. rich population of planktons dominated by members of chlorophyceae and also showing two growth peaks, one in summer and other in post-monsoon. This bimodal abundance appears to be linked with temperature since with the onset of winter and during post monsoon months their population decline dramatically.

Table 58.1: List or Chlorophycean Members in Ranchi Lake (RL) and Hatasnia Talab (HT)

Sl.No.	*Species*	*RL*	*HT*
1.	*Sphaerella lacustris*	+	+
2.	*Tetraspora gelatinosa*	+	+
3.	*Chlorococcum infusionum*	–	+
4.	*C. viliosum*	–	+
5.	*C. humicola*	+	+
6	*Nautococeus pyriformis*	–	+
7.	*Phyllobium dimorphum*	+	+
8.	*Pediastrum borryanum*	+	–
9.	*P. tetras*	+	+
10.	*Hydrodictyon reticulatum*	+	–
11.	*Tetraedron minimum*	+	+
12.	*Chlorella vulgaris*	+	+
13.	*Oocystis eclipta*	–	+
14.	*Nephrocytium obesum*	+	+
15.	*Botriococcus braunii*	+	+
16.	*Ankistrodesmus falcatus*	+	+
17.	*A. convolutus*	+	+
18.	*Actinastrum hantzschi*	–	+
19.	*Selenastrum gracile*	+	+
20.	*Kirchneriella obesa*	+	+
21.	*Coelastrum microporum*	+	+
22.	*Crucigenia triangularis*	+	+
23.	*C. tetrapedia*	+	+
24.	*C. quadrata*	+	+
25.	*Scenedesmus oblicus*	+	+
26.	*S. dimorphus*	+	+
27.	*S. bernardi*	+	+
28.	*S. bijugatus*	–	+
29.	*S. arcuatus*	–	+
30.	*S. quadricauda*	+	+
31.	*Ulothrix tennerima*	+	+
32.	*U. tenussima*	–	+
33.	*U. zonata*	+	+
34.	*Cidophora glomerata*	–	+
35.	*Stigeoclonium tenue*	+	+
36.	*S. lubricum*	+	+

Contd...

Table 58.1–Contd...

Sl.No.	*Species*	*RL*	*HT*
37.	*S. aestivale*	+	+
38.	*Oedogonium calliandrum*	+	+
39.	*O. capillare*	+	+
40.	*O. lautumniarum*	+	+
41.	*O. plagiostomum*	+	+
42.	*O. singulare*	+	+
43.	*O. sociale*	+	+
44.	*Spirogyra communis*	+	–
45.	*S. condensata*	+	–
46.	*S. formosa*	+	–
47.	*Closterium abruptum*	+	+
48.	*C. acerosum*	+	+
49.	*C. intermedium*	+	+
50.	*C. setaceum*	+	+
51.	*C. tumidum*	+	+
52.	*C. venus*	+	+
53.	*Cosmarium subcostatum*	+	+
54.	*C. punctulatum*	+	+
55.	*C. bioculatum*	+	+
56.	*C. crenatum*	+	+

During the present biosurvey of algae 26 genera belonging to 56 species have been recorded. The overwhelming dominance is shown by *Scenedesmus, Oedogonium Closterium,* and *Cosmarium.* These genera are also rich in diversity as no other genera recorded herein exhibit so rich species diversity (*Scenedesmus* 6 species, *Oedogonium* 6 species, *Closterium* 6 species and *Cosmarium* 4 species). The abundance of algae has been suggested to be indicative of nutrient pollution (De Lang 1994). The comparison of chlorophycean flora of two water bodies reveals that out of total 56 species collectively recorded 5 species are absent in Hatania Talab where as 9 species are absent in Ranchi Lake. According to Ramakrishnan (2003) the presence or absence may not be linked with environmental health of the water condition. Their absence may be due to reasons other than pollution like predation, competition or barriers which prevented them to be at the site during collection.

Another conspicuous feature of the present biosurvey is the presence of 3 species of *Spirogyra* in Ranchi Lake and their absence in Hatania Talab. The genus *Spirogyra* is considered to be inhabitant of unpolluted running clear waters (Bold and Wynne, 1978). The author is of the view that aquatic health of Hatania Talab is not conducive for the growth of this species, which may be due to its small size (1.9 acres) and excessive anthropogenic activities whereas, Ranchi Lake having a larger area (45 acres) is less affected by above mentioned factors and can still sustain the *Spirogyra* population. The author has also encountered 6 species of *Scenedesmus* from these water bodies and except one (*S. bernardi*) all other 5 species has been identified as most pollution tolerant species by Palmer (1969).

The list of pollution tolerant algae as compiled by Palmer (1969) contains 60 genera in which only 14 out of 26 genera of the present survey are enlisted. The occurrence of these genera is suggestive of the polluted status of these water bodies. The remaining 11 genera of the present survey are yet to be evaluated to assess the pollution stress, but the author believes that they can tentatively be designated as pollution tolerating genera till they are evaluated for their tolerance, as the biomonitoring approach is not static and can be modified as per the local need.

Referances

Batiuk, R.A., Orth, R.J., Moor, K.A., Dennison, W.C., Stevension, J.C., Kollar, S., Bieber, S. and Heasly, P., 1992. Cherapeak Bay submerged aquatic vegetation habitat requirements and restoration targets: A technical synthesis. *EPA*, Annapolis, M.D.

Bold, H.C. and Wynne, M.J., 1978. *Introduction to the Algae.* Prentice-Hall of India Pvt. Limited, New Delhi.

Cairns, J.Jr., McCormic, P.V. and Nitherlehner, B.R., 1994. Bioassay and field assessment of pollution effects. In: *Algae and Water Pollution,* (Eds.) Rai, L.C., Gaur, J.P. and Soider, C.J. Schwcizcrbort sch Verlagsbnchhanlung, Stuggert, pp. 267.

De Lang, E., 1994. *Manual of Simple Water Qualify Analysis.* International Water Tribunal (IWT) Foundation, Amesterdam.

Guru Sushma, D. and Sahu, R., 2006. Morphological variations in *Stigeoclonium* (Chlorophyceae) in polluted water. *Bionature,* 26(2): 55–59.

Guru Sushma, D., 2007. Effect of xenobiotic pollution on the algal (Chlorophyceae) population of some freshwater bodies of Jharkhand. *Biospectra,* 2(1): 143–147.

Hosmani, P.S., 2002. *Biological indicators of water quality.* In: *Lake 2002 Symp. on Con. Resto. Management of Aquat. Ecosystems,* (Ed.) T.V. Ramachandra. CES, IISc, Bangalore, pp 73–76.

Kanshik, S., Agarkar, M.S. and Saksena, D.N., 1993. On the planktonic algae of sewage fed Vivek Nagar Pond at Gwalior. *Poll. Res.,* 10(1): 25–32.

Kennish, M.J., 1992. *Ecology of Estuaries: Anthropogenic Effects.* CRC Press, Boca Raton.

Mohapatra, P.K. and Mohanty, R.C., 1992. Determination of water quality of the water bodies using algal bioassay method. *Phykos,* 31(1 and 2): 77–84

Palmer, C.M., 1969. Composite rating of algae tolerating organic pollution. *J. Phycol.,* 5: 78–82.

Palmer, C.M., 1980. Significance or algae. In: *Algae and Water Pollution.* Castle Home Pub., London.

Parvateesam, M. and Mishra, Maneesha, 1993. Algae of Pushkar Lake including pollution indicating forms. *Phykos,* 32(1 and 2): 27–39.

Patrick, R., 1971. *The Structure and Function of Freshwater Microbial Communities,* (Ed.) J. Cairns. *Res. Dir. Mon.* 3. Viriginia Polytechnic Institute and State Univ. Blacksburg, pp. 151–164.

Phillips, D.J.H. and Rainbow, P.S., 1993. *Biomonitoring of Trace Aquatic Contaminants.* Elsevier Applied Science, New York.

Plafkin, J.L., Barbono, M.T., Porter, K.D., Gross, S.K. and Hughes, R.M., 1989. Rapid assessment protocol for use in stream and rivers: Benthic macro-invertebrates and fish. In: *Freshwater Biomonitoring and Benthic Macroinvertebrates,* (Eds.) D.M. Rosenburg and V.H. Resh. Chapman and Hall, New York.

Rai, L.C. and Kumar, H.D., 1980. Algae and the problems of water pollution. In: *Recent Advances in the Biology of Microorganisms, Vol. 1,* (Eds.) K.S. Bilgrami and K.M. Vyas. BSMPS, pp. 127–140.

Ramkrishnan, N., 1990. Water quality assessment of two drinking water ponds using algae as indicators at Tiruvannamalai Town, Tamil Nadu. In: *Nat. Symps. on Biomonitoring Indicators in an Aquat. Ecosystem. Erod.,* Abstract No. 7.

Ramkrishnan, N., 2003. Biomonitoring approaches for water quality assessment in two water bodies at Tiruvannamalai, Tamil Nadu, India. In: *Proceeding of the Third International Conference on Environment and Health,* Chennai, India, pp. 3–14.

Siddiqui, E.N., Bazmi, S.H., Ahmad, S.M. and Ahmad, S.S., 1980. Biomonitoring of freshwater ponds at Darbhanga, North Bihar, India. In: *Environment Management in Developing Countries,* (Chief Ed.) A.K. Sinha. Gyanodaya Prakashan, Nainital, pp. 57–63.

Trainor, F.R., 1984. Indicator algal assay laboratory and field approaches. In: *Algae as Ecological Indicators,* (Ed.) Shubert. Academic Press, New York, pp. 3–14.

Tripathi, C.K.M., 1989. Phytoplanktons as indicators or water pollution in river Ganga at Varanasi. *Biol. Mem.,* 15: 65–72.

Venkatraraman, L.V., Krishi, M.K. and Suvarnalata, 1994. Algae as tool for biomonitoring and abatement of pestiside pollution in aquatic system. *Phykos,* 33(1 and 2): 171–193.

Walsh, C.E., 1975. Utilization of energy by primary producers in two ponds in north Western Florida. In: *Biostimulation and Nutrient Assessment Workshop.* Environmental Protection Agency, Ecological Research series, pp. 249–274.

Chapter 59

Antimicrobial Activity of *Merremia tridentata* (L.) Hall. f.

S. Gopalakrishnan*, M. Kamalutheen and T. Syed Ismail
Department of Chemistry, Manonmaniam Sundaranar University, Tirunelveli – 627 012, Tamil Nadu

ABSTRACT

The antibacterial and antifungal activities of petroleum ether (40°–60°C), benzene, chloroform, ethanol and water extracts of *Merremia tridentata* (L.) Hall. f. (Fam. *Convolvulaceae*) have been evaluated against seven pathogenic strains of gram-positive (*Staphylococcus aureus, Streptococcus faecalis* and *Baccillus subtilis*), gram-negative bacteria (*Escherichia coli, Pseudomonas aeruginosa* and *Klebsiella aerogenos*) and the fungus, *Candida albicans* by using zone of inhibition assay. Then these activities are confirmed by Activity Index (AI) test. Activity index values were higher for ethanolic extract against the standard ciprofloxacin for bacteria and clotrimazole for fungi followed by water extract. Among all the extracts tested the ethanolic extract of this plant showed a good antimicrobial potential against the microorganisms used in this study. The minimum inhibitory concentrations (MIC) for the ethanolic extract of this plant were determined and varied from 30 to 50 μg/ml.

Keywords: *Merremia tridentata, Antimicrobial activity, Activity index, Minimum inhibitory concentration.*

Introduction

Several antimicrobial drugs are available nowadays; but their use is limited by a number of factors, such as low potency, poor solubility, emergence of resistant strains and drug toxicity. Therefore,

* Corresponding Author; E-mail: sgkmsu@yahoo.co.in.

the search of new more effective antimicrobial agents is necessary (Penna *et al.*, 2001; Quiroga *et al.*, 2001). The plant *Merremia tridentata* commonly known as "Mudiarkunthal" or "Savulikodi" or "Thrippan Pullu" in Tamil and "Prasarini" in Sanskrit is reported to possess a number of medicinal values (Chopra *et al.*, 1956). It is a thick climbing herb with woody rootstock spreading on the walls and on the grounds. It grows in all plain districts of Tamil Nadu, Western peninsula and Bengal (Anonymous, 1962). The plant is considered bitter, astringent, tonic and used in the treatment of rheumatism, piles, swellings and urinary disorders (Chopra *et al.*, 1956). The decoction of the root is also used in toothache. Diosmetin, luteolin, diosmetin-7-O-β-D-glucoside and luteolin-7-O-β-D-glucoside have been isolated from *M. tridentata* (Tsita Tzardi *et al.*, 1992). Ergosine alkaloids have been isolated from the seeds of *M. tridentata* (Kristina *et al.*, 2005). Pyrolidine alkaloids like hygrins and nicotine have been isolated from the root and the aerial parts of *Mtridentata* (Kristina *et al.*, 2005).

A perusal of literature revealed that *M. tridentata* had not been subjected to screening for antimicrobial properties. In this chapter, we report the effect of various solvent extracts of the *M. tridentata* on the pathogenic strains of three gram-positive bacteria *viz., Staphylococcus aureus, Streptococcus faecalis,* and *Baccillus subtilis* and three gram-negative bacteria *viz., Escherichia coli, Pseudomonas aeruginosa* and *Klebsiella aerogenos* and one fungus *Candida albicans* by using zone of inhibition assay (Jain, 2003). The corresponding solvents are used as solvent control. The standard antibiotics ciproflaxacin 1 μg/disc and clotrimazole 10 μg/disc are used as standard for bacteria and for fungi respectively.

Materials and Methods

The plant was collected in the month of August from St. Xavier's College campus, Palayamkottai, Tirunelveli District, Tamil Nadu, India and identified by Dr. V. Chelladurai, Research Officer (Botany), Central Council of Research in Ayurveda and Siddha, Government Siddha Medical College, Palayamkottai, Tamil Nadu, India. A voucher specimen (MSU-034) was preserved in the Herbarium of the Department of Chemistry, Manonmaniam Sundaranar University, Tirunelveli – 627 012, Tamil Nadu, India. The fresh plants were cleaned with distilled water to remove extraneous matter, shade-dried until to get constant weight and then powdered. The dried powder of the whole plant (500 g) was successively extracted using petroleum ether (40°–60°C), benzene, chloroform, ethanol and water by using a soxhlet apparatus. The last trace of solvent was removed under reduced pressure distillation and then vacuum dried. Finally the dried crude extracts were used for the study.

Seven strains of microorganisms including three gram-positive, three gram-negative and one fungus were used in the antimicrobial assays. All the gram-positive test organisms, *Staphylococcus aureus* (NCIM 2079), *Streptococcus faecalis* (NCIM 2080) and *Baccillus subtilis* (NCIM 2063), gram-negative organisms, *Escherichia coli* (NCIM 2065), *Pseudomonas aeruginosa* (NCIM 2036) and *Klebsiella aerogenos* (NCIM 2098) and the fungus *Candida albicans* (NCIM 3102) were obtained from National Chemical Laboratory (NCL) Pune, India. The antimicrobial activity was measured by disc diffusion technique (Vanden Berghe and Vlietinck, 1991). The bacterial and fungal cultures were incubated for 24 hours at 37°C in nutrient agar slants and sabourand glucose agar slants (Himedia, Mumbai, India) respectively. Before streaking, each culture was diluted (1 : 10) with fresh sterile nutrients broth. Plates were prepared by pouring 20 ml freshly prepared No.1 medium (Himedia, Mumbai, India) into 20 mm × 100 mm Petriplates. Inoculam (5 ml) was poured directly over the surface of prepared plates to uniform depth of 4 mm and then allowed to solidify at room temperature.

Test drug solution (1000 μg) of each extract was prepared by dissolving 10 mg of each extract separately in 10 ml of the respective solvents. The test bacterial and fungal strains were inoculated into agar plates (Himedia, Mumbai) separately. Then the sterile disc containing each test drug solution of the plant extract (200 µl) was placed over the seeded agar plates in such a way that there is no overlapping of zone of inhibition. Standard antibiotics, ciproflaxacin 1 µg/disc for bacteria and clotrimazole 10 µg/disc for fungus were also incorporated as standard. The plates were kept at room temperature for two hours to allow diffusion of the test drug into the agar; they were incubated for 24 and 48 hours at 37°C for the bacterial and the yeast strains respectively. After the incubation period was over, the plates were observed for zone of inhibition (Zl) measured in millimeters (mm). From the results, the activity index was calculated using the following formula:

$$\text{Activity index (AI)} = \frac{\text{Inhibition zone of sample}}{\text{Inhibition zone of standard}}$$

The minimum inhibitory concentration (MIC) forethanolic the ethanolic extract of *M. tridentata* was determined in µg/ml (Okeke *et al.*, 2001; Okunji *et al.*, 1990). The samples of the extract were prepared at five different concentrations like 300, 200, 150, 100 and 50 µg/ml and the samples loaded into each disc with 200 µl containing 60, 40, 30, 20 and 10 µg respectively against the test organisms. The solvent 70 per cent alcohol was used as a solvent control and ciproflaxacin 1 µg/disc was used as the standard. Further the study was carried out at still lower five different concentrations of 50, 40, 30, 20 and 10 µg/ml and the samples loaded into each disc with 200 µl containing 10, 8, 6, 4 and 2 µg respectively against the test organisms.

Results and Discussion

The bioassay results for antimicrobial activity of the various solvent extracts of *M. tridentata* are presented in Table 59.1. From the results it is very clear that both the ethanol and water extracts inhibited the growth of all the tested strains of bacteria and fungus, while the chloroform extract was effective in the fungus only. But the petroleum ether and the benzene extracts were resistant to all the tested strains of bacteria and fungus. Among the extracts tested, the ethanol extract showed the maximum activity against all the tested strains used in this study followed by water extract (Table 59.1). The ethanol extract showed highest sensitivity against the *S. faecealis*, and the *B. subtilis* (gram positive) and *E. coli* (gram negative), while it showed equal sensitivity against the *P. aeruginosa* when compared with that of the standard used. In the case of the fungus, *C. albicans*, both ethanol and water extracts showed higher sensitivity when compared with that of the standard used.

Table 59.1: Antimicrobial Activity of the Various Solvent Extracts of *M. tridendata* on the Microorganisms Used

Sl.No.	*Name of the Organism*	*Zone of Inhibition (mm)*					
		Petroleum Ether (40°–60°C)	*Benzene*	*Chloroform*	*Ethanol*	*Water*	*Standards*
1.	*S. aureus*	–	–	–	18	14	22[a]
2.	*B. subtilis*	–	–	–	24	17	20[a]
3.	*S. faecalis*	–	–	–	25	16	22[a]
4.	*E. eoli*	–	–	–	22	14	18[a]
5.	*K. aerogenos*	–	–	–	17	13	20[a]
6.	*P. aeruginosa*	–	–	–	20	16	20[a]
7.	*C. albicans*	–	–	08	26	16	12[b]

a: Ciproflaxacin; b: Clotrimazole; –: No inhibitory effect.

When we consider the Activity Index (Table 59.2), it is clear from the results that the ethanol extract showed good antimicrobial potential than the water extract against the tested microorganisms. The difference in the activity may be due to the different secondary metabolites present in the ethanol and in the water extracts. This indicates that the secondary metabolites act as an antimicrobial compounds, which either inhibit or kill the bacteria by different mechanisms.

Table 59.2: Activity Index (AI) of the Crude Solvent Extracts of *M. tridendata* on the Micro-organisms Used

Sl.No.	*Name of the Organism*	*Activity Index (AI)*				
		Petroleum Ether (40°–60°C)	*Benzene*	*Chloroform*	*Ethanol*	*Water*
1.	*S. aureus*	–	–	–	0.82	0.64
2.	*B. subtilis*	–	–	–	1.20	0.85
3.	*S. faecalis*	–	–	–	1.14	0.73
4.	*E. eoli*	–	–	–	1.22	0.78
5.	*K. aerogenos*	–	–	–	0.85	0.65
6.	*P. aeruginosa*	–	–	–	1.00	0.80
7.	*C. albicans*	–	–	0.67	2.17	1.33

–: No inhibitory effect.

The results of minimum inhibitory concentration (MIC) in µg/ml for the ethanolic extract of *M. tridentata* are presented in the Tables 59.3 and 59.4. From the results, it is clear that the MIC values varied from 30 to 50 µg/ml. The ethanolic extract of *M. tridentata* was more active against the gram-positive bacteria *S. faecalis* (30 µg/ml) when compared with that of the other organisms used.

Table 59.3: MIC Determination (300 to 50 µg/ml) for Samples of the Ethanolic Extract of *M. tridendata*

Sl.No.	*Name of the Organism*	*Zone of Inhibition (mm)*				
		50	*100*	*150*	*200*	*300*
1.	*S. aureus*	10	14	15	15	17
2.	*B.subtilis*	10	15	16	16	21
3.	*S. faecalis*	10	15	16	18	22
4.	*E. coli*	13	15	15	18	21
5.	*K. aerogenos*	08	10	10	11	11
6.	*P. aeruginosa*	13	13	17	18	23
7.	*C. albicans*	11	12	12	14	18

Conclusion

The results obtained in this study demonstrated that the medicinal plant, *M. tridentata* displays antimicrobial activity *in vitro*. Thus, it has the potential to be used for medical purposes or antimicrobial natural additives in cosmetic and food industries or possibly as safe alternatives to synthetic antimicrobial drugs.

Table 59.4: MIC Determination (50 to 10 µg/ml) for Samples of the Ethanolic Extract of *M. tridendata*

Sl.No.	Name of the Organism	Zone of Inhibition (mm)				
		50	40	30	20	10
1.	*S. aureus*	11	08	–	–	–
2.	*B.subtilis*	11	10	–	–	–
3.	*S. faecalis*	11	10	08	–	–
4.	*E. coli*	12	–	–	–	–
5.	*K. aerogenos*	09	–	–	–	–
6.	*P. aeruginosa*	12	09	–	–	–
7.	*C. albicans*	11	–	–	–	–

–: No inhibitory effect.

Acknowledgement

The authors thank Dr. V. Chelladurai, Research Officer (Botany), Central Council of Research in Ayurveda and Siddha, Government Siddha Medical College, Palayamkottai, Tamil Nadu for the identification of the plant and V. Loganathan, Head, Department of Pharmaceutics, Periyar College of Pharmacy, Tiruchirappalli, Tamil Nadu for his help in the experiment.

References

Anonymous, 1962. *The Wealth of India.* Council of Scientific and Industrial Research, New Delhi, 6(L–M): 354–363.

Chopra, R.N., Nayar, S.L. and Chopra, I.C., 1956. *Glossary of Indian Medicinal Plants*. Council of Scientific and Industrial Research, New Delhi, p. 330.

Jain, N. and Sharma, M., 2003. Broad spectrum antimycotic drug for the treatment of ringworm infection in human beings. *Current Science*, 85(1): 30–34.

Kristina, J.S., Robert, W., Anke, B., Petra, M., Britta, T.R. and Sonja, C.O., Azar, G., Mald, K., Karsten, S., Ludger, W., Monika, H., Frank, M. and Eckart, E., 2005. Chemotaxonomy of the pantropical genus Merremia (Convolvulaceae) based on the distribution of tropane alkaloids. *Phytochemistry*, 66: 1448–1464.

Okeke, M.J., Iroegbu, C.U., Eze, E.N., Okoli, A.S. and Esimone, C.O., 2001. Evaluation of extracts of the root of *Landolphia owerrience* for antibacterial activity. *J. Ethanapharmacol.*, 78: 119–127.

Okunji, C.O., Okeke, C.N., Gugnani, H.C. and Jwu, M.M., 1990. Antifungal saponin from fruit pulp of *Dracaena manni. Int. J. Crude Drug Res.*, 28: 193–199.

Penna, C., Marino, S, Vivot, E, Cruanes, M.C., Munoz, J.D., Cruanes, J., Ferraro, G., Gutkind, G. and Martino, V., 2001. Antimicrobial activity of Argentine plants used in the treatment of infectious diseases: Isolation of active compounds from *Sebastiania brasiliensis. J. Ethanapharmacol.*, 77: 37–40.

Quriroga, E.N., Sampietro, A.R. and Vattuone, M.A., 2001. Screening antifungal activities of selected medicinal plants. *J. Ethanapharmacol.*, 74: 89–96.

Tsita Tzardi, E., Loukis, A. and Philianos, S., 1992. (Lab. Pharmacogn. Univ. Athens, Athens, Greece 10679), Flavonoids of *Merremia tridentata*. *Fitoterapia*, 63: 189–90.

Vanden Berghe, D.A. and Vlietinck, A.J., 1991. Screening methods for antibacterial and antiviral agents from higher plants. In: *Methods in Plant Biochemistry: Assay for Bioactivity,* (Eds.) Dey, P.M., Harborne, J.B. and Hostettman, K. Academic Press, London, pp. 47–69.

Chapter 60

In vitro Studies on Antibacterial Effect of Aquous Crude Extract of Bryophytes on *Xanthomonas citri*

G.S. Deora, Durgesh Bhati and Narendra S. Jain
Bryology Laboratory, Department of Botany, B.N. (P.G.) College, Udaipur – 313 001, Rajasthan

ABSTRACT

The present work deals with the study of antibacterial activity of *Plagiochasma articulatum* (Kash) a liverwort, *Anthoceros longii* (Steph.) a hornwort and *Fissidens bryoides* (Hedw.) a moss against test organism *Xanthomonas citri in vitro.* The various concentrations of aquous extract of these three bryophytes were prepared. The number of colonies and inhibition zone of this bacteria enriched with nutrient agar medium were examined after 24 and 48 hours. It was observed that *P. articulatum* is more active than *A. longii* and *F. bryoides* against *Xanthomonas citri.*

Keywords: *Bryophytes, Crude extract, Xanthomonas citri, Liverwort, Hornwort, Moss, Test organism, In vitro.*

Introduction

Due to antimicrobial property, bryophytes are free from the attack by parasitic microorganisms which has drawn the attention of many botanist and microbiologist in past few years. The life cycle of bryophytes involve alternation of generation between diploid sporophytic and a dominant free living haploid gametophytic generations. Hayes (1947) found that aquous extracts of *Conocephallum conicum* to be inactive against certain microorganisms. The antibacterial activity of liverwort *Lunularia crutiata* was studied by Basil, *et al.* (1998). They evaluated the action of its acetone extracts against 13 bacterial strains. Inhibition of bacterial growth was compared with that of nacefotoxine, benzyl penicillin and tetracycline.

Wakuli *et.al.* (2003) studied that the extracted pigments of bryophytes exhibited antibiotic properties against gram positive bacteria. Antibacterial potency of methanol extracts of three green lower plants *Pneumatopteries afra, Platycerium bifurcatum* and *Nephrolepsis bisserata* were determined using agar clinical strains of *Escherichia coli, Staphylococcus aureus, Klebsiella* spp. by Ojo, Ajayi and Anibijuwon (2007). Asakawa (1994) isolated several terpenoids and aromatic compounds from several liverworts (Hepaticeae) and studied their antimicrobial activities and inhibitory activities of 5-lipoxygenase and Calmodilin. The present study aims to explore antibacterial activity of some other bryophytes such as *P. articulatum, A. longii* and *F. bryoides* and to investigate their potency against *Xanthomonas citri.*

Materials and Methods

Collection and Storage of Plant Material

The plant materials namely *P. articulatum, A. longii* and *F. bryoides* were collected from the natural habitat of Mt. Abu. The materials were brought to the laboratory and were washed thoroughly with water to remove parts of other plants and soil adhered to rhizoids. They were washed with distilled water and dryed in between blotting paper to remove extra moisture. The required quantities of materials were taken in pestle and mortar to prepare aquous crude extracts. The remaining materials were wrapped in blotting paper and packed in polyethylene bags and stored in freeze to prevent from drying.

Test Organism

The test organism, *Xathomonas citri* was cultured on nutrient agar and broth in the laboratory. The pure cultures were subcultured and serially diluted to obtain single colony.

Preparation of Extract

1000 mg materials of each plant were taken in the pestle and mortar and crushed with 1000mi of distilled water repeatedly till the plant materials were properly crushed. The smooth pulp prepared was kept overnight so that all the water soluble antibacterial substances of bryophytes dissolve in the water. The extracts were filtered through Whatsman filter paper. Then the extract was centrifuged at 5000 rpm for 30 minutes. The extract so prepared was of 100000 ppm. From this extract different concentrations (10000 ppm to 90000 ppm) were prepared and stored in volumetric flask in freeze. Experiments were designed to study antimicrobial activity taking bacteria in microlitres on the culture medium.

Preparation of Medium

Nutrient agar medium and beef peptone agar medium of different pH ranging from 3 to 8.5 were prepared and autoclaved. The bacteria were then cultured over them in Petridishes. It was observed that *Xanthomonas citri* grew the best on 7.3 pH of nutrient agar. Therefore nutrient agar having pH 7.3 was selected for this study.

Assay of Antibacterial Activity

The experiments were designed in completely aseptic condition on laminar air flow bench. Two methods such as well and pour plate were adopted to assess the microbial growth under the influence of extracts of *P. articulatum, A. longii* and *F. bryoides.* In well method 10 µl of bacteria was spread over the agar film in Petri dish and well were made in it. In each well 25 µl of extract of different ppm was poured and inhibition zone of growth was measured after every 24 and 48 hours. In pour plate 10 µl bacterial cultures, 25 µl extracts of different ppm and medium were poured in Petri dishes. All these Petri dishes were sealed with parafilm and placed in an incubator at 28°C for 24 and 48 hours. Results were noted after every 24 and 48 hours.

Results and Discussion

Effect of *P. articulatum* extract on *Xanthomonas citri*

The investigations on the effect of the aquous crude extracts were carried out to assess *in vitro* potential for antibacterial activity of *P. articulatum* against test organism. The data (Table 60.1) showed that the inhibition zone of the growth of test organism was the maximum (8.9 and 10 mm) in 100000 ppm and minimum (1.5 and 2 mm) in 10000 ppm of *P. articulatum* after 24 and 48 hours respectively. Similarly it was also observed that the number of colonies (Table 60.2) were found the maximum (312 and 320) in 10000 ppm and minimum (243 and 250) in 10000 ppm after 24 and 48 hours.

Table 60.1: Effect of Aquous Crude Extract of *P. articulatum* on *Xanthomonas citri* in Well Method

Sl.No.	*Extract Concentration (ppm)*	*Inhibition Zone in mm After 24 Hours*	*Inhibition Zone in mm After 48 Hours*
1.	Control	0	0
2.	10000	1.5	2
3.	20000	3	3.5
4.	30000	4	4.5
5.	40000	4.7	4.9
6.	50000	5	5.7
7.	60000	5.2	5.9
8.	70000	6	6.4
9.	80000	6.5	7.2
10.	90000	7	8.9
11.	100000	9	10

Effect of *A. longii* Extract on *Xanthomonas citri*

The effect of aquous crude extract of *A. longii* against test organism, *Xanthmonas citri* (Table 60.3) revealed that inhibition of bacterial growth was lowest (1 and 1.5 mm) in 10000 ppm whereas it was the highest (8 and 9 mm) in 100000 ppm after 24 and 48 hours respectively. The data (Table 60.4) suggested that the number of colonies were maximum (320 and 327) in 10000 ppm while they were minimum (253 and 259) in 100000 ppm of *A. longii* after 24 and 48 hours.

Effect of *F. bryoides* Extract on *Xanthomonas citri*

The minimum (0.5 and 1 mm) inhibition of bacterial growth was reported in 10000 ppm whereas it was maximum (6.5 and 7.0) in 100000 ppm of *F. bryoides* (Table 60.5). The effect of crude extract of *F. bryoides.* (Table 60.6) indicated that the number of colonies were highest (373 and 381) in 10000 ppm whereas it was lowest (300 and 307) in 100000 ppm after 24 and 48 hours respectively.

Table 60.2: Effect of Aquous Crude Extract of *P. articulatum* on *Xanthomonas citri* in Pour Plate Method

Sl.No.	*Extract Concentration (ppm)*	*Inhibition Zone in mm After 24 Hours*	*Inhibition Zone in mm After 48 Hours*
1.	Control	390	395
2.	10000	312	320
3.	20000	303	311
4.	30000	294	299
5.	40000	288	293
6.	50000	281	285
7.	60000	273	280
8.	70000	269	273
9.	80000	260	266
10.	90000	254	259
11.	100000	243	250

Table 60.3: Effect of Aquous Crude Extract of *A. longii* on *Xanthomonas citri* in Well Method

Sl.No.	*Extract Concentration (ppm)*	*Inhibition Zone in mm After 24 Hours*	*Inhibition Zone in mm After 48 Hours*
1.	Control	0	0
2.	10000	1	1.5
3.	20000	2.5	2.7
4.	30000	3	3.7
5.	40000	4.5	4.9
6.	50000	5	5.2
7.	60000	5.9	6.4
8.	70000	6.2	7
9.	80000	7	7.9
10.	90000	7.5	8
11.	100000	8	9

On the basis of results obtained, it was concluded that the antimicrobial activity was maximum in *P. articulatum* followed by *A. longii* and *F. bryoides.* The aquous extracts of *Conocephallum conicum* to be inactive against certain microorganisms (Hayes, 1947). Observations of the present study revealed

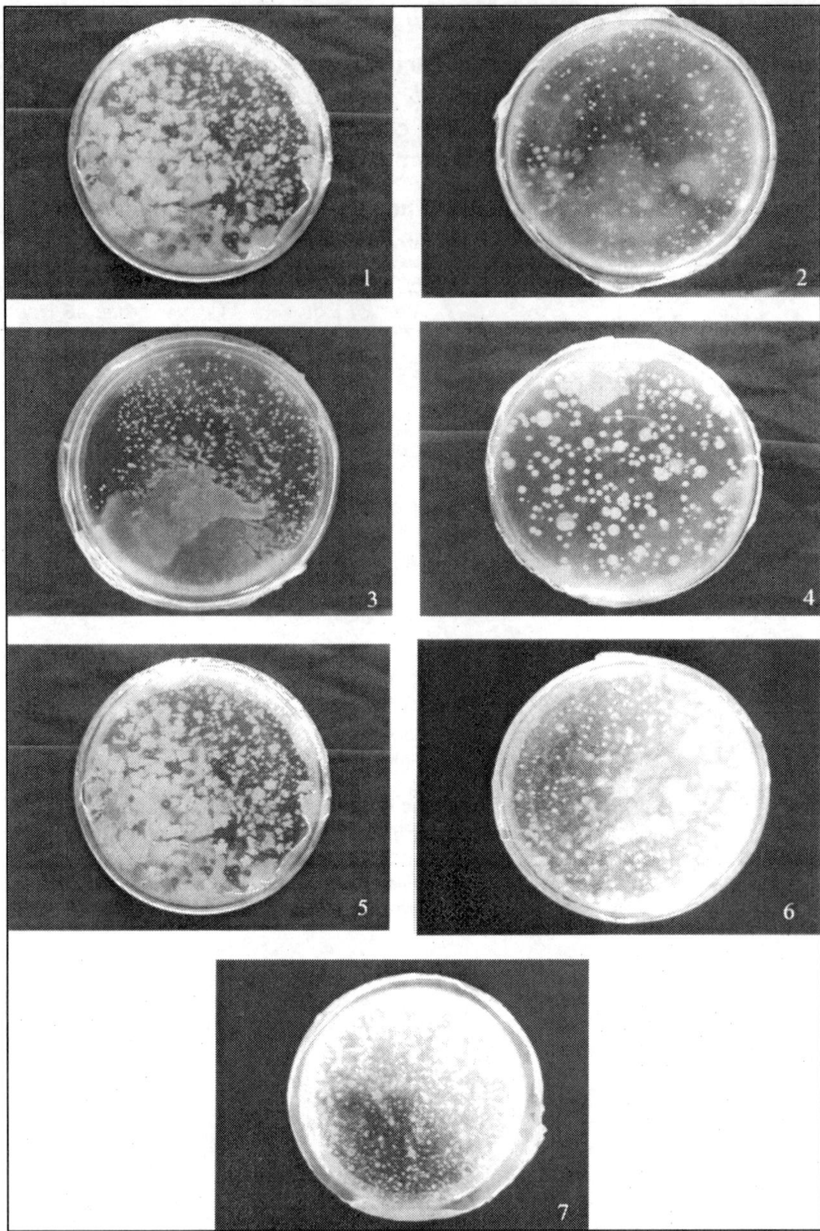

Figure 60.1: Number of Colonies in Control

Figure 60.2: Number of Colonies in 100000 ppm Extract of *P. articulatum*

Figure 60.3: Number of colonies in 10000 ppm Extract of *P. articulatum.*

Figure 60.4: Number of colonies in 100000 ppm Extract of *A. longii*

Figure 60.5: Number of colonies in 10000 ppm Extract of *A. longii*

Figure 60.6: Number of colonies in 100000 ppm Extract of *F. bryoides*

Figure 60.7: Number of colonies in 10000 ppm Extract of *F. bryoides*

that the extracts of *P. articulatum, A. longii* and *F. bryoides* (Tables 60.1, 60.3 and 60.5) inhibited the growth of test organism at 100000 ppm completely. Basil *et al.* (1998) evaluated the action of *Lunularia crudata* acetone extracts against 13 bacterial strains. Inhibition of bacterial growth was compared with that of nacefotoxine, benzyl penicillin and tetracycline.

Table 60.4: Effect of Aquous Crude Extract of *A. longii* on *Xanthomonas citri* in Pour Plate Method

Sl.No.	*Extract Concentration (ppm)*	*Inhibition Zone in mm After 24 Hours*	*Inhibition Zone in mm After 48 Hours*
1.	Control	392	397
2.	10000	320	327
3.	70000	309	315
4.	30000	301	309
5.	40000	292	300
6.	50000	288	296
7.	60000	280	289
8.	70000	273	281
9.	80000	269	277
10.	90000	262	268
11.	100000	253	259

Table 60.5: Effect of Aquous Crude Extract of *F. bryoides* on *Xanthomonas citri* in Well Method

Sl.No.	*Extract Concentration (ppm)*	*Inhibition Zone in mm After 24 Hours*	*Inhibition Zone in mm After 48 Hours*
1.	Control	0	0
2.	10000	0.5	1
3.	20000	1	1.7
4.	30000	2.2	3
5.	40000	2.9	3.5
6.	50000	3.3	3.9
7.	60000	3.9	4.3
8.	70000	4.9	5.2
9.	80000	5.4	6
10.	90000	6	6.9
11.	100000	6.5	7

The extracted pigments of bryophytes exhibited antibiotic properties against gram positive bacteria. (Wakuli, *et al.*, 2003). The present study also showed that the growth of *Xanthomonas citri* was inhibited by bryophytes due to their antimicrobial activity (Tables 60.1–60.6). Ojo, Ajayi and Anibijuwon (2007) reported that the methanol extract of three lower plants had antibacterial potency. The results of present work suggested that *P. articulatum* was more active against *Xanthomonas citri* followed by *A. longii* and *F. bryoides.*

Table 60.6: Effect of Aquous Crude Extract of *F. bryoides* on *Xanthomonas citri* in Pour Plate Method

Sl.No.	*Extract Concentration (ppm)*	*Inhibition Zone in mm After 24 Hours*	*Inhibition Zone in mm After 48 Hours*
1.	Control	395	400
2.	10000	373	381
3.	20000	367	372
4.	30000	360	368
5.	40000	353	359
6.	50000	349	353
7.	60000	342	347
8.	70000	333	341
9.	80000	321	329
10.	90000	312	317
11.	100000	300	307

References

Asakawa, T., 1994. Highlight in phytochemistry of hepaticeae biologically active terpenoids and aromatic compounds. *Pure and Appl. Chem.*, 66(10/11): 2193–2196.

Basil, A., 1998. Antibacterial activity in *Rhynchostegium riparioides* (Hedw.) Card. extract (bryophyte). *Phytotherapy Research*, 12(S1): 46–48.

Hayes, L.E., 1947. Survey of higher plants for presence of antibacterial substances. *Bot. Gaz.*, 108: 408–414.

Ojo, O.O., Ajayi, A.O. and Anibijuwon, L.L., 2007. Antibacterial potency of methanol extracts of lower plants. *J. Zhejiang Univ. Sci.*, 8, 8(3): 189–191.

Wakuli, W. *et al.*, 2003. Catenarin production by isolates of *Pyrenophora triticirepentis* and its antimicrobial activity. *Journal of Phytopathology*, 151(2): 74.

Chapter 61

Larvicidal and Ovicidal Effect of a Neem Product on Aedine and Culex Mosquitoes

Kiran Joseph, R.S. Mohanraj*, B. Dhanakkodi and K. Sathya
Department of Zoology, Kongunadu Arts and Science College (Autonomous), Coimbatore – 641 029, Tamil Nadu, India

ABSTRACT

An *in vitro* study was carried out to understand the effect of a neem based insecticide on the larvae and egg hatchability of Aedine and Culex mosquitoes. The LC_{50}/24 hour values against the different instar stages indicated larvicidal effect of the formulation. Hatchability of mosquito eggs was considerably reduced when treated with the botanical insecticide.

Keywords: *Aedes aegypti, Culex quinquefasciatus, Neem product, Toxicity, Hatchability.*

Introduction

For 260 million years, the two-winged insect has been on the face of this earth. It has outlived the dinosaurs and faced all the rough weather from volcanoes to human onslaught and emerged as a true nemesis to the human race. No other insect may have caused us so much trouble as this tiny perpetrator whom we encounter everyday. This common day foe is none other than the mosquito.

---

* Corresponding Author.

Aedes aegypti is the only known potential vector of dengue and urban yellow fever (Mekuria, 1991; Mazzarri and Georghiou, 1995). This species of mosquito was shown to be a competent laboratory vector of Chikungunya (CHIK) virus (Jupp and McIntosh, 1990). *Aedes aegypti* has also been noted to transmit filariasis and encephalitis (Dorland's illustrated Medical Dictionary, 1982). The *Culex quinquefasciatus*, a known vector of Bancroftian filariasis, is an important pest in all the tropical towns and cities in Asia (Bowers *et al.*, 1995).

Among the various control measures for mosquitoes, use of synthetic insecticides is still the most convenient and indispensable method all over the world. The extensive application of synthetic pesticides has however caused irrepairable hazards to the environment and humans. Moreover, insecticide resistance has also become the greatest problem in the control of many medically important mosquito species. *Ae. aegypti* in particular has developed resistance to a variety of organochloride, organophosphorus, carbamate and pyrethroid insecticides (Kumar *et al.*, 1991; Mekuria *et al.*, 1991; Mazzarri and Georghiou, 1995; Rawlins and Wan, 1995). Neem products are also relatively advantageous over synthetic insecticides in being eco-friendly and non-toxic to non-target organisms including humans. Purohit *et al.* (1989) demonstrated that plant products are easily biodegradable. It was also emphasized that they do not leave poisonous residues and insects do not develop resistance. Oroumchi and Lorra (1993) stated that neem products may be a source of inexpensive and environmentally sound pesticides.

Neem derivatives obtained from different parts of the plant were documented to produce diverse toxic and biological effects, on a variety of insect species (Kaur *et al.*, 2001; Ramarethinam *et al.*, 2002; Mohanraj and Dhanakkodi, 2004, 2005a, 2005b, 2006, 2007a, 2007b; Neloilya, 2007). However, investigations on the influence of neem derivatives on development aspects of mosquitoes are only a few that too fragmentary.

Reports on the toxicity of neem as mosquitocide (adulticide and larvicide) are available. Attri and Prasad (1980) observed the toxic effects of neem oil extractive on larvae of *Cx. fatigans*. Chavan and Nikam (1988) demonstrated that petroleum ether extract and alkanes from neem leaves were toxic to *Cx. pipiens fatigans*. Neem leaf extracts of polar and non-polar solvents (Deshmuhk and Renapurkar, 1992; Obomanu *et al.*, 2006) and neem oil (Das *et al.*, 1995) have been tested against larvae of *Culex* mosquito. The susceptibility of *Cx. quinquefasciatus* to acetone extract of neem seed coat (Sagar and Sehgal, 1997), NSK and neem gum extracts (Babu and Murugan, 1998) and Nimbex, Nimbecidine, Neem gold, Salotrap (Thamil Solai *et al.*, 1999) and that of *Aedes aegypti* to acetone extract fraction of the fruit of *Melia volkensii* and *Azadirachta indica* (Awala *et al.*, 1998; Umar *et al.*, 2007), resinous exudate from the tender leaves of neem (Babu *et al.*, 2000) and Econeem, a neem based insecticide (Mohanraj *et al.*, 2000) has been found. Larvicidal effect of neem products on the malarial vector, *Anopheles* have also been reported earlier (Aliero, 2003; Vatandoost and Vaziri, 2004; Okumu *et al.*, 2007). However, conducting more investigations on toxicity of different forms of neem to mosquitoes is therefore the need of the hour.

Hatchability of eggs is one of the biological features that ascertain the renewal of a species. The ovicidal effect produced by any means on egg hatching is likely to reduce the population of harmful insects. Neem derivatives were found to lessen the hatching of eggs of many species of insects (Ekesi, 2000; Sundar *et al.*, 2000, Kaur *et al.*, 2001). Although, ovicidal activities of various chemicals on mosquito eggs including *Aedes aegypti* were reported earlier (Judson, 1967; Miura *et al.*, 1976; Fournet *et al.*, 1993; Sahgal and Pillai, 1993), reports on the ovicidal effects of neem substances on mosquito eggs are meagre. It has been stated that no information is available regarding ovicidal activity of AZ and related products in mosquitoes Su and Mulla (1998a). This raises an ample chance to carry out investigations on this aspect.

Materials and Methods

Animals

Aedes and *Culex* eggs were collected, hatched, reared and maintained for many generations in the laboratory. The eggs and larvae obtained from this stock were used for different experiments.

Test Compound

A neem based insecticide, EC, containing 0.15 per cent w/w azadirachtin (1500 ppm).

Different concentrations of the neem formulation was prepared in parts per million (ppm) basis, based on the concentration of azadirachtin through serial dilution with unchlorinated filtered water.

Bioassay Test

Bioassay tests were carried out for testing the efficacy of the neem formulation *Aedes aegypti* and *Culex quinquefasciatus* at different stages of development *viz.*, I, II, III and IV instars and pupae. Instructions of WHO (1960) as detailed by Pampana (1963) for conducting bioassay experiment with mosquito larvae were carefully followed.

Different concentrations of the test compounds were prepared using unchlorinated water. Clean plastic tubs of 500 ml capacity were used as test containers. Batches of 20 larvae were exposed to 200 ml of a particular concentration of test solution. The larvae were collected with an eye dropper placed onto filter paper strips and immediately transferred to test cup containing test solution according to Cilek *et al.* (1991).

Mortality rates of larvae were recorded after 24 hours. Five or more concentrations of a test compound giving between 0 and 100 per cent mortality for larvae at different instar stages were tested. Parallel controls were maintained. Two replicates were done at each concentration. In recording the percent mortality for each concentration, the moribund and dead larvae in both replicates were combined.

For computing LC_{50}, the data were subjected to Finney's method of probit analysis as detailed by Regupathy and Dhamu (1990).

Determination of Effect of Neem Formulation on Hatching of *Aedes aegypti, Culex quinquefasciatus* Eggs

Twenty eggs of aedine were exposed to a particular concentration of the neem product tested. The hatchability was recorded until 96 hours from the initial time of the experiment. The time was fixed so, because it was demonstrated that the completion of embryogeny occurs within 4 days (Judson and Gojrati, 1967). In case of *Culex,* the egg rafts with known number of eggs were carefully exposed to different concentrations.

Hatching rate was calculated on the basis of non-hatchability of eggs according to Sahgal and Pillai (1993). To ensure non-hatchability, the eggs from any test container were collected after 96 hours. Unhatched and decapped eggs were separated and counted using dissection microscope. Five replications were conducted at each concentration of neem product. The data were statistically examined using student 't' test.

Results and Discussion

LC_{50}/24 hour values of the neem formulation to I instar larvae was 63.095 ppm (*Aedes*) 25.118 ppm (*Culex*) and this was found to gradually increase with the age of larvae. The LC_{50} values of the neem formulation to the II, III and IV instar stages were 199.526, 316.277, 501.187 ppm of (*Aedes*); 39.810, 79.432, 158.489 ppm (*Culex*) respectively. Pupae showed the highest resistance to the neem formulation as is evident from the relatively higher LC_{50} values 794.328 ppm (*Aedes*) 501.187 ppm (*Culex*) (Tables 61.1 and 61.2).

Table 61.1: LC_{50}/24 Hour Values of a Neem Formulation to the Pre-adult Stages (I, II, III, IV Instar and Pupae) of *Aedes aegypti*

Stages of Development (Instars)	*Number of Larva/ Trial*	*LC_{50}/ 24 Hour Log Dose*	*Confidence Interval*		*Variance*	*Chi-Square*	*Degree of Freedom*
			LL	*UL*			
I	20	1.8 (63.095 ppm)	1.664 (46.131 ppm)	1.936 (86.297 ppm)	0.00486	4.964	3(16.266)
II	20	2.3 (199.526ppm)	2.2362 (172.266 ppm)	2.3638 (231.100 ppm)	0.00106	1.915	3(16.266)
III	20	2.5 (316.277ppm)	2.4283 (268.101 ppm)	2.5717 (372.992 ppm)	0.00134	0.2741	3(16.266)
IV	20	2.7 (501.187 ppm)	2.6692 (466.874 ppm)	2.7308 (538.021 ppm)	0.000247	4.531	3(16.266)
Pupa	20	2.9 (794.328 ppm)	2.8562 (718.124 ppm)	2.9438 (878.617 ppm)	0.000501	7.8021	3(16.266)

Table 61.2: LC_{50}/24 Hour Values of a Neem Formulation to the Pre-adult Stages (I, II, III, IV Instar and Pupae) of *Culex quinquefasciatus*

Stages of Development (Instars)	*Number of Larva/ Trial*	*LC_{50}/ 24 Hour Log Dose*	*Confidence Interval*		*Variance*	*Chi-Square*	*Degree of Freedom*
			LL	*UL*			
I	20	1.4 (25.118 ppm)	1.3996 (25.095 ppm)	1.4003 (25.136 ppm)	0.0003749	4.032	3 (16.266)
II	20	1.6 (39.810 ppm)	1.5048 (31.974 ppm)	1.6952 (49.567 ppm)	0.002370	6.171	3 (16.266)
III	20	1.9 (79.432 ppm)	1.812 (64.868 ppm)	1.988 (97.274 ppm)	0.002022	3.871	3 (16.266)
IV	20	2.2 (158.489 ppm)	2.115 (129.270 ppm)	2.2885 (194.312 ppm)	0.002051	2.264	3 (16.266)
Pupa	20	2.7 (501.187 ppm)	2.6716 (469.461 ppm)	2.7284 (535.056 ppm)	0.0002122	0.091	3 (16.266)

Average number of eggs hatched in control medium was 18.4 out of 20 (92 per cent). This was reduced to the lowest level of 0 among eggs of *Aedes* and *Culex* placed in medium of 400 ppm, neem formulation (Tables 61.3 and 61.4).

Table 61.3: Alteration in the Hatchability of *Aedes aegypti* Eggs Exposed to Different Concentrations of a Neem Formulation and Control

Parameters	*Control*	*Concentrations (ppm)*				
		100	*150*	*200*	*300*	*400*
Number of eggs introduced	20	20	20	20	20	20
Mean number of eggs hatched*	18.4	14.0	9.6	5.2	3.0	0
S.D.	±0.894	±0.836	±0.836	±0.836	±0.707	0
Pe c ent hatchability	92	70#	48#	26#	15#	0
Per cent reduction over control		23.91	47.82	71.73	83.69	100

*: Mean (±SD) of 5 replicates.

#: Significantly different from control; P < 0.001 per cent.

Table 61.4: Alteration in the Hatchability of *Culex quinquefasciatus* Eggs Exposed to Different Concentration of a Neem Formulation and Control

Parameters	*Control*	*Concentrations (ppm)*				
		100	*150*	*200*	*300*	*400*
Number of eggs introduced/trial	20	20	20	20	20	20
Mean number of eggs hatched*	18.4	15.8	11.0	7.7	3.4	0
SD	±0.894	±0.386	±0.942	±0.833	±0.547	
Per cent hatchability	92	79#	55#	38#	17#	0
Per cent reduction over control		14.13	40.21	58.15	81.52	100

*: Mean (±SD) of 5 replicates.

#: Significantly different from control; P < 0.001 per cent.

Attri and Prasad (1980) demonstrated that neem oil extractive acted as strong toxicant to larval mosquito. Zebitz (1984) recorded 78.2 ppm as LC_{50} value of aqueous neem seed kernel extracts to fourth instar larvae of *Aedes aegypti.* Neem oil was found to kill mosquito larvae in respect to the concentration (Rawat, 1994). Mortality of 50 per cent and more occurred in larval *Aedes aegypti* (Sinniah *et al.*, 1994) and *Culex quinquefasciatus* (Amorose, 1995) exposed to relatively low concentrations (0.02 per cent as LC_{100}/24 hour as well as 0.99 and 1.2 ppm as LC_{50}/24 hour to 3rd and 4th stage larvae) of neem extract and neem oil. Acetone extract of neem seed coat at 40 ppm caused 100 per cent mortality of the first instar larvae of *Aedes aegypti* (Sagar and Sehgal, 1997). In a previous study on the toxicity of a neem formulation azadirachtin against larval stages of *Aedes aegypti* (Mohanraj *et al.*, 2000) the LC_{50}/24 hour values recorded against I, II, III, IV instar and pupal stages were 7.94, 12.365, 14.565, 14.642 and 17.38 ppm, respectively.

The difference in the values between the present and previous studies may be due to the difference in the composition of pesticides. The toxic effect may mainly be due to the active component azadirachtin present in neem. This compound was demonstrated to be the major active constituent of neem and was well known for its toxic, antifeedant, growth regulating effect on insects (Schmutterer, 1990). Xie *et al.* (1995) confirmed that azadirachtin was largely responsible for toxic actions of neem on insects. The species and formulation–dependent differences in the larvicidal activity of neem products was noticed (Mulla *et al.*, 1997 cit. Su and Mulla, 1999a). Death of mosquito larvae was probably due to the poisoning effect of neem ingredients (Friend, 1998) and deficiency of dissolved oxygen in the neem treated water (Aliero, 2003).

In the current study, a considerable increase in the LC_{50} values with the age of larvae was observed, *i.e.*, early instar larvae were more sensitive to the neem product than late instars. It was recognized that fourth stage larvae and pupae of mosquitoes were more tolerant to toxicants, than early instars (Jebanesan *et al.*, 1999). Low sensitivity with increase in age among different pre adult stages of *Aedes aegypti* exposed to various neem extracts as well as formulations (Mohanraj *et al.*, 2000). The decreased susceptibility may be attributed to increased size and weight of older larvae (Helson and Surgenoner, 1983).

Mohanraj *et al.* (2000) observed a considerable dose-based reduction in the hatchability of eggs treated with a neem formulation. One ppm of AZ found to produce hundred percent mortality in eggs (exposed for 4 hour) of *Culex tarsalis* and *Culex quinquefasciatus* (Su and Mulla, 1998a). Ovicidal activity of growth regulators/inhibitors (Miura *et al.*, 1976) and insecticides (Saghal and Pillai, 1993) in mosquitoes of different species was reported. Severe reduction in hatching of eggs from Lepidopteran and Hymenopteran insects on treatment with aqueous extracts of black pepper, neem seed and garlic bulb at concentrations 5 per cent, 10 per cent and 15 per cent and also with their volatiles was recorded (Ekesi, 2000).

Boradent and Free (1984) based on a study on the ovicidal activity of insect growth regulators, Diflubenzuron and Bay SIR 8514 opined that the chemicals diffuse into eggs and affect the vital physiological and biochemical processes associated with embryonic development, thereby inhibiting eclosion of eggs. Penetration of ovicides through chorion was demonstrated earlier (Smith and Salkerd, 1966). The mode of ovicidal action of an IGR (Dimilin) was explained as the inhibition of chitin synthesis, thus interfering with the formation of endocuticular deposition (Miura *et al.*, 1976). A similar mode of action may be suggested for neem products as they were observed to affect chitin synthesis (Sharma and Chander Sheikher, 1999a).

It is therefore suggested that the neem derivatives can also be considered for mosquito control operation, besides their use against agricultural pests.

References

Aliero, B.L., 2003. Larvicidal effects of aqueous extracts of *Azadirachta indica* (neem) on the larvae of *Anopheles* mosquito. *African Journal of Biotechnology*, 2: 325–327.

Amorose, T., 1995. Larvicidal efficacy of neem (*Azadirachta indica*) oil and defatted cake on *Culex quinquefasciatus* Say. *Geobios*, 22: 169–173.

Attri, B.S. and Prasad, R., 1980. Neem oil extractive: An effective mosquito larvicide. *Indian J. Ent.*, 42: 371–374.

Awala, P., Mwangi, R.M. and Irungu, L.W., 1998. Larvicidal activity of a granular formation of a *Melia volkensii* (Gurke) acetone extract against *Aedes aegypti* L. *Insect Sci. Applic.*, 18: 225–228.

Babu, R. and Murugan, K., 1998. Interactive effect of neem seed kernel and neem gum extracts on the control of *Culex quinquefasciatus* Say. *Neem Newsletter*, 15: 9–10.

Babu, R., Murugan, K. and Mohan, P.S., 2000. Larvicidal effect of resinous exudate from the tender leaves of *Azadirachta indica. Neem Newsletter*, 17: 1–3.

Boradbent, A.B. and Free, D.J., 1984. Effects of dyflobenzuron and Bay SIR 8514 on the oriental fruit moth (Lepidotera : Olethreutidae) and the oblique banded leaf roller (Lepidotera : Tortricidae). *J. Econ. Entomol.*, 77: 194–197.

Bowers, W.S., Sener, H., Evans, F., Bingal, A. and Erdogon, J., 1995. Activity of Turkish medical plant against mosquitoes *Aedes aegypti* and *Anopheles gambiae. Insect Sci. Applic.*, 16: 339–342.

Chavan, S.R. and Nikam, S.T., 1988. Investigation of alkanes from neem leaves and their mosquito larvicidal activity. *Pesticides*, 22: 32–33.

Cilek, J.W., Webb, J.D. and Knapp, F.W., 1991. Residual concentration and efficacy of three temphos formulations for control of larval *Aedes aegypti. J. Am. Mosq. Control Assoc.*, 7: 310–312.

Das, T.K., Somchoudhury, A.K., Saha, K. and Sarkar, P.K., 1995. Persistant toxicity of modern insecticides, germicides and plant extracts against mosquito larvae. *Pestology*, 19: 5–12.

Deshmukh, P.B. and Renapurkar, D.M., 1992. Insect repellent activity of *Azadirachta indica* Juss against mosquito *Culex pipiens fatigans. Pestology*, 16: 33–36.

Dorland's Illustrated Medical Dictionary, 1982, 26th Edition. W.B. Saunder Company, Igaku-Shoin/ Saunders, London, p. 35.

Ekesi, S., 2000. Effect of volatiles and crude extracts of different plant materials on egg viability of *Maruca vitrata* and *Clavigralla tomentosicollis. Phytoparasitica*, 28: 305–310.

Fournet, F., Sannier, C. and Monteny, N., 1993. Effects of the insect growth regulators OMS 2017 and diflubenzuron on the reproductive potential of *Aedes aegypti. J. Am. Mosq. Control Assoc.*, 9: 426–430.

Friend, J.A., 1998. Twenty years of horticulture research with neem against diamond backmoth, *Plutella xylostella* L. *Pestology*, 22: 33–40.

Helson, B.V. and Surgeoner, G.A., 1983. Effect of temperature and stage of development on susceptibility of *Aedes euedes* and *Aedes stimulans* (Diptera : Culicidae) larvae to temephos. *Can. Ent.*, 115: 623–628.

Jebanesan, A., Venkatachalam, M.R. and Jagadeesan, G., 1999. Toxicity evaluation of certain neem based bio–insecticides against the larval mosquitoes of *Culex quinquefasciatus* Say. *Pestology*, 23: 72–75.

Judson, C.L., 1967. Alternation of feeding behaviour and fertility in *Aedes aegypti* by the chemosterilant apholate. *Entomol. Exp. Appl.*, 10: 387–394.

Judson, C.L. and Gojrati, H.A.N., 1967. The effects of various oxygen tensions on embryogeny and larval responses of *Aedes aegypti. Entomol. Exp. Appl.*, 10: 181–188.

Jupp, P.G. and McIntosh, B.M., 1990. *Aedes furcifer* and other mosquitoes as vectors of Chikungunya virus at mica, north-eastern transvaal, South Africa. *J. Am. Mosq. Control Assoc.*, 12: 415–419.

Kaur, J.J., Rao, D.K., Sehgal, S.S. and Seth, R.K., 2001. Effect of hexane extract of neem seed kernel on development and reproductive behaviour of *Spodoptera litura* Fabr. *Ann. Pl. Protec. Sci.*, 9: 171–178.

Kumar, S., Thomas, S. and Pillai, M.K.K., 1991. Involvement of mono-oxygenases as a major mechanism of deltamethrin resistance in larvae of three species of mosquitoes. *Indian J. Exp. Biol.,* 29: 379–384.

Mazzarri, M.B. and Georghiou, P., 1995. Characterization of resistance to organophosphate, carbamate and pyrethroid insecticides in field populations of *Aedes aegypti* from Venezuela. *J. Am. Mosq. Control Assoc.,* 11: 315–322.

Mekuria, Y., Gwinn, T.A., Williams, D.C. and Tidwell M.A., 1991. Insecticide susceptibility of *Aedes aegypti* from Santo Domingo, Dominican Republic. *J. Am. Mosq. Control Assoc.,* 7: 69–72.

Miura, T., Schaefer, H., Takahashi, R.M. and Mulligan, F.S., 1976. Effects of the insects growth inhibitor, dimilin on hatching of mosquito egg. *J. Econ. Entomol.,* 69: 655–658.

Mohanraj, R.S. and Dhanakkodi, B., 2004. Effect of neem extracts on ovarian length and protein of *Aedes aegypti. J. Exp. Zool., India,* 7: 125 – 128.

Mohanraj, R.S. and Dhanakkodi, B., 2005a. Ovicidal effects of neem products on mosquito eggs. *J. Ecotoxicol. Environ. Monit.,* 15: 73–77.

Mohanraj, R.S. and Dhanakkodi, B., 2005b. Growth inhibiting effect of crude neem seed kernel extract of *Aedes aegypti. Indian J. Ent.,* 67(3): 241–243.

Mohanraj, R.S. and Dhanakkodi, B., 2006. Influence of neem extracts of ovipositional activity of *Aedes aegypti. Pestology,* 30: 38–41.

Mohanraj, R.S. and Dhanakkodi, B., 2007a. Delayed effect of neem extracts on fitness parameters of *Aedes aegypti. Journal of Current Sciences,* 10: 161–165.

Mohanraj, R.S. and Dhanakkodi, B., 2007b. Interference of botanical extracts on oviposition of *Aedes aegypti. Adv. Bio. Sci.,* 6: 11–17.

Mohanraj, R.S., Saraswathy, S., Thangavel, K., Logaswamy, S. and Dhanakkodi, B., 2000. Effect of neem formulation on pre-adult stages and egg hatchability of *Aedes aegypti. Neem Newsletter,* 17: 4–6.

Neoliya, N.K., Singh, D. and Sangwan, R.S., 2007. Azadirachtin-based insecticides induce alteratioin in *Helicoverpa armigera* Hub. Head polypeptides. *Current Science,* 92: 94–99.

Obomanu, F.G., Ogbalu, O.K., Gabriel, U.U., Fekarurhobo, G.K. and Adediran, B.I., 2006. Larvicidal properties of *Lepiadagathis alopecuroides* and *Azadirachta indica* on *Anopheles gambiae* and *Culex quinquefasciatus. African Journal of Biotechnology,* 5: 761–765.

Okumu, F.O., Knols, B.G.J. and Fillinger, U., 2007. Larvicidal effects of a neem (*Azadirachta indica*) oil formulation on the malaria vector *Anopheles gambiae. Malaria Journal,* 6: 63–73.

Oroumchi, S. and Lorra, C., 1993. Investigation on the effects of aqueous extracts of neem and Chinaberry on development and mortality of the alfalfa weevil *Hypera postica* Gyllenh (Col., Curculionidae). *J. Appl. Ent.,* 116: 345–351.

Pampana, E., 1963. *A Textbook of Malaria Eradication.* Oxford University Press, London, pp. 459–462.

Purohit, P., Jyotsna, D. and Srimannarayana, G., 1989. Anti-feedant activity of indigenous plant extracts against larvae of castor semi-looper. *Pesticides,* 25: 23–26.

Ramarethinam, S., Loganathan, S., Marimuthu, S. and Murugesan, N.V., 2002. Studies on the evaluation of neem oil based EC formulation (0.03 per cent Azadirachtin) on the semilooper caterpillar, *Achaeajanata* L. (Lepidoptera : Noctuidae) infesting castor plant, *Ricinus communis* L. *Pestology,* 26: 9–14.

Rawat, S., 1994. Neem products and their pesticidal characteristics. *Everyman's Sciences.* December–January, 1993–94.

Rawlins, S.C. and Wan, J.O.H., 1995. Resistance in some caribbean populations of *Aedes aegypti* to several insecticides. *J. Am. Mosq. Control Assoc.,* 11: 59–65.

Regupathy, A. and Dhamu, K.P., 1990. *Statistic Workbook for Insecticide Toxicology.* Surya Desktop Publishers, India, pp. 4–7, 16–22.

Sagar, S.K. and Sehgal, S.S., 1997. Toxicity of neem seed coat extract against mosquitoes. *Indian J. Ent.,* 59: 215–223.

Sahgal, A. and Pillai, M.K.K., 1993. Ovicidal activity of permethrin and deltamethrin on mosquitoes. *Entomon.,* 17: 149–154.

Schmutterer, H., 1990. Properties and potential of natural pesticides from the neem tree, *Azadirachta indica. Ann. Rev. Ent.,* 35: 271–297.

Sharma, P.L. and Chander Sheikher, 1999a. Toxic and morphogenic effects of neem leaf extracts on *Helicoverpa armigera* Hubner. *Pest Management and Economic Zoology,* 5: 95–100.

Sinniah, B., Sinniah, D. and Ibrahim, J., 1994. Effect of neem oil on mosquito larvae. *Mosq. Borne. Dis. Bull.,* 11: 90–94.

Smith, E.H. and Salkerd, E.H., 1966. The use and action of ovicides. *A. Rev. Ent.,* 11: 331–368.

Su, T.Y. and Mulla, M.S., 1998a. Ovicidal activity of neem products (Azadirachtin) against *Culex tarsalis* and *Culex quinquefasciatus* (Diptera : Culicidae). *J. Am. Mosq. Control Assoc.,* 14: 204–209.

Su, T.Y. and Mulla, M.S., 1999a. Oviposition bioassay responses of *Culex tarsalis* and *Culex quinquefaciatus* to neem products containing azadirachtin. *Entomol. Exp. Appl.,* 91: 337 – 345.

Sundar, M.S., Krishnayya, P.V. and Rao, P.A., 2000. Evaluation of lethal concentrations of azadirachtin based proprietary formulation for larvicidal and ovicidal effect against *Spodoptera litura* Fab. *J. Ent. Res.,* 24: 147–150.

Thamil Solai, P.K., Selvaraj Pandian, R. and Charles Manoharan, A., 1999. Toxicity evaluation of the pesticides Salotrap, Neemgold, Nimbecidine and Nimbex against the larvae and pupae of *Culex quinquefasciatus* Say. In: *Biopesticides in Insect Pest Management,* (Eds.) S. Ignacimuthu and Alok Sen. Phoneix Publishing House Pvt. Ltd., New Delhi, pp. 46–49.

Umar, A., Kela, S.L. and Ogidi, J.A., 2007. Effects of extraction solvents on the toxicity of *Azadirachta indica* A. Juss (Neem) seed kernel extracts to *Aedes aegypti* (Diptera : Culicidae) larvae. *Int. Jor. P. App. Scs.,* 1: 32–38.

Vatandoost, H. and Haziri, V.M., 2004. Larvicidal activity of a neem tree extract (Neemarin) against mosquito larvae in the Islamic Republic of Iran. *Eastern Mediterranean Health Journal,* 10: 573–581.

Xie, Y.S., Fields, P.G. and Isman, M.B., 1995. Repellency and toxicity of azadirachtin and neem concentrates to three stored-product beetles. *J. Econ. Entomol.,* 80: 1024–1031.

Zebitz, C.P.W., 1984. Effect of some crude and azadirachtin enriched neem (*Azadirachta indica*) seed kernel extracts on larvae of *Aedes aegypti. Entomol. Exp. Appl.,* 35: 11–16.

Chapter 62

Comparative Effect and Efficacy of Fenbendazole and Tetramisole on *Ascaridia galli*

K. Surender Reddy, V. Bikshapathi and G. Benarjee*
Department of Zoology, Kakatiya University, Warangal – 506 009, A.P.

ABSTRACT

Comparative therapeutic trials for two anthelminthic drugs were conducted on domestic fowl, *Gallus gallus* which were naturally infected with the nematode parasite, *Ascaridia galli.* Two anthelmintic drugs *i.e.,* Fenbendazole and Tetramisole were selected in the present study, to assess the efficacy of these drugs in deworming of parasites in the domestic fowl. The Tetramisole and Fenbendazole were administered to the birds @ 5 mg/bird/kg weight/single dose and there by the result shows that the Tetramisole was more effective in expulsion of Parasitic ova. In another experiment the birds were treated with Tetramisole and Fenbendazole @ 5 mg/bird/kg rate/single dose to study the efficacy of these anthelminth drugs on the direct effect on parasitic worms. The studies shows that there is a decline of worm count in the birds treated with Tetramisole over Fenbendazole, the Tetramisole was more effective over Fenbendazole, it is concluded.

Keywords: *Gallus gallus, Ascaridia galli, Fenbendazole, Tetramisole.*

Introduction

The domestic fowl, *Gallus gallus* has been assumed an important place for meat due to high

* Corresponding Author; E-mail: gbgsss@yahoo.co.in.

nutritional supplement to all sections of people. This meat contains almost all essential amino acids and low amount of cholesterol. The eggs of domestic hen is considered as unadulterated balanced diet. The infection of *Ascaridia galli* in domestic hen results in heavy morbidity and ultimately leads to mortality and causing extensive loss to the poultry industry. The parasite infestation results tremendous adverse effect on body weight and egg production in hen. It has been estimated that the annual loss goes to the tune of Rs. 301.96 million in our country (Chakrabarti, 1995). Several anthelminth drugs have been screened (Andrews and Thomas, 1979; Agarwal *et al.*, 1989; Blair and Campbell, 1978; Chevis, 1998; Maki and Yanagisawa, 1985) and they were proved effective on several parasites (Bennet *et al.*, 1977; Gerald and Elward, 1971; Coles *et al.*, 1975; Friedman and Pletzer, 1980; Matta, 1980; McCracen, 1978; Wolfe and Wershing, 1974). Earlier studies have been made by several investigators on effect of certain Anthelminthic drugs (Altaif; 1972; Amer *et al.*, 1976; Kovolenki, 1986; Kumar *et al.*, 2001; Kumar *et al.*, 2003 and Maqbool *et al.*, 1998). But the comparative account of the anthelmenth drugs on the efficacy of production of ova by the worms, and the effect of these drugs on expulsion of parasites is the subject to be probed and this may be done directly by the worm count. Therefore an attempt has been made to evaluate the comparative efficacy of two anthelmenth drugs Fenbendazole and Tetramisole on ova production and effect of these two drugs on efficacy of expulsion of parasites in the domestic fowl has been investigated.

Materials and Methods

In the present study a separate batch of birds have been maintained for Fenbendazole and Tetramisole. In each group 20 birds of 3 to 4 months old age were used. To examine the presence of ova (*Ascaridia galli*) in the bird, the method of solusby, (1982) was followed. Fenbendazole and Tetramisole was administered to the birds @ 5 mg/bird/kg weight/single dose each. The efficacy of the drugs was evaluated on the basis of no. of ova (egg) present/gram of faecal matter (egg per gram). The eggs in faecal sample of the birds was counted according to stoll's technique. The faecal matter of the birds was examined before administration of drugs and after completion of the 4th, 7th and 10th day of treatment and 24 hours composite sample of the faecal matter was collected to count the ova of the parasites.

In another experimental batch two groups of birds, each consisting of 20 nos. were maintained and Fenbendazole and Tetramisole was administered @ 5 mg/bird/kg/body weight/single dose. Four birds from each group were sacrificed on 4th, 7th and 10th day of treatment. The birds were cut and opened the digestive tract to count the worms. The evaluation of efficacy of the drugs was done on the basis of the worm count in each bird. The percentage of efficacy of drugs was evaluated by the following formula:

$$\text{Per cent efficacy} = \frac{\text{Mean epg/worms of pre-medication}}{\text{Mean epg/worms post-medication}} \times 100$$

To count the no. of ova, the statistical methods as described by Snedecor and Cochran (1989) was adopted. The 'p' values were calculated with the student "t" test (*epg*: egg per gram).

Results and Discussion

In the present study, anthelminth drugs Fenbendazole and Tetramisole was ad ministered on the domestic fowl *Gallus gallus* already infected with *Ascaridia galli*.. The efficacy of release of ova by these two parasites was studied and the results are presented in Table 62.1. The results reveals that after administered anthelminth drugs, on the first day the no. of ova found in the bird faecal matter were 13

51.27, in F enbendazole treated birds, they were 1600 and Tetramisole treated birds were 1500. It shows that the effect of anthelminth drug, Fenbendazole was more effective than the Tetramisole (Figure 62.1). The subsequent expulsion of ova by the birds treated with Fenbendazole and Tetramisole shows that the worm expulsion rate was also more (86.0 per cent) with the Tetramisole and less (73.75 per cent) with the Fenbendazole on 4th day of treatment. On the 7th day of treatment too the Tetramisole shown higher with 99.97 per cent when compared to Fenbendazole (94.37 per cent). However, on the 10th day of treatment Fenbendazole and Tetramisole were almost equal with 99.99 per cent and 99.9 per cent respectively. It is noticed that at all the stages *i.e.*, after 4 days, 7 days and 10 days, the results obtained (Table 62.1) on the expulsion rate of ova by the treated birds were significant (P<0.05).

Table 62.1: The Efficacy of Fenbendazole and Tetramisole on the Expulsion Rate of Ova by *Ascaridia galli* in Domestic Fowl

Drug	*4 Days No. of Eggs*	*7 Days No. of Eggs*	*10 Days No. of Eggs*	*P-value*
Control	1420.3±0.95	1496.5±1.0	1550.56±0.55	<0.05
Fenbendazole	420.1±2.1 (73.75 per cent)	90.0±3.2 (94.37 per cent)	6.0±4.5 (99.99 per cent)	<0.05
Tetramisole	210±2.2 (86.0 per cent)	30.31±2.6 (99.97 per cent)	1.5±0.01 (99.9 per cent)	<0.05

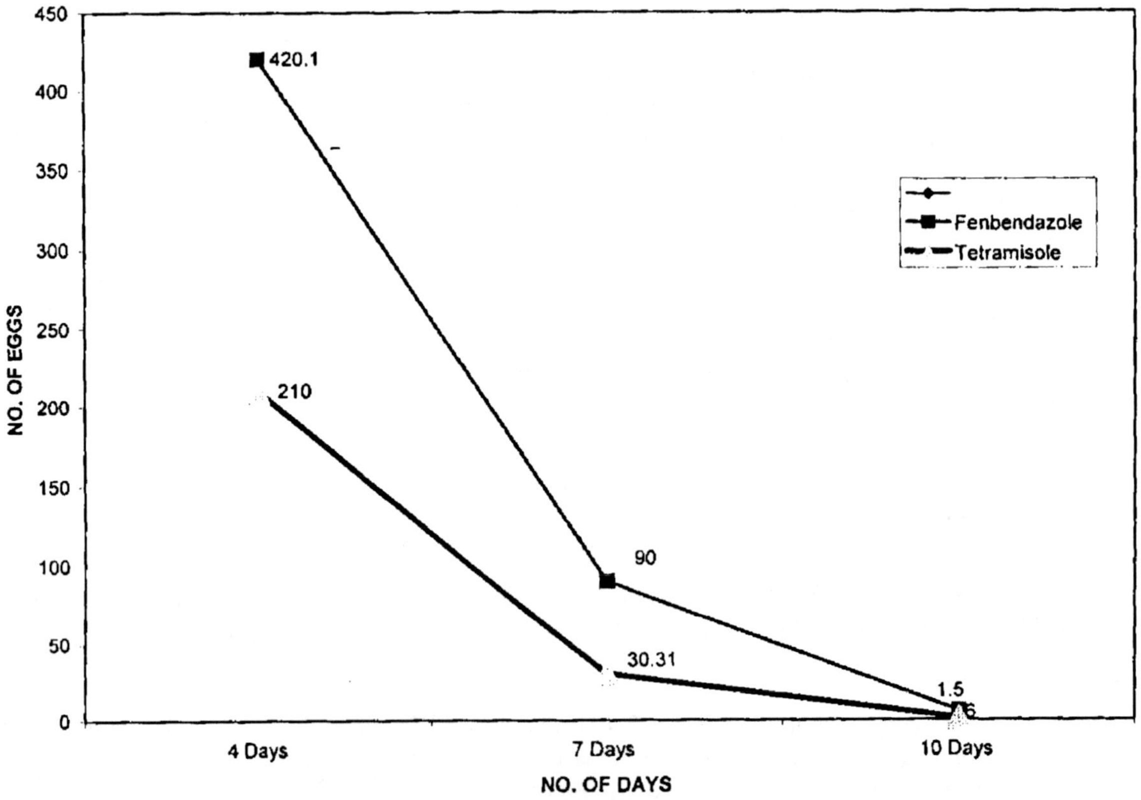

Figure 62.1: The Graph Shows the Efficacy of Fenbendazole and Tetramisole of the Expulsion Rate of Ova by *Ascaridia galli*

Brody and Elward (1971) screened certain anthelminth drugs against certain helminth species in mice have shown the similar results. The efficacy of Tetramisole was found more after 4th and 7th days as compared to Fenbendazole and the reason may be attributed that certain anthelminth drugs exhibit rapid action (Chevis, 1980). The Fenbendazole and Tetramisole seems to be effective anthleminthic drugs against *Ascaridia galli* like other anthelminth drugs such as Mebendazole (Matt, 1980) and Fenbendazole (Maki and Yamagisawa, 1985). The effect of Fenbendazole and Tetramisle on the worms (*Ascaridia galli*) has been studied. The effect of anthelminth drugs can be assessed by the expulsion of worms by the host (McCracen, 1978) after anthelminth administration. In the present study the worm expulsion by Tetramisole was more (46.33 per cent) and Fenbendazole was less (30.66 per cent) which is a clear indication that the Tetramisole was more effective than Fenbendazole against the infection of *Ascardia galli,* therefore the Tetramisole could expel more worms from the host rather than Fenbendazole (Table 62.2 and Figure 62.2). Both the Fenbendazole and Tetramisole have shown their efficacy (during 7th and 10th day), which also shows that the Tetramisole was potent antheminthic drug compared to Fenbendazole. However, the mean worm expulsion count at various stages of intervals (4th day, 7th day and 10th day) are significant (Table 62.2). The oral administration of anthelminth drugs is the only

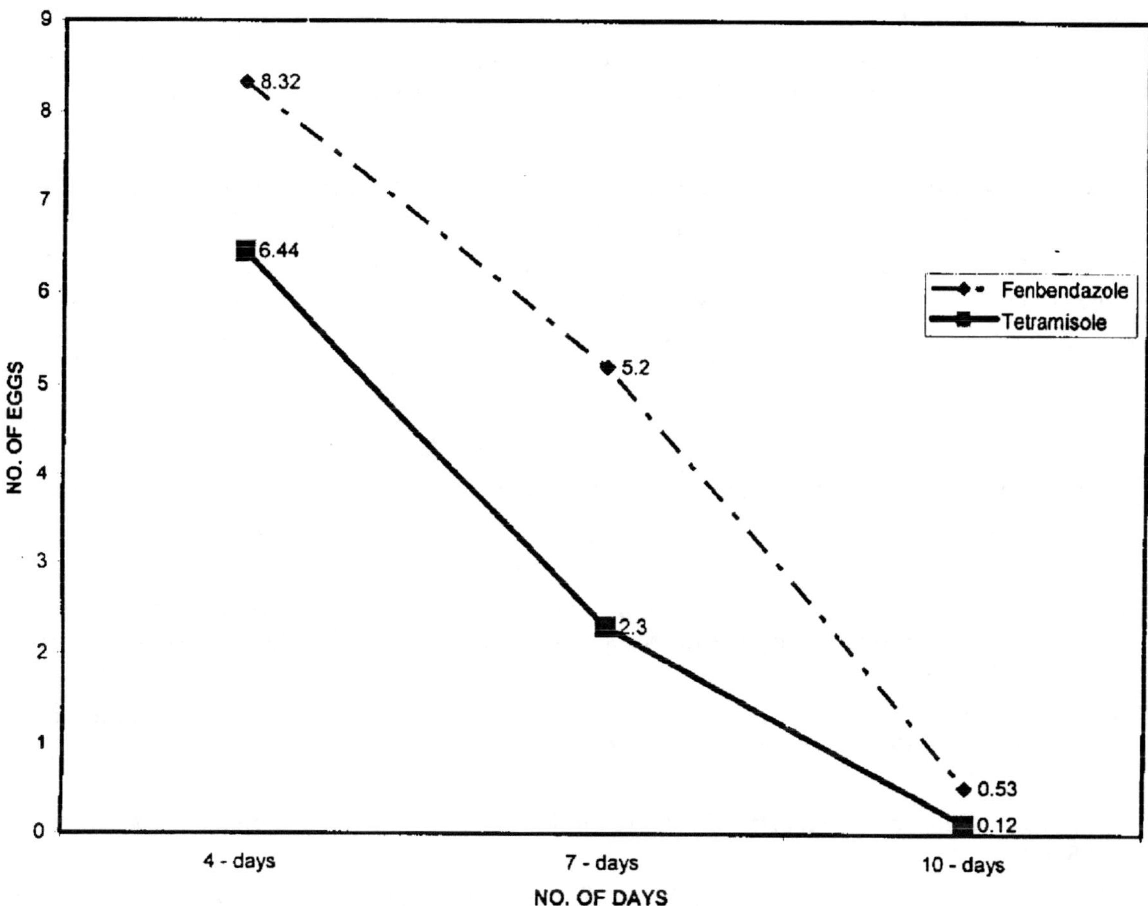

Figure 62.2: The Graph Shows the Effect of Fenbendazole and Tetramisole on Mean Worm Count

suitable method of treating the host against helminth infestation (Heath and Lawrance, 1978). However, when compared to Mebandazole (Matta, 1980) Tetramisole and Fenbendazole are more effective. But the comparative efficacy of Tetramisole and Fenbendazole shows that Tetramisole is more potent Anthelminth drug than Fenbendazole. Therefore, Tetramisole may be prescribed often to treat the domestic fowl especially in severe infections of *Ascaridia galli.*

Table 62.2: The Effect of Fenbendazole and Tetramisole on the Mean Worm Count (MWC) of *Ascaridia galli* in Domestic Fowl

Drug Administered	*4th Day*	*7th Day*	*10th Day*	*P-Value*
Control	NIL	NIL	NIL	–
Fenbendazole	8.32 (30.66 per cent)	5.2 (61.16 per cent)	0.53 (95.58 per cent)	<0.05
Tetramisole	6.44 (46.33 per cent)	2.3 (80.83 per cent)	0.12 (99.0 per cent)	<0.05

Acknowledgements

The Authors thank Head, Department of Zoology, Kakatiya Univeristy for providing facilities to carryout this work and the author (KSR) thank CSIR for providing financial assistance.

References

Agarwal, Abha, Tekwani, B.L., Shukla, O.P. and Ghatak, S., 1979. Effect of anthelminthics and phenothiazines on ATPase of filarial parasite *Setaria cervi. Ind. J. Expt. Bio.,* 28(3): 245–248.

Andrews, P. and Thomas, H., 1979. The effect of praziquantel on *Hymenolepis diminuta in vitro. Tropenmedzin and Parasitologic.,* 30: 391–400.

Blair, L.S. and Campbell, W.C., 1978. Trial of avermectin A, B, mebendazole and melarsoprol against pre-cardiac *Dirofilaric immitis* in the ferret. *J. Parasitology,* 64(6): 1032–1033.

Brugmans, J.P., Thienpont, D.C., I. Van Wijngaarden, vanparij S.O.E, Schuermans, V.L. and Lauwers, H.L., 1971. Effect of mebendazole in enterobiasis. *J. Ame, Med. Asso.,* 217: 313–316.

Bueding, E., 1962. Effect of benzylic diamines on *Schistosoma mansoni. Biochem. Pharmacol.,* 11: 17–28.

Callear, J.F.F. and Neave, R.M.S., 1971. The clinical use of the anthelmintic mebendazole. *Br. Vet. J.,* 127(9): 41–43.

Campbell, W.C., McCracken, R.O. and Blair, Lyndia S., 1975. Effect of parenterally injected benzinidazole compounds on *E. multilocularis* and *Taenia crassiceps* metacestodes in laboratory animals. *J. Parasit.,* 61(5): 844–852.

Chandrasekharan, K., Ramdas, K., Cheeran, J.Y. and Pairly, E.P., 1979. A clinical note on the use of Mebendazole against Ancylostomiasis and toxocariasis in dogs, Kerala. *J. Vet. Sci.,* 10(1): 159–162.

Chevis, R.A.F., 1980. The speed of action of anthelmintics: Mebendazole. *Veterinary Record,* 107(17): 398–399.

Colglazier, M.L., Enzie, E.D. and Kates, K.C., 1977. Critical anthelimithic trail in ponies with four Benzimidazoles. *J. Parasit.,* 63(4): 724–727.

Craig, T.M. and Ronald, N.C., 1978. Preliminary studies on the effect of Albendazole on cysticerci of *Taenia saginata*. *Southwest Vet*., 31: 121–124.

Eva-Maria, Bennet, Behm, Carolyn and Bryant, 1977. Effect of mebendazole and Levamisole on Tetrathyridia of *Mesocestoides corti* in the mouse. *Int. J. Parasitology*, 8: 463–466.

Friedman, P.A. and Pletzer, E.G., 1980. Interaction of anthelmintic benzimidazole with *A. suum* embryonic tubulin. *Bio. Bioph. Acta*, 630: 271–278.

Gerald, Brody and Elward, Thomas E., 1971. Comparative activity of 29 known anthelmintics under standardized drug-diet and gavage medication regimens against four helminthes species in mice. *J. Parasit.*, 57(3): 1068–1077.

Heath, D.D. and Lawrance, S.B., 1978. A single oral treatment with Mebendazole for the control of *T. crassiceps* larval infections in rats. *Int. J. Parasit*., 9: 73–76.

Maki, Jun and Yanagisawa, Toshio, 1985. Antheliminitic effect of bithinol, aramomycin sulphate, flubendazole and Mebendazole on mature and immature *H. nana* in mice. *Journal of Helminthology*, 59: 211–216.

Matta, S.C., 1980. Efficacy of Mebendazole against some common cestodes of poultry. *Indian Journal of Poultry Science*, 15(3): 207 210.

McCracen, O. Richard, 1978. Efficacy of Mebendazole and Albendazole against *T. spiralis* in mice. *Journal of Parasitology*, 64(2): 214–219.

Westhuizen, B. Vander, New Comb, K. and Guerrero, J., 1984. Antheliminitic efficacy of Mebendazole suspension against induced helminth infection in S. African sheep and cattle. *Am. J. Vet. Res.*, 45(4): 779–782.

Wolfe and Wershing, 1974. Effect of Mebendazole on *T. trichura* and *A. lubricoides*. *JAMM*, 230(10): 1408.

Chapter 63

Postharvest Fungal Rot of Chilli (*Capsicum annuum* L.) Fruits

V. Bali Reddy, G. Bindu Madhavi and D. Venkata Reddy
Regional Agricultural Research Station, Lam, Guntur – 522 034, A.P.

ABSTRACT

Survey was conducted in five districts of Andhra Pradesh *viz.*, Guntur, Warangal, Khammam, Prakasam and Krishna to observe the postharvest fungal fruit rot causing fungal pathogens in chilli. It was observed that *Drechslera tetramera, Bipolaris tetramera* and *Alternaria alternata* were causing major fruit rot diseases during pre-monsoon period. *Alternaria alternata, Colletotrichum capsici* were prevalent in monsoon period. Postharvest fungal fruit rot damage during post-monsoon was mainly due to *Fusarium semitectum, Aspergillus flavus, Aspergillus niger, Rhizopus* spp. and *Mucor*.

Introduction

Chilli fruits (*Capsicum annuum* L.) is one of the spice crop famous for its pungent aromatic flavour. Among the spices dry chillies contributes a major share in India growth for both home and export market. The mature green fruits are used as vegetable or in fish and meat preparations. The green fruits are very often damaged by species of *Alternaria, Fusarium, Aspergillus, Colletotrichum, Rhizopus* and *Mucor*. An attempt was made to study the post harvest fungal rots caused by various pathogens in different districts of Andhra Pradesh.

Materials and Methods

Chilli fruits were collected from vegetable market from seven districts of Andhra Pradesh during pre- monsoon, monsoon and post-monsoon periods. The infected fruits were collected and pathogens

Table 63.1: Postharvest Fruit Rot Causing Fungi Observed in Chilli Fruits Collected from Markets of Different Districts

District	*Variety Grown in Major Area*	*Fungi Observed**										
		Drechslera tetra-mera	*Bipolaris tetra-mera*	*Alter-naria alternata*	*Colletot-richum capsici*	*Fusarium semi-tectum*	*Asper-gillus flavus*	*A. niger*	*A. ohraceus*	*Curvularia lunata*	*Rhizopus spp.*	*Mucor*
Guntur	LCA–334	+++	++	+++	+++	+++	+++	+++	++	+	+	+
Khammam	LCA-335	+++	++	+++	+++	+	++	++	+	–	–	–
Warangal	G–4	++	++	+++	+++	+	+++	+++	+	+	–	–
Prakasam	LCA–206	+	++	+++	+++	+	+	+	–	–	–	–
Krishna	CA–960	++	++	+++	+++	++	+	+	++	++	++	+

*: Cumulative of pre-monsoon, monsoon and post-monsoon periods; +++: Severe; ++: Moderate; +: Low.

were isolated and identified (Ellis, 1971,1976). Fungi associated were cultured in Potato Dextrose Agar medium. Fresh chilli fruit samples were surface sterilized with mercuric chloride (1 per cent) and plated on PD agar and incubated at 32±1°C for 5 days and confirmed pathogenecity by following Koch's postulates.

Results and Discussion

Fruit rot due to *Alternaria alternata* (Fr.) Keissl. was recorded in fruits collected during the monsoon period. The temperature was as low as 26°C during the period. This disease was characterized by the production of small dark brown spots on fruits. These spots later coalesced and covered large necrotic areas on fruit. The fruits finally rotted completely. The fungi produced abundant conidia *in vivo* and *in vitro*. This disease has been reported earlier (Wall and Biles, 1993; Prabhavathy and Reddy, 1995; Vijayalakshmi and Rao, 2003) from different parts of India. Fruit rot caused by *Fusarium semitectum* Berk and Rav. was recorded during post-monsoon season. The fungus produced white hairy hyphae on the stalk, calyx and the stalk end in the early stages of infection. Later the whole surface of the fruit was covered by the fungal mycelium. The fungus produced microconidia and macroconidia on the fruit surface, the microconidia were aseptate and macroconidia were 3–5 septate and measured 24.5–34.5 × 3.5–4.0 µm and were curved with a wedge shaped but not pedicillate basal cell curved with a wedge shaped but not pedicillate basal cell and pointed apex. Micro-conidia measured 7.0–11.0 × 3.0–3.5 µm species of *Fusarium* have been found to infect chilli fruits (Wall and Biles, 1993; Prabhavathy and Reddy, 1995; Vijayalakshmi and Rao, 2003; Bilgrami *et al.*, 1979, 1981, 1991; Hosagoudar *et al.*, 1996; Mukerji and Bhasin, 1986). *Aspergillus niger*, caused rot of chilli fruits during the monsoon season. Water soaked brown areas developed near the blossom end or sometimes near the stem end. As the disease advanced further, conidiophores bearing black, catenulate and spiny conidia covered the whole fruit. This disease was found to spread rapidly and affected fresh fruits in storage. Infection of dried fruits by *A. niger* was observed. Anthracnose was observed in chilli fruits during monsoon season, caused by *Colletotrichum capsici* (Syd.) Bud. and Bisby. This disease was characterized by dark brown or dirty black, oval to circular depressed spots over the fruits. The mycelium penetrated the inner tissues and even the seeds and produced a peculiar pungent odour. The pathogen produced setae and conidia *in vivo* and *in vitro*. *C. capsici* infection was recorded earlier from Andhra Pradesh (Vijayalakshmi and Rao, 2003) and Hosagoudar *et al.* (1996) from Kerala Rot due to *Drechslera tetramera* (McKinney) Subram. and Jain [=*Bipolaris tetramera* (McKinney) shoemaker] was observed during the pre-monsoon season. The infected fruits developed brown areas on the fruit wall which later turned into deep brown or black and covered by sparse white mycelia. The conidiophores were brown and produced grey to brown 3-septate conidia. Colonies effuse, grey to brown, hairy and velvety. Mycelium submerged 4–9 µm wide. Conidiophores were brown, macronematous, straight or flexuous, geniculate, unbranched and smooth. Conidia solitary, four celled, straight, brown, smooth, cylindrical with rounded ends. They measured 22–38 µm long and 9–13 µm thick. *D. rostrate* was reported earlier from other parts of India (Vijayalakshmi and Rao, 2003). These fungi are pathogenic and produced toxins in chilli (Dadar, 1998; Dubey *et al.*, 1999). Seunghum *et al.* (1997) reported that *A. alternata* causing fruit rot of chilli produced toxin metabolites such as altemariol and alternariol monomethyl ether. Mathur (1995) observed the production of phytotoxin metabolites by *C. capsici* in chilli fruits.

References

Bilgrami, K.S., Jamaluddin and Rizwi, M.A., 1979. *Fungi of India*. Today and Tomorrow's Printers and Publishers, New Delhi, India.

Bilgrami, K.S., Jamaluddin and Rizwi, M.A., 1981. *Fungi of India*. Today and Tomorrow's Printers and Publishers, New Delhi, India.

Bilgrami, K.S., Jamaluddin and Rizwi, M.A., 1991. *Fungi of India*. Today and Tomorrow's Printers and Publishers, New Delhi, India.

Datar, V.V., 1998. *Indian J. Mycol. and Pl. Pathol.*, 26: 239–242.

Dubey, L.N., Das, K.K. and Jha, R.R. 1999. *Indian J. Mycol. and Pl. Pathol.*, 29: 276.

Ellis, M.B., 1976. *Dermataceous Hyphomycetes*. CMI, Kew, England.

Hosagoudar, V.B., Abraham, T.K. and Pushpangadan, P., 1996. *Fungi of Kerala*. TBGRI, Palode, Kerala, pp. 121.

Mathur, K. 1995. *Indian J. Mycol. and Pl. Pathol.*, 29: 276.

Mukerji, K.G. and Bhasin, J., 1986. *Plant Diseases of India*. Tata McGraw Hill Co., Ltd., New Delhi, pp. 468.

Prabhavathy, K.G. and Reddy, S.M., 1995. *Indian Phytopath.*, 4: 492.

Seunghun, Y., Sohyoung, O., Hyangburn, L., Byungrgum, K., Minill, C. and Subong, P., 1997. *Koean J. Pl. Pathol.*, 13: 323–330.

Vijayalakshmi, M. and Rao, V. Umamaheswara, 2003. *Indian J. Mycol. and Pl. Pathol.*, 33(3): 451–452.

Wall, M.M. and Biles, C.L., 1993. *Phytopath.*, 83: 324–328.

Chapter 64

Evaluation of Toxicity of Yellow Oleander (*Thevetia peruviana*) and Tomato (*Lycopersicon esculentum*) as Fish Toxicants on some Important Blood Parameters of an Air Breathing Fish (*Heteropneustes fossilis*)

Tapan Chandra Das and Samir Banerjee
Aquaculture Research Unit, Department of Zoology, University of Kolkata, 35 Ballygunge Circular Road, Kolkata – 700 019

ABSTRACT

The present studies are made on the efficiency of *Thevetia peruvian a* seed dust (TSD) and *Lycopersicon esculentum* leaf dust (LLD) as fish toxicants. TSD is more effective as fish toxicant than LLD. Both of these toxicants are haemotoxic. Haematological variables in fish blood *viz.*, Total Erythrocyte Count (TEC), Haemoglobin (Hb per cent) content, Haematocrit (Ht) values, Mean Cell Volume (MCV), Mean Cell Haemoglobin (MCH) and Mean Cell Haemoglobin Concentration (MCHC) were observed and recorded following exposure to TSD and LLD in the air breathing predator fish *Heteropneustes fossilis* to assess the level of toxic transfer in the blood.

***Keywords**: Plant toxicants, Thevetia peruviana, Lycopersicon esculentum, Fish toxicants.*

Introduction

Plant origin toxicants are extensively used all over the world for the eradication of unwanted weed fish, predator animals during pre-stocking pond management. But use of such plant origin toxicants is still in infancy in our country (Chakraborty, 1999; Mallick, 1999). In Indian fresh water aquaculture system, mahua oil cake (*Bassia latifolia*) and tamarind (*Tamarindus indicus*) seed husk are mainly used as plant toxicants. Both of these toxins are with haemolytic properties and toxic to fish (Misra *et al.*, 1986).

With the rapid demand in fishery management it has now become necessary to find out other plant origin piscicides. In this context, the present investigation has been envisaged to assess the mode of action of two selected toxins of plant origin *i.e.*, Tomato (*Lycopersicon esculentum* Mill.) leaf dust (LLD) and Yellow Oleander (*Thevetia peruviana* Pers.) seed dust (TSD) on fish and their environment. The main aim of this work is to assess the mode of action and possibility of these plants parts to use as potential piscicide. These plants are widely distributed in West Bengal, Bihar Orissa, Uttar Pradesh and also other provinces of India though their use as piscicide is not fully exploited. The present study is planned to evaluate the mode of action of the plant toxins on some important haematological parameters of an air breathing fish *Heteropneustes fossilis* (Bloch). The active ingredient of Yellow Oleander is thevetin a kind of Glycosides (Chopra *et al.*, 1965; Saravanapavananthan *et al.*, 1988) and that of Tomato is tomatine which is a kind of Gluco-alkaloid. (The Wealth of India, Vol. VI, 1962).

Materials and Methods

For the present study the toxin materials selected are seed dust of *Thevetia peruviana* (TSD) and leaf dust of *Lycopersicon esculentum* (LLD). Different doses of the toxicants were applied to find out the minimum lethal dose. In case of LLD, 100 ppm, 150 ppm, 200 ppm, 250 ppm, 300 ppm, 400 ppm and 500 ppm were applied. But in case of TSD 25 ppm, 50 ppm, 75 ppm, 100 ppm, 150 ppm, 200 ppm, and 250 ppm were applied. Then the dose for further study is selected as 250 ppm in case of LLD and 100 ppm in case of TSD. Three glass aquaria containing 50L of water were chosen to perform the experiments. One aquarium is used as control. The other two were treated with phytotoxins. One aquarium was treated with 100 ppm of TSD and the other was treated with 250 ppm of LLD.

The study was carried out on an air breathing fish species *H. fossilis* of same weight (27–30 g) and length (14–16 cm) groups. A number of live specimens of *H. fossilis* were collected from freshwater pond. Equal number of the specimens were given in each of the aquarium. The fish of the control and treated aquaria were constantly observed to detect any behavioural changes due to poisoning.

Haematological parameters *viz.* total erythrocyte count (TEC), Haemoglobin (Hb) concentration were estimated following the methods by Hesser (1960), Blaxhall and Daisley (1973), Dacie and Lewis (1984). Macrohaematocrit (Ht) value was measured following Wintrobe (Wintrobe, 1967) method. The blood smears were prepared and stained with Graham-Knoll's benzidine method (Ghaham-Knoll, 1918) for observation of qualitative changes of blood cells.

The mean cell volume (MCV), mean cell haemoglobin (MCH) and mean cell haemoglobin concentration (MCHC) are regarded as absolute values. The values were calculated from the red cell count, haemoglobin and haemotocrit values. Absolute values are essential for classification of anaemic condition of the studied animal.

$$MCV = \frac{\text{Haematocrit (\%)} \times 10}{\text{RBC number in million/cu. mm.}}$$

$$MCH = \frac{\text{Haematocrit (g\%)} \times 10}{\text{RBC number in million/cu. mm.}}$$

$$MCHC = \frac{\text{Haematocrit (g\%)} \times 10}{\text{Haematocrit volume (\%)}}$$

Results

The control value of RBC of *H. fossilis* was 3.10×10^6. But after 14 hours of exposure to LLD the value decreased significantly to 2.48×10^6 (Table 64.1, Figure 64.1). But after 5 hours of exposure to TSD the control value decreased significantly from 3.06×10^6 to 1.58×10^6 (Table 64.2, Figure 64.3). The haemoglobin content of blood of *H. fossilis* decreased significantly from a control value of 12.98 (g per cent) to 11.56 (g per cent) after 14 hours of exposure to LLD (Table 64.1, Figure 64.1). But after 5 hours of exposure to TSD the control value of haemoglobin content decreased significantly from 12.9 (g%) to 8.96 (g%) (Table 64.2, Figure 64.3). The haematocrit value of *H. fossilis* decreased significantly from a control value of 35.70 (per cent) to 29.36 (per cent) after 14 hours of exposure to LLD. (Table 64.1, Figure 64.1). But after 5 hours of exposure to TSD the control haematocrit value decreased significantly from 34.50 (per cent) to 19.76 (per cent) (Table 64.2, Figure 64.3).

Table 64.1: Alterations in the Erythrocycte Related Parameters (mean value) of *Heteropneustes fossilis* Treated with *Lycopersicon esculentum* Leaf Dust

No. of Observations	*Dosages (ppm)*	*Duration of Treatment*	*TEC × 10⁶ Mean ±SD*	*Hb (g%) Mean ±SD*	*Ht (%) Mean ±SD*	*MCV (fl) Mean ±SD*	*MCH (Pg) Mean±SD*	*MCHC (%) Mean±SD*
5	250	Control	3.10 ±0.14	12.98 ±0.17	35.70 ±0.19	115.38 ±4.69	41.97 ±2.21	36.36 ±0.59
		4 hours	2.78 ±0.07	12.36 ±0.08	34.10 ±0.32	122.77 I4.05	44.49 ±0.97	36.25 ±0.55
		9 hours	2.58 * ±0.07	11.86* ±0.16	31.90* ±0.14	123.75 ±3.79	44.49 ±0.95	37.2 ±0.49
		14 hours	2.48* ±0.07	11.56* ±0.08	29.36* to.39	120.88 ±2.60	46.66 ±1.60	38.51 ±0.48

*: Significant at 5 per cent level

TEC: Total Erythrocyte count; Hb: Haemoglobin; H: Haematocrit value; MCH: Mean Cell Haemoglobin; MCV: Mean Cell Volume; MCHC: Mean Cell Haemoglobin Concentration

The MCV of control *H. fossilis* increased insignificantly from a control value of 115.38 (n) to 120.88 (fl) after 14 hours of exposure to LLD (Table 64.1 Figure 64.2). But the MCV of *H. fossilis* also increased significantly from a control value of 112.91 (fl) to 124.54 (fl) after 5 hours of exposure to TSD (Table 64.2, Figure 64.4).

The MCH of fish also increased insignificantly from a control value of 41.97 (pg) to 46.66 (pg) after 14 hours exposure to LLD (Table 64.1, Figure 64.2). But after 5 hours exposure to TSD the MCH of the fish increased significantly from a control value of 42.21 (pg) to 56.63 (pg) (Table 64.2, Figure 64.4).

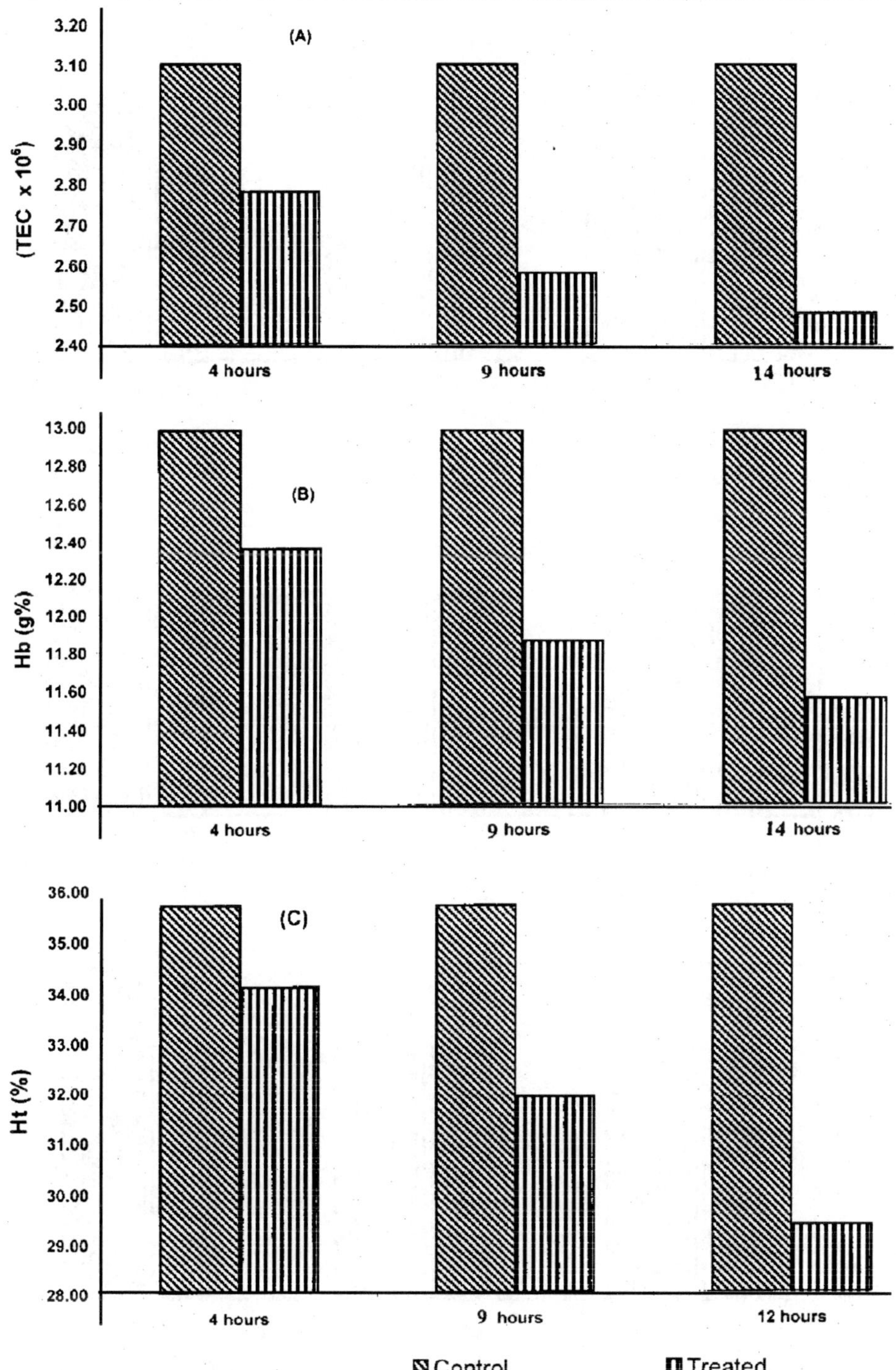

Figure 64.1: Alterations in the Erythrocyte Related Parameters of *H. fossilis* Treated with *Lycopersicon* Leaf Dust (A: TEC; B: Hb; C: Ht)

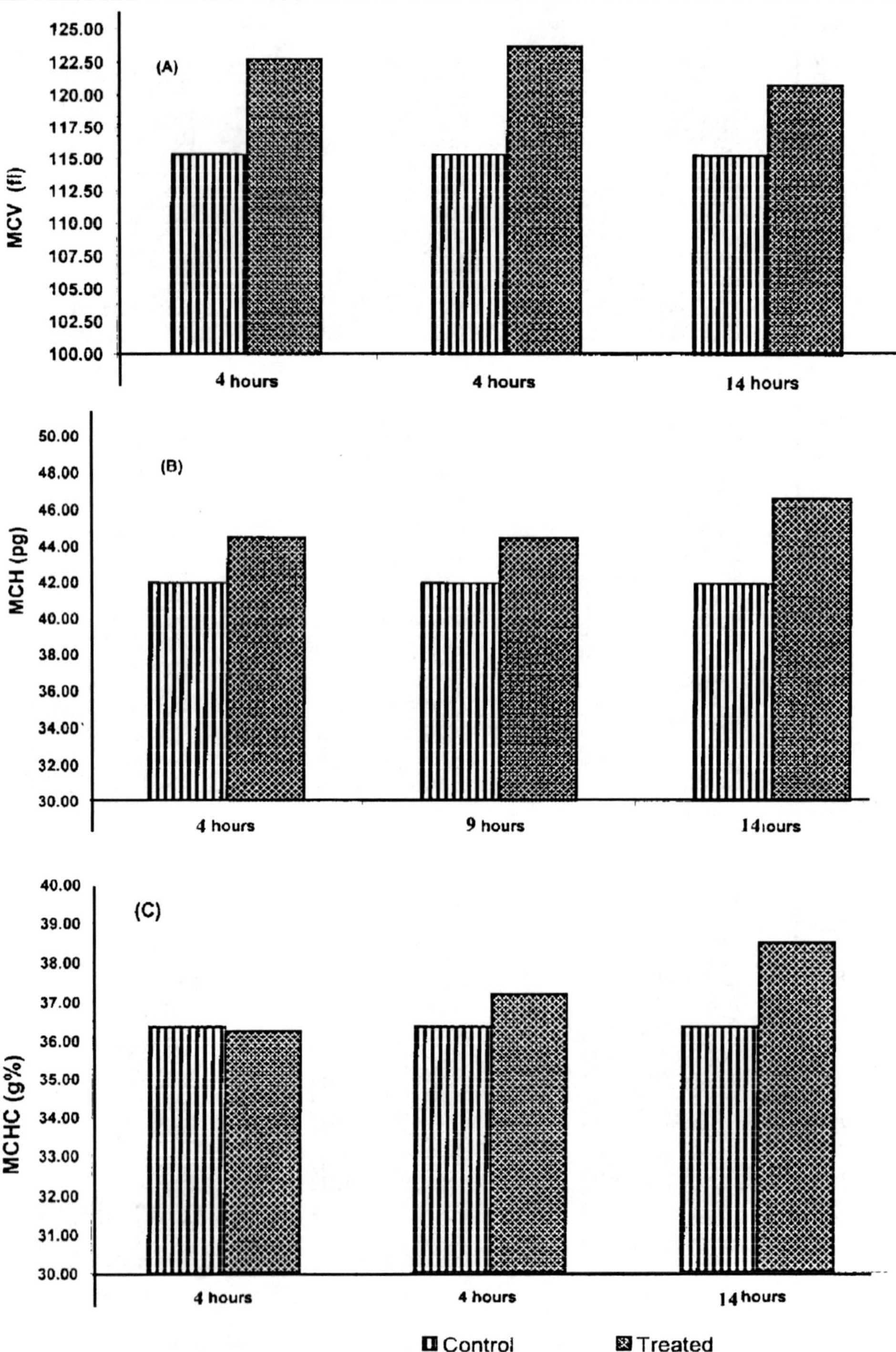

Figure 64.2: Alterations in the Erythrocyte Related Parameters of *H. fossilis* treated with *Lycopersicon* Leaf Dust (A: MCV; B: MCH; C: MCHC)

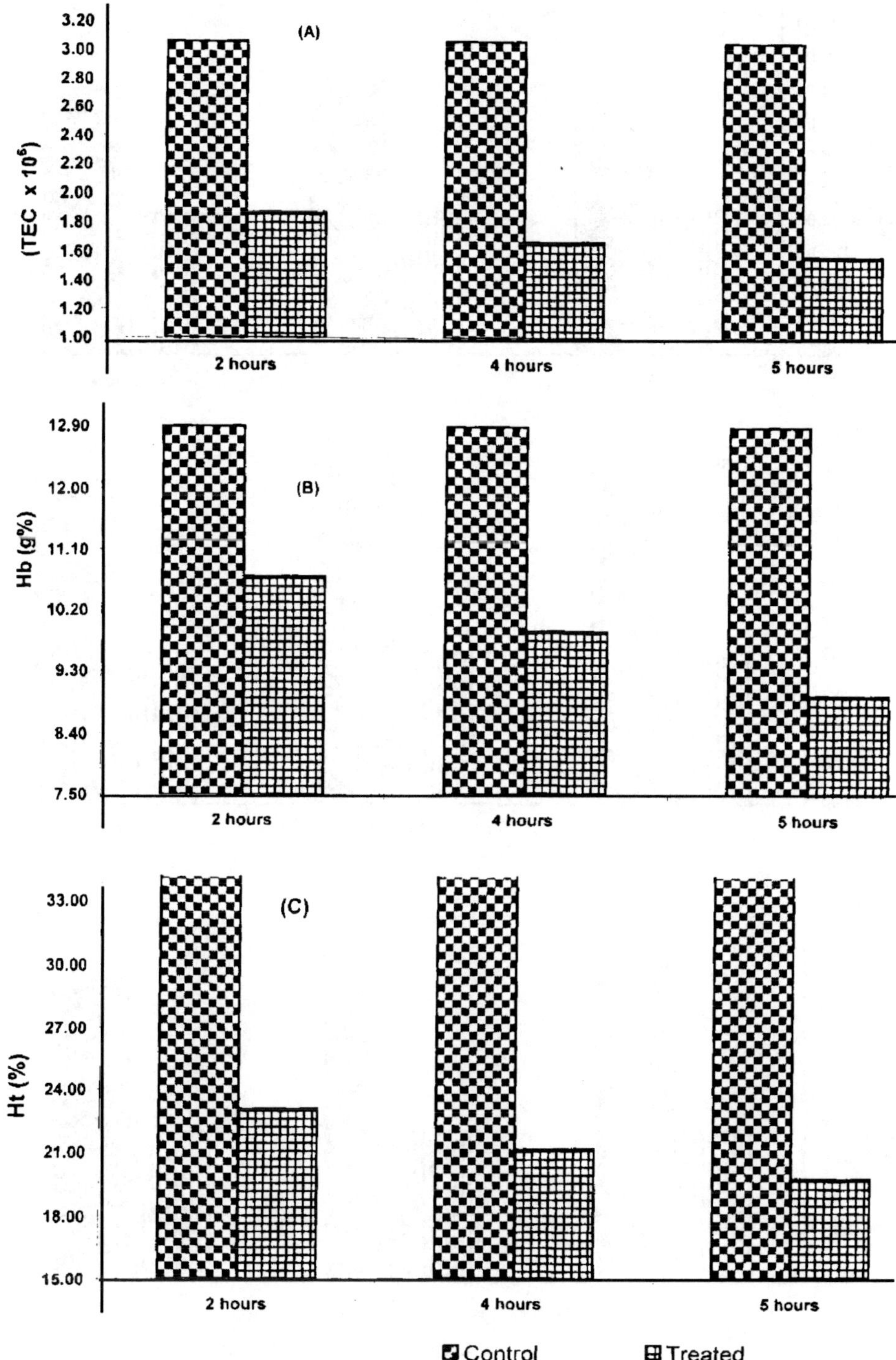

Figure 64.3: Alterations in the Erythrocyte Related Parameters of *H. fossilis* treated with *Thevetia* Seed Dust (A: TEC; B: Hb; C: Ht)

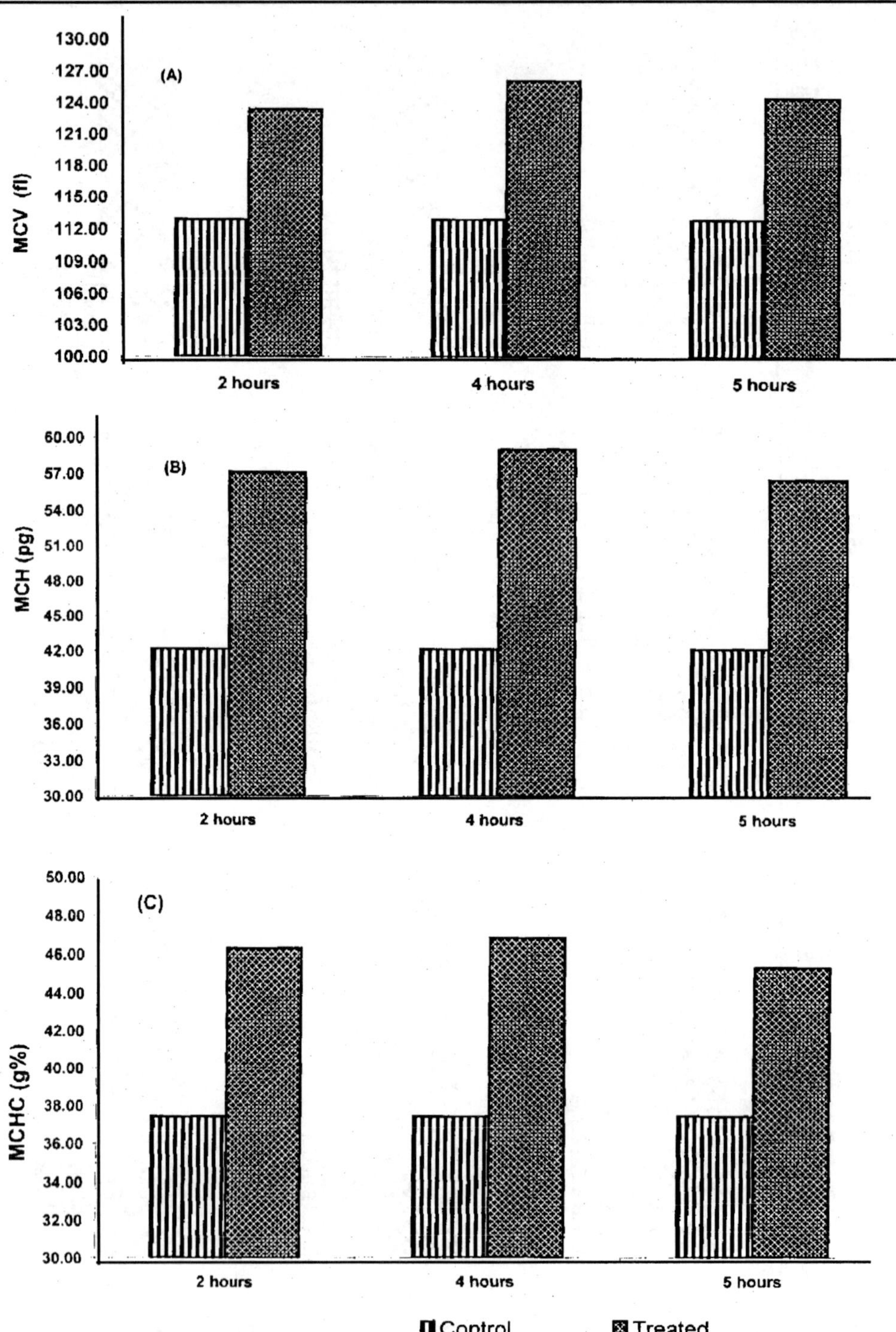

Figure 64.4: Alterations in the Erythrocyte Related Parameters of *H. fossilis* treated with *Thevetia* Seed Dust (A: MCV; B: MCH; C: MCHC)

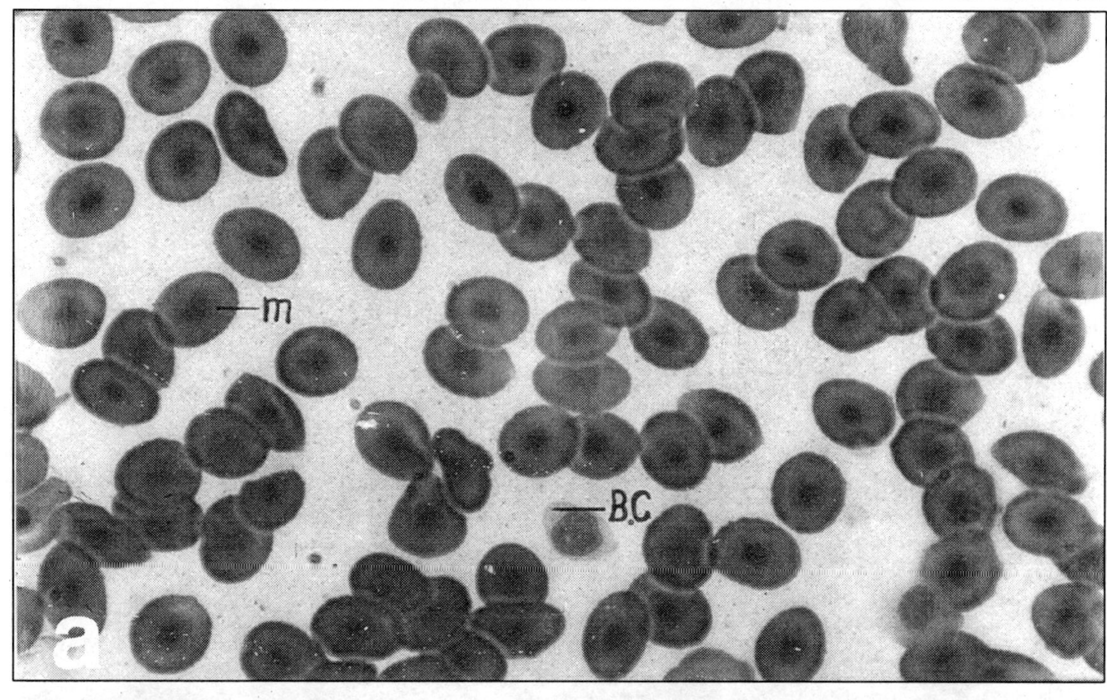

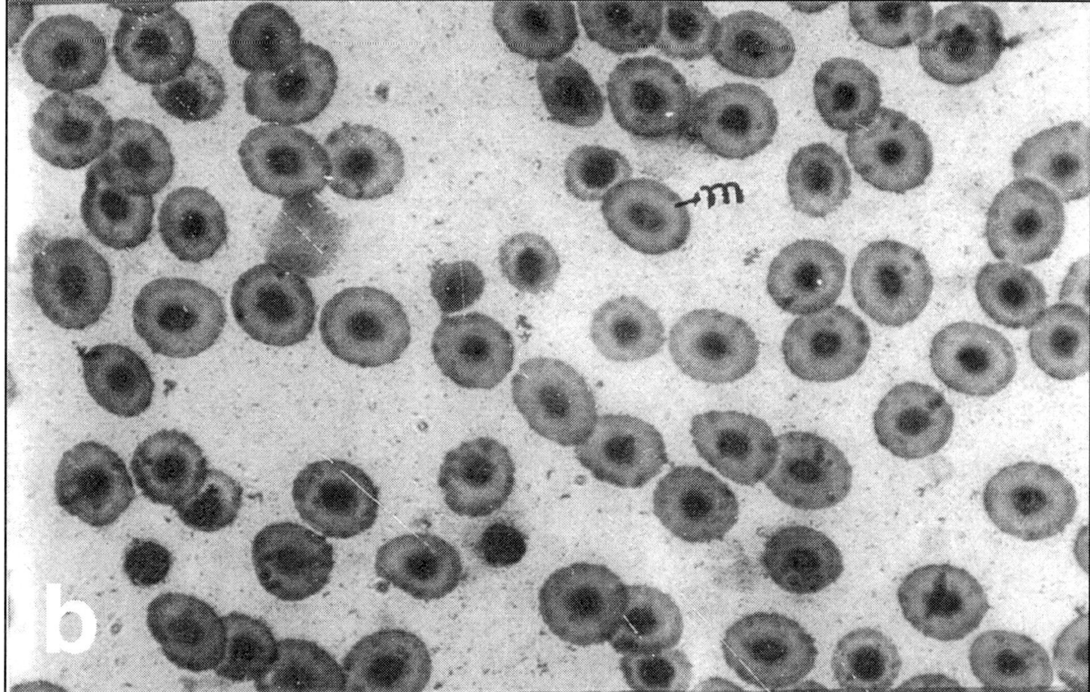

Plate 64.1: Light Microscopic Observations of the Blood Cells of *H. fossilis* (X 1000)
a: Control; b: Treated with TSD; m: Mature red blood corpuscles; B.C.: Blast Cell

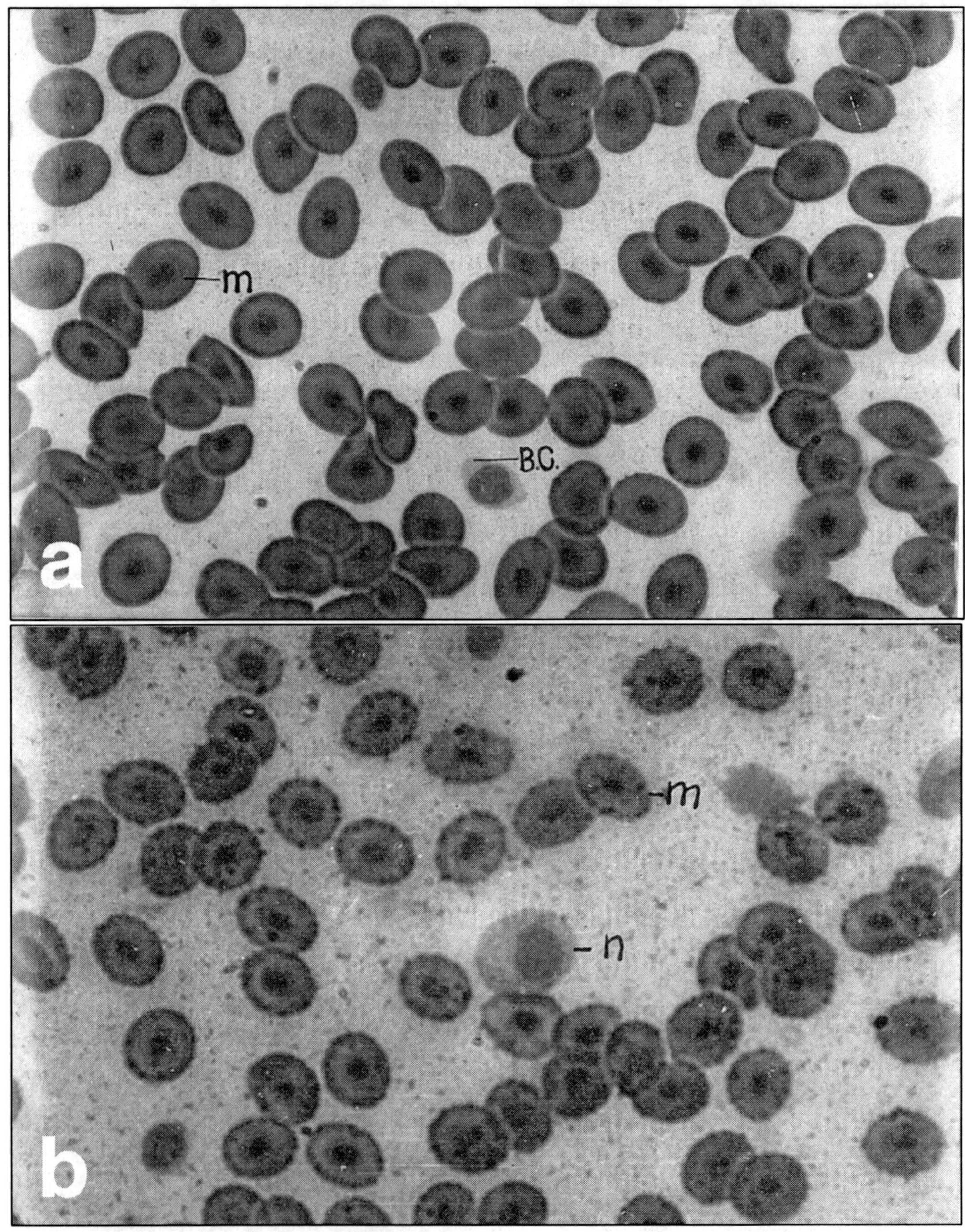

Plate 64.2: Light Microscopic Observations of the Blood Cells of *H. fossilis* (X 1000)
a: Control; b: Treated with LLD; m: Mature red blood corpuscles; B.C.: Blast Cell; n: Neutrophil.

Table 64.2: Alterations in the Erythrocyte Related Parameters (Mean Value) of *Heteropneustes fossilis* Treated with *Thevetia peruviana* Seed Dust

No. of Observations	*Dosages (ppm)*	*Duration of Treatment*	*TEC x 10^6 Mean ±SD*	*Hb (g%) Mean ±SD*	*Ht (%) Mean ±SD*	*MCV (fl) Mean ±SD*	*MCH (Pg) Mean±SD*	*MCHC (%) Mean±SD*
5	100	Control	3.06 ±1 8	12.9 ±0.21	34.50 ±1.33	112.91 ±2.93	42.21 ±1.87	37.42 ±0.92
		2 hours	1.87* ±0.06	10.70* ±0.29	23.06* ±0.36	123.37* ±4.40	57.14* ±1.22	46.31* ±0.88
		4 hours	1.67* ±0.02	9.90* ±0.22	21.14* ±0.27	126.13* ±0.38	59.07* ±0.89	46.83* to.77
		5 hours	1.58* ±0.02	8.96* ±0.17	19.76* ±0.22	124.54* ±3.52	56.63* ±0.47	45.34* ±0.44

*: Significant at 5 per cent level

TEC: Total Erythrocyte count; Hb: Haemoglobin; H: Haematocrit value; MCH: Mean Cell Haemoglobin; MCV: Mean Cell Volume; MCHC: Mean Cell Haemoglobin Concentration

The MCHC of *H. fossilis* increased insignificantly from a control value of 36.36 (per cent) to 38.51 (per cent) after 14 hours of exposure to LLD (Table 64.1, Figure 64.2). But after 5 hours of exposure to TSD the control value of MCHC of the fish increased significantly from 37.42 per cent to 45.34 per cent (Table 64.2, Figure 64.4).

The flood smear of control and treated *H. fossilis* was studied. The low stainability of TDS treated blood film clearly exhibited the leakage of Hb from the RBC into the plasma through the broken cell membrane (Plate 64.1).

The blood smear of LLD treated *H. fossilis* showed adverse effect on the erythrocyte population. The erythrocytic membrane was disrupted after the exposure and Hb also come out into the plasma through the broken cell membrane (Plate 64.2).

Behavioural Responses of Test Fishes Exposed

It was observed that the fishes exposed to TSD were affected within a period of 15 to 20 minutes. The test fishes exhibited various behavioural responses such as fast swimming, rapid ventilation, rolling of eyes, restlessness, and erratic swimmmg movement under the exposure to TSD. Occasionally they tried to jump out of the aquarium and finally leading to death.

But in exposure to LLD the fishes were affected within a period of 30 to 60 minutes and became initially inactive. The opercular movement decreased and the fishes became sluggish, came to the surface and float more or less perpendicular to the water surface and gasped in the same way as fish in low oxygen tension gasped for air. The fishes also exhibited excessive mucus secretion, leading to gill clogging ultimately causing death.

Discussion

Thevetin a kind of glucosides extracted from the seed dusts of *Thevetia peruviana* has got a unique mode of action on circulatory system (Chopra *et al.*, 1965). Thevetin has digitialis like action on the amphibian and mammalian hearts (Chopra *et al.*, 1940). The basis for the physiological action of Oleander cardenolides is similar to that of the classic digitalis glycosides, *i.e.*, inhibition of plasmalemma

Na^+, K^+ ATPase (Longford *et al.*, 1996). Tomatine, a kind of glucoalkaloid is extracted from the leaf dust of *Lycopersicon esculentum* Tomatine contains α-tomatine, a glycoside increases the fluorescence measured in the membrane permeability of frog embryos by about 600 per cent as compared to control. α-tomatine also diminished sodium active transport in frog skin by about 16 per cent compared to the control values. Animal and human cell membranes maybe disrupted following changes in ion fluxes due to the activity of tomatine (Blankemeyer *et al.*, 1996).

The present haematological investigation showed that TEC, Hb and Ht values decreased significantly in exposure to TSD. But MCV, MCH and MCHC increased significantly in treated fish blood. Thevetin is a highly water soluble biotoxin and quickly diffuses through the gill surface membrane of the ventilating area causing disturbance in O_2 carriage mechanism due to the rapid reduction of RBC and haemoglobin. The results of the present study clearly highlight the entry of the phytotoxin component into the blood stream of the fishes exposed to TSD exposure which appear to be haemolytic in nature. In exposure to LLD, TEC, Hb and Ht values also decreased significantly but the MCV, MCH, and MCHC increased insignificantly in the treated fish blood. TSD and LLD enter the circulation through gills and buccal epithelium. Generally RBC may undergo premature destruction by intravascular haemolysis in the circulation. It may be caused by trauma to the RBC or by exogenous toxins (Cooper and Bunn, 1983). The haemolysis of RBC and subsequent alteration of the blood values of *H. fossilis* could be understood from this fact. Alteration of blood values is more drastic in case of TSD treatment than that of LLD treatment. It suggest its high toxicity on fish through changes of blood profile. Haemoglobin may be released into the plasma through the disrupted cell membrane as a result of haemolysis (Cheung *et al.*, 1989). The following may happen.

1. The viscosity of blood may rise.
2. Loss of haemoglobin will critically reduce the oxygen carrying capacity of blood, thus producing anoxia and acidiosis (James and Sampath, 1995).
3. Possibility of haemoglobinuria condition where the Hb would be excreted through urin, as free haemoglobin has easy access into the glomerular membrane of kidney tubules as a result of which a serious disturbance in the kidney functioning will be caused (Cooper and Bunn, 1983).
4. Acute anaemia is a distinct possibility (Goel and Kalpana, 1985).
5. Cell membrane of RBC may be disrupted due to the presence of glycoside (Blankemeyer *et. al.*, 1996).
6. The seeds of *Thevetia peruviana* contain, cardiac glycosides that may cause vomiting, dizziness and dysrhythmiasis such as conduction block affecting sinus and AV nodes (Eddleston *et al.*, 1999).

All these deleterious effects depend upon the number of RBC haemolysed and the rapidity of haemolysis. It can be assumed from the experimental results that the toxicity of TSD and LLD affect very rapidly on the blood of all the fish specimens and lead to all these abnormalities which became fatal. The rapid disintegration or RBC might have contributed the poor value of RBC count, Hb content and Ht value.

Acknowledgement

The authors would like to pay their gratitude to Head, Department of Zoology, University of Calcutta for necessary permission to do this work.

References

Blankemeyer, J.T., White, I.B., Stringer, B.K. and Friedman, M., 1996. Effect of α-tomatine and tomalidine on membrane potential of frog embryos and active transport of ions in frog skin. Bibliographic Citation, *Food-Chem. Toxicol.*, 35: 639–646.

Blaxhall, P.C. and Daisely, K.W., 1973. Routine haematological methods for use with blood. *J. Fish. Biol.*, 4: 593–604.

Chakraborti, P.S., 1999. Studies on the impacts of toxicants of plant origin on fishes and their environment. *Ph.D. Thesis*, University of Calcutta.

Cheung, K., Hinds, J.A. and Duffy, P., 1989. Detection of poisoning by plant origin cardiac glycoside with the abbott-TDX, analyzen. Department of Clinical Chemistry, Prince Charles Hospital Chermiside, Australia, Bibliographic Citation, *Clin. Chem.*, 35: 295–297.

Chopra, R.N., Badhawar, R.L. and Ghosh, S., 1940. *Poisonous Plant of India*, Vol. 1. Govt. of India Press, pp. 16–19.

Chopra, R.N., Badhawar, R.L. and Ghosh S., 1965. *Poisonous Plant of India*, Vol. 1. Govt. of India Press, pp. 566–568.

Cooper, R.A. and Bunn, H.F., 1983. In: *Principles of International Medicines.* McGraw Hill International Book Company, 324: 1862–1964.

Dacie, J.V. and Lewis, S.M., 1984. *Practical Haematology*, 6th edn. Churchill Livingstone, Edinburgh, London, Melbourre and New York, pp. 453.

Eddlestom, M., Ariaratnam, C.A., Meyer, W.P., Perera, G., Kulartne, A.M., Attapattus, S., Sheriff, M.H. and Warrella, D.A., 1999. Epidemic of self poisoning with the seeds of the yellow oleander tree (*Thevetia peruviana*) in North Sri Lanka. *Trop. Med. Int. Health*, 4: 266–273.

Goel, K.A. and Kalpana, G., 1985. Haematological characteristics of *Heteropneustes fossilis* under the stress of zinc. *Indian J. Fish*, 36: 256–259.

Graham-Knoll, 1918. *Atlas of Blood Cytology*, (Ed.) F.G. Boyer. Grune and Stratton, New York, London, p. 28.

Hesser, E.F., 1960. Methods for routine haematology. *Prog. Fish Cult.*, 22: 164–171.

James, R. and Sampath, K., 1995. Sublethal effects of mixtures of copper and ammonia on selected biochemical and physiological parameters in cat fish *Heteropneustes fossilis* (Bloch). *Bull. Environ. Contam. Toxicol.*, 55: 187–194.

Longford, S.D. and Boor, P.J., 1996, Oleander toxicity: an examination of human and animal toxic exposures. *Toxicology*, 109(1): 1–13.

Mallick, A., 1999. Piscicidal potency of plant origin toxicants with respect to the effect on fish and aquatic environment. *Ph.D. Thesis*, University of Calcutta.

Misra, S., Chaudhuri, S.M. and Banerjee, S., 1986. Evaluation of toxicity of mahua oil cake and tamarind seed husk as fish toxicants. *Environment and Ecology*, 4(2): 388–390.

Saravanapavananthan, N. and Ganeshamoorthy, J., 1988. Yellow oleander poisoning: A study of 170 cases. Faculty of Medicine, University of Jaffna. Bibliographic Citation, *Forensic Sri. Int.*, 36: 247–250.

The Wealth of India, Vol. VI, 1962. CSIR, New Delhi, pp. 187–197.

Index

A

Aedes aegypti, 478

Agaricus bitorquis, 404-409

Agroecosystem, 84-97

Agrostis capillaries, 235

Albino mice, 120-123

Alnus nepalensis, 189-199

Alternaira alternata, 494

Alternaria, 492

Amphora, 258

Anand, K. Vijay, 296

Annaduri, R., 182

ANOVA, 69

Antimicrobial activity of *Merremia tridentate*, 464-469

Aphanocapsa, 258

Arcella, 324

Ascaridia galli, 486-491

Aspergillus, 492

Azotobacter, 87

B

Baccillus subtilis, 465

Bacillus, 36

Bacteriological assessment of the marine environment of Kollam, 205-211

Banerjee, Samir, 496

Barik, K.L., 442

Baruah, Debojit, 66

Baskar, M., 216

Behera, D.K., 288

Belloo Beel, 280-287

Benarjee, G., 421, 486

Bhakat, R.K., 275, 280

Bhandarkar, S.V., 343, 347, 352

Bhandarkar, W.R., 352

Bhaskar, C. Vijaya, 430

Bhat, Gh. Mohi-u-din, 212

Bhati, Durgesh, 470

Bhatnagar, Pradeep, 310

Bhoomika, H.R., 129, 133

Bikshapathi, V., 486

Bindu, L., 200, 321, 389

Binukumari, S., 35

Biomedical waste management, 421-429

- introduction, 422
- materials and methods, 422-424
- results and discussion, 425-428

Biosuppression of *Scirphophaga incertulas*, 364-371

- introduction, 365
- materials and methods, 364
- results and discussion, 366-371

Biradar, S.L., 129, 133

Borah, Sabita, 66

Botia Dario, 72

C

C. decipiens, 51
Cajanus cajan, 339
Callosobruchus chinensis, 161
Candida albicans, 465
Capsicum annuum, 492
Cassia hybrida, 92
Ceratophyllum demersum, 285
Chaca chaca, 72
Chakraborty, Anita, 275, 280
Chanda nama, 72
Channa barca, 81
Chavan, R.N., 343
Chela cachius, 72
Chitlala chitala, 81
Chrysopogon zeylanicus, 451
Cicer arietinum, 261, 265
Citrus aurantifolia, 139
Closterium, 258
Colletotrichum, 492
Correlation and path analysis of some economic traits in potato, 129-132
Cryopreservation of spermatozoa of an endangered fish *Osphronemus goramy* and a common table fish *Puntius sarana*, 200-204
  introduction, 200-201
  materials and methods, 201
  collection and extension of semen, 201
  thawing and observation, 202
  results and discussion, 202-203
Culex quinquefasciatus, 478
Cyprinus carpio, 283
Cyril, Wilma, 47, 205
Cyrtorrhinus lividipennis, 360

D

Dalvi, U.S., 261, 265
Damame, S.V., 261, 265
Das, P.K., 288
Das, Tapan Chandra, 496
Dash, A.K., 288
Dash, B.B., 288
Deora, G.S., 470
Department of Ancient History and Archaeology, 167
Deshmukh, G.P., 339
Development Assistance Committee, 2
Devi, A. Premila, 155
Devi, Manimala, 410
Devi, N. Sangbanbi, 410
Dhanakodi, B., 477
Dhevagi, P., 84
Dudhade, D.D., 339
Dwivedi, S.C., 377

E

Ecology and biodiversity of belloo beel, 280-287
  benthos, 282-283
  fish fauna, 283-284
  introduction, 280-281
  macrophytes, 284-286
  materials and methods, 281
  plankton, 281-282
  results and discussion, 281
  study area, 281
Economic traits in potato, 129-132
Effect of water activity on the germination of certain fungi isolated from working environments in Imphal, 155-160
Effect of wheat straw and sugarcane trash on yield and nutrient status of *Agaricus bitorquis*, 404-409
Electricity generation and effect of catholyte in a dual chambered microbial fuel cell using Klebsiella pneumoniae as biocatalyst, 174-181
  MFC construction and operation, 176
  chemical analysis, 176
Electron spin resonance spectroscopic technique for dating of fossil bones, 166-173
  introduction, 166-167
  sample preparation and dose determination, 167
  results and discussion, 167-172
Enterococus faecalis, 106
Environment and sustainable development indicators, 1-8
  baselines, 5
  benefits of sustainable development indicators, 4
  biodiversity indicators, 3
  comparators, 6-7
  introduction, 1-2
  natural resource management indicators, 2
  pollution indicators, 2
  problems of sustainable development indicators, 5

sustainable development indicators, 2-3
targets, 6
thresholds, 5-6
Environmental impact of mine spoils, 216-247
action plan, 240-241
biological costs of not reclaiming mine spoils, 236
environmental consequences of mine spoils, 218
environmental impact of specific mine spoils, 228-236
factors influencing hydraulic characteristics, 222-228
groundwater flow systems, 221-222
hydrogeologic characteristics, 220-221
introduction, 216-217
mine spoil site development problems, 217
mine-related landslides site development problems, 217
mining guidelines and norms, 240
natural recovery of mine spoils, 238-240
nutrient deficiencies, 219
postmining care, 241
surface mining control and reclamation act of 1977 (U.S.). 241-242
Erecthetis pusilas, 72
Escherichia coli, 465, 471

F

FAO, 85
Farm yard manure and mime on soil adsorption and plant uptake of phosphorus, 288-295
Fenbendazole and tetramisole on *Ascaridia galli*, 486-491
Festuca rubra, 235
Field culture of fairy shrimp *Streptocephalus dichotomus*, 256-260
discussion, 258-259
feeding method, 257-258
introduction, 256-257
materials and methods, 257
recommended feeding level, 257-258
results, 258
abiotic factors influencing population density, 258
Flavobacterium, 99
Fusarium, 492

G

Gallus gallus, 486, 487
Genetic variability studies in potato, 133-138
Gopalakrishnan, S., 464
Gopale, G.N., 347
Groundnut under the influence of fluctuating groundwater table and levels of irrigation, 248-255
crop growth rate, 252
introduction, 249
materials and methods, 249
relative growth rate, 252
results and discussion, 249
Growth and biomass production of *Alnus nepalensis*, as affected by crop and mulches, 189-199
effect of mulch on biomass distribution, 191-198
effect of mulch on growth increments, 191
effect of species moisture on growth increments, 191
introduction, 190
materials and methods, 190
results and discussion, 191
Gulati, J.M.L., 248
Guru, Sushma Das, 458

H

H. fossilis, 505
Harer, P.N., 339
Hazarika, Lakhi Prasad, 66
HIV, 272
HIV / AIDS, 271
Human activity on mangrove ecosystem south east coast of India, 437-441
aquaculture, 439
anthropogenic factors, 440
cattle grazing, 438-439
clear felling, 439
dumping of garbage, 439
introduction, 438
study area, 438
siltation, 439-440
Hydrilla verticillata, 285

I

Identification of molecular marker associated with chickpea for drought tolerance using RAPD analysis, 265-269

introduction, 265-266
materials and methods, 266
results and discussion, 266-268
Impact of paper mill effluent on agroecosystem, 84-97
combined use of effluents and amendments, 92
effect of effluent on
soil properties, 88
soil physico-chemical properties, 88-89
soil microbial properties, 89-90
plants, 90-92
effluent characteristics, 85-88
introduction, 84-85
Impact of weeds on some of the morphological attributes of a little millet crop, 442-447
introduction, 442-443
study site, environment and soil, 443
materials and methods, 443
results and discussion, 443-446
In vitro studies on antibacterial effect of aquous crude extract of bryophytes on *Xanthomonas citri*, 470-476
materials and methods, 471
results and discussion, 472-476
Indira Gandhi wildlife sanctuary: introduction, 449
Indiresh, K.M., 133
Influence of silicate solubilizing bacterial inoculation to rice in different soils, 328-338
biometric observation, 336
effect of silicate solubilizing bacteria on rice, 330
identification and characterization of the bacterial isolates, 330
introduction, 329
materials and methods, 329
results and discussion, 332-335
International Development Goals (IDGs), 2
Irrigation water quality parameters of certain tank located around Tumkur, 430-436
Ismail, T. Syed, 464

J

Jagne, A.B., 352
Jain, Narendra S., 470
Janaki, K., 166
Jasmine, P., 382
Jayakumar, R., 120
Jha, B.C., 275, 280
Joseph, Kiran, 477
Joseph, S.T. Shibu, 47

K

Kalaiselvi, P., 328, 394
Kamalutheen, M., 464
Kandla Port Trust, 298
Kennedy, P., 437
Khan, M.A., 359, 364
Khapra beetle, 377-381
Klebsiella, 207, 471
Klebsiella aerogenos, 465
Klebsiella pneumoniae, 174-181
Kollam, 205-211
Kulshreshta, Shweta, 310
Kumar, Ankit, 104
Kumar, Arvind, 1
Kumar, Ashwini, 147
Kumar, M. Sumanth, 98
Kumar, Prince Prakash Jeba, 296
Kumar, Purushottum, 104

L

Labeo pungusia, 81
Lake Bhandara, 347-351
Lakshmi, S. Rama, 9
Larvicidal and ovicidal effect of a neem product on aedine and culex mosquitoes, 477-485
introduction, 477-479
materials and methods, 479
results and discussion, 480-482
Lenka, D., 248
Leucaena leucocephala, 92
Limnogonus niditus, 360
Limnological studies of Ooty lake in Tamil Nadu, 35-46
Logankumar, K., 150
Lokhande, P.K., 261, 265

M

Macrophytes in Bhandardaha Beel, 275-279
Madhaveelatha, P., 120
Madhavi, G. Bindu, 139, 372, 404, 492
Mahadevaiah, 430
Mahajan, R.T., 161
Maharashtra State of the Zoological Survey of India, 389
Manian, S., 448
Manimegalai, M., 35

Manna, Monalisa, 113
Marine environment of Kollam, 205-211
Mathur, Nupur, 310
Mavarkar, N.S., 129, 133
Merremia tridentate, 464-469
Meyyappan, Rm., 20, 174
Mhase, L.B., 261, 265
Microbial fuel cells, 20-34
applications and recent developments, 30-31
basic concepts, 21
biofuel cell-mechanism, 22
enzymatic fuel cells, 26-27
fuel cell-functions, 21
mediator less microbial fuel cells, 25-26
mediator-coupled microbial fuel cells, 25
requirements for a mediator, 25
type of fuel cells, 21-22
types of biological fuel cells, 22
design of microbial fuel cells, 27
electrical measurements, 29
introduction, 21
membrane less MFC, 29
Millennium Development Goals (MDGs), 3
Mine spoils, 216-247
Mineral characterization studies of sediment samples of Tuna, Gujarat by spectroscopic studies, 296-309
carbonate class, 302-304
clay minerals, 304
description of the study area, 297-298
feldspar, 302
geology of the study area, 298
introduction, 297
materials and methods, 298-299
mineral-wise analysis, 306-307
quartz, 300-302
results and discussion, 300
sample analysis, 300
XRD analysis, 305-306
Mir, A.A., 212
Mishra, J.N., 270
Mishra, M.M., 248
Mishra, Onkar, 147
Modeling on risk of vulnerability to STDs/HIV infection among pre-marital female migrants in urban area, 270-274
autonomous female migration, 271
data, 273
estimation of parameters, 272
family migration, 271-272
introduction, 270-271
model, 272
relay migration, 271
results and discussion, 273
Mohanraj, R.S., 477
Mondal, Hossain Ali, 113
Monocrotophos stress, 120-123
Mughal, A.H., 212
Mugloo, J.A., 212
Muthukumar, T.S., 216

N

N. seriata, 51
Nagendrappa, G., 430
Naik, R.M., 261, 265
National AIDS Control Organization (NACO), 271
Nawake, S.S., 339
Neem product on aedine and culex mosquitoes, 477-485
Neem, 477-485
Nitzchia longissima, 51
Non-keratinizing carcinoma in the sputum of occupational workers as an indicator of health hazard, 124-128
introduction, 124-125
materials and methods, 125-126
mineralogical and chemical analysis of stone, 126
results and discussion, 126-127
Nutrient dynamics and phytoplankton abundance in the marine environment of Kollam, 47-65
cyanophyta, 52
diatoms, 51-52
dinoflagellates, 52
discussion, 52-64
materials and methods, 48-50
results, 50-51
Nutrient dynamics in an annual fire influencing humid subtropical grassland of Indira Gandhi wildlife sanctuary, 448-457

introduction, 449
materials and methods, 450
results and discussion, 450-455
study area, 449-450
Nutrient molecules of albino mice under monocrotophos stress, 120-123

O

O. goramy, 201
Oblisami, G., 84
Observation on the avifaunal diversity in and around Shringarbandh lake, 343-346
On some ciliates of bor wildlife sanctuary, 389-393
Ooty lake in Tamil Nadu, 35-46
Organization for Economic Cooperation and Development, 2
Oryza sativa, 329
Oscillatoria, 87, 148
Osphronemus goramy, 200-204

P

P. sarana, 201
Pandey, Himanshu, 270
Paneer, 105
Panicum sumatrense, 442
Paramecium caudatum, 324
Parida, Puspanjali, 124
Patole, S.S., 161
Paulsamy, S., 448
PCP degradation by *Pseudomonas aeruginosa* and *Pseudomonas putida*, 98-103
chemicals and reagents, 100
flask culture, 100
influence of pH of the medium on PCP degradation, 100
introduction, 99
materials and methods, 99
results and discussion, 100-102
Physico-chemical properties with zooplankton in sager lake Bhandara, 347-351
Platanus occidentalis, 235
Population fluctuations and abundance of predators fauna in rice ecosystem, 359-363
Post-harvest fungal rot of chilli fruits, 492-495
Potato, 129-132, 133-138
Prakash, D.J., 382
Prasad, G. Laxman, 421
Pre-impact studies of the 2000MW lower Subansiri dam on certain hydrobiological aspects of down stream of river Subansiri with special reference to plankton and fishes, 66-83
discussion, 72-81
fish diversity, 69
fish diversity, 71-72
hydrobiological parameters, 67
introduction, 66-67
materials and methods, 67
planktonic diversity, 71
planktonic study, 67-68
results, 69
river hydrology, 69
temperature, 69-70
transparency, 71
Prospect of biomonitoring of pollution in aquatic system by algal assessment, 458-463
Proteus, 207
Protozoa, 148
Pseudomonas aeruginosa, 98-103, 465
Pseudomonas putida, 98-103
Pseudomonas, 87
Pulse beetle, 161-165
introduction, 162
materials and methods, 162
method, 163
results and discussion, 163-164
Puntius sarana, 200-204

Q

Quercus robur, 212-215

R

Raj, S. Anthoni, 328, 394
Rajalakshmi, A., 296
Ramesh, P.T., 216
Ramu, G., 421
Ramulu, K. Narasimha, 421
Rao, K. Jayantha, 120
Rao, K. Jeevan, 9
Rao, Narasimha, 87
RAPD analysis of F1 diallel cross population of chickpea for drought tolerant trait, 261-264
introduction, 261-262
materials and methods, 262
DNA extraction and RAPD analysis, 262
results and discussion, 262-264

Rather, T.A., 212
Raveendran, S., 437
Ravisankar, R., 296
Reddy, D. Venkata, 139, 372, 404, 492
Reddy, K. Gaurava, 372, 404
Reddy, K. Surender, 486
Reddy, M. Chandra Sekhar, 372, 404
Reddy, T. Byragi, 98
Reddy, V. Bali, 139, 372, 404, 492
Reddy, V. Chenga, 139, 372, 404
Rhizobium, 90
Rhizopus, 492
Rhizosolenia, 51
Rice under different land situations, 113-119
  coefficient of variability, 117
  heritability and genetic advance, 117-118
  materials and methods, 114
  pattern of variability, 114-117
  results and discussion, 114
Ricinus communis, 379
River Ayroor, 321-322
River Subansiri, 66-83
Rotifer diversity in two water bodies of Brahmapuri, 352-358

S

Sahoo, U.K., 189
Salmonella, 207
Sankaraiyah, K., 120
Santosh, K.G., 129, 133
Sasmal, Biman Gopal, 113
Sathya, K., 477
Scirphophaga incertulas, 365
Screening of eight plants for their repellent action against adult Khapra beetle, 377-381
Seasonal analysis of sanganer handmade paper and cardboard industries effluent, 310-320
Seasonal periodicity and pollution level of a LNMU campus pond, 147-149
Seed stratification and nursery raising of *Quercus robur*, 212-215
Selection of potential site for solid waste disposal through GIS technique, 182-188
Selvarani, B. Jesi, 256
Selvi, D. Sahayathna, 296
Senapati, H.K., 288
Shaji, C., 321
Sharma, B.S., 47, 147, 205
Shelf life and quality of acid lime at room temperature, 139-146
Shelf life and quality of acid lime at room temperature: chemical constituents, 141-143
Shewanella putrefaciens, 25
Shigella, 207
Shukla, Umashankar, 270
Silicate solubilizing bacterial application along with rice residues on growth and yield attributes of rice, 394-403
  field testing of silicate solubilizing bacteria I on rice, 396-397
  identification and characterization of the bacterial isolates, 395-396
  introduction, 395
  isolation of silicate solubilizing bacteria, 395
  materials and methods, 395
  results and discussion, 398-400
Silicate, 328-338
Singh, Ajay Kumar, 104
Singh, Birendra Kumar, 104
Singh, N. Iranbanta, 155
Singh, Nirmala, 104
Singh, S. Mukta, 155
Singh, S.R., 410
Sivasubramanian, C., 437
Sofi, P.A., 212
Soil solarization on damping off disease management in vegetable nursery, 372-376
  effect of nursery diseases, 374
  effect of soil fungi, 374-375
  polythene thickness, 373
Solaimalai, A., 216
Solanum tuberosum, 129
Spoilage causing pathogenic microorganisms in Indian cheese sold in Allahabad city, 104-112
  collection of samples, 105
  evaluation of microbiological quality of paneer, 105
  introduction, 105
  materials and methods, 105

results and discussion, 106-111

Srikanth, K., 421

Srinu, B., 139

STD/HIVs, 272

Streptocephalus dichotomus, 256-260

Sundar, I., 1

Surendran, M., 382

Suresh, D., 448

Sustainable development of environmental conservation on natural land resources and population through carrying capacity and footprints, 410-420

Suvidha, S., 321

T

T. japonicum, 365

Tesdinella, 324

Thalassiothrix frauenfeldii, 51

Thevetia peruviana, 505

Toxicity, 161-165

Toxicity of yellow oleander and tomato as fish toxicants on some important blood parameters of an air breathing fish, 496-507

discussion, 505-506

introduction, 497

materials and methods, 497-498

results, 498-505

Trichogramma chilonis, 365

Trigonella foenumgraecum, 91

Trogoderma granarium, 377

U

Udayasoorian, C., 84

Umavathi, S., 150

Umavathi, S., 35

United Nations Environment Programme (UNEP), 411

United Nations, 2

Urban and agricultural waste composts on yield and nutrient uptake by different crops, 9-19

dry mater production, 11-12

materials and methods, 10

plant uptake, 12-16

results and discussion, 11

uptake of nutrients by edible parts of different crops, 16-19

yield, 12

V

Vahini, K. Swapna, 98

Vanlalhluna, P.C., 189

Variations in the primary productivity of Singanallur pond in Coimbatore, 150-154

Velraj, G., 166

Verma, K., 359, 364

Verma, Om Prakash, 104

Vijay, M., 20, 174

Vision in relation to light in birds, 382-388

analysis of nerve fibres, 383-384

conduction velocity, 387

discussion, 386

introduction, 382-383

materials and methods, 383

results, 384

results: diameter of afferent fibres, 384

results: myelin thickness, 384-386

results: number of fascicles, 384

W

Water quality assessment of Ayroor river with special reference to zooplankton diversity, 321-322

World Bank, 2

World Conservation Strategy, 411

World Conservation Union, 411

World Health Organization, 2, 427

World Wide Fund for Nature, 411

X

Xanthomonas citri, 470-476

Y

Yadav, Santosh, 377

Yield and economics of intercropping of pulses in pigeonpea, 339-342

materials and methods, 340

results and discussion, 340-341

Z

Zinnia angustifolia, 379

Zooplankton diversity, 321-322

Lightning Source UK Ltd.
Milton Keynes UK
UKOW03n1401040615

252905UK00003B/25/P

9 789351 24